SIGNAL TRANSDUCTION
AND HUMAN DISEASE

SIGNAL TRANSDUCTION AND HUMAN DISEASE

Edited by

TOREN FINKEL, M.D., Ph.D.
National Heart, Lung, and Blood Institute
National Institutes of Health
Bethesda, Maryland

J. SILVIO GUTKIND, Ph.D.
National Institute of Dental and Craniofacial Research
National Institutes of Health
Bethesda, Maryland

A JOHN WILEY & SONS, INC., PUBLICATION

Library of Congress Cataloging-in-Publication Data:

Signal transduction and human disease / edited by Toren Finkel,
  J. Silvio Gutkind.
       p.   ;   cm.
  Includes bibliographical references and index.
  ISBN 0-471-02011-7 (cloth : alk. paper)
  1. Pathology, Molecular.   2. Cellular signal transduction.   3. Drugs—Design.
  [DNLM: 1. Signal Transduction.   2. Drug Design. QH 601 S5773 2003]
  I.   Finkel, Toren.   II. Gutkind, J. Silvio.
  RB113 .S525 2003
  616.07—dc21   2002152366

Printed in the United States of America.

10 9 8 7 6 5 4 3 2 1

To my three muses
Beth, Kira and Nadia

TF

To my four pillars
Silvia, Sarah, Naomi, and Juanita

JSG

CONTENTS

ACKNOWLEDGMENTS

We are grateful to our numerous contributors who have brought their extensive experience and expertise to help craft this volume. In addition, the staff at Wiley Publishing have been extraordinary helpful throughout this effort. We are particularly grateful to Luna Han, a Senior Editor at Wiley, who initially proposed this project and who expertly guided it along, providing innumerable insights and suggestions. We are also grateful to members of our own laboratory who have provided many of the insights that we now write about. Finally, a special thanks to our families who have joined us on this journey and whose support and love give meaning to the destination.

CONTRIBUTORS

MICHAEL A. BEAVEN, Ph.D., Laboratory of Molecular Immunology, National Heart, Lung, and Blood Institute, National Institutes of Health, Bethesda, Maryland

PETER D. BURBELO, Ph.D., Lombardi Cancer Center, Georgetown University Medical Center, Washington, D.C.

FABIO CANDOTTI, M.D., Genetics and Molecular Biology Branch, National Human Genome Research Institute, National Institutes of Health, Bethesda, Maryland

JOYA CHANDRA, Ph.D., Division of Oncology Research, Mayo Graduate School, Rochester, Minnesota

JING DU, M.D., Ph.D., Laboratory of Molecular Pathophysiology, National Institute of Mental Health, National Institutes of Health, Bethesda, Maryland

TOREN FINKEL, M.D., Ph.D., Cardiovascular Branch, National Heart, Lung, and Blood Institute, National Institutes of Health, Bethesda, Maryland

NEILL A. GIESE, Ph.D., Millennium Pharmaceuticals, Inc., South San Francisco, California

TODD D. GOULD, M.D., Laboratory of Molecular Pathophysiology, National Institute of Mental Health, National Institutes of Health, Bethesda, Maryland

J. SILVIO GUTKIND, Ph.D., Cell Growth Regulation Section, Oral and Pharyngeal Cancer Branch, National Institute of Dental and Craniofacial Research, National Institutes of Health, Bethesda, Maryland

JONATHAN M. HILL, M.A., M.R.C.P., Cardiovascular Branch, National Heart, Lung and Blood Institute, National Institutes of Health, Bethesda, Maryland

KEITH M. HULL, M.D., Ph.D., Office of the Clinical Director, National Institute of Arthritis and Musculoskeletal and Skin Diseases, National Institutes of Health, Bethesda, Maryland

THOMAS R. HUNDLEY, Ph.D., Laboratory of Molecular Immunology, National Heart, Lung, and Blood Institute, National Institutes of Health, Bethesda, Maryland

DANIEL L. KASTNER, M.D., Ph.D., Genetics and Genomics Branch, National Institute of Arthritis and Musculoskeletal and Skin Diseases, National Institutes of Health, Bethesda, Maryland

SCOTT H. KAUFMANN, M.D., Ph.D., Division of Oncology Research, Mayo Clinic, and Department of Molecular Pharmacology, Mayo Graduate School, Rochester, Minnesota

DEREK LE ROITH, M.D., Ph.D., Clinical Endocrinology Branch, National Institutes of Health, Bethesda, Maryland

STEWART J. LEVINE, M.D., Pulmonary-Critical Care Medicine Branch, National Heart, Lung, and Blood Institute, National Institutes of Health, Bethesda, Maryland

HUSSEINI K. MANJI, M.D., Laboratory of Molecular Pathophysiology, National Institute of Mental Health, National Institutes of Health, Bethesda, Maryland

SILVIA MONTANER, Ph.D., Cell Growth Regulation Section, Oral and Pharyngeal Cancer Branch, National Institute of Dental and Craniofacial Research, National Institutes of Health, Bethesda, Maryland

JOEL MOSS, M.D., Ph.D., Pulmonary-Critical Care Medicine Branch, National Heart, Lung, and Blood Institute, National Institutes of Health, Bethesda, Maryland

JOHN J. O'SHEA, M.D., Molecular Immunology and Inflammation Branch, National Institute of Arthritis and Musculoskeletal and Skin Diseases, National Institutes of Health, Bethesda, Maryland

WALTER A. PATTON, Ph.D., Department of Chemistry, Lebanon Valley College, Annville, Pennsylvania

JEREMY W. PECK, M.S., Lombardi Cancer Center, Georgetown University Medical Center, Washington, D.C.

MICHAEL J. QUON, M.D., Ph.D., Cardiology Branch, National Heart, Lung, and Blood Institute, National Institutes of Health, Bethesda, Maryland

ILSA I. ROVIRA, M.S., Cardiovascular Branch, National Heart, Lung and Blood Institute, National Institutes of Health, Bethesda, Maryland

WILLIAM F. SIMONDS, M.D., Metabolic Diseases Branch, National Institute of Diabetes and Digestive and Kidney Diseases, National Institutes of Health, Bethesda, Maryland

AKRIT SODHI, Ph.D., Cell Growth Regulation Section, Oral and Pharyngeal Cancer Branch, National Institute of Dental and Craniofacial Research, National Institutes of Health, Bethesda, Maryland

DORA C. STYLIANOU, M.S., Lombardi Cancer Center, Georgetown University Medical Center, Washington, D.C.

JAMES N. TOPPER, M.D., Ph.D., Millennium Pharmaceuticals, Inc., South San Francisco, California

MARTHA VAUGHAN, M.D., Pulmonary-Critical Care Medicine Branch, National Heart, Lung, and Blood Institute, National Institutes of Health, Bethesda, Maryland

ROBERTA VISCONTI, M.D., Ph.D., Istituto di Endocrinologia ed Oncologia Sperimentale "G. Salvatore" del Consiglio Nazionale delle Ricerche, Napoli, Italy

BENJAMIN WOLOZIN, M.D., Ph.D., Department of Pharmacology, Loyola University Medical Center, Maywood, Illinois

YEHIEL ZICK, Ph.D., Department of Molecular Cell Biology, The Weizmann Institute of Science, Rehovot, 76100, Israel

INTRODUCTION

Flower in the crannied wall,
I pluck you out of the crannies,
I hold you here, root and all, in my hand,
Little flower—but if I could understand
What you are, root and all, and all in all,
I should know what God and man is.

> From "Flower in the Crannied Wall"
> Alfred Lord Tennyson
> 1809–1892

Pick up any newspaper or turn on any television set, and undoubtedly you will be confronted by the dizzying array and breathtaking speed of scientific and medical advances. Future historians will certainly note that a mere 50 years separated the initial discovery of the structure of DNA from the description of the complete sequence of the human genome. Similarly, the pace of scientific discovery has forever altered our expectations and perspectives. For instance, in the past, deciphering the causative mutations for conditions such as sickle cell anemia or familiar hypercholesterolemia would take years of meticulous planning and painstaking work and, in the end, the isolation of the culpable gene would shake the very foundation of science and medicine. In contrast, these days the genetic bases for diseases are reported with such frequency that their discovery is often treated with the indifference one reserves for stories on insurance premiums or crop forecasts.

Despite the pace of medical research, the sad fact remains, however, that the incidence of many fatal diseases continues to increase. In addition, although new treatments are continually discovered and tested, it is also safe to say that today the life expectancy following the diagnosis of an advanced solid tumor or end-stage congestive heart failure remains exceedingly short. What then is the impact of our increasing knowledge of human biology on our ability to treat the most severe and crippling of human diseases? The short answer is that although it is too early to know for sure, certain promising signs are emerging. Indeed, there appears to be a growing list of drugs being tested in early clinical trials that translate insight garnered from basic laboratory research to specifically target molecular pathways fueling disease. For example, once-fatal leukemias can now be successfully treated with drugs such as the newly described agent Gleevac, which inhibits a kinase specifically activated in the process of malignant transformation. Similarly, other novel agents that target receptor tyrosine kinases, such as the epidermal growth factor (EGF) receptor, appear to be promising drugs to treat a number of solid tumors.

When surveying the field, we could find no text that straddled the productive interface between modern biology and modern medicine. Indeed, we began to feel that a laboratory researcher working in the field of asthma might be very conversant with the intricate molecular signaling pathway of NF-κB and its myriad intervening components and target genes, yet he or she might never have been exposed to the simple clinical tool of flow-volume loops or seen graphically the effects of bronchodilators on airway resistance. Conversely, a rheumatologist might be quite adept at examining a joint and developing an appropriate differential diagnosis but be quite unaware of the details surrounding TNF signaling. In the first edition of this book, we have attempted to bring these two complementary approaches into one volume. Together with our contributors, we have labored to describe a host of disease processes from common conditions such as atherosclerosis and cancer to disorders such as TRAPS, a rare rheumatological syndrome characterized by periodic fevers and rashes. Within many of the chapters, where appropriate, we have first tried to give the reader a sense of the disease process, what it affects, how it presents, how common it is, and what the current treatments are. These clinical descriptions are not meant to be exhaustive but rather to serve as an outline to the reader regarding the disease's manifestations and current treatment options. After this introduction, we usually present a more in depth discussion of one or two signal transduction pathways or biological process relevant to the disease. Throughout these fourteen chapters we have endeavored to cover most of the major signaling pathways using a variety of different human diseases as our framework and point of embarkation.

The book is divided like many medical textbooks into subspecialty areas. In our case this includes sections in cardiopulmonary disease, oncology, endocrinology, infectious disease, allergy/rheumatology, and neurology/psychiatry. Diseases discussed include among others cancer, asthma, atherosclerosis, diabetes, rheumatoid arthritis, Parkinson disease, and depression. In addition, we outline the current understanding of diverse pathways from MAPK activation in cancer to the role of NF-κB in asthma and arthritis, from JAK/STAT signaling in immune deficiencies to the molecular basis of dysentery.

We begin with cardiology, discussing the basis of atherosclerosis and the role that small diffusible radical species such as nitric oxide and superoxide have on the vessel wall. Rather than viewing them simply as toxic molecules, we show that these reactive oxygen species play an important role in vascular homeostasis. Pharmacological manipulation of these pathways has in fact been known for a century or more, as nitroglycerin (a nitric oxide generator) has been widely used by symptomatic patients for treating chest pain (i.e., angina pectoris). Indeed, Alfred Noble, the Swedish benefactor of the Noble Prizes, used nitroglycerin as a starting point for his discovery of dynamite in the 1860s. Close to 150 years later, three scientists would share a Nobel Prize for the understanding that nitric oxide regulates vascular tone, with pharmacological agents such as nitroglycerin deriving their clinical benefit by mimicking these effects. The further description of other agents such as Viagra, which prolong the half-life of nitric oxide in certain, shall we say, critical organs, have reenforced the importance of this pathway in health and disease.

After this description of atherosclerosis we discuss the growing epidemic of asthma, a disease that affects both children and adults. We use this condition to discuss an essential regulator of the inflammatory process, namely, the NF-κB pathway. In particular, we discuss activation of the NF-κB pathway through cytokine receptors such as the tumor necrosis factor (TNF) receptor.

We next delve into cancer biology. One chapter in this section is a general review of the molecular mechanisms of cancer, focusing on the key biochemical pathways involved in cell cycle regulation and the acquisition of the malignant phenotype. Among the areas discussed are the Ras-MAPK pathway, small GTPases and exchange factors, p53, Rb, and other tumor suppressors, as well as receptor tyrosine kinases. After this discussion, we discuss in a separate chapter the biology of programmed cell death including caspases, the Bcl-2 family of pro- and antiapoptotic proteins, and the Akt kinase. Our goal is to demonstrate how these pathways and their intricate interplay relate to tumor progression and ultimately how they will shape future treatment modalities.

We next move into the area of endocrinology with two separate chapters. The first chapter deals with the molecular basis for diabetes. This chapter primarily discusses the basis for insulin resistance and discusses downstream signaling from the insulin receptor and other relevant receptor tyrosine kinases. The following chapter deals with G protein-coupled receptors (GPCRs) and in particular the multiple endocrine manifestations resulting from inappropriate GPCR activity.

After the section on endocrinology we move on to infectious diseases. The first chapter discusses the interaction of bacteria with the cell and in particular the lessons these interactions have taught us regarding dynamic regulation of the cytoskeleton. The next chapter deals with the molecular basis underlying the diarrhea associated with infectious agents such as cholera or *Escherichia coli* that result in a staggering amount of mortality each year in the developing world. The toxins from these organisms have provided a number of valuable lessons in cell biology, and the authors provide an in-depth description of an interesting posttranslational modification, ADP-ribosylation.

The next chapters are concerned with allergy and rheumatology. We begin with a primer on severe combined immunodeficiencies, a constellation of over 95 different syndromes that impact the immune system. This syndrome is a natural starting point to discuss the world of cytokine signaling and the downstream pathway regulated by JAK and STAT proteins. We next discuss the basis for allergic reaction, from the devastating forms of anaphylaxis to milder syndromes such as hay fever, paying particular attention to the mast cell as the underlying cell type responsible for these allergic responses. Finally, we discuss two rheumatological conditions, the rare periodic fever syndrome TRAPS and the more common rheumatoid arthritis. These two syndromes allow for a discussion of TNF signaling and a look at NF-κB signaling in another disease context.

In the last major section we turn to the brain to discuss both neurological diseases and mood disturbances. In the first chapter, we discuss a variety of debilitating diseases characterized histologically by neurological degeneration. This section allows for a discussion of protein aggregation and the various intracellular processes stimulated by pathological protein aggregates. In the next-to-last chapter we discuss syndromes such as depression and bipolar disease. These disorders, which can in their severe form be life threatening, provide the impetus to discuss signaling through the neurotrophic receptors and the regulation of the CREB transcription factor. The last chapter is devoted to novel drug development and in particular how one goes from candidate target to candidate drug, in essence, how one translates the emerging knowledge of the basic scientific advance into a practical and useful medicine.

As you can see from these brief descriptions, we have, for the benefit of clarity, limited each section to covering only a handful of relevant pathways. Clearly, for instance, the MAPK pathway affects a host of diseases besides cancer and would be just as relevant to talk about in the context of diabetes or a number of neurological conditions. Similar arguments could undoubtedly be made for other signaling pathways such as NF-κB or nitric oxide that have important manifestations in a number of diseases. Therefore, the reader is cautioned that these fourteen chapters are meant as an overview and guide for future explorations. Although we have chosen to discuss important pathways for disease initiation or progression, signal transduction is an integrated subject and no single pathway can or should be viewed in total isolation.

The worlds of laboratory science and clinical medicine are both moving at breakneck speed. As they grow, the tools, techniques, and language of these two areas invariably become more specialized and unique to each discipline. We hope that we have managed in this volume to provide the reader a footing in both camps, in essence, to provide both a big picture as well as giving a sense of the individual brush strokes. We believe that this holistic approach will allow the reader to conveniently integrate both the important clinical and molecular aspects of a number of important disease processes. Such a range of knowledge will undoubtedly be essential if we are to be successful in creating the next generation of molecular therapies.

CHAPTER I

ATHEROSCLEROSIS: SIGNAL TRANSDUCTION BY OXYGEN AND NITROGEN RADICALS

JONATHAN M. HILL, ILSA I. ROVIRA, and TOREN FINKEL
Cardiovascular Branch, National Heart, Lung, and Blood Institute,
National Institutes of Health, Bethesda, Maryland

INTRODUCTION

We are faced with a growing pandemic of cardiovascular disease and stroke at the start of the third millennium. According to World Health Organization estimates, in 1999, cardiovascular disease contributed to one-third of all deaths, with 78% of those deaths occurring in low- and middle-income countries. Atherosclerosis, a disease affecting large arteries, is the underlying cause of most of these deaths. In developed societies, despite access to complex drug therapy and invasive treatment, it remains the number one killer, contributing to nearly one-half of all deaths, while in the developing world, economic transition and industrialization appear to be bringing about lifestyle changes destined to create a new generation of cardiovascular disease victims. Indeed, by 2010 it is estimated that in the developing world, cardiovascular disease will be the leading cause of death. The majority of this mortality burden appears to be at least partly preventable and controllable.

Although there were early descriptions of atherosclerosis in Egyptian mummies, the first careful anatomic and physiological descriptions of atherosclerosis date from the mid-eighteenth century. In recent years, a number of important studies have allowed for a more fundamental understanding of disease mechanisms, with some of these studies providing the first dissection of the relevant intracellular signaling pathways. These studies have pointed the way for the development of new pharmacologic therapies and novel risk reduction strategies. In this chapter, we outline the epidemiological and clinical aspects of atherosclerotic disease from the early stages of endothelial dysfunction and plaque for-

Signal Transduction and Human Disease, Edited by Toren Finkel and
J. Silvio Gutkind
ISBN 0-471-02011-7   Copyright © 2003 John Wiley & Sons, Inc.

mation to eventual plaque rupture. In an overview of signal transduction mechanisms in atherosclerosis, we focus on just one aspect of the disease process by describing the biology of nitric oxide and other reactive oxygen species (ROS) in the arterial wall. In a review of modern treatment approaches we underscore how the understanding of signaling pathways has led to better therapeutic options.

ATHEROSCLEROTIC LESION DEVELOPMENT AND CLINICAL PRESENTATIONS

Large arteries are comprised of three distinct layers. The intima is the endoluminal layer and is lined with endothelial cells bound to a sheet of connective tissue made up predominantly of collagen and proteoglycans. It is in this layer that many of the initial and predominant changes of atherosclerosis occur. The media consists of smooth muscle cells, whereas the adventitia consists mostly of connective tissue elements such as fibroblasts. In general, it is the outermost endothelial layer and the underlying smooth muscle cell layer (i.e., the intima and media) that are thought to be the most important in maintaining overall vascular tone.

In normal individuals, physiological increases in blood flow are caused in large part by endothelium-mediated vasodilatation. This enables blood flow to increase in line with tissue oxygen demands. The major endothelium-derived relaxing factor (EDRE) was discovered by Furchott and Zawadzki (Furchgott and Zawadzki, 1980) and was subsequently identified chemically as nitric oxide (NO) (Palmer et al., 1987; Ignarro et al., 1986; Ignarro et al., 1987). In addition to NO there are a number of other vasodilator and vasoconstrictor substances that regulate vascular tone and homeostasis including endothelin-1, prostacyclin, prostaglandin H_2, and the endothelium-derived hyperpolarizing factor EDHF. In the coronary circulation there is evidence that NO is constantly released from the endothelium (Quyyumi et al., 1995) to maintain a basal state of vasodilatation and to counteract the vasoconstricting effects of substances such as noradrenalin, angiotensin, and endothelin.

Clinical assessment of the endothelial function of the coronary and peripheral circulations can be measured by monitoring the vasodilator response to endothelium-dependent agonists such as acetylcholine. Dysfunctional endothelium is characterized by reduced vasodilatation in response to agents such as acetylcholine. It should be noted that acetylcholine, besides stimulating NO release, also stimulates release of other vasodilating substances such as EDHF. More recently, an ultrasound technique measuring brachial artery flow-mediated vasodilatation allows the repetitive and noninvasive measurement of endothelial function in human subjects (Celermajer et al., 1992).

The Framingham Study (Stokes et al., 1987; Kannel, 2000; D'Agostino et al., 2000) is probably the best-known large-scale epidemiological study

TABLE 1.1. Established and Emerging Risk Factors for Coronary Artery Disease

Risk Factors with Genetic Component	Risk Factors Triggered by the Environment
Lipid abnormalities ↑LDL/VLDL, ↓HDL, or ↑Lp(a) Abnormal hemostatic factors ↑Fibrinogen, ↑PAI-1 Depression Diabetes mellitus and insulin resistance Family history Homocysteinemia Hypertension Male gender Obesity	High-fat diet Smoking Lack of exercise Infection (e.g., *Chlamydia pneumoniae*, CMV)

Risk factors are divided into those that have some genetic basis and those that are purely environmental.

that generated the idea of specific "risk factors." Data began emerging from this study in the 1960s showing the relative contributions of multiple risk factors to the pathogenesis of atherosclerosis and its numerous clinical manifestations. They can be divided into factors with a predominant genetic component and those that are largely environmental (Table 1.1). Individual risk factors can interact with each other and may synergistically affect the progression of the disease. Data from the original Framingham Study were extremely important in the identification of a number of classic risk factors such as smoking, diabetes, and hypertension. Recently, in addition to these conventional risk factors, there are a number of emerging novel atherosclerotic risk factors such as homocysteine levels, fibrinogen levels, and potentially infectious agents.

The most common way for atherosclerotic disease to present clinically is the development of angina. This is experienced by patients as a tightness or pain across the chest and sometimes down the arm. It is the result of a narrowing in a coronary artery supplying the heart muscle, reducing the blood flow and causing myocardial ischemia (Fig. 1.1). Anginal symptoms may be precipitated by situations requiring increased myocardial blood flow, such as during exercise, anxiety, and cold weather and after heavy meals. They are often associated with a feeling of breathlessness. Atherosclerotic disease affecting the peripheral arteries presents in the same way when the narrowed arteries cannot supply enough blood to meet the tissue oxygen demands. The symptoms for patients with peripheral vascular disease is often described as a tightness or aching in the calf muscles after exercise. This syndrome is called intermittent claudication. As with most symptoms associated with vascular

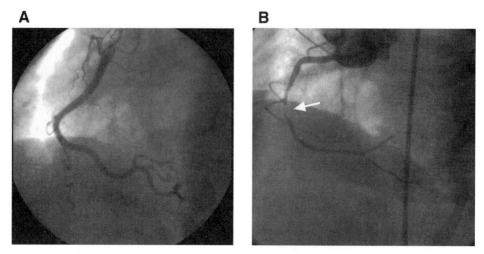

Figure 1.1. Representative coronary angiograms from a normal individual (**A**) and a patient with coronary artery disease (**B**). These studies are performed within a cardiac catheterization laboratory, where a tiny tube is threaded from the patient's groin area to the arteries of the heart. Dye is then injected and an X-ray camera then takes a series of pictures to assess for blockages in the coronary arteries. The arrow indicates an area of severe arterial blockage.

disease, the disease tends to be slowly progressive, reflecting the chronic nature of plaque progression.

The most common presentations of atherosclerosis in the acute setting are myocardial infarction, unstable angina, and stroke. The precipitating event is a result of the acute instability of an atherosclerotic plaque, the surface of which may rupture, causing acute thrombosis and vessel occlusion. Myocardial infarction produces irreversible necrosis of part of the heart muscle and is often fatal before the patient reaches the hospital. It can be treated with drugs targeting the thrombotic cascade and clot formation or by opening the closed artery with a small balloon (angioplasty). At present, little clinical information is available to guide patients or physicians as to when a plaque will convert from a stable lesion to the much more dangerous unstable plaque.

REDOX SIGNALING PATHWAYS IN ATHEROSCLEROSIS

The concept that endothelial injury is the initiating factor in atherosclerosis dates back to the observations of Virchow, who suggested that atherosclerosis developed after mechanical irritation to the intima, which in turn caused degenerative and inflammatory responses leading to local cellular proliferation (Virchow, 1856). It is now generally believed that, in addition to these mechanical forces, risk factors such as smoking, diabetes, and hypercholesterolemia function as continuous endothelial damaging agents. It is also thought by many, but certainly not

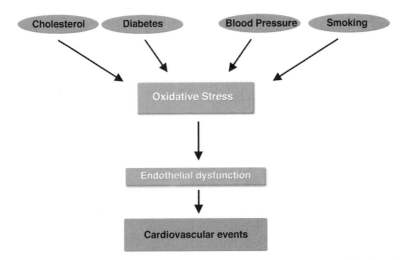

Figure 1.2. *Model for atherosclerotic disease progression.* In this widely hypothesized, but by no means universally accepted model, various cardiovascular risk factor induce a prooxidant stress within the vessel wall. This increase in reactive oxygen species inhibits NO activity and results in the clinical syndrome of endothelial dysfunction, one of the earliest markers of disease susceptibility. This model suggests that a variety of seemingly disparate cardiovascular risk factors may be unified mechanistically by their ability to induce a prooxidant state.

all investigators, that oxidative stress is the common mediator of a host of environmental and genetic cardiovascular risk factors (Fig. 1.2). As such, risk factors are thought to act in large part by promoting vascular oxidative stress. Because superoxide can readily inactivate NO, a rise in oxidative stress, particularly in superoxide levels, can counteract the biological activity of NO. Hence, an understanding of NO and other ROS is thought to be critical to understanding the initiation and progression of cardiovascular disease.

Consistent with their ability to induce oxidative stress, atherosclerotic risk factors appear to modulate normal physiological signal transduction pathways within the vessel wall to induce a syndrome termed "endothelial dysfunction." For the purposes of this review, we define endothelial dysfunction as an impairment of endothelial vasodilator function principally related to the bioavailability of NO. This is most often detected as an impairment of the vascular response to agents such as acetylcholine. The presence of endothelial dysfunction even without overt macroscopic atherosclerotic disease was recently demonstrated to predict adverse cardiovascular events and long-term outcome (Schachinger et al., 2000). As such, the clinical syndrome of endothelial dysfunction may be one of the earliest markers of the atherogenic process.

Early pioneering experiments by Furchgott and colleagues determined that the ability of acetylcholine to induce vasorelaxation required a functional endothelial layer. Later experiments demonstrated that acetycholine induced the synthesis of NO. That a small, diffusible gas could be purposely produced within the vessel wall and have an important physiological role was a remarkable departure from conventional thinking with regards to all forms of ROS. The major discoverers of this concept, Furchgott, Ignarro, and Murad, would go on to share the Nobel Prize for Medicine or Physiology in 1998. The production of NO is now known to occur via the action of nitric oxide synthase (NOS), an enzyme family that catalyzes the conversion of L-arginine to L-citrulline in the presence of molecular oxygen and NADPH to yield NO. In addition to its principal role in stimulating vasorelaxation by the production of cGMP in smooth muscle cells, NO and its derivatives play a key role in the development of atherosclerosis by regulating monocyte and platelet adhesion, altering endothelial permeability, and inhibiting vascular smooth muscle cell proliferation and migration (Garg and Hassid, 1989; Cornwell et al., 1994).

There are three distinct isoforms of NOS, arising from three separate genes, with variations in their structure reflecting their specific in vivo functions (Stuehr, 1997). Each enzyme is a highly complex system with distinct functional domains and a multitude of cofactors and prosthetic groups. The enzyme generally functions as a homodimer of identical subunits each bearing two major functional domains: an N-terminal oxygenase, which binds heme and tetrahydrobiopterin (BH4) as well as the substrate L-arginine, and a C-terminal reductase, which contains the binding sites for NADPH, FAD, and FMN. The enzyme is similar to the cytochrome P-450 family of enzymes, especially in its ability to catalyze flavin-mediated electron transport from the electron donor NADPH to a prosthetic heme group. The calmodulin binding domain (CaM) lies between these two functional regions of NOS and is integral to structure and enzymatic function. In the absence of appropriate levels of the substrate L-arginine or BH4, NOS enzymes can produce superoxide and H_2O_2 (NOS uncoupling). The physiological role of this uncoupling is not completely understood, although some recent reports suggest that NOS-produced superoxide can also function as a signaling molecule (Wang et al., 2000).

The main endothelial isoform, eNOS (also called NOS3), differs from the other isoforms with a unique subcellular localization. This localization is achieved because only eNOS is acylated by both palmitate and myristate. Specific residues are modified with Cys-15 and Cys-26 undergoing palmitoylation while an N-terminal glycine undergoes myristoylation (Shaul et al., 1996). Although the myristoylation is an irreversible modification, the palmitoylation step is reversible and subject to physiological regulation by a host of agonists that increase intracellular calcium. One end result of this complex and unique posttranslational modification is that eNOS is not uniformly distributed throughout the endothelial cell membrane but is instead confined to plasmalemmal microdomains known as calveolae. These structures are becoming increasingly important in signal transduction and represent areas in which signaling proteins and their downstream effectors appear to be substantially enriched. A number of caveolin proteins have been defined, and it appears that eNOS can directly bind to both caveolin-1 and caveolin-3 (Feron et al., 1996), with binding appearing to inhibit NOS activity. Recent studies supporting the importance of these interactions come from mice with targeted deletions of caveolin-1, in which it has been observed that there is a major alteration in NO-mediated vasorelaxation (Drab et al., 2001).

Given that eNOS is the gene product responsible for producing the EDRF described by Furchgott, it is not surprising that a number of reports have examined the physiological regulation of the enzyme at both the transcriptional and posttranslational levels. These studies demonstrated that a number of important physiological stimuli regulate eNOS gene expression, such as shear stress, oxidized LDL, and exercise training (Uematsu et al., 1995; Sessa et al., 1994). Some reports suggested that at early stages of atherosclerotic lesion development eNOS expression may be downregulated through a decrease in transcription and a destabilization of mRNA, whereas as the lesion matures the overall level of eNOS expression may actually increase. The physiological significance of these observations is unclear. In addition, there is not always a clear relationship between mRNA level, protein levels, and enzymatic activity, suggesting the possibility of additional layers of complexity and regulation by yet undefined posttranscriptional and posttranslational mechanisms.

Another emerging important form of regulation of eNOS activity appears to be protein phosphorylation. The enzyme has a number of consensus sequence sites for phosporylation by protein kinase A (PKA), protein kinase B (Akt), protein kinase C (PKC), and calmodulin kinase II. There is now evidence that in addition to the phosphorylation of serine residues (Michel et al., 1993) eNOS can also be tyrosine phosphorylated (Garcia-Cardena et al., 1996). Most evidence suggests that tyrosine phosphorylation appears to regulate the interaction of eNOS with caveolin-1 and hence its subcellular localization. Physiologically relevant stimuli such as shear stress appear to stimulate eNOS phosphorylation and increase NO production, in agreement with the known capacity of blood vessels to dilate in response to increased flow. Recently, several studies demonstrated that Ser-1179 of the protein is phosphorylated by protein kinase B/Akt (Dimmeler et al., 1999; Fulton et al., 1999). Again, this phosphorylation was noted to increase NO production. Interestingly, other reports have suggested that HMG-CoA reductase inhibitors, widely used drugs that so effectively lower serum cholesterol, appear to significantly increase the activity of endothelial Akt (Kureishi et al., 2000). These agents, including such widely prescribed agents such as lovastatin and simvastatin, have been shown to have a dramatic effect on cardiovascular mortality. Indeed, their effects on patients' overall mortality appear to exceed what one would expect from simply lowering cholesterol. As such, a considerable amount of effort has been expended to understand what other potential cardiovascular effects statin therapy might provide. One potential mechanism to alter plaque

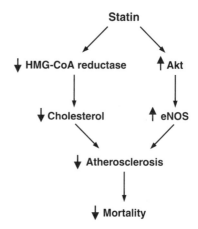

Figure 1.3. *Cholesterol-dependent and -independent effects of statin therapy.* The widely prescribed class of cholesterol-lowering agents referred to as statins appear to lower death rates from cardiovascular disease to a greater extent than would be predicted by the amount of cholesterol lowering obtained. The search for these cholesterol independent effects include augmenting NO levels by modulating the Akt kinase.

progression would be that by raising Akt activity, statins are increasing NO output in the vessel wall and thus potentially providing an additional, cholesterol-independent, benefit to patients. (Fig. 1.3).

Besides the eNOS isoform, two other NOS isoforms have been described and extensively studied. Initially thought to be confined to the nervous system, nNOS (NOS1) was actually the first of the isoforms to be purified and cloned (Bredt et al., 1991). Despite its name, it is clear that nNOS is expressed outside the central and peripheral nervous system and, more specific to our discussion, nNOS is expressed in endothelial and vascular smooth muscle cells (Papapetropoulos et al., 1997; Boulanger et al., 1998) as well as in human atherosclerotic lesions (Wilcox et al., 1997). Nonetheless, the functional significance and role of nNOS in atherosclerosis remain unclear.

The expression of the inducible form of NOS, iNOS (NOS2), is best characterized in inflammatory cells. These cells are very abundant in atherosclerotic lesions. In addition, iNOS upregulation in smooth muscle cells contributes to an overall increase in production of NO in atherosclerosis. A recent study has shown colocalization of this upregulated iNOS with epitopes of oxidized LDL and peroxynitrite-modified proteins (Luoma et al., 1998). It is important to note that the level of NO production from the iNOS isoform is several orders of magnitude higher than from either the eNOS or nNOS isoforms.

The notion that a diffusible gas such as NO could function as a physiological regulator of vascular tone suggests that there are direct and specific protein targets of NO. Evidence suggests that the principal target of endothelium-produced NO is the inactive form of guanylate cyclase located in the underlying vascular smooth muscle cells. The NO-

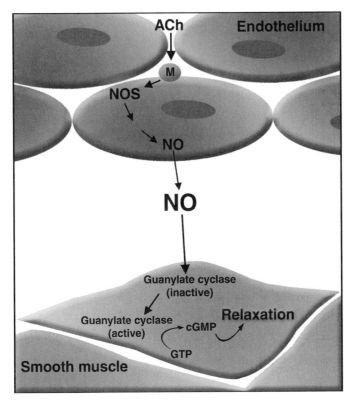

Figure 1.4. *The release of NO from the endothelium regulates vascular tone.*
Agents such as acetylcholine bind to their cognate receptor (M) and stimulate
the calcium-dependent activation of nitric oxide synthase (NOS). The subsequent production of NO diffuses to the underlying smooth muscle cells, activating guanylate cyclase and resulting in vessel relaxation.

dependent conversion from the inactive to active form of guanylate
cyclase in turn catalyzes the production of cGMP from GTP. This causes
the relaxation of the smooth muscle cell and subsequent vasodilatation
(Fig. 1.4). Although guanylate cyclase represents an important target of
NO, many other proteins containing transition metals such as iron, zinc,
or copper can also be regulated by NO. The molecular basis for this
regulation differs for each target. In the case of guanylate cyclase, NO
attacks the bond between His-105 and the ferrous iron associated with
the enzyme. This leads to the activation of the enzyme. Other metal-
containing proteins that serve as important NO targets include hemo-
globin, which appears to be a major intravascular carrier of both molecular oxygen and NO. In addition, a host of transcription factors such as
the large family of zinc finger proteins also can be functionally altered
by NO exposure. In general, such exposure leads to a decrease in DNA
binding, which stands in contrast to the case of guanylate cyclase, where
NO exposure activates the enzyme. Finally, other important targets of
NO are the enzymes involved in aerobic respiration that contain Fe-S

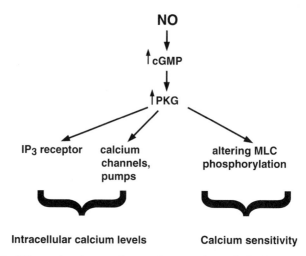

Figure 1.5. *NO-regulated smooth muscle tone through multiple mechanisms.* Included among these effects are the activation of protein kinase G (RKG), which in turn alters intracellular calcium levels and myofilament calcium sensitivity.

centers. These enzymes such as aconitase are also subject to inactivation by other ROS and in particular are rapidly inactivated by exposure to superoxide anions.

The mechanisms by which a rise in cGMP and the subsequent activation of the cGMP-activated protein kinase G (PKG) produce a change in vascular tone are the subject of considerable interest and debate (Lincoln et al., 2001). There are at least two major pathways that contribute to the NO-induced vasorelaxation (Fig. 1.5). The first mechanism involves a reduction in intracellular calcium concentrations in vascular smooth muscle. This reduction in calcium is achieved through a number of distinct mechanisms. Recent evidence has demonstrated that the IP_3 receptor is a direct target of PKG. Phosphorylation of the smooth muscle IP_3 receptor results in a decrease in calcium release from the sarcoplasmic reticulum. In addition, calcium levels are also modulated by the effect of PKG on a number of calcium pumps and voltage-gated channels.

Besides altering calcium levels, the rise in cGMP and the activation of PKG directly alters actin-myosin kinetics. A number of studies have addressed the regulation of myosin light chain (MLC) phosphorylation at Ser-19. Two enzymes with antagonistic functions are primarily responsible for the degree of MLC phosphorylation. These enzymes are myosin light chain kinase (MLCK) and MLC phosphatase. The level of regulatory MLC phosphorylation in turn determines the level of force production and hence the degree of vascular tone. Most available evidence suggests that PKG functions to increase MLC phosphatase activity, producing a decrease in MLC phosphorylation. This decrease in MLC phosphorylation contributes to vascular relaxation. Interestingly, another signaling pathway that impacts vascular tone is the agonist-induced activation of the small GTPase Rho and the downstream effector Rho kinase (Pfitzer et al., 2001). Activation of Rho kinase leads to the inhibition of the MLC phosphatase. Thus, NO-dependent, PKG-mediated vasorelaxation acts in an antagonistic fashion to agonist-mediated vasoconstriction mediated by the Rho/Rho kinase pathway. The myosin-binding subunit (MBS) of the MLC phosphatase appears to be the direct molecular target of PKG because the two proteins can form a direct molecular complex (Surks et al., 1999).

In addition to guanylate cyclase, NO has a number of important other cellular targets that may be important in vascular homeostasis and atherosclerotic disease progression. The caspase family of cysteine proteases, critical to the execution of apoptosis, appear to be direct protein targets. In particular, caspase-1, -2, -3, -4, -6, -7, and -8, can be reversibly inhibited in vitro by NO-mediated nitrosylation of the active cysteine residue. The nitrosylation of caspase-1 was recently shown to block the ability of the cytokine TNF-α to trigger apoptosis of endothelial cells (Dimmeler et al., 1997). Interestingly, these effects were seen predominantly with low concentrations of NO whereas a proapoptotic effect was seen with higher NO levels. It is presently unclear to what degree, if any, endothelial apoptosis contributes to atherogenesis, although a number of studies suggest that an increase in apoptosis accompanies plaque formation (Mallat and Tedgui, 2000). Therefore, one potential protective effect of NO may be by directly modifying vascular caspase activity and therefore altering the propensity of endothelial cells to undergo cell death. Although the work in endothelial cells has lagged behind that in other cell types, in lymphocytes it was noted that a significant proportion of caspase-3 is nitrosylated under basal conditions. Apoptotic stimuli such as engagement of the Fas receptor result in denitrosylation of caspase-3, allowing for subsequent increases in caspase activity and cell death (Mannick et al., 1999).

In addition to the interaction of NO with metal-containing proteins such as guanylate cyclase, the interaction with cysteine residues in caspases, NO-derived species can also directly modify tyrosine residues. The exact in vivo pathway for tyrosine nitration remains controversial (Davis et al., 2001). Many studies have suggested that the interaction of NO with superoxide leads to peroxynitrite formation. This radical species is capable in vitro of nitrating tyrosine residues. Other potential pathways involve the interaction of peroxynitrite with carbon dioxide or the action of myeloperoxidase on oxidized NO to form other reactive nitrogen species (RNS). Interestingly, a number of studies have demonstrated that atherosclerotic plaque appears to have a significant increase in the level of nitrotyrosine-containing material (Buttery et al., 1996). This observation led to the belief that these areas have an increase in peroxynitrite formation. One potential direct target of peroxynitrite that may be important in vascular tone is prostacyclin synthase, the enzyme responsible for producing the vasorelaxing substance prostacyclin. Tyrosine nitrosylation of prostacyclin synthase has been demonstrated to inhibit enzymatic activity, suggesting that peroxynitrite formation may increase vasoconstriction not only by directly destroying NO, but also by inhibiting other key regulatory enzymes required for vascular tone (Zou et al., 1999). Interestingly, the ability of RNS to modify tyrosine residues suggests that these species may modify important regulatory amino acids linked to the tyrosine kinases/phosphatases. Although in vitro studies suggest that some cross-talk might exist between NO and classic tyrosine kinases such as c-src (Gow et al., 1996), to date, few in vivo data are available for such processes (MacMillan-Crow et al., 2000) and therefore the overall physiological importance remains unclear.

The observation that atherosclerotic plaque has increased levels of nitrotyrosine and by implication increased levels of peroxynitrite suggests that regions within a plaque have elevated levels of NO, superoxide, or both radical species. A variety of evidence suggests that in fact, areas of plaque have an increase in superoxide production (Warnholtz et al., 1999). Because the balance of NO and superoxide represents a critical element in vascular tone we will next explore where superoxide, hydrogen peroxide, and other ROS are produced and how they may also contribute to specific alterations in signaling pathways.

In phagocytic cells such as neutrophils or macrophages the production of superoxide is required for the specialized function of these cells to provide for host defense. In these cells, production of ROS requires the assembly of a specialized enzyme system, the NADPH oxidase. This enzyme complex is composed of two plasma membrane components, gp91phox and p22phox, as well as three cytoplasmic components, p47phox, p67phox, and the small GTPase rac2. Activation of the neutrophil causes a series of events leading to the recruitment of cytosolic components to the membrane to create a fully assembled oxidase complex and the subsequent high-level production of ROS. Many of the components of the classic NADPH oxidase appear to be present in cells of the vascular wall. Consistent with a role for this oxidase in vascular biology, an animal model with a targeted deletion in p47phox had reduced levels of atherosclerotic plaque formation (Barry-Lane et al., 2001). Besides the components of the well-described neutrophil NADPH oxidase complex, novel cell specific components have recently been described (Lambeth et al., 2000). In particular, proteins with significant homology to gp91phox have been recently described in smooth muscle cells (Suh et al., 1999). In addition, in a variety of other cells, homologs of gp91phox appear to function in oxidase complexes (see Fig. 1.6). To date, the exact mechanisms by which these novel oxidases function is incompletely understood. Indeed, it is unclear whether these oxidases require the participation of the other known classic NADPH oxidase members such as p47phox or p67phox. It is tempting, however, to speculate on a parallel between the NO- and superoxide-generating systems. In cells involved in host defense, the production of NO is produced by the iNOS (NOS2) enzyme. This enzyme produces a considerable amount of RNS to provide an immune surveillance function. Similarly, the pro-

EXPANDING FAMILY OF NADPH OXIDASES

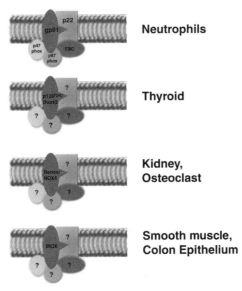

Neutrophils

Thyroid

Kidney,
Osteoclast

Smooth muscle,
Colon Epithelium

Figure 1.6. *An expanding family of NADPH oxidases.* The classic NADPH oxidase found in neutrophils is composed of membrane bound subunits (gp91phox and p22phox) and several other components recruited from the cytosol (p47phox, p67phox and rac2). Novel homologs of the well-characterized gp91phox component have recently been isolated, which appear to have specific and unique tissue distribution.

duction of superoxide and other ROS by neutrophils, macrophages, and other immune cells by the classic NADPH oxidase results in high-level oxidant production necessary for clearance of microorganisms. Indeed, individuals with mutations in any of the components of the classic NADPH oxidase complex have a condition known as chronic granulomatous disease. These unfortunate individuals usually die in early childhood or adolescence from their inability to fight a wide range of bacterial and fungal pathogens. In contrast to the high-level production of NO or superoxide restricted to cells performing a host defense function, in other cases, the production of ROS or RNS appears to be produced for a signaling function. We detailed above the evidence for this signaling function for NO and will also do so shortly for ROS. In these cases, a different set of enzymes is required for low levels of oxidant production. For the case of NO these enzymes are related in structure, namely, eNOS and nNOS (low-level NO production) compared to iNOS (high-level NO production). In the case of ROS, the description of novel NADPH oxidases suggests that a similar theme may emerge in which stucturally related enzyme systems will exist that share overall functional homology but which produce differing amounts of ROS depending on whether the function is host defense or signal transduction (Fig. 1.7).

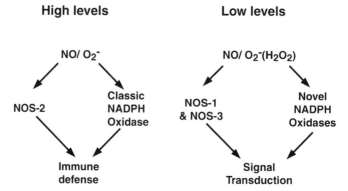

Figure 1.7. *The divergent role of high and low levels of RNS and ROS.* A potential interesting parallel between the production of high and low levels of reactive nitrogen (RNS) and reactive oxygen (ROS) species in vivo. High level of RNS and ROS may be required for host defense, whereas low levels of these compounds are used for signaling pathways. Different but related enzymes are used to accomplish high- or low-level radical production.

Although the production of ROS by the classic or novel NADPH oxidase represents one enzyme system that is potentially responsible for vascular ROS production, a number of other important systems exist. Included among these are enzymes such as xanthine oxidase as well as the mitochondrial respiratory chain. Increased activity of these or other superoxide-generating systems would result in an increase in ROS levels leading to a decrease in bioactive NO and a reduction in vasorelaxation. Presently, the relative contribution of these enzyme systems to vascular reactivity is unknown. One potential strategy to resolve these issues is to explore the phenotype of animals with targeted deletions in the various components and enzyme systems. Initial studies with animals with a knockout of p47phox suggest that these animals have an approximately 50% reduction in production of superoxide within the vessel wall (Hsich et al., 2000). Presumably the production from other non-NADPH oxidases contributes the other half of vascular superoxide production.

Although enzyme systems such as the NADPH oxidase produce superoxide and it is superoxide that can react and inactivate NO, most studies have concentrated on hydrogen peroxide as the ROS that mediates intracellular signaling. Superoxide is in fact rapidly dismutated to hydrogen peroxide. This phenomenon occurs spontaneously but is significantly accelerated by the enzyme superoxide dismutase (SOD). In humans, there are three forms of SOD, a copper- and zinc-containing enzyme that is cytosolic, a manganese-containing form that localizes to the mitochondria, and a secreted form that functions extracellularly. The reason for three separate gene products that all appear to have similar enzymatic function is not entirely clear.

A number of studies have demonstrated that ligand stimulation of a variety of cells including endothelial and vascular smooth muscle cells results in a burst of ROS (Thannickal and Fanburg, 2000). Most of these studies have measured ROS production with membrane-permeant dyes such as dichlorofluoroscein diacetate (DCF), an agent whose fluorescence is dependent in part on intracellular hydrogen peroxide levels. With these methods, studies have demonstrated that this burst in oxidants occurs within minutes of ligand addition and gradually returns to baseline within 30 minutes. In one early study in vascular smooth muscle cells, platelet-derived growth factor (PDGF) caused a rapid increase in DCF fluorescence (Sundaresan et al., 1995). This burst of DCF fluorescence could be blocked by increasing the level of intracellular catalase, an enzyme responsible for hydrogen peroxide degradation. Interestingly, examination of PDGF-dependent tyrosine phosphorylation also demonstrated that blocking the burst of hydrogen peroxide resulted in suppression of downstream signaling. These results suggested that the production of hydrogen peroxide is actually required for downstream PDGF signaling in this cell type. Subsequent studies have significantly extended these observations to a variety of cell types and a variety of different ligands. Included among these are other ligands transduced by tyrosine kinase receptors such as EGF, as well as ligands transduced by G-coupled receptors, such as angiotensin II (Thannickal et al., 2000). In each of these cases, scavenging ROS and, in particular, scavenging hydrogen peroxide, appears to block downstream signal transduction, demonstrating an essential redox-dependent aspect of signaling.

It remains unclear why so many cells produce a burst of ROS after ligand stimulation. One particularly attractive protein target for intracellularly generated hydrogen peroxide is the tyrosine phosphatase class of enzymes. These enzymes all contain cysteine residues in their active site. Similar to the ability of NO to modify the active site of cysteine proteases such as caspases, hydrogen peroxide could potentially directly modify tyrosine phosphatases and thereby modify signaling. In this case, hydrogen peroxide would transiently inactivate tyrosine phosphatases, leading to unopposed tyrosine kinase activity and perhaps allowing for the burst of tyrosine phosphorylation that is observed after the addition of growth factors such as PDGF. Such a mechanism (see Fig. 1.8) could explain how this burst of tyrosine phosphorylation occurs even though ligand binding appears to simultaneously activate both kinases and phosphatases. In such a scenario, hydrogen peroxide would transiently oxidize and hence inactivate tyrosine phosphatases, allowing, at least for a brief time, the unopposed action of tyrosine kinases. The subsequent reduction of redox-modified tyrosine phosphatases, by as yet unclear mechanisms, would allow for the eventual return to baseline of intracellular tyrosine kinase substrate phosphorylation. Given that tyrosine phosphatases are 2–3 orders of magnitude more efficient enzymes than tyrosine kinases, the ability of ROS to transiently inactivate tyrosine phosphatases may be an essential and general mechanism for growth

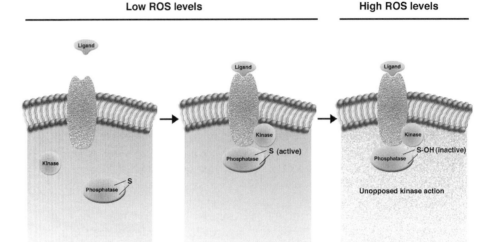

Figure 1.8. *A model for ROS as regulators of receptor signaling.* In this model, ligand binding recruits both tyrosine kinases and phosphatases to the receptor. The activity of phosphatases is, however, reduced by the rise in intracellular ROS that inactivate the critical cysteine residue in the active site of the enzyme. This allows for unopposed action of tyrosine kinases and the burst of tyrosine phosphorylation. When ROS levels fall, phosphatase activity is restored and the increased specific activity of phosphatases, compared to kinases, predominates.

factor signaling. Recent evidence consistent with such a mechanism has begun to emerge. In particular, PTP-1B, a ubiquitous tyrosine phosphatase, was shown to have a redox-dependent reduction in activity following growth factor addition (Lee et al., 1998).

In the case of tyrosine phosphatases it appears that the active cysteine residue essential for enzymatic activity is the cysteine residue directly affected by hydrogen peroxide. Interestingly, a presumably ancient bacterial precedent exists for such redox-dependent cysteine modification. A series of studies have suggested that hydrogen peroxide regulates gene expression in *Escherichia coli* and that this is mediated by a redox-dependent transcription factor, OxyR (Choi et al., 2001). Detailed mechanistic studies have demonstrated that hydrogen peroxide directly regulates OxyR function by oxidation of a specific, reactive cysteine in the molecule. Emerging from these and other studies is the concept of a critical role for reactive cysteine residues. Although most cysteines are not ionized to a thiolate anion at physiological pH, certain residues, presumably because of their unique local environment, appear substantially more reactive. The exact basis for such reactivity is not well understood; however, these cysteines appear often to be surrounded by highly charged basic amino acids in either the primary or tertiary structure. These reactive cysteines therefore represent a general and important class of targets for intracellular ROS. In many ways, this is analogous to the established paradigm that specific tyrosine, serine, or threonine residues are uniquely targeted by intracellular kinases based on both the specific target amino acid and the surrounding amino acid recognition motifs. Phosphorylation on such specific tyrosine, serine, or threonine residues provides in turn a reversible means to alter protein function. Emerging evidence suggests that a similar reversible modification by oxidants can occur again by the covalent modification of unique amino acids, in this case reactive cysteine residues. Specific cysteine residues targeted by both ROS may also be targeted by RNS. The oxidation of these critical cysteines in turn would presumably alter protein function. In the case of tyrosine phosphatases, this alteration leads to inactivation. Nevertheless, as long as the oxidant burst is not too excessive and the irreversible sulfinic species is avoided, the oxidation of the reactive cysteine is thought to be reversible. The exact basis for the intracellular reduction of the now oxidized cysteine residue is not established; however, the participation of glutathione, a three-amino acid reducing agent present intracellularly in millimolar concentrations, represents an attractive candidate. In such a scenario (see Fig. 1.9), after oxidant stress

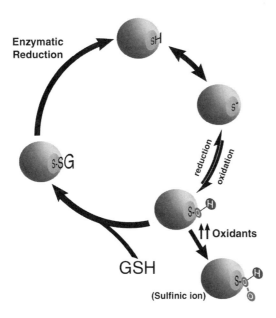

Figure 1.9. *A putative cycle of redox modification of target proteins.* The signature of specific targets for hydrogen peroxide in cells may be the presence of a reactive cysteine residue. This amino acid can be ionized at physiological pH and is therefore a target for ROS-mediated signaling. Oxidation of the cysteine to the sulfenic form is reversible, potentially by interaction with glutathione (GSH) forming a mix-disulfide intermediate. Stronger oxidants could result in further oxidation to the sulfonic ion, which is generally not viewed as a reversible modification.

a mixed disulfide would occur between the molecule with a reactive cysteine and glutathione. The cycle outlined in Figure 1.9 represents a specific and reversible redox cycle analogous, as described above, to the specific and reversible means to modify protein function by phosphorylation. Evidence to support such a scheme is beginning to emerge, including the observation of the transient redox-dependent glutathiolation of PTP-1B induced by growth factor addition (Barrett et al., 1999). In addition, recent methods designed to trap the mixed disulfide form have allowed for a proteomic approach to potentially identify direct targets of ROS (Sullivan et al., 2000).

Finally, it is important to note that the pharmacologic treatment of atherosclerotic disease has in many instances been effective at altering the availability or increasing the activity of NO within the vessel wall. This increase in NO bioavailability by pharmacological agents happens through both direct and indirect methods. The widespread use of antioxidants is conceived to work by lowering the levels of cellular ROS and thereby providing an atheroprotective effect. Although some studies have shown effective risk reduction with agents such as vitamin E, on balance, most randomized studies have seen either small effects or none at all (Meagher and Rader, 2001). The characterization of the enzymatic source of vascular oxidant production should hopefully lead to the development of specifically targeted agents that will represent a new class of antioxidant agents that do not act as scavengers of ROS but actually inhibit production.

Manipulation of vascular NO levels is also the basis of several widely used pharmacological agents. We have already discussed how cholesterol-lowering statin therapy may lead to increased Akt activity, leading in turn to higher eNOS activity and more NO production. Perhaps more direct is the action of nitroglycerin and other longer-acting nitrates long known to be effective in relieving anginal symptoms. It is now appreciated that nitrates provide an exogenous source of NO, inducing coronary vasodilatation. The antianginal mechanism is produced by dilating large coronary arteries and arterioles, relieving myocardial ischemia especially induced by exercise. This vasodilator effect is independent of the endothelium but exerts its effects by stimulating guanylate cyclase. Clinically, it has been noted that the ability of nitrates to relieve anginal symptoms disappears with continuous usage. This phenomena, termed "nitrate tolerance," was thought to arise after depletion of the intracellular sulfhydryl groups (Fung and Bauer, 1994); however, more recently it has been postulated that excess nitrate could lead to the formation of peroxynitrite, which can directly inhibit guanylate cyclase. Finally, one other therapeutic agent related to the NO pathway is sildenafil, commonly known as Viagra. This agent works not by increasing NO levels but instead by inhibiting phosphodiesterases that are responsible for the breakdown of cGMP in smooth muscle cells. This in turn allows for a NO mimetic effect by maintaining high levels of cGMP. The success of this drug underlies how understanding of the signal transduction pathways leading to vascular relaxation may continue to provide new agents leading to improved health, not to mention happiness.

In summary, both RNS and ROS appear to play an important role in vascular homeostasis. The idea that free radicals can be purposely produced and that they can have specific targets represents an important paradigm shift. The concept that ROS and RNS can act specifically will undoubtably have significant implications in other disease processes such as aging or neurodegenerative diseases, in which a variety of evidence suggests free radicals play a role. Further studies understanding these redox-regulated signaling pathways will hopefully provide a variety of new treatment strategies to combat the growing pandemic of cardiovascular disease.

REFERENCES

Barrett, W. C., DeGnore, J. P., Keng, Y. F., Zhang, Z. Y., Yim, M. B., and Chock, P. B. (1999). Roles of superoxide radical anion in signal transduction mediated by reversible regulation of protein-tyrosine phosphatase 1B. J Biol Chem 274, 34543–34546.

Barry-Lane, P. A., Patterson, C., Van der Merwe, M., Hu, Z. Y., Holland, S. M., Yeh, E. T. H., and Runge, M. S. (2001). p47phox is required for atherosclerotic lesion progression in ApoE(−/−) mice. J Clin Invest 108, 1513–1522.

Boulanger, C. M., Heymes, C., Benessiano, J., Geske, R. S., Levy, B. I., and Vanhoutte, P. M. (1998). Neuronal nitric oxide synthase is expressed in rat vascular smooth muscle cells: activation by angiotensin II in hypertension. Circ Res 83, 1271–1278.

Bredt, D. S., Hwang, P. M., Glatt, C. E., Lowenstein, C., Reed, R. R., and Snyder, S. H. (1991). Cloned and expressed nitric oxide synthase structurally resembles cytochrome P-450 reductase. Nature 351, 714–718.

Buttery, L. D., Springall, D. R., Chester, A. H., Evans, T. J., Standfield, E. N., Parums, D. V., Yacoub, M. H., and Polak, J. M. (1996). Inducible nitric oxide synthase is present within human atherosclerotic lesions and promotes the formation and activity of peroxynitrite. Lab Invest 75, 77–85.

Celermajer, D. S., Sorensen, K. E., Gooch, V. M., Spiegelhalter, D. J., Miller, O. I., Sullivan, I. D., Lloyd, J. K., and Deanfield, J. E. (1992). Non-invasive detection of endothelial dysfunction in children and adults at risk of atherosclerosis. Lancet 340, 1111–1115.

Choi, H., Kim, S., Mukhopadhyay, P., Cho, S., Woo, J., Storz, G., and Ryu, S. (2001). Structural basis of the redox switch in the OxyR transcription factor. Cell 105, 103–113.

Cornwell, T. L., Arnold, E., Boerth, N. J., and Lincoln, T. M. (1994). Inhibition of smooth muscle cell growth by nitric oxide and activation of cAMP-dependent protein kinase by cGMP. Am J Physiol 267, C1405–C1413.

D'Agostino, R. B., Russell, M. W., Huse, D. M., Ellison, R. C., Silbershatz, H., Wilson, P. W., and Hartz, S. C. (2000). Primary and subsequent coronary risk appraisal: new results from the Framingham study. Am Heart J 139, 272–281.

Davis, K. L., Martin, E., Turko, I. V., and Murad, F. (2001). Novel effects of nitric oxide. Annu Rev Pharmacol Toxicol *41*, 203–236.

Dimmeler, S., Fleming, I., Fisslthaler, B., Hermann, C., Busse, R., and Zeiher, A. M. (1999). Activation of nitric oxide synthase in endothelial cells by Akt-dependent phosphorylation. Nature *399*, 601–605.

Dimmeler, S., Haendeler, J., Nehls, M., and Zeiher, A. M. (1997). Suppression of apoptosis by nitric oxide via inhibition of interleukin-1β-converting enzyme (ICE)-like and cysteine protease protein (CPP)-32-like proteases. J Exp Med *185*, 601–607.

Drab, M., Verkade, P., Elger, M., Kasper, M., Lohn, M., Lauterbach, B., Menne, J., Lindschau, C., Mende, F., Luft, F. C., Schedl, A., Haller, H., and Kurzchalia, T. V. (2001). Loss of caveolae, vascular dysfunction, and pulmonary defects in caveolin-1 gene-disrupted mice. Science *293*, 2449–2452.

Feron, O., Belhassen, L., Kobzik, L., Smith, T. W., Kelly, R. A., and Michel, T. (1996). Endothelial nitric oxide synthase targeting to caveolar. Specific Interactions with caveolin isoforms in Cardiac Myocytes and Endothelial Cells. J Biol Chem *271*, 22810–22814.

Fulton, D., Gratton, J. P., McCabe, T. J., Fontana, J., Fujio, Y., Walsh, K., Franke, T. F., Papapetropoulos, A., and Sessa, W. C. (1999). Regulation of endothelium-derived nitric oxide production by the protein kinase Akt. Nature *399*, 597–601.

Fung, H. L., and Bauer, J. A. (1994). Mechanisms of nitrate tolerance. Cardiovasc Drugs Ther *8*, 489–499.

Furchgott, R. F., and Zawadzki, J. V. (1980). The obligatory role of endothelial cells in the relaxation of arterial smooth muscle by acetylcholine. Nature *288*, 373–376.

Garcia-Cardena, G., Fan, R., Stern, D. F., Liu, J., and Sessa, W. C. (1996). Endothelial nitric oxide synthase is regulated by tyrosine phosphorylation and interacts with caveolin-1. J Biol Chem *271*, 27237–27240.

Garg, U. C., and Hassid, A. (1989). Nitric oxide-generating vasodilators and 8-bromo-cyclic guanosine monophosphate inhibit mitogenesis and proliferation of cultured rat vascular smooth muscle cells. J Clin Invest *83*, 1774–1777.

Gow, A. J., Duran, D., Malcolm, S., and Ischiropoulos, H. (1996). Effects of peroxynitrite-induced protein modifications on tyrosine phosphorylation and degradation. FEBS Lett *385*, 63–66.

Hsich, E., Segal, B. H., Pagano, P. J., Rey, F. E., Paigen, B., Deleonardis, J., Hoyt, R. F., Holland, S. M., and Finkel, T. (2000). Vascular effects following homozygous disruption of p47 (phox): An essential component of NADPH oxidase. Circulation *101*, 1234–1236.

Ignarro, L. J., Buga, G. M., Wood, K. S., Byrns, R. E., and Chaudhuri, G. (1987). Endothelium-derived relaxing factor produced and released from artery and vein is nitric oxide. Proc Natl Acad Sci USA *84*, 9265–9269.

Ignarro, L. J., Harbison, R. G., Wood, K. S., and Kadowitz, P. J. (1986). Activation of purified soluble guanylate cyclase by endothelium-derived relaxing factor from intrapulmonary artery and vein: stimulation by acetylcholine, bradykinin and arachidonic acid. J Pharmacol Exp Ther *237*, 893–900.

Kannel, W. B. (2000). The Framingham Study: Its 50-year legacy and future promise. J Atheroscler Thromb *6*, 60–66.

Kureishi, Y., Luo, Z., Shiojima, I., Bialik, A., Fulton, D., Lefer, D. J., Sessa, W. C., and Walsh, K. (2000). The HMG-CoA reductase inhibitor simvastatin activates the protein kinase Akt and promotes angiogenesis in normocholesterolemic animals. Nat Med *6*, 1004–1010.

Lambeth, J. D., Cheng, G., Arnold, R. S., and Edens, W. A. (2000). Novel homologs of gp91phox. Trends Biochem Sci *25*, 459–461.

Lee, S. R., Kwon, K. S., Kim, S. R., and Rhee, S. G. (1998). Reversible inactivation of protein-tyrosine phosphatase 1B in A431 cells stimulated with epidermal growth factor. J Biol Chem *273*, 15366–15372.

Lincoln, T. M., Dey, N., and Sellak, H. (2001). cGMP-dependent protein kinase signaling mechanisms in smooth muscle: from the regulation of tone to gene expression. J Appl Physiol *91*, 1421–1430.

Luoma, J. S., Stralin, P., Marklund, S. L., Hiltunen, T. P., Sarkioja, T., and Yla-Herttuala, S. (1998). Expression of extracellular SOD and iNOS in macrophages and smooth muscle cells in human and rabbit atherosclerotic lesions: colocalization with epitopes characteristic of oxidized LDL and peroxynitrite-modified proteins. Arterioscler Thromb Vasc Biol *18*, 157–167.

MacMillan-Crow, L. A., Greendorfer, J. S., Vickers, S. M., and Thompson, J. A. (2000). Tyrosine nitration of c-SRC tyrosine kinase in human pancreatic ductal adenocarcinoma. Arch Biochem Biophys *377*, 350–356.

Mallat, Z., and Tedgui, A. (2000). Apoptosis in the vasculature: mechanisms and functional importance. Br J Pharmacol *130*, 947–962.

Mannick, J. B., Hausladen, A., Liu, L., Hess, D. T., Zeng, M., Miao, Q. X., Kane, L. S., Gow, A. J., and Stamler, J. S. (1999). Fas-induced caspase denitrosylation. Science *284*, 651–654.

Meagher, E., and Rader, D. J. (2001). Antioxidant therapy and atherosclerosis: animal and human studies. Trends Cardiovasc Med *11*, 162–165.

Michel, T., Li, G. K., and Busconi, L. (1993). Phosphorylation and subcellular translocation of endothelial nitric oxide synthase. Proc Natl Acad Sci USA *90*, 6252–6256.

Palmer, R. M., Ferrige, A. G., and Moncada, S. (1987). Nitric oxide release accounts for the biological activity of endothelium-derived relaxing factor. Nature *327*, 524–526.

Papapetropoulos, A., Desai, K. M., Rudic, R. D., Mayer, B., Zhang, R., Ruiz-Torres, M. P., Garcia-Cardena, G., Madri, J. A., and Sessa, W. C. (1997). Nitric oxide synthase inhibitors attenuate transforming-growth-factor-β1-stimulated capillary organization in vitro. Am J Pathol *150*, 1835–1844.

Pfitzer, G., Sonntag-Bensch, D., and Brkic-Koric, D. (2001). Thiophosphorylation-induced Ca^{2+} sensitization of guinea-pig ileum contractility is not mediated by Rho-associated kinase. J Physiol *533*, 651–664.

Quyyumi, A. A., Dakak, N., Andrews, N. P., Husain, S., Arora, S., Gilligan, D. M., Panza, J. A., and Cannon, R. O., III (1995). Nitric oxide activity in the human coronary circulation. Impact of risk factors for coronary atherosclerosis. J Clin Invest *95*, 1747–1755.

Schachinger, V., Britten, M. B., and Zeiher, A. M. (2000). Prognostic impact of coronary vasodilator dysfunction on adverse long-term outcome of coronary heart disease. Circulation *101*, 1899–1906.

Sessa, W. C., Pritchard, K., Seyedi, N., Wang. J., and Hintze, T. H. (1994). Chronic exercise in dogs inreases coronary vascular nitric oxide production and endothelial cell nitric oxide synthase gene expression. Circ Res *74*, 349–353.

Shaul, P. W., Smart, E. J., Robinson, L. J., German, Z., Yuhanna, I. S., Ying, Y., Anderson, R. G., and Michel, T. (1996). Acylation targets endothelial nitric-oxide synthase to plasmalemmal caveolae. J Biol Chem *271*, 6518–6522.

Stokes, J., III, Kannel, W. B., Wolf, P. A., Cupples, L. A., and D'Agostino, R. B. (1987). The relative importance of selected risk factors for various manifestations of cardiovascular disease among men and women from 35 to 64 years old: 30 years of follow-up in the Framingham Study. Circulation *75*, V65–V73.

Stuehr, D. J. (1997). Structure-function aspects in the nitric oxide synthases. Annu Rev Pharmacol Toxicol *37*, 339–359.

Suh, Y. A., Arnold, R. S., Lassegue, B., Shi, J., Xu, X., Sorescu, D., Chung, A. B., Griendling, K. K., and Lambeth, J. D. (1999). Cell transformation by the superoxide-generating oxidase Mox1. Nature *401*, 79–82.

Sullivan, D. M., Wehr, N. B., Fergusson, M. M., Levine, R. L., and Finkel, T. (2000). Identification of oxidant-sensitive proteins: THF-α induces protein glutathiolation. Biochemistry *39*, 11121–11128.

Sundaresan, M., Yu, Z. X., Ferrans, V. J., Irani, K., and Finkel, T. (1995). Requirement for generation of H_2O_2 for platelet-derived growth factor signal transduction. Science *270*, 296–299.

Surks, H. K., Mochizuki, N., Kasai, Y., Georgescu, S. P., Tang, K. M., Ito, M., Lincoln, T. M., and Mendelsohn, M. E. (1999). Regulation of myosin phosphatase by a specific interaction with cGMP-dependent protein kinase Iα. Science *286*, 1583–1587.

Thannickal, V. J., Day, R. M., Klinz, S. G., Bastien, M. C., Larios, J. M., and Fanburg, B. L. (2000). Ras-dependent and -independent regulation of reactive oxygen species by mitogenic growth factors and TGF-β1. FASEB J *14*, 1741–1748.

Thannickal, V. J., and Fanburg, B. L. (2000). Reactive oxygen species in cell signaling. Am J Physiol *279*, L1005–L1028.

Uematsu, M., Ohara, Y., Navas, J. P., Nishida, K., Murphy, T. J., Alexander, R. W., Nerem, R. M., and Harrison, D. G. (1995). Regulation of endothelial cell nitric oxide synthase mRNA expression by shear stress. Am J Physiol *269*, C1371–C1378.

Virchow, R. (1856). *Gesammelte Abhandhungen zur Wissenchaftlichen Medicin.* (Frankfurt-am-Main, Germany: Meidinger Sohn), p. 458.

Wang, W., Wang, S., Yan, L., Madara, P., Del Pilar, C. A., Wesley, C. A., and Danner, R. L. (2000). Superoxide production and reactive oxygen species signaling by endothelial nitric-oxide synthase. J Biol Chem *275*, 16899–16903.

Warnholtz, A., Nickenig, G., Schulz, E., Macharzina, R., Brasen, J. H., Skatchkov, M., Heitzer, T., Stasch, J. P., Griendling, K. K., Harrison, D. G., Bohm, M., Meinertz, T., and Munzel, T. (1999). Increased NADH-oxidase-mediated superoxide production in the early stages of atherosclerosis: evidence for involvement of the renin-angiotensin system. Circulation *99*, 2027–2033.

Wilcox, J. N., Subramanian, R. R., Sundell, C. L., Tracey, W. R., Pollock, J. S., Harrison, D. G., and Marsden, P. A. (1997). Expression of multiple isoforms of nitric oxide synthase in normal and atherosclerotic vessels. Arterioscler Thromb Vasc Biol *17*, 2479–2488.

Zou, M. H., Leist, M., and Ullrich, V. (1999). Selective nitration of prostacyclin synthase and defective vasorelaxation in atherosclerotic bovine coronary arteries. Am J Pathol *154*, 1359–1365.

CHAPTER 2

NF-κB: A KEY SIGNALING PATHWAY IN ASTHMA

STEWART J. LEVINE
Pulmonary-Critical Care Medicine Branch, National Heart, Lung, and Blood Institute, National Institutes of Health, Bethesda, Maryland

INTRODUCTION

Asthma is a chronic inflammatory disease of the airways with a substantial public health and economic impact. It is the ninth most common chronic illness in the United States, affecting at least 14 to 15 million individuals (Collins, 1997; Mannino et al., 1998). Furthermore, asthma prevalence and morbidity rates have increased over the past two decades, as evidenced by increases in the number of asthma-related hospitalizations and deaths (Mannino et al., 1998; Elias, 1999). The effects of asthma also have a disproportionate impact on children. Asthma is the most common chronic illness of childhood and was responsible for an estimated 198,000 pediatric hospitalizations in 1993 (CDC, 1996; Mannino et al., 1998). There is, in addition, a substantial cost related to asthma care, which was estimated to total $14.5 billion dollars in the United States during 2000 (CDC, 1997).

Signal transduction pathways play a critical role in the recruitment and activation of inflammatory cells, such as lymphocytes and eosinophils, in the asthmatic airway. Transcription factors that have been implicated in the pathogenesis of asthma include signal transducers and activators of transcription (STAT), activator protein-1 (AP-1), nuclear factor of activated T cells (NFAT), cyclic AMP response-element binding proteins (CREB), guanine-adenine and thymine-adenine repeats (GATA), Ets family proteins, and nuclear factor-κB (NF-κB) (Rahman and MacNee, 1998; Busse and Lemanske, 2001; Finotto et al., 2001).

The NF-κB pathway plays an important role in the pathogenesis of several important human inflammatory diseases, including cancer, dia-

Signal Transduction and Human Disease, Edited by Toren Finkel and
J. Silvio Gutkind
ISBN 0-471-02011-7   Copyright © 2003 John Wiley & Sons, Inc.

betes, AIDS, rheumatoid arthritis, atherosclerosis, multiple sclerosis, inflammatory bowel disease, *Helicobacter pylori*-associated gastritis, chronic inflammatory demyelinating polyradiculoneuritis, euthyroid sick syndrome, and the systemic inflammatory response syndrome (Baldwin, 2001; Tak and Firestein, 2001). Activation of the NF-κB signaling pathway in response to inflammation, stress, infection, or allergy induces the targeted phosphorylation and degradation of the inhibitory protein IκB and subsequent translocation of NF-κB into the nucleus (Baldwin, 1996; Rahman and MacNee, 1998; Tak and Firestein, 2001). Binding of NF-κB dimers to target promoters initiates the transcriptional activation of a wide variety of proinflammatory and immunomodulatory genes (Baldwin, 1996; Rahman and MacNee, 1998). Activation of NF-κB also plays an important role in the regulation of apoptosis, cellular proliferation, responsiveness to cancer chemotherapeutic agents, and viral (e.g., HIV) transcription and replication (Baldwin, 2001; Tak and Firestein, 2001).

Recent investigations have also identified NF-κB as an essential transcription factor in the pathogenesis of asthma. This chapter reviews the clinical manifestations and pathogenesis of asthma and provides an overview of the NF-κB/Rel family of transcription factors. Regulation of the NF-κB signaling pathway by IκB proteins and IκB kinases as well as the important role of the NF-κB signaling pathway in the pathogenesis of asthmatic airway inflammation are considered.

ASTHMA—CLINICAL MANIFESTATIONS

Asthma is not a single disease entity, but rather a syndrome resulting from multiple pathogenetic mechanisms that produce common signs and symptoms (Fish and Peters, 1998). Asthma frequently occurs in the context of atopy, with aeroallergens commonly acting as inducers of asthmatic symptoms. Furthermore, allergen exposure plays a key role in both initiating and sustaining airway inflammatory responses that predispose susceptible atopic individuals to heightened airway hyperreactivity (McFadden and Gilbert, 1992; Fish and Peters, 1998; Busse and Lemanske, 2001). Nonspecific stimuli, such as viral upper respiratory tract infections, exercise, cold air, smoke, oxidant air pollutants, strong vapors, noxious fumes, and particulate matter, can then trigger airway hyperresponsiveness in asthmatic patients with underlying airway inflammation (McFadden and Gilbert, 1992; Fish and Peters, 1998). Sensitivity to nonsteroidal inflammatory medications and occupational exposures can also induce the asthmatic phenotype in the absence of atopy (Fish and Peters, 1998).

Asthma can be diagnosed utilizing the following criteria: (1) symptoms of episodic airflow obstruction, (2) the presence of airflow obstruction that is at least partially reversible, and (3) the exclusion of alternative diagnoses (e.g., vocal cord dysfunction, vascular rings, foreign bodies, upper airway malignancies) (Murphy et al., 1997). Symptoms consistent with asthma include cough, shortness of breath (e.g., dyspnea),

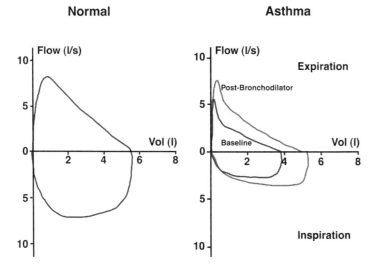

Figure 2.1. *Airflow Obstruction.* The presence of airflow obstruction can be demonstrated by a decrease in the FEV_1. The FEV_1 represents the forced expiratory volume of gas that can be exhaled in 1 s. The FEV_1 is measured by having the patient perform a maximal inhalation and then exhaling as rapidly as possible [e.g., the forced vital capacity (FVC)]. The FVC is displayed as flow (in liters per second) vs. volume (liters) with expiratory flow plotted above and inspiratory flow plotted below the horizontal axis. Both the peak expiratory flow and FEV_1 are decreased in the patient with asthma at baseline (graph on **right**). Furthermore, the expiratory flow curve is characteristically concave in asthmatic airflow limitation. There is a significant improvement in both the peak expiratory flow and the FEV_1 after administration of a short-acting β_2 agonist (e.g., albuterol). The flow-volume loop of a normal individual is on the **left** for comparison.

wheezing, and chest tightness. Because asthma symptoms are nonspecific, the diagnosis of asthma requires the documentation of reversible airflow obstruction, which can be assessed by pulmonary function testing. The most reproducible test for the presence of airflow obstruction is the FEV_1. The FEV_1 is the forced expiratory volume or the volume of gas that can be exhaled in 1 second after a maximal inhalation. An FEV_1 of less than 80% of predicted is consistent with the presence of airflow obstruction. To establish reversible airflow limitation, the FEV_1 should increase by more than 12% and at least 200 ml after inhalation of a short-acting β_2-agonist, such as albuterol (see Fig. 2.1). Methacholine challenge testing can also be utilized to assess airway hyperresponsiveness. Asthmatic patients with airway hyperreactivity will demonstrate at least a 20% decrease in FEV_1 after inhalation of small amounts of methacholine, a synthetic derivative of the neurotransmitter acetylcholine (Crapo et al., 2000).

A stepwise approach to the classification of asthma severity was developed by a National Heart, Lung, and Blood Institute (NHLBI)

Expert Panel to provide guidelines for the management of asthma (Murphy et al., 1997). Asthma severity can be classified as mild intermittent, mild persistent, moderate persistent, or severe persistent on the basis of the frequency of asthma symptoms, FEV_1, and variability in peak expiratory flow rates. This classification scheme can then be utilized to design a stepwise approach for the management of asthma symptoms. For example, adult patients with mild intermittent symptoms (e.g., Step 1) can be treated with a short-acting inhaled β_2-agonist as required for symptoms. In contrast, patients with mild, moderate, or severe persistent asthma (e.g., Steps 2, 3, and 4) require daily, long-term anti-inflammatory medications, as well as usually inhaled corticosteroids, to help reduce inflammation and stabilize asthma symptoms. At present, inhaled corticosteroids are the most effective medication for long-term control of asthma and have been demonstrated to reduce asthma symptoms, reduce the occurrence of severe exacerbations, decrease the frequency of short-acting inhaled β_2-agonist use, reduce airway hyperresponsiveness, and improve lung function. Therefore, the goal of asthma therapy should be the administration of optimal anti-inflammatory medications to prevent acute exacerbations and chronic asthma symptoms, while avoiding drug toxicity.

PATHOGENESIS OF ASTHMATIC AIRWAY INFLAMMATION

It is clear that the pathogenetic mechanism underlying the initiation, perpetuation, and modulation of asthma is the presence of airway inflammation. The presence of airway inflammation was initially recognized in postmortem studies of fatal asthma that demonstrated bronchial wall infiltration by increased numbers of eosinophils, lymphocytes, and neutrophils, in addition to the classic findings of luminal mucus plugging, goblet cell and mucus gland hyperplasia, bronchial smooth muscle hyperplasia and hypertrophy, epithelial cell desquamation, basement membrane thickening, and lung hyperinflation (Fig. 2.2) (James and Carroll, 1998; Busse and Lemanske, 2001). Mast cells and plasma cells may also participate in the pathogenesis of asthmatic airway inflammation (James and Carroll, 1998). Examination of bronchial biopsies subsequently led to the recognition that airway inflammation, epithelial shedding, and basement membrane thickening are also present in patients with mild, stable asthma (Azzawi et al., 1990; Laitinen et al., 1993; Vignola et al., 1998; Busse and Lemanske, 2001). Furthermore, airway structural remodeling, resulting in fixed or only partially reversible airflow obstruction, may develop as a consequence of chronic airway inflammation (Elias et al., 1999). Pathological manifestations of airway remodeling include airway wall thickening, subepithelial fibrosis, mucus metaplasia, myofibroblast hyperplasia, myocyte hyperplasia and hypertrophy, and epithelial cell hypertrophy (Elias et al., 1999). Mediators that have been implicated in the pathogenesis of airway remodeling

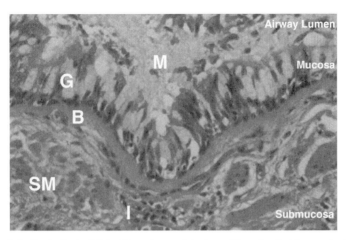

Figure 2.2. *Pathological Manifestations of Asthma.* The classic histopathological manifestations of asthma are present in the airway of this asthmatic patient. The airway lumen is occluded with mucus (M) as a consequence of bronchial mucosal goblet cell (G) and mucus gland hyperplasia. The basement membrane (B) is thickened, and increased numbers of inflammatory cells (I) are present in the submucosa. Bronchial smooth muscle (SM) hyperplasia and hypertrophy are also present. (Photograph courtesy of William D. Travis, M.D., Armed Forces Institute of Pathology, Washington, DC.)

and fibrosis include TGF-β, interleukin (IL)-6, IL-11, and IL-13 (Elias et al., 1999).

Allergic airway inflammation is the most common etiologic factor associated with the initiation of asthma and is thought to develop as a consequence of inappropriate IgE-mediated immune responses in patients with a genetic predisposition to atopy (Wills-Karp, 1999). Recent investigations have provided evidence that the origins of the asthmatic diathesis in genetically susceptible, atopic individuals may arise early in life, perhaps in utero, and are strongly correlated with the presence of allergy (Elias, 1999; Gern et al., 1999). Evidence that allergen sensitization occurs in utero includes the demonstration of allergen-specific proliferative responses in cord blood lymphocytes (Gern et al., 1999). Similarly, the house dust mite antigen, Der p 1, has been detected in both amniotic fluid and fetal cord blood samples (Holloway et al., 2000). Subsequent patient-environment interactions during infancy and childhood, such as viral upper respiratory tract infections, diet, and toxin exposure, may then contribute to the establishment of the asthmatic phenotype (Gern et al., 1999).

T_H2-type CD4$^+$ T cells have been recognized as key players in initiating and perpetuating allergic asthmatic responses (Elias et al., 1999; Wills-Karp, 1999; Busse and Lemanske, 2001). Induction of T_H2-type CD4$^+$ T cell responses requires antigen presentation by dendritic cells (Fig. 2.3) (Wills-Karp, 1999). Once activated, T_H2-type CD4$^+$ T cells

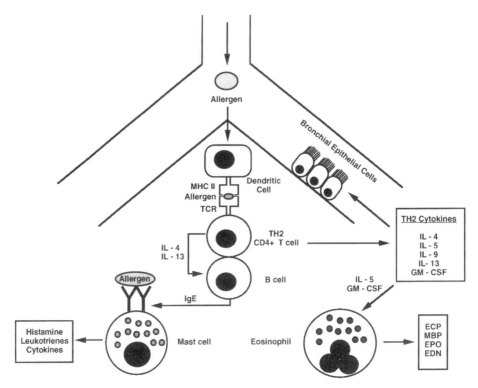

Figure 2.3. *Pathogenesis of Atopic Asthma.* Inhaled allergen is presented by MHC class II molecules on airway dendritic cells to T_H2-type $CD4^+$ T cells. The activation of T_H2-type $CD4^+$ T cells induces the production of a characteristic set of cytokines that are essential for the pathogenesis of atopic airway inflammation. The T_H2-type cytokines, IL-4 and IL-13, provide the first signal to B cells for isotype switching to IgE synthesis. Cross-linking of IgE bound to mast cells by allergen induces the immediate release of preformed granule components that mediate the immediate- or early-phase allergic asthmatic response. IL-5 and granulocyte-macrophage colony-stimulating factor (GM-CSF) induce the proliferation, activation, and enhanced survival of eosinophils in the asthmatic airway. IL-5 is also an eosinophil chemotactic factor. IL-9 mediates goblet cell hyperplasia and enhanced airway epithelial cell mucin gene expression. Activated eosinophils cause epithelial cell injury as a result of the generation of cytotoxic proteins, such as eosinophilic cationic protein (ECP), major basic protein (MBP), eosinophil peroxidase (EPO), and eosinophil-derived neurotoxin (EDN).

produce a characteristic set of cytokines, including IL-4, IL-5, IL-9, IL-10, and IL-13, that promote allergic inflammatory airway responses (Elias et al., 1999; Wills-Karp, 1999). Transgenic mice overexpressing IL-4, IL-5, IL-9, or IL-13 demonstrate many of the features associated with the asthmatic phenotype, such as lymphocytic and eosinophilic airway inflammation and mucus metaplasia (Elias et al., 1999). In addition, individual T_H2-type cytokines mediate important biological functions that contribute to asthmatic airway inflammation. For example, IL-4 is required for the induction of T_H2-type $CD4^+$ T cells, whereas IL-13 signaling through the IL-4Rα chain is an important effector pathway, capable of mediating airway hyperreactivity, eosinophilia, and mucus production (Grunig et al., 1998; Vogel, 1998; Wills-Karp et al., 1998; Ray and Cohn, 1999). IL-5 is essential for eosinophil activation, chemotaxis, proliferation, and survival, whereas IL-9 mediates goblet cell hyperplasia and mastocytosis (Ray and Cohn, 1999; Townsend et al., 2000; Busse and Lemanske, 2001). Of note, other cell types, such as $CD8^+$ T cells, γ/δ T cells, eosinophils, and mast cells also produce T_H2-type cytokines and enhance airway inflammation and hyperreactivity in animal models (Ray and Cohn, 1999). In addition, other regulatory pathways, such as components of the innate immune system (e.g., complement factor 5), may also contribute to the pathogenesis of asthmatic airway inflammation (Karp et al., 2000).

The T_H2-type cytokines IL-4 and IL-13 deliver the first signal for B cell isotype switching to IgE synthesis, which is followed by a second signal generated via the interaction between T cell CD40 ligand and B cell CD40, as well as interactions between other costimulatory molecules, such as CD28 and B7 (Busse and Lemanske, 2001). IgE then binds to high-affinity IgE receptors (FcεRI) on tissue mast cells and low-affinity IgE receptors (FcεRII, CD23) on eosinophils, lymphocytes, platelets, and macrophages, to mediate cellular activation and mediator release when cross-linked by allergen (Busse and Lemanske, 2001). Bridging of mast cell FcεRI receptors by allergens induces the immediate release of preformed granule contents, such as histamine, prostaglandins, leukotrienes, and platelet-activating factor. These mediators produce bronchoconstriction, microvascular leak, and mucus production as part of the early-phase asthmatic response, which occurs within minutes (Wills-Karp, 1999; Busse and Lemanske, 2001). A late-phase response occurs several hours later. The late-phase response is manifested by airway inflammation, airflow obstruction, and airway hyperreactivity occurring as a consequence of inflammatory cell recruitment (Fig. 2.4). Both inflammatory cells and resident airway cells can generate cytokines and chemokines that upregulate allergic airway inflammation (Busse and Lemanske, 2001). For example, chemokines, such as eotaxin, RANTES (regulated on activation, normal T cell expressed and secreted), and macrophage inflammatory protein-1α, play an important role in eosinophil chemotaxis to the asthmatic airway (Busse and Lemanske, 2001). Eosinophils are important effector cells in asthmatic airway inflammation that cause epithelial cell injury via the generation of cytotoxic proteins, such as eosinophilic cationic protein (ECP) and major basic protein (MBP) (Elias et al., 1999; Wills-Karp, 1999; Busse and Lemanske, 2001).

THE NF-κB/Rel FAMILY OF TRANSCRIPTION FACTORS

Since its initial description by Sen and Baltimore in 1986 as a DNA-binding protein capable of interacting with B cell immunoglobulin κ light

Pre-Allergen Challenge   **Post-Allergen Challenge**

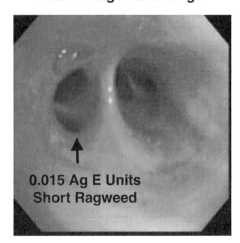

Figure 2.4. *Late-Phase Asthmatic Response.* The late-phase asthmatic response is demonstrated after a segmental allergen challenge. With fiberoptic bronchoscopy, 0.015 antigen E units of short ragweed were instilled in a subsegmental bronchus of an atopic asthmatic patient. Twenty-four hours after the bronchoscopic segmental allergen challenge, the subsegmental bronchus is edematous and 50% occluded. In addition, increased airway mucus secretions are present. This asthmatic patient gave informed consent for participation in a protocol involving a bronchoscopic segmental allergen challenge that was approved by the National Heart, Lung, and Blood Institute Institutional Review Board for the protection of human subjects.

chain enhancer sequences, NF-κB has been recognized as one of the key signaling molecules in pathways that lead to transcriptional activation (Sen and Baltimore, 1986). The NF-κB/Rel family of transcription factors has five mammalian members, each of which contains a highly conserved, approximately 300-amino acid, N-terminal Rel homology domain with two immunoglobulin-like motifs (Baeuerle and Baltimore, 1996; Baldwin, 1996; Karin and Ben-Neriah, 2000). The NF-κB Rel homology domain mediates dimerization of NF-κB family members, DNA binding, and binding to inhibitory IκB proteins and also contains a nuclear localization signal (Baeuerle and Baltimore, 1996; Ghosh et al., 1998; Karin and Ben-Neriah, 2000).

Among members of the mammalian NF-κB/Rel family are NF-κB1 (p50), NF-κB2 (p52), c-Rel, RelA (p65), and RelB (Fig. 2.5) (Baldwin, 1996; Ghosh et al., 1998; Karin and Ben-Neriah, 2000). Other Rel family members include the *Drosophila* proteins *dorsal*, *dif*, and *relish*, which regulate morphogenesis, immunity, and host defense (Ghosh et al., 1998). c-Rel, RelA (p65), and RelB contain C-terminal transactivation domains that are responsible for the recruitment of the transcriptional apparatus with resultant increases in gene transcription (Makarov, 2000). c-Rel, RelA (p65), and RelB do not require proteolytic processing to generate

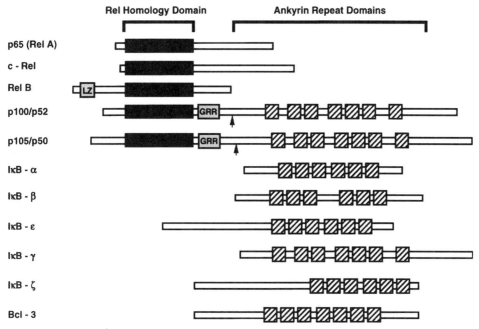

Figure 2.5. *NF-κB and IκB Family Members.* NF-κB family members contain a highly conserved, approximately 300-amino acid, N-terminal Rel homology domain (black box). The Rel homology domain mediates dimerization of NF-κB family members, DNA binding, and binding to inhibitory IκB proteins and also contains a nuclear localization signal. IκB family members contain multiple ankyrin repeats, which mediate specific protein associations with the dimerized immunoglobulin-like folds of the NF-κB Rel homology domain. GRR denotes the glycine-rich region of p100 and p105; LZ denotes the leucine zipper domain of RelB. This diagram is adapted from Baldwin (1996), Ghosh et al. (1998), and Karin and Ben-Neriah (2000).

an active form (Perkins, 2000). In contrast, NF-κB1 (p50) and NF-κB2 (p52) are generated from precursors of 105 kDa and 100 kDa, respectively. p105 and p100 contain multiple copies of ankyrin-repeat domains within their C-terminal IκB-like domains that are typical of the inhibitory IκB proteins (Baldwin, 1996). The ankyrin-repeat domains bind active NF-κB1 (p50) and RelA (p65) subunits, which results in the sequestration of inactive p105 and p100 precursors in the cytoplasm and inhibits their processing to NF-κB1 (p50) and NF-κB2 (p52) (Hatada et al., 1992; Cohen et al., 2001). As discussed in greater detail below, the processing of p105 and p100 precursors to NF-κB1 (p50) and NF-κB2 (p52) is an ATP-dependent process that requires polyubiquitination and limited proteasome-mediated degradation of the C-terminal ankyrin-repeat domains (Fan and Maniatis, 1991; Ghosh et al., 1998; Cohen et al., 2001).

The active DNA-binding forms of NF-κB are homo- or heterodimers, each of which contains one-half of the DNA binding site (Ghosh et al.,

1998). The first NF-κB to be identified was a heterodimer composed of NF-κB1 (p50) and RelA (p65), which binds the DNA consensus sequence 5′-GGGRNNYYCC-3′ (R denotes an unspecified purine, Y denotes an unspecified pyrimidine, and N denotes any nucleotide) (Baldwin, 1996; Ghosh et al., 1998; Karin and Ben-Neriah, 2000). Crystal structure analysis of p50/p65 NF-κB heterodimers and p50 homodimers bound to DNA revealed a butterfly-like structure (Chen et al., 1998; Ghosh et al., 1998; Karin and Ben-Neriah, 2000). The Rel homology domains of each subunit contain two immunoglobulin-like folds that form the wings of the butterfly and are connected by a flexible linker region. Subunits dimerize via formation of a β-sheet sandwich within the C-terminal dimerization domains. DNA binding is mediated by loops originating from the edges of the N- and C-terminal domains. The N-terminal fold is responsible for sequence-specific recognition, whereas the C-terminal fold forms an interface with the DNA that localizes the NF-κB dimer to the major groove.

Although multiple combinations of Rel family heterodimers have been identified, not all combinations exist (Baldwin, 1996). p50/p65 heterodimers and homodimers of p50 or p65 are the most abundant NF-κB isoforms (Phelps et al, 2000). Heterodimer formation confers specificity to NF-κB-mediated transcriptional activation by modulating binding to variations in the DNA consensus binding sequence (Baldwin, 1996). For example, among members of the mammalian NF-κB/Rel family, only RelA (p65) and c-Rel are able to induce robust transcriptional activation, whereas homodimers of NF-κB1 (p50) or NF-κB2 (p52) are transcriptionally repressive (Baldwin, 1996; Karin and Ben-Neriah, 2000). Unlike c-Rel, RelA (p65), and RelB, both NF-κB1 (p50) and NF-κB2 (p52) lack a transactivation domain in the C terminus that is required for transcriptional activation of NF-κB-inducible genes (Ghosh et al., 1998). Additional mechanisms by which NF-κB dimers determine specificity include distinct patterns of cellular expression, selective interactions with IκB isoforms, and differential activation (Baldwin, 1996).

THE INHIBITORY IκB PROTEINS

Activation of NF-κB is prevented by binding to inhibitory IκB proteins that maintain NF-κB in an inactive state in the cytoplasm (Baeuerle, 1998). Formation of a stable cytosolic complex with IκB sequesters the NF-κB nuclear localization signal, thereby preventing translocation from the cytoplasm into the nucleus (Fig. 2.6). IκB proteins are characterized by an ankyrin-repeat domain, containing six or seven repeats, that forms a stacked structural unit. The ankyrin-repeat domain mediates specific protein associations with the dimerized immunoglobulin-like folds of the Rel homology domains of NF-κB subunits (Baeuerle, 1998; Jacobs and Harrison, 1998).

Members of the IκB family in higher eukaryotes include IkBα, IkBβ, IkBε, IkBγ, Bcl-3, and the recently identified IκBζ (Fig. 2.5). The p105

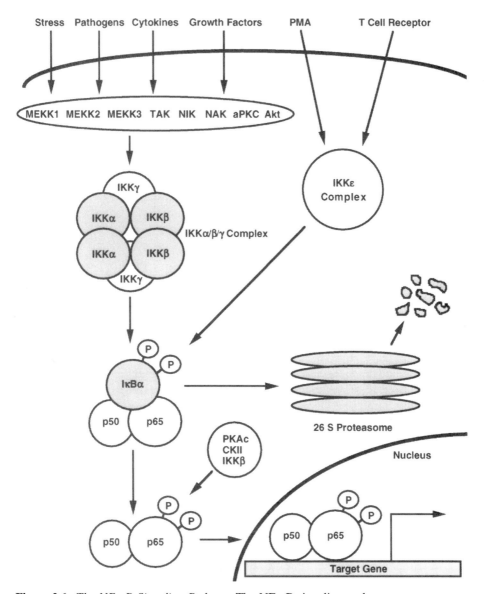

Figure 2.6. *The NF-κB Signaling Pathway.* The NF-κB signaling pathway coordinates stress responses to a variety of diverse stimuli. The IKKα/β/γ complex can be activated by members of the mitogen-activated protein kinase kinase kinase (MAP3K) family, as well as by atypical protein kinases (aPKC) and Akt (protein kinase B). In contrast, the recently identified IKKε complex can be activated in response to phorbol ester (PMA) or T cell receptor signaling. The IKK complexes then phosphorylate IκB, which undergoes ubiquitination and subsequent degradation by the 26S proteasome. Liberated NF-κB dimers can be phosphorylated by p65 kinases, which increases the transactivation potential. NF-κB then translocates to the nucleus and binds DNA κB sites, thereby inducing the transcriptional activation of multiple proinflammatory target genes.

NF-κB1 and p100 NF-κB2 precursor proteins also contain ankyrin repeats that are degraded during their conversion to the active form and are therefore included in the IκB family (Ghosh et al., 1998). IκBγ is identical to the C-terminal region of p105 and is generated via an alternative p105 promoter (Inoue et al., 1992; Ghosh et al., 1998). IkBα, IkBβ, and IkBε have N-terminal regulatory domains and function as stimulus-dependent regulators of NF-κB activation (Baeuerle, 1998; Karin and Ben-Neriah, 2000). IκBζ is also induced by proinflammatory stimuli, such as lipopolysaccharide (LPS) and IL-1β, but not TNF-α (Yamazaki et al., 2001). In addition, a subclass of evolutionarily conserved Ras-like proteins, named κB-Ras1 and κB-Ras2, can bind to IκBα or IκBβ and decrease their rate of degradation (Fenwick et al., 2000). κB-Ras1 and κB-Ras2 mediate this interaction via binding to IκBα and IκBβ C-terminal PEST domains (see below).

Members of the IκB family interact with specific NF-κB subunits. For example, the majority of p50-p65 and p50-c-Rel complexes are regulated by IκBα and IκBβ (Ghosh et al., 1998; Heissmeyer et al., 1999), whereas IκBε exclusively binds to RelA (p65) homodimers or p65/c-Rel heterodimers (Whiteside et al., 1997; Ghosh et al., 1998). Unlike IκBα and IκBβ, which are cytoplasmic, Bcl-3 is localized to the nucleus and demonstrates specificity toward NF-κB1 (p50) and NF-κB2 (p52) homodimers (Nolan et al., 1993; Ghosh et al., 1998). Similarly, IκBζ localizes to the nucleus via its N-terminal region (Yamazaki et al., 2001). IκBζ preferentially associates with NF-κB1 (p50) rather than RelA (p65), and inhibits DNA binding by p65/p50 heterodimers and p50 homodimers.

IκBα, the best characterized member of the IκB family, is a 37-kDa protein containing an N-terminal signal-receiving domain (SRD) that mediates inducible signal-dependent degradation, an acidic C-terminal PEST-like domain (e.g., enriched in proline, glutamic acid, serine, and threonine) that mediates constitutive phosphorylation by casein kinase II, and the central ankyrin-repeat domain (Ghosh et al., 1998; Jacobs and Harrison, 1998; Phelps et al., 2000). The affinity of IκBα for binding p50/p65 heterodimers is 20- and 2.5-fold that for p50 and p65 homodimers, respectively (Phelps et al., 2000).

Analysis of crystal structures of the IκBα-NF-κB complex has provided insight into its mechanism of action and revealed extensive interactions between all six IκBα ankyrin-repeat domains and the NF-κB1 (p50)/RelA (p65) immunoglobulin-like domains (Baeuerle, 1998). Crystallography has revealed that the first and second ankyrin repeats bind to a long region that contains the nuclear localization signal, thereby retaining NF-κB in the cytoplasm. Ankyrin-repeats 3, 4, and 5 provide an extensive area for interface with the dimerized immunoglobulin-like domain of NF-κB1 (p50)/RelA (p65) (Baeuerle, 1998; Huxford et al., 1998; Jacobs and Harrison, 1998; Phelps et al., 2000). In addition, ankyrin repeat 6 binds the N-terminal immunoglobulin-like domain and may be important in modifying the DNA-binding capabilities of NF-κB (Baeuerle, 1998). Analysis of crystallographic structures demonstrated that the DNA binding is inhibited by IκB-induced changes in the RelA (p65) conformational structure. Formation of the IκB-RelA (p65) complex results in the 180° rotation of the flexible N-terminal immunoglobulin-like domain of RelA (p65), which sequesters amino acids important for DNA binding (Baeuerle, 1998; Phelps et al., 2000). A second interaction between the acidic C-terminal PEST-like region and the RelA (p65) N-terminal domain contributes to this complex and is enhanced by casein kinase II-catalyzed IκBα phosphorylation (Phelps et al., 2000).

An important physiological function of IκBα is its participation with NF-κB in an autoregulatory negative feedback loop (Ghosh et al., 1998). After stimulus-induced IκBα degradation, NF-κB translocates into the nucleus and initiates the transcriptional activation of the IκBα gene. NF-κB induction of IκBα synthesis is mediated by multiple κB sites within the proximal IκBα promoter (Le Bail et al., 1993). IκBα is then imported into the nucleus via an energy-dependent process that involves the ankyrin-repeat regions and requires importins α and β as well as the GTPase Ran (Turpin et al., 1999). After transport into the nucleus, newly synthesized IκBα can bind to and cause dissociation of DNA-bound NF-κB, thereby terminating its transactivating activity (Zabel and Baeuerle, 1990; Han and Brasier, 1997; Han et al., 1999). After capturing NF-κB subunits, IκBα is transported out of the nucleus into the cytoplasm via a N-terminal nuclear localization signal that binds to the nuclear export receptor CRM1 (Sachdev and Hannink, 1998; Johnson et al., 1999; Tam et al., 2000). Therefore, binding of RelA (p65) by IκBα in the nucleus terminates NF-κB transactivating activity and subsequently resequesters NF-κB in the cytoplasm (Brown et al., 1993). This autoregulatory negative feedback loop ensures that NF-κB activation is transient.

Recent studies have demonstrated that reversible acetylation of RelA (p65) also regulates the duration of nuclear NF-κB action (Chen et al., 2001). Acetylation of RelA (p65) prevents IκBα binding, thereby allowing p50/p65 heterodimers to be retained in the nucleus. The acetylation of RelA (p65) can be induced by TNF and is likely mediated by the p300 and CBP (CREB-binding protein) coactivators, which have histone acetyltransferase activity. Acetylated RelA is deacetylated by histone deacetylase 3 (HDAC3), which permits binding to IκBα and nuclear export via a CRM1-dependent pathway. Therefore, deacetylation of RelA (p65) by HDAC3 may serve as an intranuclear molecular switch that regulates the duration of nuclear NF-κB activity.

IκBβ may also be important in the persistent activation of NF-κB that follows chronic stimulation, despite elevated levels of newly synthesized IκBα (Karin and Ben-Neriah, 2000). After persistent stimulation with lipopolysaccharide (e.g., LPS), IκBβ is resynthesized and accumulates in the cytoplasm in an unphosphorylated form (Suyang et al., 1996). The resynthesized IκBβ forms a stable complex with cytoplasmic NF-κB but fails to mask the nuclear localization signal and the DNA-binding domain, thereby allowing the IκBβ-NF-κB complex to enter the nucleus and bind DNA. The binding of unphosphorylated IκBβ to NF-κB shields these complexes from IκBα, thereby functioning as a chaperone that allows nuclear translocation and prevents inhibition of NF-κB activity by newly synthesized IκBα protein. Nuclear degradation of IκBα is another mechanism by which chronic stimulation mediates prolonged NF-κB activation (Renard et al., 2000). It was recently demonstrated that proteasome-dependent degradation of newly synthesized, phosphorylated, and ubiquitinated IκBα occurs in the nucleus, thereby permitting persistent NF-κB activity during chronic stimulation. Therefore, intranuclear proteolysis of IκBα appears to be necessary to prevent the self-termination of NF-κB activity during prolonged activation.

Bcl-3 has a unique mechanism of action whereby binding to NF-κB subunits leads to transcriptional activation rather than repression (Ghosh et al., 1998). Bcl-3 is a putative protooncogene that was identified in a t(14;19) translocation in a subgroup of B cell lymphocytic leukemias (Ohno et al., 1990; Brasier et al., 2001). Bcl-3 is abundant in the nucleus and is not degraded after activation of NF-κB stimulating pathways (Heissmeyer et al., 1999). Bcl-3 can positively regulate NF-κB activity via several mechanisms. Homodimers of NF-κB1 (p50) or NF-κB2 (p52) are the known targets for Bcl-3. Binding of Bcl-3 to NF-κB1 (p50) homodimers can induce their dissociation from NF-κB binding sites, thereby allowing transactivating NF-κB complexes to bind (Franzoso et al., 1993). Because neither NF-κB1 (p50) nor NF-κB2 (p52) contains transactivation domains, this represents a mechanism by which Bcl-3 can antagonize p50-mediated inhibition (Franzoso et al., 1992; Heissmeyer et al., 1999). Furthermore, stimuli that activate p50/p65 dimers, such as TNF-α, IL-1β, and phorbol ester (PMA), also trigger the rapid formation of Bcl-3-p50 complexes, with the same kinetics of activation as p50-p65 complexes, in a process that requires IκB kinase (IKK)-mediated p105 degradation (Heissmeyer et al., 1999). A second mechanism by which Bcl-3 positively regulates NF-κB activity is by formation of a ternary complex with NF-κB1 (p50) and NF-κB2 (p52) homodimers that are bound to DNA κB sites (Bours et al., 1993; Fujita et al., 1993). This association allows Bcl-3 to transactivate gene transcription directly through two cooperating domains located N- and C-terminal to the ankyrin-repeat domains (Bours et al., 1993). Therefore, Bcl-3 ternary complexes can act as transcriptional coactivators, whereas p50 and p52 homodimers alone cannot. Furthermore, Bcl-3 can bind to other nuclear proteins, such as Jab1, Pirin, Tip60, and Bard1, through its ankyrin-repeat domains (Dechend et al., 1999). For example, formation of a quaternary complex containing the histone acetylase Tip60 can enhance Bcl-3/p50-mediated transcription. Thus Bcl-3 can act as a bridging factor or adaptor with Bcl-3 interacting proteins that modulate NF-κB-mediated gene transcription.

Generation of IκB knock-out mice further elucidated the function of the IκB family and suggested that individual members have unique roles in regulating NF-κB activity (Cheng et al., 1998). IkBα-deficient mice have elevated basal NF-κB activity in hematopoietic tissues with consequent extensive granulopoiesis, dermatitis and death in the early postnatal period (Beg et al., 1995; Klement et al., 1996; Cheng et al., 1998). This phenotype was reversed in knock-in mice generated by replacing the IκBα gene with the IκBβ gene. This suggests that IκBβ can functionally replace IκBα and that these molecules have acquired distinct functions as a result of differential tissue expression and activation (Cheng et al., 1998). Mice with a homozygous deletion of the C-terminal ankyrin repeats of p100, but still containing functional NF-κB2 (p52), demonstrate marked gastric hyperplasia, lymph node enlargement, and death in the early postnatal period, associated with significant increased κB-binding complexes containing NF-κB2 (p52) (Ishikawa et al., 1997). This indicates that the p100 precursor is required for the proper regulation of NF-κB2 (p52)-containing NF-κB complexes and can not be compensated by IκBα or IκBβ. Finally, Bcl-3 deficient mice demonstrate impaired antigen-specific T and B cell responses as well as a partial deficit of B cells (Franzoso et al. 1997).

NF-κB ACTIVATION VIA IKK-CATALYZED PHOSPHORYLATION OF IκB

The critical step in NF-κB activation is the signal-induced phosphorylation of inhibitory IκB proteins by IκB kinases, which triggers ubiquitin-mediated IκB degradation and translocation of cytoplasmic NF-κB to the nucleus (Stancovski and Baltimore, 1997; Karin, 1999; Mercurio and Manning, 1999; Karin and Ben-Neriah, 2000). Stimuli that cause NF-κB activation initiate signaling cascades that culminate in the specific phosphorylation of serines at positions 32 and 36 in IκBα and 19 and 23 in IκBβ, resulting in their rapid degradation (Chen et al., 1996; Woronicz et al., 1997). This IKK activity is mediated by a multimeric protein complex or signalsome, with a molecular mass of approximately 700 to 900 kDa, and is composed of several proteins, including IKKα, IKKβ, and IKKγ (Mercurio et al., 1997; Zandi et al., 1997; Karin and Ben-Neriah, 2000). The IKK signalsome is likely composed of a dimer of IKKα/IKKβ dimers, plus a dimer or trimer of IKKγ (Rothwarf et al., 1998; Karin and Ben-Neriah, 2000).

IKKα (IKK1, CHUK) is an 85-kDa protein that was identified in a yeast two-hybrid screen by its ability to interact with NF-κB-inducing kinase (NIK), a mitogen-activated protein kinase kinase kinase (MAP3K) (DiDonato et al., 1997; Regnier et al., 1997; Karin and Ben-Neriah, 2000). IKKβ (IKK2) is an 87-kDa protein that is 52% identical with IκBα (Woronicz et al., 1997; Zandi et al., 1997). Shared structural characteristics include an N-terminal protein kinase catalytic domain and C-terminal leucine-zipper and helix-loop-helix domains that participate in protein interactions (Stancovski and Baltimore, 1997; Karin and Ben-Neriah, 2000). Both IKKα and IKKβ can phosphorylate IκBα and IκBβ, although the ability of IKKα to phosphorylate S19 in IκBβ is limited (Woronicz et al., 1997). IKKα and IKKβ, which are the catalytically active subunits of the IKK signalsome, normally exist as leucine-zipper linked heterodimers (Woronicz et al., 1997; Karp et al., 2000).

IKKγ/NEMO/FIP3 (NF-κB essential modulator) is the third component of the IKK complex (Rothwarf et al., 1998). IKKγ is a 48-kDa regulatory subunit that preferentially binds IKKβ and is required for activation of the IKK complex (Rothwarf et al., 1998; Karin and Ben-Neriah, 2000). IKKγ-deficient cells fail to assemble the IKK complex and cannot generate IKK or NF-κB activity after stimulation with LPS, IL-1, double-stranded RNA, or TNF (Yamaoka et al., 1998; Karin, 1999). An N-terminal α-helical region, identified as the NEMO-binding domain (NBD), is required for association with the C-terminal regions of IKKα and IKKβ (May et al., 2000). IKKγ facilitates the association of IKKβ with IκB, thereby increasing IKKβ kinase activity (Yamamoto et al., 2001). IKKγ has a predominantly helical structure with two coiled-coil domains and a leucine-zipper motif at the C terminus (Rothwarf et al., 1998; Karin and Ben-Neriah, 2000). The C terminus of IKKγ is a binding site for upstream activators of the IKK complex (Rothwarf et al., 1998). For example, activation of the p55 type I TNF receptor by TNF induces binding of the kinase receptor-interacting protein (RIP) and the NF-κB inhibitory protein A20 to IKKγ, thereby linking the IKK signalsome to the p55 TNF receptor complex (Zhang et al., 2000). Other proteins that bind IKKγ include the viral transactivator Tax and CIKS (connection to IKK and SAPK/JNK), a protein that interacts with and activates both IKK and stress-activated protein kinase (SAPK)/Jun kinase (JNK) signaling complexes (Harhaj and Sun, 1999; Leonardi et al., 2000).

Another IKK complex has recently been identified that can phosphorylate serines 32 and 36 of IκBα and thereby activate NF-κB (Peters et al., 2000; Peters and Maniatis, 2001). IKKε, which is identical to IKK-i, was identified in a database search as having homology to IKKα and IKKβ, all proteins of approximately 85 kDa with a characteristic kinase domain near the N terminus and a helix-loop-helix domain near the C terminus (Shimada et al., 1999; Peters et al., 2000). In contrast to IKKα and IKKβ, IKKε lacks a leucine-zipper motif and contains a coiled-coil multimerization domain near the C terminus that is thought to mediate self-association (Peters et al., 2000). IKKε is present in a novel IKK complex, which is distinct from the IKKα/β/γ complex and is activated by both PMA and the T cell receptor but not by TNF-α or IL-1. Recombinant IKKε can directly phosphorylate serine 36, but not serine 32, in IκBα, whereas the PMA-activated IKKε complex can phosphorylate both serine residues (Peters and Maniatis, 2001). IKKε/IKK-i can also bind and phosphorylate I-TRAF/TANK (TRAF-interacting protein, TRAF family member-associated NF-κB activator), which may also participate in the regulation of NF-κB activation (Nomura et al., 2000; Peters and Maniatis, 2001).

The importance of IKKs was further demonstrated by the creation of mice with targeted IKK deletions. IKKα −/− mice can activate IKK complexes in response to TNF, IL-1, or LPS, thereby demonstrating that IKKα is not required for IKK activation in response to proinflammatory stimuli (Hu et al., 1999; Karin and Ben-Neriah, 2000). Furthermore, this function can be replaced by IKKβ. In contrast, mice deficient in IKKβ have low basal NF-κB and IKK activity and impaired cytokine-induced NF-κB activation (Li et al., 1999a; Li et al., 1999b; Tanaka et al., 1999). Thus, IKKβ is required for IKK activation in response to most inducers of NF-κB activity, whereas IKKα appears to be dispensable (Senftleben et al., 2001). In addition, IKKβ –/– mice demonstrate embryonic lethality due to severe liver degeneration and TNF-induced apoptosis as a consequence of impaired NF-κB activation (Li et al., 1999a; Li et al., 1999b; Tanaka et al., 1999). Mice that are doubly deficient in IKKα and IKKβ demonstrate a complete lack of NF-κB activation in response to TNF, IL-1, and LPS, as well as defective neural tube development as a consequence of enhanced neuroepithelial cell apoptosis (Li et al., 2000).

IKKα –/– mice display multiple morphologic defects, manifested by the absence of limbs, tails, and ears, as well as severe craniofacial deformities, and they die shortly after birth. The most striking defect is the inability to develop a stratified, well-differentiated epidermis because of a failure to activate the keratinocyte terminal differentiation program (Hu et al., 2001). This activity of IKKα is independent of the classic IKK complex or signalosome and appears to involve a separate pathway that controls the production of a keratinocyte differentiation-inducing factor (Hu et al., 2001). It was recently reported that IKKα kinase activity is also required for B cell maturation, formation of secondary lymphoid organs, and increased expression of a subset of NF-κB target genes (Senftleben et al., 2001). This function of IKKα requires its phosphorylation by upstream kinases, such as NIK, and is exerted via processing of the NF-κB2 p100 precursor. Thus, IKKα can activate a distinct NF-κB pathway that involves regulated p100 processing rather than IκB degradation.

Targeted disruption of the X-linked IKKγ gene results in male embryonic lethality due to severe liver degeneration secondary to TNF-induced apoptosis. IKKγ-deletion also prevents NF-κB activation in response to proinflammatory cytokines and LPS (Makris et al., 2000; Rudolph et al., 2000; Schmidt-Supprian et al., 2000). IKKγ-deficient mice are a model for incontinentia pigmenti or Bloch–Sulzberger syndrome, an X-linked genetic disorder characterized by male lethality and females with skin lesions characterized by massive granulocyte infiltration (Makris et al., 2000; Schmidt-Supprian et al., 2000). Mutations in the human IKKγ/NEMO gene are associated with defective NF-κB activation and familial incontinentia pigmenti (Smahi et al., 2000). In addition, mutations affecting a putative zinc finger domain at the C terminus of the IKKγ protein are associated with an X-linked primary immunodeficiency characterized by the hyper-IgM syndrome, defective CD40-dependent NF-κB activation and B cell immunoglobulin class switching, as well as hypohydrotic ectodermal dysplasia (Jain et al., 2001).

UPSTREAM SIGNALS FOR IKK ACTIVATION

Signals leading to NF-κB activation are most frequently initiated by stress or exposure to pathogens (Tak and Firestein, 2001). For example, stimuli capable of activating NF-κB include bacteria, viruses, parasites, physiological stress (e.g., hemorrhage, hyperoxia, hyperosmotic shock, ischemia, shear force), physical stress (e.g., UV irradiation), oxidative stress (e.g., ozone, hydrogen peroxide), environmental toxins (e.g., cigarette smoke, asbestos, heavy metals, PCBs), therapeutic agents (e.g., AZT, cisplatinum, doxorubicin, etoposide, haloperidol, tamoxifen), physiological mediators (e.g., complement, fMLP, leukotriene B4, platelet-activating factor), growth factors [insulin, platelet-derived growth factor (PDGF), basic fibroblast-derived growth factor, TGF-α], and inflammatory cytokines (e.g., IL-1, IL-2, IL-12, IL-15, IL-17, IL-18, TNF-α, TNF-β, LIF) (Pahl, 1999; Makarov, 2000). Therefore, NF-κB functions as a central regulator of the stress response via its ability to activate multiple proinflammatory genes in response to a variety of stimuli.

Individual receptors that activate NF-κB signaling pathways utilize a variety of distinct serine kinases (Fig. 2.6) (Yin et al., 2001). Members of the MAP3K family that have been identified as direct mediators of cytokine-induced IKK activation include MEKK1, MEKK2, MEKK3, TGF-β-activating kinase (TAK), NIK, and NF-κB-activating kinase (NAK) (Zhao and Lee, 1999; Karin and Ben-Neriah, 2000; Tojima et al., 2000; Lee et al., 2001; Yang et al., 2001). Furthermore, the activating phosphorylation sites of IKKα and IKKβ are similar to those of the MAP2Ks (Karin and Ben-Neriah, 2000).

Consistent with the central role of NF-κB in coordinating the stress response is the fact that multiple signaling pathways converge on the IκBα complex, leading to NF-κB activation. The signaling pathways involved in TNF-induced NF-κB activation are well characterized (Fig. 2.7). Binding of TNF-α to TNFRI recruits TNFRI-associated death-domain protein (TRADD) via death domain association (Wallach et al., 1999). TRADD then facilitates the recruitment of the serine-threonine kinase RIP, again by death domain association, as well as the recruitment of the adapter protein TNF receptor-associated factor 2 (TRAF2) to the TNFRI complex (Hsu et al., 1996a; Hsu et al. 1996b; Wallach et al., 1999; Tsao et al., 2000). TRAF2 also binds to the intermediary domain of RIP, thereby forming a trimolecular complex composed of TRADD, RIP, and TRAF2 (Hsu et al., 1996a). The chaperone protein Hsp90 also binds RIP and prevents its proteasomal degradation (Lewis et al., 2000). In contrast, RIP3 can be recruited to the TNFRI signaling complex by binding to RIP and serves to attenuate both RIP- and TNFRI-induced NF-κB activation (Sun et al., 1999). In this way, RIP3 may function as a proapoptotic molecule that binds RIP and inhibits its activation of NF-κB in response to TNFRI signaling.

TRAF2 and RIP play distinct roles in the TNFRI-initiated activation of IKK; TRAF2 recruits IKK to TNFRI, whereas RIP mediates IKK activation independent of its kinase activity (Devin et al., 2000). Either

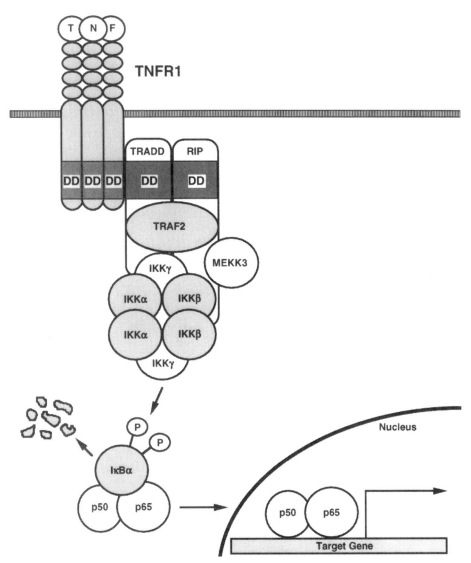

Figure 2.7. *TNFRI-Mediated NF-κB Signaling.* The pathway by which the type I, 55-kDa tumor necrosis factor receptor (TNFRI) activates NF-κB signaling is well characterized. Binding of TNF-α to TNFRI recruits TRADD via death domain (DD) association, which in turn recruits RIP, also via a death domain interaction, as well as TRAF2. TRAF2 then recruits the IKKα/β/γ complex to TNFRI, while RIP binds MEKK3. MEKK3 directly phosphorylates and activates IKK, which then phosphorylates IκBα. IκBα is ubiquitinated and degraded, thereby liberating NF-κB to translocate to the nucleus and induce the transcriptional activation of target genes.

IKKα or IKKβ is sufficient for the recruitment of IKK to TNFRI, which is mediated via the leucine-zipper motif common to IKKα, IKKβ, and the RING finger domain of TRAF2 (Devin et al., 2001). Although IKKγ is not essential for TNF-induced IKK recruitment to TNFRI, IKKγ binds RIP in a TRAF2-dependent fashion and may stabilize the IKK complex (Zhang et al., 2000; Devin et al., 2001). Furthermore, MEKK3 has recently been identified as a key signaling molecule in TNF-induced NF-κB activation (Yang et al., 2001). MEKK3 acts downstream of RIP and TRAF2 but upstream of IKK. MEKK3 binds to RIP and directly phosphorylates IKK, thereby linking RIP and IKK in TNF-mediated NF-κB activation. Inhibitory regulatory proteins can also be recruited to the TNF receptor to regulate NF-κB activation. A20, a zinc finger protein that inhibits TNF-induced NF-κB activation, can also be recruited to TNFRI via binding to IKKγ (Zhang et al., 2000). A20 acts downstream of RIP and TRAF2 and may exert its inhibitory activity via an interaction with the A20-binding inhibitor of NF-κB activation (ABIN) (Heyninck et al., 1999). CARDINAL, a recently identified caspase recruitment domain protein, can also bind IKKγ and inhibit TNF-induced NF-κB activation (Bouchier-Hayes et al., 2001).

The signaling pathways mediating IL-1-induced NF-κB activation have also been recently characterized (Fig. 2.8). Signaling from the type I IL-1 receptor (IL-1RI) and the closely related Toll-like receptors 2 (TLR2) and 4 (TLR4) activate NF-κB via TRAF6-dependent pathways (Karin and Ben-Neriah. 2000; Zhang and Ghosh. 2001). After IL-1 binding, IL-1RI complexes with the IL-1 receptor accessory protein (IL-1RacP), which facilitates the recruitment of the adapter MyD88, a functional analog of TRADD, and the serine/threonine kinase IL-1 receptor-associated kinase (IRAK) (Ninomiya-Tsuji et al., 1999; Zhang and Ghosh, 2001). IRAK is autophosphorylated and dissociates from the IL-1 receptor complex, allowing it to interact with TNF receptor-associated factor 6 (TRAF6). Activation of IKK by TRAF6 requires two intermediary complexes, TRAF6-regulated IKK activator 1 (TRIKA1) and 2 (TRIKA2) (Wang et al., 2001). TRIKA1 is a dimeric ubiquitin-conjugating enzyme complex that is comprised of the ubiquitin-conjugating enzyme Ubc13 and the Ubc-like protein, Uev1A (Deng et al., 2000; Wang et al., 2001). Oligomerization of TRAF6 triggers its own polyubiquitination by the TRIKA1 complex via lysine (K63) of ubiquitin. Formation of K63-linked polyubiquitin chains directly activates the MAP3K TAK1 in a proteasome-independent fashion (Deng et al., 2000; Wang et al., 2001). IL-1 also stimulates translocation of the adapter protein TAB2 from the membrane to the cytosol, where it mediates the interaction between TRAF6 and TAK1 via the formation of the TRIKA2 complex composed of TAK1, TAB1, and TAB2 (Takaesu et al., 2000; Wang et al., 2001). TAK1 then directly phosphorylates and activates IKK, without the requirement for NIK (Wang et al., 2001; Yin et al., 2001). TAK1 had previously been thought to activate IKK via an intermediary kinase such as NIK (Wang et al., 2001). However, studies in NIK-deficient mice have recently demonstrated that NIK does not represent a common upstream kinase that regulates IKK activation after stimulation with TNF or IL-1,

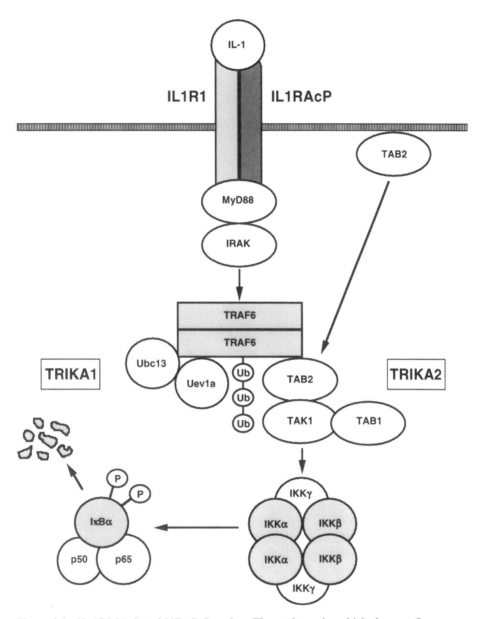

Figure 2.8. *IL-1RI-Mediated NF-κB Signaling.* The pathway by which the type I interleukin-1 receptor (IL-1RI) activates NF-κB signaling has also been characterized. After IL-1 binding, the IL-1RI complexes with the IL-1 receptor accessory protein (IL-1RacP), thereby facilitating the recruitment of the adapter protein MyD88 and the serine/threonine kinase IL-1 receptor-associated kinase (IRAK). IRAK is autophosphorylated and dissociates from the IL-1 receptor complex, allowing it to interact with TNF receptor-associated factor 6 (TRAF6). TRAF6-dependent activation of IKK requires two intermediary complexes, TRAF6-regulated IKK activator 1 (TRIKA1), composed of Ubc13 and Uev1a, and TRIKA2, composed of TAK1, TAB1, and TAB2. Oligomerization of TRAF6 triggers its own polyubiquitination by the TRIKA1 complex, which mediates the activation of the MAP3K TAK1 in a proteasome-independent fashion. IL-1 also stimulates translocation of the adapter protein TAB2 from the membrane to the cytosol, where it mediates the interaction between TRAF6 and TAK1 via the formation of the TRIKA2 complex. TAK1 then directly phosphorylates and activates IKK.

but rather regulates NF-κB transcriptional activity in a receptor-restricted fashion (Yin et al., 2001). Therefore, TAK1 is an ubiquitin-dependent IKK kinase capable of directly catalyzing IKK activation.

Protein kinases that do not belong to the MAP3K family, such as the atypical protein kinases (aPKCs) ζPKC and λ/ιPKC and Akt (protein kinase B), are also putative activators of the IKK signalsome. This is consistent with the concept that multiple signaling pathways are capable of mediating IKK activation (Perkins, 2000). The aPKC-binding protein p62 is reported to bind RIP and thereby allow aPKCs to phosphorylate and activate IKKβ in response to TNF stimulation (Lallena et al., 1999; Sanz et al., 1999). p62 also can interact with the TRAF domain of TRAF6 and serve as an adapter connecting the aPKCs to TRAF6 during IL-1-mediated NF-κB activation (Sanz et al., 2000). The serine/threonine kinase Akt (protein kinase B) has also been reported to increase NF-κB activity via activation of the IKK signalsome in response to either TNF or PDGF (Ozes et al., 1999; Romashkova and Makarov, 1999). Phosphatidylinositol-3-OH kinase (PI(3)K) is an upstream activator of Akt in this pathway (Ozes et al., 1999; Romashkova and Makarov, 1999). Akt then directly associates with and phosphorylates IKKα, leading to the nuclear translocation of NF-κB and the subsequent transcription activation of antiapoptotic genes.

Alternative or atypical IKK activation pathways have also been identified (Karin and Ben-Neriah, 2000). One atypical IKK activation pathway involves the tyrosine phosphorylation of IκBα at tyrosine 42 with subsequent NF-κB activation (Mukhopadhyay et al., 2000). Stimuli that induce IκBα tyrosine phosphorylation include hypoxia, reoxygenation, and the PTPase inhibitor pervanadate (Beraud et al., 1999; Mukhopadhyay et al., 2000). The responsible tyrosine kinases have not been definitively identified but may include members of the Src family (Beraud et al., 1999; Karin and Ben-Neriah, 2000). It is reported that tyrosine phosphorylation induces the dissociation of IκB from NF-κB as a result of binding of the p85a regulatory subunit of phosphotidylinositol 3-kinase (PI3-kinase) through its Src homology 2 domains (Imbert et al., 1996; Beraud et al., 1999). Although this process was initially thought not to involve the proteolytic degradation of IκBα, a recent report has demonstrated that pervanadate can mediate IκBα degradation after tyrosine phosphorylation (Imbert et al., 1996; Mukhopadhyay et al., 2000). A second atypical IKK activation pathway involves NF-κB activation in response to ionizing radiation (Karin and Ben-Neriah, 2000). Short UV irradiation has been demonstrated to induce IκB degradation by the 26S proteasome independent of its N-terminal serine phosphorylation via an unknown mechanism (Li and Karin, 1998).

REGULATION OF NF-κB ACTIVITY DOWNSTREAM FROM IκB

Inducible NF-κB phosphorylation promotes DNA binding and transactivation and thereby serves as an additional regulatory mechanism downstream of IKK activation and IκB degradation (Fig. 2.6) (Schmitz et al., 2001). Inducible NF-κB phosphorylation occurs in response to a variety of stimuli, including TNF-α, IL-1, LPS, phorbol ester (PMA), phytohemagglutinin, and hydrogen peroxide (Karin and Ben-Neriah, 2000; Schmitz et al., 2001). NF-κB-mediated transactivation requires its association with the transcriptional coactivators CBP/p300 via two sites that are blocked in unphosphorylated RelA (p65) by intramolecular masking of the N terminus by the C-terminus (Zhong et al., 1998). Phosphorylation of serine 276 in RelA (p65) by the catalytic subunit of protein kinase A (PKAc) weakens the interaction between the N- and C-terminal regions, thereby facilitating the interaction between RelA (p65) and CBP/p300 with resultant enhanced NF-κB-mediated transcription. Furthermore, PKAc, but not the PKA regulatory subunit, is bound to IκBα or IκBβ and is maintained in an inactive state in an NF-κB-IκB-PKAc complex (Zhong et al., 1997). Degradation of IκB allows the activation of PKAc in a cAMP-independent fashion, thereby facilitating p65 phosphorylation. Casein kinase II (CKII) also can phosphorylate RelA (p65), on serine 529 in the C-terminal transactivation domain, after stimulation with TNF or IL-1 (Bird et al., 1997; Wang et al., 2000). Similar to PKAc, the association between RelA (p65) and IκBα inhibits CKII-mediated phosphorylation. IκBα degradation permits CKII to phosphorylate RelA (p65) and increase the transactivation potential of NF-κB without affecting nuclear translocation or DNA binding affinity (Wang and Baldwin, 1998). IKKβ can also phosphorylate RelA (p65) on serine 536 located in the C-terminal transactivation domain (Sakurai et al., 1999).

Additional signaling pathways have been demonstrated to mediate NF-κB phosphorylation. Akt has been reported to stimulate the transactivation domain I of RelA (p65) in response to IL-1β or oncogenic Ras in a PI3K-dependent fashion (Sizemore et al., 1999; Madrid et al., 2000; Madrid et al., 2001). Activated Akt mediates the phosphorylation of RelA (p65) on serines 529 and 536 in an indirect fashion, functioning through IKKβ and the mitogen-activated protein kinase p38 (Madrid et al., 2000; Madrid et al., 2001). Similarly, a pathway involving activation of protein kinase C ζ and p21[ras] leads to phosphorylation of RelA (p65) RHD in primary endothelial cells (Anrather et al., 1999; Schmitz et al., 2001). Glycogen synthase-3β (GSK-3β) may also regulate NF-κB activity at the level of the transcriptional complex (Hoeflich et al., 2000; Schmitz et al., 2001). Finally, TNF-α -induced phosphorylation of c-Rel on serine 471 in the transactivation domain appears to be necessary for TNF-α-induced c-Rel activation (Martin and Fresno, 2000).

The ability of proinflammatory stimuli to induce binding of NF-κB subunits to NF-κB-inducible promoters before their rapid extrusion from the nucleus represents an additional regulatory mechanism by which NF-κB activity can be regulated (Saccani et al., 2001). It was recently demonstrated by chromatin immunoprecipitation that acute LPS stimulation induces two distinct waves of NF-κB recruitment to target promoters. A fast wave of recruitment occurs within 20 minutes to constitutively and immediately accessible (CIA) genes, whereas a late wave occurs after 90 minutes to promoters that require stimulus-dependent hyperacetylation that renders NF-κB sites accessible [promoters with regulated and late accessibility (RLA)]. Therefore, chromatin accessibility to individual NF-κB-target genes after cellular activation can provide an additional mechanism by which NF-κB activity is regulated.

UBIQUITIN-MEDIATED IκB DEGRADATION

Induction of NF-κB activity typically requires the IKK-regulated degradation of IkBα, IκBβ, and IκBε by the ubiquitin-proteasome pathway (Karin and Ben-Neriah, 2000). After IκB phosphorylation, ubiquitination occurs via a three-stage process (Maniatis, 1999). First, ubiquitin (an 8.6-kDa highly conserved protein) is attached in an ATP-dependent manner via a high-energy thioester bond to an ubiquitin-activating enzyme (E1) (Maniatis, 1999; Tanaka et al., 2001). The ubiquitin is then transferred to an ubiquitin-conjugating enzyme (E2), which in conjunction with the E3 protein, catalyzes the covalent linkage of ubiquitin to lysine residues of the targeted protein. The E3 protein functions as an adaptor molecule that confers specificity to the recruitment of proteins to the complex containing the E2 enzyme. Repetition of the ubiquitination process results in the formation of a polyubiquitin chain via the attachment of multiple ubiquitin molecules to specific lysine residues in the conjugated ubiquitin. Polyubiquitination then serves as a signal for attack by the 26S proteasome, leading to the degradation or processing of the targeted protein.

IKK activation results in phosphorylation of N-terminal serines 32 and 36 on IκBα and serines 19 and 23 on IκBβ, which leads to their selective ubiquitination and subsequent degradation (Baldwin, 1996; Ghosh et al., 1998; Mercurio and Manning, 1999; Tanaka et al., 2001). Ubiquitination of IκBα is mediated by an E3 pIκBα-ubiquitin ligase complex, SCF$^{\beta\text{-}TrCP}$, that is specific for IκBs. Although the SCF complex does not possess inherent catalytic activity, it functions to direct the ubiquitin-conjugating enzyme (E2) to phosphorylated IκB (Yaron et al., 1998). After IκB binding by the E3 pIκBα-ubiquitin ligase SCF$^{\beta\text{-}TrCP}$ the E2 enzyme transfers ubiquitin from the ubiquitin-E1 complex to IκBα (Maniatis, 1999).

SCF complexes, initially identified in studies of the yeast cell cycle, are composed of Skp1, Cdc53/Cul-1, Roc1, and the F-box protein (Fig. 2.9) (Maniatis, 1999; Tanaka et al., 2001). The F-box protein is the SCF component that confers specificity to the polyubiquitination process by binding to the protein targeted for degradation (Maniatis, 1999; Tanaka et al., 2001). WD40-repeat domains of the F-box proteins β-TrCP1 (Fbw1a) and β-TrCP2 (Fbw1b) specifically bind phosphorylated IκB via the conserved recognition motif, DS(P)GLDS(P), and thereby promote its ubiquitination in the presence of E1 and E2 (Ubch5 or Ubch7) (Yaron et al., 1998; Maniatis, 1999; Shirane et al., 1999; Spencer et al., 1999;

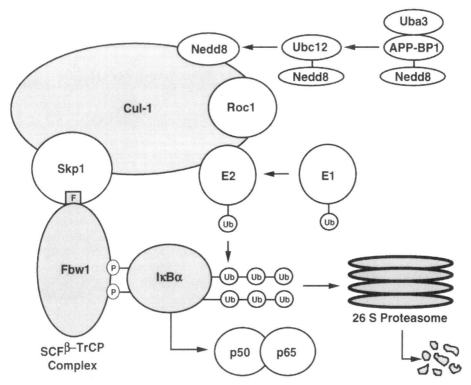

Figure 2.9. *The E3 pIκBα-Ubiquitin Ligase, SCF^{β-TrCP}.* The SCF^{β-TrCP} complex has four components, Skp1, Cdc53/Cul-1, Roc1, and an F-box protein. The F-box proteins β-TrCP1 (Fbw1a) and β-TrCP2 (Fbw1b) confer specificity to the polyubiquitination process by binding IκBα via the conserved recognition motif DS(P)GLDS(P). Skp1 functions as an adaptor linking the F-box proteins to Cul-1. Cul-1 acts as a scaffolding molecule that links the SCF complex to the E2 ubiquitin-conjugating enzyme. Roc1 contains a RING-H2 finger domain and binds the C-terminal region Cul-1 and promotes its nuclear import. Covalent conjugation of NEDD8/Rub1, an ubiquitin-like protein, to the C-terminal region of Cul-1 accelerates the formation of the E2-E3 complex and the subsequent polyubiquitination of pIκBα by enhancing the recruitment of the E2-ubiquitin complex to the SCF complex. Nedd8-modification is catalyzed by the APP-BP1/Uba3 heterodimer (E1) and Ubc12 (E2).

Suzuki et al., 1999; Winston et al., 1999; Wu and Ghosh, 1999; Suzuki et al., 2000). Skp1 binds to the F-box proteins via the F-box domain and also interacts with the N terminus of Cul-1, thereby functioning as an adaptor that links the F-box proteins to Cul-1 (Tanaka et al., 2001). Cul-1 functions as a scaffolding molecule that links the SCF complex to the E2 ubiquitin-conjugating enzyme. In addition, covalent conjugation of NEDD8/Rub1, an ubiquitin-like protein, to the C-terminal region of Cul-1 accelerates the formation of the E2-E3 complex and the subsequent polyubiquitination of pIκBα by enhancing the recruitment of the E2-ubiquitin complex to the SCF complex (Read et al., 2000; Kawakami et al., 2001; Tanaka et al., 2001). Nedd8 modification is catalyzed by the APP-BP1/Uba3 heterodimer (E1) and Ubc12 (E2) (Tanaka et al., 2001). Roc1/Rbx1/Hrt1, which contains a RING-H2 finger domain, is another essential subunit of the SCF complex that binds the C-terminal region of Cul-1 and promotes its nuclear import (Furukawa et al., 2000; Karin and Ben-Neriah, 2000; Tanaka et al., 2001).

The ubiquitin-proteasome system is also responsible for the limited processing of the p105 and p100 precursors to their NF-κB1 (p50) and NF-κB2 (p52) mature forms (Karin and Ben-Neriah, 2000; Orian et al., 2000; Xiao et al., 2001). Limited processing of precursor proteins is unusual, because ubiquitination and subsequent proteasome activity generally result in the complete degradation of targeted proteins (Karin and Ben-Neriah, 2000; Orian et al., 2000). Before processing, both p105 and p100 function as IκB molecules (Heissmeyer et al., 2001). Masking of the nuclear localization signals by the C-terminal ankyrin-repeat domain sequesters the precursor proteins in the cytosol. Posttranslational processing of p105 to p50 requires prior ubiquitination and degradation of the C-terminal ankyrin-repeat domain (Cohen et al., 2001). A 23-amino-acid, glycine-rich domain (amino acids 376–404) functions as both a processing signal and a "stop" signal that terminates 26S proteasomal digestion, thereby stabilizing the p50 molecule (Lin and Ghosh, 1996; Ghosh et al., 1998; Orian et al., 1999; Orian et al., 2000; Cohen et al., 2001). Processing of p105 is predominantly constitutive, although it can be enhanced in response to stimuli such as phorbol ester, okadaic acid, TNF-α, or IL-1β (MacKichan et al., 1996; Heissmeyer et al., 1999; Karin and Ben-Neriah, 2000; Heissmeyer et al., 2001).

p105 is targeted for processing and/or degradation by two ubiquitin-recognition motifs that are recognized by two different E3 ubiquitin ligases (Cohen et al., 2001). Amino acid residues 441–454 in p105 appear to be important for constitutive processing. Two essential lysines (positions 441 and 442) are major ubiquitination sites, whereas an acidic region (positions 446–454) may function as an E3-recognition motif for constitutive processing and/or degradation (Orian et al., 1999; Cohen et al., 2001). A second C-terminal E3-recognition motif appears to be important for signal-induced p105 processing and/or degradation (Cohen et al., 2001). Stimulus-induced IKK-catalyzed phosphorylation of the p105 C-terminal domain creates a recognition motif for the SCF[β-TrCP] E3 ubiquitin ligase, which results in accelerated processing to p50 as well as complete degradation (Orian et al., 2000; Heissmeyer et al., 2001; Salmeron et al., 2001). The DS[923]VCDS[927] β-TrCP recognition motif is similar to the targeting motifs in IκBα, -β, and -ε as well as β-catenin and HIV-Vpu (Karin and Ben-Neriah, 2000; Cohen et al., 2001; Heissmeyer et al., 2001). Although the processing of p100 to p52 appears to be limited, it is mediated by a mechanism similar to that used for p105 (Betts and Nabel, 1996; Orian et al., 1999; Cohen et al., 2001). Processing of p100 was recently demonstrated to be positively regulated by NIK. NIK catalyzes the site-specific phosphorylation and ubiquitination of p100 and serves as a molecular trigger for p100 processing (Xiao et al.,

2001). In addition, p100 contains a C-terminal processing-inhibitory domain (PID) that suppresses its constitutive processing.

NF-κB IS AN IMPORTANT TRANSCRIPTION FACTOR IN THE PATHOGENESIS OF ASTHMATIC AIRWAY INFLAMMATION

Multiple lines of evidence show that NF-κB plays a critical role in the pathogenesis of asthmatic airway inflammation. First, mediators capable of inducing asthmatic airway inflammation can function as activators of the NF-κB signaling pathway. Second, activation of the NF-κB signaling pathway can induce the transcription of many proinflammatory genes that contribute to asthmatic airway inflammation. Third, increased NF-κB activation has been demonstrated both in experimental models of asthmatic allergic inflammation and in asthmatic patients. Finally, the essential role of NF-κB in the pathogenesis of atopic asthma has been definitively established in mice harboring targeted deletions in NF-κB subunits.

ACTIVATORS OF THE NF-κB SIGNALING PATHWAY IN ASTHMA

A variety of proinflammatory stimuli and mediators that contribute to the pathogenesis of asthma, such as allergens, respiratory viruses, cytokines, interleukins, lipid mediators, and reactive gases, are capable of activating the NF-κB pathway (Fig. 2.10) (Barnes and Karin, 1997; Rahman and MacNee, 1998). Allergens, histamine, and respiratory viruses represent common stimuli that induce asthmatic airway inflammation via a mechanism that includes direct NF-κB activation. Der p1, the major allergen of the dust mite *Dermatophagoides pteronyssinus*, can directly induce NF- κB activation in human bronchial epithelial cells (Stacey et al., 1997). In addition, NF-κB activation was implicated in the generation of IL-16 and T cell chemotactic activity by bronchial explants from asthmatic patients after ex vivo stimulation with *D. pteronyssinus* (Hidi et al., 2000). Similarly, infections by respiratory viruses, such as rhinovirus, adenovirus, and influenza, can directly activate pulmonary epithelial cell NF-κB and thereby induce the expression of proinflammatory genes (Pahl and Baeuerle, 1995; Pahl et al., 1996). Rhinovirus infection directly activates RelA (p65), NF-κB1 (p50), and NF-κB2 (p52) in human bronchial epithelial cells (Zhu et al., 1996; Thomas et al., 1998; Papi and Johnston, 1999). Rhinovirus-stimulated NF-κB activation can then induce the expression of adhesion molecules (e.g., ICAM-1 and VCAM-1), cytokines (e.g., IL-6 and IL-11), and chemokines (e.g., RANTES) (Zhu et al., 1996; Bitko et al., 1997; Thomas et al., 1998; Papi and Johnston, 1999a; Papi and Johnston, 1999b). ICAM-1 and VCAM-1 have important roles in eosinophil transmigration to asthmatic airways, and ICAM-1 also serves as the rhinovirus cellular receptor. Histamine,

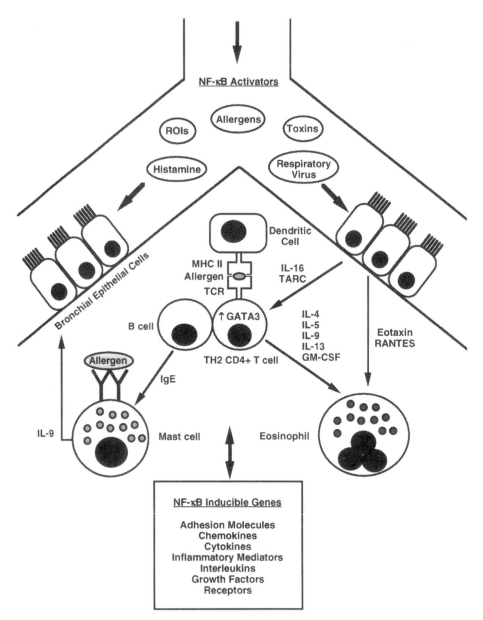

Figure 2.10. *NF-κB signaling is essential for the pathogenesis of asthma.* Multiple stimuli that trigger asthmatic airway inflammation can also activate the NF-κB signaling pathway. NF-κB signaling plays an important role in the differentiation and activation of dendritic cells, which are the primary airway antigen-presenting cell that mediates T cell activation. Activated T_H2-type $CD4^+$ T cells generate a specific set of cytokines that promote allergic airway inflammation. Furthermore, NF-κB is essential for the differentiation of T_H2-type $CD4^+$ T cells, which requires NF-κB-mediated GATA3 gene transcription. NF-κB activation also promotes B cell IgE class switching. Secreted IgE can bind mast cell high-affinity IgE receptors and cause mast cell activation after cross-linking of IgE by allergen. IL-9 generated by activated mast cells and T_H2-type $CD4^+$ T cells

which is released by mast cells and mediates the early response to allergen, can also induce human bronchial epithelial cell NF-κB activation and IL-8 gene transcription (Adcock et al., 1994; Aoki et al., 1998).

Reactive oxygen species can also activate NF-κB and exacerbate asthma (Adcock et al., 1994; Allen and Tresini, 2000; Finkel and Holbrook, 2000). Ozone (O_3), the most common outdoor air pollutant in the United States, is a highly reactive oxidant gas capable of generating reactive oxygen intermediates (ROIs) and inducing airway inflammation with associated increases in NF-κB binding activity in airway epithelial cells (Jaspers et al., 1997; Nichols et al., 2001). ROIs can also augment IL-1-mediated NF-κB activation in human bronchial epithelial cells (Jany et al., 1995). Hydrogen peroxide, which is elevated in the exhaled gases of asthmatic patients in parallel with increased levels of lipid peroxidation products, has also been reported to stimulate IκB degradation and NF-κB activation (Antczak et al., 1997; Allen and Tresini, 2000; Antczak et al., 2000). Although H_2O_2 increases airway epithelial cell IKK activity and IκBα ubiquitination, it fails to induce IκBα degradation, which suggests that H_2O_2 functions as a general inhibitor of proteasome function in the airway epithelium (Jaspers et al., 2001). Therefore, the effect of ROIs on NF-κB activation may be cell type specific. Finally, diesel exhaust particles, another important air pollutant that induces asthmatic airway hyperreactivity and eosinophilia, can also activate human airway epithelial cell NF-κB (Takizawa et al., 1999).

NF-κB–MEDIATED TRANSACTIVATION OF PROINFLAMMATORY GENES IN ASTHMA

Transcription of many proinflammatory genes that participate in the regulation of asthmatic airway inflammation can be activated by NF-κB. NF-κB-inducible genes include those for adhesion molecules (e.g., ICAM-1, VCAM-1, E-selectin), cell surface receptors (e.g., CCR5, MHC class I and II, CD86, T cell receptor, IL-2 receptor α chain, platelet-activating factor receptor), chemokines [e.g., eotaxin, RANTES, macrophage inflammatory protein (MIP)-1α, monocyte chemoattractant protein-1 (MCP-1), Gro-α/β/γ], growth factors (e.g., GM-CSF, M-CSF, G-CSF), enzymes (iNOS, cyclooxygenase-2, 5-lipoxygenase, cytosolic phospholipase A_2), and interleukins/cytokines (e.g., IL-1β, IL-2, IL-6, TNF-α) (Barnes and Karin, 1997; Rahman and MacNee, 1998).

In particular, the NF-κB-mediated transcriptional activation of the CC-type chemokines, eotaxin, RANTES, and TARC, as well as IL-9, is

---

can induce airway mucus cell hypertrophy and enhanced mucin gene expression. Furthermore, NF-κB activation in bronchial epithelial cells induces the generation of the T cell chemotactic factors IL-16 and TARC as well as chemokines with chemotactic activity for eosinophils, such as eotaxin and RANTES. Multiple additional NF-κB-inducible genes also participate in the initiation, perpetuation, and amplification of asthmatic airway inflammation.

relevant to the pathogenesis of asthma. Eotaxin is overexpressed in asthmatic airways and functions as a potent eosinophilic chemotactic factor via binding to the CCR3 receptor (Matsukura et al., 1999; Lilly et al., 2001). The eotaxin promoter contains overlapping consensus binding sites for NF-κB and STAT6 that share four base pairs and mediate responsiveness to TNF-α and the T_H2-type cytokine IL-4, respectively (Matsukura et al., 1999). In studies using human airway epithelial cells, the NF-κB and STAT6 sites acted synergistically to enhance eotaxin gene transcription in response to TNF-α and IL-4, an effect that was completely abrogated when both sites were mutated. As discussed below, cooperative activity between NF-κB and STAT6 also regulates B cell IgE-isotype class switching. IL-1β also induces eotaxin gene transcription in human airway epithelial cells via an NF-κB-dependent mechanism (Jedrzkiewicz et al., 2000). However, it should be noted that STAT6 and NF-κB can act in an antagonistic fashion in other promoters. For example, the E-selectin promoter contains NF-κB and STAT6 sites that overlap by five base pairs, thereby allowing STAT6 binding to act in a competitive fashion and to antagonize NF-κB binding and transcriptional activation (Bennett et al., 1997). Thus differences in DNA motifs and promoter structure may differentially influence NF-κB and STAT6 binding and promoter activation (Matsukura et al., 1999). Inhibition of NF-κB-dependent transcription by STAT6 can also result from the competition for limited amounts of transcriptional coactivator proteins, such as CBP, as described for the interferon regulatory factor-1 promoter (Ohmori and Hamilton, 2000). This represents an alternative mechanism by which IL-4-induced STAT6 activation can suppress NF-κB-dependent transcriptional activation without altering NF-κB DNA binding. Furthermore, during osteoclastogenesis, IL-4 can inhibit IκB phosphorylation and degradation (Abu-Amer, 2001).

RANTES is also expressed in the airways of asthmatic patients. RANTES acts via the CCR3 receptor and plays a central role in eosinophil chemotaxis, as well as the recruitment of mast cells and CD45 RO+ memory T cells (Ray et al., 1997; Busse and Lemanske, 2001). Furthermore, RANTES is an NF-κB-inducible gene that is upregulated in response to viral infection in a cooperative fashion by NF-κB and interferon-regulatory factor-3 (Moriuchi et al., 1997; Genin et al., 2000). In lung epithelial cells, TNF-α and IFN-γ mediate increases in RANTES gene expression (Stellato et al., 1995). Overexpression of IκB-related protein (IκBR) in human airway epithelial cells can augment the TNF-α-induced increases in RANTES gene expression via the sequestration of inhibitory p50 homodimers (Ray et al., 1997). IκBR was initially cloned from human alveolar epithelial cells and shows significant similarity to the *Drosophila* protein Cactus, an inhibitor of the NF-κB-like protein Dorsal (Ray et al., 1995).

TARC binds to the CCR4 chemokine receptor, which is preferentially expressed on T_H2-type CD4+ T cells, and functions as a T cell chemotactic factor (Berin et al., 2001). Human bronchial epithelial cells produce TARC, and TARC levels in bronchoalveolar lavage fluid (BALF) are increased after segmental allergen challenge in atopic asthmatics. Furthermore, induction of TARC gene expression by TNF-α and IL-4 is NF-κB dependent. Therefore, NF-κB-mediated TARC expression may contribute to the chemotactic recruitment of T_H2-type CD4$^+$ T cells in the pathogenesis of asthma.

IL-9, a product of T_H2-type CD4$^+$ T cells and mast cells, has been identified as a candidate gene for asthma (Nicolaides et al., 1997). In models of atopic asthma, mice overexpressing IL-9 demonstrate eosinophilic and lymphocytic airway inflammation, mast cell hyperplasia, elevated serum IgE, and bronchial hyperresponsiveness (McLane et al., 1998; Temann et al., 1998). IL-9 also mediates airway mucus cell hypertrophy and enhanced mucin gene (MUC5AC) expression (Temann et al., 1998; Longphre et al., 1999). Mast cells that have been activated by crosslinked IgE and LPS demonstrate enhanced IL-9 gene expression that is mediated by three NF-κB binding sites within the IL-9 promoter (Stassen et al., 2001). IL-9 also regulates T helper and mast cell function via increased expression of Bcl-3 and increased p50 homodimer DNA-binding activity (Richard et al., 1999).

NF-κB also plays a key role in the regulation of dendritic cell and B lymphocyte function. Dendritic cells represent the primary antigen-presenting cell in the airway and play an important role in T cell activation during both primary and recall responses to antigen. Dendritic cells constitutively express NF-κB. Furthermore, NF-κB activation mediates dendritic cell differentiation and activation (Verhasselt et al., 1999; Lyakh et al., 2000). In particular, RelB is essential for the development of myeloid-derived CD8α-DEC-205$^-$ dendritic cells (Wu et al., 1998).

NF-κB activation also promotes B cell IgE class switching in a cooperative fashion with STAT6 during activation of the human IgE germline promoter by IL-4 (Messner et al., 1997; Stutz and Woisetschlager, 1999). B cell stimulation by IL-4 synergizes with the requisite second signal provided by CD40 to mediate strong NF-κB activation and IgE production (Iciek et al., 1997; Jeppson et al., 1998). This synergy is mediated by nuclear complexes containing both STAT6 and NF-κB proteins that bind to and activate a CD40L/IL-4 responsive element in the germline ε promoter (Iciek et al., 1997). Secreted IgE can bind to both high-affinity IgE receptors (FcεRI) on mast cells and basophils or low-affinity IgE receptors (FcεRII) on eosinophils, macrophages, lymphocytes, and platelets (Busse and Lemanske, 2001). Molecular bridging of FcεRI molecules by allergen results in mast cell activation and secretion of preformed and newly generated mediators. Although the significance of the low-affinity IgE receptor in the pathogenesis of asthma is incompletely defined, IgE-dependent activation of alveolar macrophages from atopic asthmatic patients is associated with secretion of enhanced amounts of TNF-α and IL-6 (Gosset et al., 1992). p50/p65 heterodimers represent the major transcription factor mediating gene activation after IgE binding to the alveolar macrophage FcεRIIb receptor (CD23) (Ten et al., 1999). The signal transduction pathway mediating CD23-dependent NF-κB activation involves a tyrosine kinase activity upstream of IKK, with subsequent

IKK activation and phosphorylation of serines 32 and 36 in the IκBα N-terminal domain (Ten et al., 1999).

NF-κB ACTIVATION IN ASTHMA

NF-κB activation has been demonstrated in asthmatic airways in experimental models of asthmatic allergic inflammation as well as in asthmatic patients. Increased nuclear binding of NF-κB transcription factors including NF-κB1 (p50), RelA (p65), and c-Rel subunits was demonstrated in thoracic lymphocytes after ovalbumin sensitization and aerosolization challenge in a murine model of asthma (Donovan et al., 1999). In a naturally occurring model of asthma, horses afflicted with heaves have bronchial epithelial cells that contain high levels of p50/p65 heterodimers (Bureau et al., 2000).

Enhanced NF-κB activation has also been demonstrated in asthmatic patients. Increased nuclear NF-κB1 (p50) and RelA (p65) subunits were demonstrated in sputum macrophages and in bronchial biopsies from asthmatic individuals (Hart et al., 1998). Interestingly, in a follow-up study, the anti-inflammatory effects of inhaled corticosteroids (e.g., fluticasone) were not mediated by decreases in either RelA (p65) subunit expression or NF-κB DNA-binding in alveolar macrophages and bronchial biopsies, despite clinical improvement in airway eosinophilia and bronchial hyperreactivity (Hart et al., 2000). Furthermore, an increase in airway epithelial cell nuclear RelA (p65) expression was noted after fluticasone therapy, which was reproduced in the A549 epithelial cell line after dexamethasone treatment. Similarly, studies conducted in A549 and BEAS-2B pulmonary epithelial cell lines demonstrated that the repressive effects of glucocorticoids are not mediated by upregulation of IκBα, decreased p50/p65 gene expression, or inhibition of NF-κB DNA binding (Newton et al., 1998). Therefore, these studies suggest that the anti-inflammatory effects of corticosteroids in asthmatic airway epithelial cells may not be mediated by inhibition of RelA (p65).

NF-κB ACTIVATION IS ESSENTIAL FOR THE PATHOGENESIS OF ASTHMA

The most compelling evidence that NF-κB plays an essential role in the pathogenesis of asthmatic airway inflammation comes from recent studies utilizing p50 and c-Rel knock-out mice (Yang et al., 1998; Das et al., 2001). Although mice harboring targeted p50 deletions do not display any developmental abnormalities, they demonstrate defective B cell proliferative and antibody responses as well as marked reductions in immunoglobulin class switching to IgE, IgG3, and IgA (Sha et al., 1995; Snapper et al., 1996; Yang et al., 1998). After allergen sensitization and challenge, p50$^{-/-}$ mice do not develop eosinophilic airway inflammation, thereby demonstrating a key role for p50 in the pathogenesis of asthma (Yang et al., 1998; Das et al., 2001). Furthermore, p50$^{-/-}$ mice demonstrate impaired production of T_H2-type cytokines such as IL-5 and, to a lesser extent, IL-4 and IL-13. $p50^{-/-}$ mice also have impaired production of chemokines such as eotaxin and macrophage inflammatory proteins (MIP-1α and MIP-1β).

These defects in $p50^{-/-}$ mice did not occur as a consequence of impaired T cell chemotaxis but instead reflected impaired differentiation of T_H2-type CD4$^+$ cells secondary to reduced GATA-3 expression (Das et al., 2001). In contrast, induction of the T-bet gene, which plays an important role in T_H1 lineage development, was not impaired. NF-κB-mediated transcriptional activation of the GATA3 gene is mediated via two κB-binding sites, one located between residues –301 and –310 and another between +746 and +755 (Das et al., 2001). GATA-3 is a key transcription factor that is expressed in a T_H2 cell-specific manner and induces T_H2 cell differentiation as well as the expression of T_H2-type cytokines (Zhang et al., 1997; Zheng and Flavell, 1997; Lee et al., 2001). In addition, GATA-3 has been demonstrated to play a key role in the pathogenesis of allergic asthma. Expression of a dominant-negative GATA-3 mutant in a T cell-specific fashion in a murine model of allergic asthma attenuates airway inflammation, as evidenced by reductions in T_H2-type cytokines IL-4, IL-5, and IL-13, eosinophilia, mucus hyperproduction, and IgE synthesis (Zhang et al., 1999). Similarly, inhibition of GATA-3 utilizing antisense oligonucleotides inhibits airway eosinophilia, T_H2-type cytokine production, and airway hyperresponsiveness in a separate model of murine allergic asthma (Finotto et al., 2001). Therefore, these studies highlight the importance of the p50 subunit, most likely in the context of a p50/p65 heterodimer, in the induction of GATA-3 expression and subsequent T_H2 cell differentiation in asthmatic airways (Das et al., 2001).

The NF-κB c-Rel subunit has also been identified as playing an important role in the pathogenesis of asthmatic airway inflammation. c-Rel is primarily expressed in lymphocytes and is required for normal lymphocyte activation and proliferation (Donovan et al., 1999). c-Rel-deficient mice display normal lymphocyte development but manifest defective T cell proliferation in response to T cell receptor activation as a consequence of defective production of IL-2, IL-3, and GM-CSF (Kontgen et al., 1995; Donovan et al., 1999; Liou et al., 1999). $c-Rel^{-/-}$ mice also have defects in humoral immunity, as manifested by impaired mitogen-induced B cell survival and cell cycle progression (Tumang et al., 1998). After allergen sensitization and challenge, $c-Rel^{-/-}$ mice do not develop pulmonary inflammation, BALF eosinophilia, airway hyperreactivity, or increases in total serum IgE (Donovan et al., 1999). $c-Rel^{-/-}$ mice also demonstrate defective expression of the chemokine MCP-1, which contains two NF-κB binding sites capable of binding c-Rel/p65 heterodimers and p65 homodimers (Ueda et al., 1997; Donovan et al., 1999). MCP-1 may play an important role in the pathogenesis of airway inflammation, because inhibition of MCP-1 markedly diminishes bronchial hyperreactivity, as well as BALF eosinophilia, IgE, IL-4, and IL-5 levels (Gonzalo et al., 1998). Therefore, both the p50 and c-Rel subunits have been identified as critical factors in the pathogenesis of asthmatic airway inflammation.

NF-κB AS A MOLECULAR TARGET FOR DRUG DEVELOPMENT

Because of its central role in mediating inflammatory responses in asthma and other disorders, the NF-κB signaling pathway represents an important target for the future development of anti-inflammatory medications. Indeed, many of the currently utilized anti-inflammatory agents, such as glucocorticoids, nonsteroidal anti-inflammatory drugs, and immunosuppressive agents (e.g., cyclosporin A, FK-506) mediate their function, at least in part, via the inhibition of the NF-κB pathway (Yamamoto and Gaynor, 2001).

Efforts are in progress to develop novel, specific inhibitors of the NF-κB signaling pathway. These inhibitors may target different protein complexes in the NF-κB signaling cascade, such as IκB, the IKK signalsome, and the proteasome. For example, an IκBα superrepressor has been developed that functions as a dominant negative as a result of its ability to sequester NF-κB in the cytoplasm and thereby prevent its activation of NF-κB-inducible genes (Yamamoto and Gaynor, 2001). The IκBα superrepressor is an IκBα protein that is resistant to IKK-mediated phosphorylation and subsequent proteasomal degradation because of mutations in serine residues 32 and 36 (Yamamoto and Gaynor, 2001). Furthermore, adenoviral delivery of the IκBα superrepressor has been demonstrated to sensitize chemotherapy-resistant tumors to TNF-α-mediated and chemotherapy-mediated apoptosis in a murine model (Wang et al., 1999). Therefore, inhibition of the NF-κB pathway may be effective in increasing the efficacy of cancer chemotherapy. Similarly, a cell-permeant peptide corresponding to the IKK-γ NBD has been developed that blocks the association of IKK-γ with the IKK complex. The NBD peptide can inhibit both TNF-α-induced NF-κB activation and the activation of NF-κB-inducible genes while slightly increasing basal levels of NF-κB activity (May et al., 2000). Furthermore, the NBD peptide is capable of blocking inflammatory responses in animal models of inflammation. This suggests that the NBD peptide may be efficacious via its ability to block cytokine-induced IKK activity while maintaining basal levels of NF-κB activity. Inhibitors of proteasome function have also been developed that prevent the proteasome-mediated degradation of IκB (Yamamoto and Gaynor, 2001). However, these proteasome inhibitors are not specific for IκB and may be associated with unanticipated adverse consequences.

Therefore, the development of specific inhibitors of the NF-κB signaling pathway represents a promising approach for the future treatment of inflammatory disorders, including asthma.

REFERENCES

Abu-Amer, Y. (2001). IL-4 abrogates osteoclastogenesis through STAT6-dependent inhibition of NF-κB. J Clin Invest *107*(11), 1375–1385.

Adcock, I. M., Brown, C. R. et al. (1994). Oxidative stress induces NF κB DNA binding and inducible NOS mRNA in human epithelial cells. Biochem Biophys Res Commun *199*(3), 1518–1524.

Allen, R. G., and Tresini, M. (2000). Oxidative stress and gene regulation. Free Radic Biol Med *28*(3), 463–99.

Anrather, J., Csizmadia, V. et al. (1999). Regulation of NF-κB RelA phosphorylation and transcriptional activity by p21(ras) and protein kinase Cζ in primary endothelial cells. J Biol Chem *274*(19), 13594–13603.

Antczak, A., Kurmanowska, Z. et al. (2000). Inhaled glucocorticosteroids decrease hydrogen peroxide level in expired air condensate in asthmatic patients. Respir Med *94*(5), 416–421.

Antczak, A., Nowak, D. et al. (1997). Increased hydrogen peroxide and thiobarbituric acid-reactive products in expired breath condensate of asthmatic patients. Eur Respir J *10*(6), 1235–1241.

Aoki, Y., Qiu, D. et al. (1998). Leukotriene B4 mediates histamine induction of NF-κB and IL-8 in human bronchial epithelial cells. Am J Physiol *274*(6 Pt 1), L1030–L1039.

Azzawi, M., Bradley, B. et al. (1990). Identification of activated T lymphocytes and eosinophils in bronchial biopsies in stable atopic asthma. Am Rev Respir Dis *142*(6 Pt 1), 1407–1413.

Baeuerle, P. A. (1998). IκB-NF-κB structures: at the interface of inflammation control. Cell *95*(6), 729–731.

Baeuerle, P. A., and Baltimore, D. (1996). NF-κB: ten years after. Cell *87*(1), 13–20.

Baldwin, A. S., Jr. (1996). The NF-κB and I κB proteins: new discoveries and insights. Annu Rev Immunol *14*, 649–683.

Baldwin, A. S., Jr. (2001). Series introduction: the transcription factor NF-κB and human disease. J Clin Invest *107*(1), 3–6.

Barnes, P. J., and Karin, M. (1997). Nuclear factor-κB: a pivotal transcription factor in chronic inflammatory diseases. N Engl J Med *336*(15), 1066–1071.

Beg, A. A., Sha, W. C., et al. (1995). Constitutive NF-κB activation, enhanced granulopoiesis, and neonatal lethality in IκBα-deficient mice. Genes Dev *9*(22), 2736–2746.

Bennett, B. L., Cruz, R., et al. (1997). Interleukin-4 suppression of tumor necrosis factor α-stimulated E- selectin gene transcription is mediated by STAT6 antagonism of NF-κB. J Biol Chem *272*(15), 10212–10219.

Beraud, C., Henzel, W. J., et al. (1999). Involvement of regulatory and catalytic subunits of phosphoinositide 3-kinase in NF-κB activation. Proc Natl Acad Sci USA *96*(2), 429–434.

Berin, M. C., Eckmann, L. et al. (2001). Regulated production of the T helper 2-type T-cell chemoattractant TARC by human bronchial epithelial cells in vitro and in human lung xenografts. Am J Respir Cell Mol Biol *24*(4), 382–389.

Betts, J. C., and Nabel, G. J. (1996). Differential regulation of NF-κB2(p100) processing and control by amino-terminal sequences. Mol Cell Biol *16*(11), 6363–6371.

Bird, T. A., Schooley, K. et al. (1997). Activation of nuclear transcription factor NF-κB by interleukin-1 is accompanied by casein kinase II-mediated phosphorylation of the p65 subunit. J Biol Chem *272*(51), 32606–32612.

Bitko, V., Velazquez, A. et al. (1997). Transcriptional induction of multiple cytokines by human respiratory syncytial virus requires activation of NF-κB and is inhibited by sodium salicylate and aspirin. Virology *232*(2), 369–378.

Bouchier-Hayes, L., Conroy, H. et al. (2001). CARDINAL, a novel caspase recruitment domain protein, is an inhibitor of multiple NF-κB activation pathways. J Biol Chem *276*(47), 44069–44077.

Bours, V., Franzoso, G. et al. (1993). The oncoprotein Bcl-3 directly transactivates through κB motifs via association with DNA-binding p50B homodimers. Cell *72*(5), 729–739.

Brasier, A. R., Lu, M. et al. (2001). NF-κB-inducible BCL-3 expression is an autoregulatory loop controlling nuclear p50/NF-κB1 residence. J Biol Chem *276*(34), 32080–32093.

Brown, K., Park, S. et al. (1993). Mutual regulation of the transcriptional activator NF-κB and its inhibitor, IκB-α. Proc Natl Acad Sci USA *90*(6), 2532–2536.

Bureau, F., Bonizzi, G. et al. (2000). Correlation between nuclear factor-κB activity in bronchial brushing samples and lung dysfunction in an animal model of asthma. Am J Respir Crit Care Med *161*(4 Pt 1), 1314–1321.

Busse, W. W., and Lemanske, R. F., Jr. (2001). Asthma. N Engl J Med *344*(5), 350–362.

CDC (1996). Asthma mortality and hospitalization among children and young adults—United States, 1980–1993. MMWR *45*(17), 350–353.

CDC (1997). Facts about asthma. CDC.

Chen, F. E., Huang, D. B. et al. (1998). Crystal structure of p50/p65 heterodimer of transcription factor NF-κB bound to DNA. Nature *391*(6665), 410–413.

Chen, L. F., Fischle, W. et al. (2001). Duration of nuclear NF-κB action regulated by reversible acetylation. Science *293*, 1653–1657.

Chen, Z. J., Parent, L. et al. (1996). Site-specific phosphorylation of IκBα by a novel ubiquitination-dependent protein kinase activity. Cell *84*(6), 853–862.

Cheng, J. D., Ryseck, R. P. et al. (1998). Functional redundancy of the nuclear factor κB inhibitors IκBα and IκBβ. J Exp Med *188*(6), 1055–1062.

Cohen, S., Orian, A. et al. (2001). Processing of p105 is inhibited by docking of p50 active subunits to the ankyrin repeat domain, and inhibition is alleviated by signaling via the carboxyl-terminal phosphorylation/ubiquitin-ligase binding domain. J Biol Chem *276*(29), 26769–26776.

Collins, J. G. (1997). Prevalence of selected chronic conditions: United States, 1990–1992. National Center for Health Statistics. Vital Health Stat *10*(194).

Crapo, R. O., Casaburi, R. et al. (2000). Guidelines for methacholine and exercise challenge testing—1999. This official statement of the American Thoracic Society was adopted by the ATS Board of Directors, July 1999. Am J Respir Crit Care Med *161*(1), 309–329.

Das, J., Chen, C. H. et al. (2001). A critical role for NF-κB in GATA3 expression and TH2 differentiation in allergic airway inflammation. Nat Immunol *2*(1), 45–50.

Dechend, R., Hirano F., et al. (1999). The Bcl-3 oncoprotein acts as a bridging factor between NF-κB/Rel and nuclear co-regulators. Oncogene *18*(22), 3316–3323.

Deng, L., Wang, C. et al. (2000). Activation of the IκB kinase complex by TRAF6 requires a dimeric ubiquitin-conjugating enzyme complex and a unique polyubiquitin chain. Cell *103*(2), 351–361.

Devin, A., Cook, A. et al. (2000). The distinct roles of TRAF2 and RIP in IKK activation by TNF-RI: TRAF2 recruits IKK to TNF-RI while RIP mediates IKK activation. Immunity *12*(4), 419–429.

Devin, A., Lin, Y. et al. (2001). The α and β subunits of IκB kinase (IKK) mediate TRAF2- dependent IKK recruitment to tumor necrosis factor (TNF) receptor 1 in response to TNF. Mol Cell Biol *21*(12), 3986–3994.

DiDonato, J. A., Hayakawa, M. et al. (1997). A cytokine-responsive IκB kinase that activates the transcription factor NF-κB. Nature *388*(6642), 548–554.

Donovan, C. E., Mark, D. A. et al. (1999). NF-κB/Rel transcription factors: c-Rel promotes airway hyperresponsiveness and allergic pulmonary inflammation. J Immunol *163*(12), 6827–6833.

Elias, J. (1999). Inspirations on asthma. J Clin Invest *104*(7), 827–828.

Elias, J. A., Zhu, Z. et al. (1999). Airway remodeling in asthma. J Clin Invest *104*(8), 1001–1006.

Fan, C. M., and Maniatis, T. (1991). Generation of p50 subunit of NF-κB by processing of p105 through an ATP-dependent pathway. Nature *354*(6352), 395–358.

Fenwick, C., Na, S. Y. et al. (2000). A subclass of Ras proteins that regulate the degradation of IκB. Science *287*(5454), 869–873.

Finkel, T., and Holbrook, N. J. (2000). Oxidants, oxidative stress and the biology of ageing. Nature *408*(6809), 239–247.

Finotto, S., De Sanctis, G. T. et al. (2001). Treatment of allergic airway inflammation and hyperresponsiveness by antisense-induced local blockade of GATA-3 expression. J Exp Med *193*(11), 1247–1260.

Fish, J. E., and Peters, S.P. (1998). Asthma: clinical presentation and management. *Fishman's Pulmonary Diseases and Disorders,* 3rd Ed. Fishman, A. P., Elias, J.A., Fishman, J.A., Grippi, M.A., Kaiser, L.R., Senior, R.M. New York, McGraw-Hill. *1:* 757–776.

Franzoso, G., Bours, V. et al. (1992). The candidate oncoprotein Bcl-3 is an antagonist of p50/NF-κB- mediated inhibition. Nature *359*(6393), 339–342.

Franzoso, G., Bours, V. et al. (1993). The oncoprotein Bcl-3 can facilitate NF-κ B-mediated transactivation by removing inhibiting p50 homodimers from select κB sites. EMBO J *12*(10), 3893–3901.

Franzoso, G., Carlson, L. et al. (1997). Critical roles for the Bcl-3 oncoprotein in T cell-mediated immunity, splenic microarchitecture, and germinal center reactions. Immunity *6*(4), 479–490.

Fujita, T., Nolan, G. P. et al. (1993). The candidate proto-oncogene bcl-3 encodes a transcriptional coactivator that activates through NF-κB p50 homodimers. Genes Dev *7*, 1354–1363.

Furukawa, M., Zhang, Y. et al. (2000). The CUL1 C-terminal sequence and ROC1 are required for efficient nuclear accumulation, NEDD8 modification, and ubiquitin ligase activity of CUL1. Mol Cell Biol *20*(21), 8185–8197.

Genin, P., Algarte, M. et al. (2000). Regulation of RANTES chemokine gene expression requires cooperativity between NF-κB and IFN-regulatory factor transcription factors. J Immunol *164*(10), 5352–5361.

Gern, J. E., Lemanske, R. F., Jr. et al. (1999). Early life origins of asthma. J Clin Invest *104*(7), 837–843.

Ghosh, S., May, M. J. et al. (1998). NF-κB and Rel proteins: evolutionarily conserved mediators of immune responses. Annu Rev Immunol *16*, 225–260.

Gonzalo, J. A., Lloyd, C. M. et al. (1998). The coordinated action of CC chemokines in the lung orchestrates allergic inflammation and airway hyper-responsiveness. J Exp Med *188*(1), 157–167.

Gosset, P., Tsicopoulos, A. et al. (1992). Tumor necrosis factor αand interleukin-6 production by human mononuclear phagocytes from allergic asthmatics after IgE-dependent stimulation. Am Rev Respir Dis *146*(3), 768–774.

Grunig, G., Warnock, M. et al. (1998). Requirement for IL-13 independently of IL-4 in experimental asthma. Science *282*(5397), 2261–2263.

Han, Y., and Brasier, A. R. (1997). Mechanism for biphasic rel A. NF-κB1 nuclear translocation in tumor necrosis factor α-stimulated hepatocytes. J Biol Chem *272*(15), 9825–9832.

Han, Y., Meng, T. et al. (1999). Interleukin-1-induced nuclear factor-κB-IκBα autoregulatory feedback loop in hepatocytes. A role for protein kinase Cα in post-transcriptional regulation of IκBα resynthesis. J Biol Chem *274*(2), 939–947.

Harhaj, E. W., and Sun, S. C. (1999). IKKγ serves as a docking subunit of the IκB kinase (IKK) and mediates interaction of IKK with the human T-cell leukemia virus Tax protein. J Biol Chem *274*(33), 22911–22914.

Hart, L., Lim, S. et al. (2000). Effects of inhaled corticosteroid therapy on expression and DNA-binding activity of nuclear factor κB in asthma. Am J Respir Crit Care Med *161*(1), 224–231.

Hart, L. A., Krishnan, V. L. et al. (1998). Activation and localization of transcription factor, nuclear factor-κB, in asthma. Am J Respir Crit Care Med *158*(5 Pt 1), 1585–1592.

Hatada, E. N., Nieters, A. et al. (1992). The ankyrin repeat domains of the NF-κB precursor p105 and the protooncogene bcl-3 act as specific inhibitors of NF-κB DNA binding. Proc Natl Acad Sci USA *89*(6), 2489–2493.

Heissmeyer, V., Krappmann, D. et al. (1999). NF-κB p105 is a target of IκB kinases and controls signal induction of Bcl-3-p50 complexes. EMBO J *18*(17), 4766–4778.

Heissmeyer, V., Krappmann, D. et al. (2001). Shared pathways of IκB kinase-induced SCF(βTrCP)-mediated ubiquitination and degradation for the NF-κB precursor p105 and IκBα. Mol Cell Biol *21*(4), 1024–1035.

Heyninck, K., De Valck, D. et al. (1999). The zinc finger protein A20 inhibits TNF-induced NF-κB-dependent gene expression by interfering with an RIP- or TRAF2-mediated transactivation signal and directly binds to a novel NF-κB-inhibiting protein ABIN. J Cell Biol *145*(7), 1471–1482.

Hidi, R., Riches, V. et al. (2000). Role of B7-CD28/CTLA-4 costimulation and NF-κB in allergen-induced T cell chemotaxis by IL-16 and RANTES. J Immunol *164*, 412–418.

Hoeflich, K. P., Luo, J. et al. (2000). Requirement for glycogen synthase kinase-3β in cell survival and NF-κB activation. Nature *406*(6791), 86–90.

Holloway, J. A., Warner, J. O. et al. (2000). Detection of house-dust-mite allergen in amniotic fluid and umbilical-cord blood. Lancet *356*(9245), 1900–1902.

Hsu, H., Huang, J. et al. (1996a). TNF-dependent recruitment of the protein kinase RIP to the TNF receptor-1 signaling complex. Immunity *4*(4), 387–396.

Hsu, H., Shu, H. B. et al. (1996b). TRADD-TRAF2 and TRADD-FADD interactions define two distinct TNF receptor 1 signal transduction pathways. Cell *84*(2), 299–308.

Hu, Y., Baud, V. et al. (1999). Abnormal morphogenesis but intact IKK activation in mice lacking the IKKα subunit of IκB kinase. Science *284*(5412), 316–320.

Hu, Y., Baud, V. et al. (2001). IKKα controls formation of the epidermis independently of NF-κB. Nature *410*(6829), 710–714.

Huxford, T., Huang, D. B. et al. (1998). The crystal structure of the IκBα/NF-κB complex reveals mechanisms of NF-κB inactivation. Cell *95*(6), 759–770.

Iciek, L. A., Delphin, S. A. et al. (1997). CD40 cross-linking induces Ig ε germline transcripts in B cells via activation of NF-κB: synergy with IL-4 induction. J Immunol *158*(10), 4769–4779.

Imbert, V., Rupec, R. A. et al. (1996). Tyrosine phosphorylation of IκB-α activates NF-κB without proteolytic degradation of IκB-α. Cell *86*(5), 787–798.

Inoue, J., Kerr, L. D. et al. (1992). IκBγ, a 70 kd protein identical to the C-terminal half of p110 NF-κB: a new member of the IκB family. Cell *68*(6), 1109–1120.

Ishikawa, H., Carrasco, D. et al. (1997). Gastric hyperplasia and increased proliferative responses of lymphocytes in mice lacking the COOH-terminal ankyrin domain of NF-κB2. J Exp Med *186*(7), 999–1014.

Jacobs, M. D., and Harrison, S. C. (1998). Structure of an IκBα/NF-κB complex. Cell *95*(6), 749–758.

Jain, A., Ma, C. A. et al. (2001). Specific missense mutations in NEMO result in hyper-IgM syndrome with hypohydrotic ectodermal dysplasia. Nat Immunol *2*(3), 223–228.

James, A. L., and Carroll, N. (1998). The pathology of fatal asthma. *Inflammatory mechanisms in asthma.* S. T. Holgate, Busse, W.W. New York, Marcel Dekker: 1–26.

Jany, B., Betz, R.et al. (1995). Activation of the transcription factor NF-κB in human tracheobronchial epithelial cells by inflammatory stimuli. Eur Respir J *8*(3), 387–391.

Jaspers, I., Flescher, E. et al. (1997). Ozone-induced IL-8 expression and transcription factor binding in respiratory epithelial cells. Am J Physiol *272*(3 Pt 1), L504–511.

Jaspers, I., Zhang, W. et al. (2001). Hydrogen peroxide has opposing effects on IKK activity and IκBα breakdown in airway epithelial cells. Am J Respir Cell Mol Biol *24*(6), 769–777.

Jedrzkiewicz, S., Nakamura, H. et al. (2000). IL-1β induces eotaxin gene transcription in A549 airway epithelial cells through NF-κB. Am J Physiol *279*(6), L1058–1065.

Jeppson, J. D., Patel, H. R. et al. (1998). Requirement for dual signals by anti-CD40 and IL-4 for the induction of nuclear factor-κB, IL-6, and IgE in human B lymphocytes. J Immunol *161*(4), 1738–1742.

Johnson, C., Van Antwerp, D. et al. (1999). An N-terminal nuclear export signal is required for the nucleocytoplasmic shuttling of IκBα. EMBO J *18*(23), 6682–6693.

Karin, M. (1999). The beginning of the end: IκB kinase (IKK) and NF-κB activation. J Biol Chem *274*(39), 27339–27342.

Karin, M., and Ben-Neriah, Y. (2000). Phosphorylation meets ubiquitination: the control of NF-κB activity. Annu Rev Immunol *18*, 621–663.

Karp, C. L., Grupe, A. et al. (2000). Identification of complement factor 5 as a susceptibility locus for experimental allergic asthma. Nat Immunol *1*(3), 221–226.

Kawakami, T., Chiba, T. et al. (2001). NEDD8 recruits E2-ubiquitin to SCF E3 ligase. EMBO J *20*(15), 4003–4012.

Klement, J. F., Rice, N. R. et al. (1996). IκBα deficiency results in a sustained NF-κB response and severe widespread dermatitis in mice. Mol Cell Biol *16*(5), 2341–2349.

Kontgen, F., Grumont, R. J. et al. (1995). Mice lacking the c-rel proto-oncogene exhibit defects in lymphocyte proliferation, humoral immunity, and interleukin-2 expression. Genes Dev *9*(16), 1965–1977.

Laitinen, L. A., Laitinen, A. et al. (1993). Airway mucosal inflammation even in patients with newly diagnosed asthma. Am Rev Respir Dis *147*(3), 697–704.

Lallena, M. J., Diaz-Meco, M. T. et al. (1999). Activation of IκB kinase β by protein kinase C isoforms. Mol Cell Biol *19*(3), 2180–2188.

Le Bail, O., Schmidt-Ullrich, R. et al. (1993). Promoter analysis of the gene encoding the IκB-α/MAD3 inhibitor of NF-κB: positive regulation by members of the rel/NF-κB family. EMBO J *12*(13), 5043–5049.

Lee, G. R., Fields, P. E. et al. (2001). Regulation of IL-4 gene expression by distal regulatory elements and GATA-3 at the chromatin level. Immunity *14*(4), 447–459.

Leonardi, A., Chariot, A. et al. (2000). CIKS, a connection to IκB kinase and stress-activated protein kinase. Proc Natl Acad Sci USA *97*(19), 10494–10499.

Lewis, J., Devin, A. et al. (2000). Disruption of hsp90 function results in degradation of the death domain kinase, receptor-interacting protein (RIP), and blockage of tumor necrosis factor-induced nuclear factor-κB activation. J Biol Chem *275*(14), 10519–10526.

Li, N., and Karin, M. (1998). Ionizing radiation and short wavelength UV activate NF-κB through two distinct mechanisms. Proc Natl Acad Sci USA *95*(22), 13012–13017.

Li, Q., Estepa, G. et al. (2000). Complete lack of NF-κB activity in IKK1 and IKK2 double-deficient mice: additional defect in neurulation. Genes Dev *14*(14), 1729–1733.

Li, Q., Van Antwerp, D. et al. (1999a). Severe liver degeneration in mice lacking the IκB kinase 2 gene. Science *284*(5412), 321–325.

Li, Z. W., Chu, W. et al. (1999b). The IKKβ subunit of IκB kinase (IKK) is essential for nuclear factor κB activation and prevention of apoptosis. J Exp Med *189*(11), 1839–1845.

Lilly, C. M., Nakamura, H. et al. (2001). Eotaxin expression after segmental allergen challenge in subjects with atopic asthma. Am J Respir Crit Care Med *163*(7), 1669–1675.

Lin, L., and Ghosh, S. (1996). A glycine-rich region in NF-κB p105 functions as a processing signal for the generation of the p50 subunit. Mol Cell Biol *16*(5), 2248–2254.

Liou, H. C., Jin, Z. et al. (1999). c-Rel is crucial for lymphocyte proliferation but dispensable for T cell effector function. Int Immunol *11*(3), 361–371.

Longphre, M., Li, D. et al. (1999). Allergen-induced IL-9 directly stimulates mucin transcription in respiratory epithelial cells. J Clin Invest *104*(10), 1375–1382.

Lyakh, L. A., Koski, G. K. et al. (2000). Bacterial lipopolysaccharide, TNF-α, and calcium ionophore under serum-free conditions promote rapid dendritic cell-like differentiation in CD14$^+$ monocytes through distinct pathways that activate NK-κB. J Immunol *165*(7), 3647–3655.

MacKichan, M. L., Logeat, F. et al. (1996). Phosphorylation of p105 PEST sequence via a redox-insensitive pathway up-regulates processing of p50 NF-κB. J Biol Chem *271*(11), 6084–6091.

Madrid, L. V., Mayo, M. W. et al. (2001). Akt stimulates the transactivation potential of the RelA/p65 subunit of NF-κB through utilization of the IκB kinase and activation of the mitogen-activated protein kinase p38. J Biol Chem *276*(22), 18934–18940.

Madrid, L. V., Wang, C. Y. et al. (2000). Akt suppresses apoptosis by stimulating the transactivation potential of the RelA/p65 subunit of NF-κB. Mol Cell Biol *20*(5), 1626–1638.

Makarov, S. S. (2000). NF-κB as a therapeutic target in chronic inflammation: recent advances. Mol Med Today *6*(11), 441–448.

Makris, C., Godfrey, V. L. et al. (2000). Female mice heterozygous for IKK gamma/NEMO deficiencies develop a dermatopathy similar to the human X-linked disorder incontinentia pigmenti. Mol Cell *5*(6), 969–979.

Maniatis, T. (1999). A ubiquitin ligase complex essential for the NF-κB, Wnt/Wingless, and Hedgehog signaling pathways. Genes Dev *13*(5), 505–510.

Mannino, D. M., Homa, D. M. et al. (1998). Surveillance for asthma—United States, 1960–1995. MMWR *47(SS-1)*, 1.

Martin, A. G., and Fresno, M. (2000). Tumor necrosis factor-α activation of NF-κB requires the phosphorylation of Ser-471 in the transactivation domain of c-Rel. J Biol Chem *275*(32), 24383–24391.

Matsukura, S., Stellato, C. et al. (1999). Activation of eotaxin gene transcription by NF-κB and STAT6 in human airway epithelial cells. J Immunol *163*(12), 6876–6883.

May, M. J., D'Acquisto, F. et al. (2000). Selective inhibition of NF-κB activation by a peptide that blocks the interaction of NEMO with the IκB kinase complex. Science *289*(5484), 1550–1554.

McFadden, E. R., Jr., and Gilbert, I. A. (1992). Asthma. N Engl J Med *327*(27), 1928–1937.

McLane, M. P., Haczku, A. et al. (1998). Interleukin-9 promotes allergen-induced eosinophilic inflammation and airway hyperresponsiveness in transgenic mice. Am J Respir Cell Mol Biol *19*(5), 713–720.

Mercurio, F., and Manning, A. M. (1999). Multiple signals converging on NF-κB. Curr Opin Cell Biol *11*(2), 226–232.

Mercurio, F., Zhu, H. et al. (1997). IKK-1 and IKK-2: cytokine-activated IκB kinases essential for NF-κB activation. Science *278*(5339), 860–866.

Messner, B., Stutz, A. M. et al. (1997). Cooperation of binding sites for STAT6 and NF κB/rel in the IL-4- induced up-regulation of the human IgE germline promoter. J Immunol *159*(7), 3330–3337.

Moriuchi, H., Moriuchi, M. et al. (1997). Nuclear factor-κB potently up-regulates the promoter activity of RANTES, a chemokine that blocks HIV infection. J Immunol *158*(7), 3483–3491.

Mukhopadhyay, A., Manna, S. K. et al. (2000). Pervanadate-induced nuclear factor-κB activation requires tyrosine phosphorylation and degradation of

IκBα. Comparison with tumor necrosis factor-α. J Biol Chem *275*(12), 8549–8555.

Murphy, S., Bleecker, E.R., Boushey, H., Buist, A .S., Busse, W., Clark, N. M., Eigen, H., Ford, J. G., Janson, S., Kelly, H. W., Lemanske, R .F., Lopez, C. C., Martinez, F., Nelson, H. S., Nowak, R., Platts-Mills, T. A. E., Shapiro, G. G., Stoloff, S., and Weiss, K. (1997). Practical Guide for the diagnosis and management of asthma. Based on the Expert Panel Report 2: Guidelines for the Diagnosis and Management of Asthma., National Heart, Lung, and Blood Institute, National Institutes of Health, Public Health Service, U.S. Department of Health and Human Services.

Newton, R., Hart, L. A. et al. (1998). Effect of dexamethasone on interleukin-1β-(IL-1β)-induced nuclear factor-κB (NF-κB) and κB-dependent transcription in epithelial cells. Eur J Biochem *254*(1), 81–89.

Nichols, B. G., Woods, J. S. et al. (2001). Effects of ozone exposure on nuclear factor-κB activation and tumor necrosis factor-α expression in human nasal epithelial cells. Toxicol Sci *60*(2), 356–362.

Nicolaides, N. C., Holroyd, K. J. et al. (1997). Interleukin 9: a candidate gene for asthma. Proc Natl Acad Sci USA *94*(24), 13175–13180.

Ninomiya-Tsuji, J., Kishimoto, K. et al. (1999). The kinase TAK1 can activate the NIK-I κB as well as the MAP kinase cascade in the IL-1 signalling pathway. Nature *398*(6724), 252–256.

Nolan, G. P., Fujita, T. et al. (1993). The bcl-3 proto-oncogene encodes a nuclear I κB-like molecule that preferentially interacts with NF-κB p50 and p52 in a phosphorylation-dependent manner. Mol Cell Biol *13*(6), 3557–3566.

Nomura, F., Kawai, T. et al. (2000). NF-κB activation through IKK-i-dependent I-TRAF/TANK phosphorylation. Genes Cells *5*(3), 191–202.

Ohmori, Y., and Hamilton, T. A. (2000). Interleukin-4/STAT6 represses STAT1 and NF-κB-dependent transcription through distinct mechanisms. J Biol Chem *275*(48), 38095–38103.

Ohno, H., Takimoto, G. et al. (1990). The candidate proto-oncogene bcl-3 is related to genes implicated in cell lineage determination and cell cycle control. Cell *60*(6), 991–997.

Orian, A., Gonen, H. et al. (2000). SCF(β)(-TrCP) ubiquitin ligase-mediated processing of NF-κB p105 requires phosphorylation of its C-terminus by IκB kinase. EMBO J *19*(11), 2580–2591.

Orian, A., Schwartz, A. L. et al. (1999). Structural motifs involved in ubiquitin-mediated processing of the NF-κB precursor p105: roles of the glycine-rich region and a downstream ubiquitination domain. Mol Cell Biol *19*(5), 3664–3673.

Ozes, O. N., Mayo, L. D. et al. (1999). NF-κB activation by tumour necrosis factor requires the Akt serine-threonine kinase. Nature *401*(6748), 82–85.

Pahl, H. L. (1999). Activators and target genes of Rel/NF-κB transcription factors. Oncogene *18*(49), 6853–6866.

Pahl, H. L., and Baeuerle, P. A. (1995). Expression of influenza virus hemagglutinin activates transcription factor NF-κB. J Virol *69*(3), 1480–1484.

Pahl, H. L., Sester, M. et al. (1996). Activation of transcription factor NF-κB by the adenovirus E3/19K protein requires its ER retention. J Cell Biol *132*(4), 511–522.

Papi, A., and Johnston, S. L. (1999a). Respiratory epithelial cell expression of vascular cell adhesion molecule-1 and its up-regulation by rhinovirus infection via NF-κB and GATA transcription factors. J Biol Chem *274*(42), 30041–30051.

Papi, A., and Johnston, S. L. (1999b). Rhinovirus infection induces expression of its own receptor intercellular adhesion molecule 1 (ICAM-1) via increased NF-κB-mediated transcription. J Biol Chem *274*(14), 9707–9720.

Perkins, N. D. (2000). The Rel/NF-κB family: friend and foe. Trends Biochem Sci *25*(9), 434–440.

Peters, R. T., Liao, S. M. et al. (2000). IKKε is part of a novel PMA-inducible IκB kinase complex. Mol Cell *5*(3), 513–522.

Peters, R. T., and Maniatis, T. (2001). A new family of IKK-related kinases may function as I κB kinase kinases. Biochim Biophys Acta *2*(62), M57–M62.

Phelps, C. B., Sengchanthalangsy, L. L. et al. (2000). Mechanism of I κBα binding to NF-κB dimers. J Biol Chem *275*(38), 29840–29846.

Rahman, I., and MacNee, W. (1998). Role of transcription factors in inflammatory lung diseases. Thorax *53*(7), 601–612.

Ray, A., and Cohn, L. (1999). Th2 cells and GATA-3 in asthma: new insights into the regulation of airway inflammation. J Clin Invest *104*(8), 985–993.

Ray, P., Yang, L. et al. (1997). Selective up-regulation of cytokine-induced RANTES gene expression in lung epithelial cells by overexpression of IκBR. J Biol Chem *272*(32), 20191–20197.

Ray, P., Zhang, D. H. et al. (1995). Cloning of a differentially expressed I κB-related protein. J Biol Chem *270*(18), 10680–10685.

Read, M. A., Brownell, J. E. et al. (2000). Nedd8 modification of cul-1 activates SCF(β(TrCP))-dependent ubiquitination of IκBα. Mol Cell Biol *20*(7), 2326–2333.

Regnier, C. H., Song, H. Y. et al. (1997). Identification and characterization of an IκB kinase. Cell *90*(2), 373–383.

Renard, P., Percherancier, Y. et al. (2000). Inducible NF-κB activation is permitted by simultaneous degradation of nuclear IκBα. J Biol Chem *275*(20), 15193–15199.

Richard, M., Louhahed, J. et al. (1999). Interleukin-9 regulates NF-κB activity through BCL3 gene induction. Blood *93*, 4318–4327.

Romashkova, J. A., and Makarov, S. S. (1999). NF-κB is a target of AKT in anti-apoptotic PDGF signalling. Nature *401*(6748), 86–90.

Rothwarf, D. M., Zandi, E. et al. (1998). IKK-γ is an essential regulatory subunit of the IκB kinase complex. Nature *395*(6699), 297–300.

Rudolph, D., Yeh, W. C. et al. (2000). Severe liver degeneration and lack of NF-κB activation in NEMO/IKKγ-deficient mice. Genes Dev *14*(7), 854–862.

Saccani, S., Pantano, S. et al. (2001). Two waves of nuclear factor κB recruitment to target promoters. J Exp Med *193*(12), 1351–1359.

Sachdev, S., and Hannink, M. (1998). Loss of IκBα-mediated control over nuclear import and DNA binding enables oncogenic activation of c-Rel. Mol Cell Biol *18*(9), 5445–5556.

Sakurai, H., Chiba, H. et al. (1999). IκB kinases phosphorylate NF-κB p65 subunit on serine 536 in the transactivation domain. J Biol Chem *274*, 30353–30356.

Salmeron, A., Janzen, J. et al. (2001). Direct phosphorylation of NF-κB1 by p105 by the IκB kinase complex on serine 927 is essential for signal-induced p105 proteolysis. J Biol Chem *276*, 22215–22222.

Sanz, L., Diaz-Meco, M. T. et al. (2000). The atypical PKC-interacting protein p62 channels NF-κB activation by the IL-1-TRAF6 pathway. EMBO J *19*(7), 1576–1586.

Sanz, L., Sanchez, P. et al. (1999). The interaction of p62 with RIP links the atypical PKCs to NF-κB activation. EMBO J *18*(11), 3044–3053.

Schmidt-Supprian, M., Bloch, W. et al. (2000). NEMO/IKKγ-deficient mice model incontinentia pigmenti. Mol Cell *5*(6), 981–992.

Schmitz, M. L., Bacher, S. et al. (2001). IκB-independent control of NF-κB activity by modulatory phosphorylations. Trends Biochem Sci *26*(3), 186–190.

Sen, R., and Baltimore, D. (1986). Multiple nuclear factors interact with the immunoglobulin enhancer sequences. Cell *46*(5), 705–716.

Senftleben, U., Cao, Y. et al. (2001). Activation by IKKα of a second, evolutionary conserved, NF-κB signaling pathway. Science *293*, 1495–1498.

Sha, W. C., Liou, H. C. et al. (1995). Targeted disruption of the p50 subunit of NF-κB leads to multifocal defects in immune responses. Cell *80*(2), 321–330.

Shimada, T., Kawai, T. et al. (1999). IKK-i, a novel lipopolysaccharide-inducible kinase that is related to IκB kinases. Int Immunol *11*(8), 1357–1362.

Shirane, M., Hatakeyama, S. et al. (1999). Common pathway for the ubiquitination of IκBα, IκBβ, and IκBε mediated by the F-box protein FWD1. J Biol Chem *274*(40), 28169–28174.

Sizemore, N., Leung, S. et al. (1999). Activation of phosphatidylinositol 3-kinase in response to interleukin-1 leads to phosphorylation and activation of the NF-κB p65/RelA subunit. Mol Cell Biol *19*(7), 4798–4805.

Smahi, A., Courtois, G. et al. (2000). Genomic rearrangement in NEMO impairs NF-κB activation and is a cause of incontinentia pigmenti. The International Incontinentia Pigmenti (IP) Consortium. Nature *405*(6785), 466–472.

Snapper, C. M., Zelazowski, P. et al. (1996). B cells from p50/NF-κB knockout mice have selective defects in proliferation, differentiation, germ-line CH transcription, and Ig class switching. J Immunol *156*(1), 183–191.

Spencer, E., Jiang, J. et al. (1999). Signal-induced ubiquitination of IκBα by the F-box protein Slimb/β-TrCP. Genes Dev *13*(3), 284–294.

Stacey, M. A., Sun, G. et al. (1997). The allergen Der p1 induces NF-κB activation through interference with IκBα function in asthmatic bronchial epithelial cells. Biochem Biophys Res Commun *236*(2), 522–526.

Stancovski, I., and Baltimore, D. (1997). NF-κB activation: the IκB kinase revealed? Cell *91*(3), 299–302.

Stassen, M., Muller, C. et al. (2001). IL-9 and IL-13 production by activated mast cells is strongly enhanced in the presence of lipopolysaccharide: NF-κB is decisively involved in the expression of IL-9. J Immunol *166*(7), 4391–4398.

Stellato, C., Beck, L. A. et al. (1995). Expression of the chemokine RANTES by a human bronchial epithelial cell line. Modulation by cytokines and glucocorticoids. J Immunol *155*(1), 410–418.

Stutz, A. M., and Woisetschlager, M. (1999). Functional synergism of STAT6 with either NF-κB or PU.1 to mediate IL-4-induced activation of IgE germline gene transcription. J Immunol *163*(8), 4383–4391.

Sun, X., Lee, J. et al. (1999). RIP3, a novel apoptosis-inducing kinase. J Biol Chem *274*(24), 16871–16875.

Suyang, H., Phillips, R. et al. (1996). Role of unphosphorylated, newly synthesized IκBβ in persistent activation of NF-κB. Mol Cell Biol *16*(10), 5444–5449.

Suzuki, H., Chiba, T. et al. (1999). IκBα ubiquitination is catalyzed by an SCF-like complex containing Skp1, cullin-1, and two F-box/WD40-repeat proteins, βTrCP1 and βTrCP2. Biochem Biophys Res Commun *256*(1), 127–132.

Suzuki, H., Chiba, T. et al. (2000). Homodimer of two F-box proteins βTrCP1 or βTrCP2 binds to IκBα for signal-dependent ubiquitination. J Biol Chem *275*(4), 2877–2884.

Tak, P. P., and Firestein, G. S. (2001). NF-κB: a key role in inflammatory diseases. J Clin Invest *107*(1), 7–11.

Takaesu, G., Kishida, S. et al. (2000). TAB2, a novel adaptor protein, mediates activation of TAK1 MAPKKK by linking TAK1 to TRAF6 in the IL-1 signal transduction pathway. Mol Cell *5*(4), 649–658.

Takizawa, H., Ohtoshi, T. et al. (1999). Diesel exhaust particles induce NF-κB activation in human bronchial epithelial cells in vitro: importance in cytokine transcription. J Immunol *162*(8), 4705–5711.

Tam, W. F., Lee, L. H. et al. (2000). Cytoplasmic sequestration of rel proteins by IκBα requires CRM1- dependent nuclear export. Mol Cell Biol *20*(6), 2269–2284.

Tanaka, K., Kawakami, T. et al. (2001). Control of IκBα proteolysis by the ubiquitin-proteasome pathway. Biochimie *83*(3-4), 35135–6.

Tanaka, M., Fuentes, M. E. et al. (1999). Embryonic lethality, liver degeneration, and impaired NF-κB activation in IKK-β-deficient mice. Immunity *10*(4), 421–429.

Temann, U. A., Geba, G. P. et al. (1998). Expression of interleukin 9 in the lungs of transgenic mice causes airway inflammation, mast cell hyperplasia, and bronchial hyperresponsiveness. J Exp Med *188*(7), 1307–1320.

Ten, R. M., McKinstry, M. J. et al. (1999). Signal transduction pathways triggered by the FcεRIIb receptor (CD23) in human monocytes lead to nuclear factor-κB activation. J Allergy Clin Immunol *104*(2 Pt 1), 3763–87.

Ten, R. M., McKinstry, M. J. et al. (1999). The signal transduction pathway of CD23 (FcεRIIb) targets IκB kinase. J Immunol *163*(7), 3851–3857.

Thomas, L. H., Friedland, J. S. et al. (1998). Respiratory syncytial virus-induced RANTES production from human bronchial epithelial cells is dependent on nuclear factor-κB nuclear binding and is inhibited by adenovirus-mediated expression of inhibitor of κBα. J Immunol *161*(2), 1007–1016.

Tojima, Y., Fujimoto, A. et al. (2000). NAK is an IκB kinase-activating kinase. Nature *404*(6779), 778–782.

Townsend, J. M., Fallon, G. P. et al. (2000). IL-9-deficient mice establish fundamental roles for IL-9 in pulmonary mastocytosis and goblet cell hyperplasia but not T cell development. Immunity *13*(4), 573–583.

Tsao, D. H., McDonagh, T. et al. (2000). Solution structure of N-TRADD and characterization of the interaction of N-TRADD and C-TRAF2, a key step in the TNFRI signaling pathway. Mol Cell *5*(6), 1051–1057.

Tumang, J. R., Owyang, A. et al. (1998). c-Rel is essential for B lymphocyte survival and cell cycle progression. Eur J Immunol *28*(12), 4299–4312.

Turpin, P., Hay, R. T. et al. (1999). Characterization of IκBα nuclear import pathway. J Biol Chem *274*(10), 6804–6812.

Ueda, A., Ishigatsubo Y., et al. (1997). Transcriptional regulation of the human monocyte chemoattractant protein-1 gene. Cooperation of two NF-κB sites and NF-κB/Rel subunit specificity. J Biol Chem *272*(49), 31092–31099.

Verhasselt, V., Vanden Berghe, W. et al. (1999). *N*-acetyl-L-cysteine inhibits primary human T cell responses at the dendritic cell level: association with NF-κB inhibition. J Immunol *162*(5), 2569–2574.

Vignola, A. M., Chanez, P. et al. (1998). Airway inflammation in mild intermittent and in persistent asthma. Am J Respir Crit Care Med *157*(2), 403–409.

Vogel, G. (1998). Interleukin-13's key role in asthma shown. Science *282*(5397), 2168.

Wallach, D., Varfolomeev, E. E. et al. (1999). Tumor necrosis factor receptor and Fas signaling mechanisms. Annu Rev Immunol *17*, 331–367.

Wang, C., Deng, L. et al. (2001). TAK1 is a ubiquitin-dependent kinase of MKK and IKK. Nature *412*(6844), 346–351.

Wang, C. Y., Cusack, J. C. et al. (1999). Control of inducible chemoresistance: Enhanced anti-tumor therapy through increased apoptosis by inhibition of NF-κB. Nat Med *4*, 412–417.

Wang, D., and Baldwin, A. S., Jr. (1998). Activation of nuclear factor-κB-dependent transcription by tumor necrosis factor-α is mediated through phosphorylation of RelA/p65 on serine 529. J Biol Chem *273*(45), 29411–29416.

Wang, D., Westerheide, S. D. et al. (2000). Tumor necrosis factor α-induced phosphorylation of RelA/p65 on Ser529 is controlled by casein kinase II. J Biol Chem *275*(42), 32592–32597.

Whiteside, S. T., Epinat, J. C. et al. (1997). IκBε, a novel member of the IκB family, controls RelA and cRel NF-κB activity. EMBO J *16*(6), 1413–1426.

Wills-Karp, M. (1999). Immunologic basis of antigen-induced airway hyperresponsiveness. Annu Rev Immunol *17*, 255–281.

Wills-Karp, M., Luyimbazi, J. et al. (1998). Interleukin-13: central mediator of allergic asthma. Science *282*(5397), 2258–2261.

Winston, J. T., Strack, P. et al. (1999). The SCFbeta-TRCP-ubiquitin ligase complex associates specifically with phosphorylated destruction motifs in IκBα and β-catenin and stimulates IκBα ubiquitination in vitro. Genes Dev *13*(3), 270–283.

Woronicz, J. D., Gao, X. et al. (1997). IκB kinase-β: NF-κB activation and complex formation with IκB kinase-α and NIK. Science *278*(5339), 866–869.

Wu, C., and Ghosh, S. (1999). β-TrCP mediates the signal-induced ubiquitination of IκBβ. J Biol Chem *274*(42), 29591–2954.

Wu, L., D'Amico, A. et al. (1998). RelB is essential for the development of myeloid-related CD8α– dendritic cells but not of lymphoid-related CD8α⁺ dendritic cells. Immunity *9*(6), 839–847.

Xiao, G., Harhaj, E. W. et al. (2001). NF-κB-inducing kinase regulates the processing of NF-κB2 p100. Mol Cell *7*(2), 401–409.

Yamamoto, Y., and Gaynor, R. B. (2001). Therapeutic potential of inhibition of the NF-κB pathway in the treatment of inflammation and cancer. J Clin Invest *107*, 135–142.

Yamamoto, Y., Kim, D. W. et al. (2001). IKKγ/NEMO facilitates the recruitment of the IκB proteins into the IκB kinase complex. J Biol Chem *24*, 24.

Yamaoka, S., Courtois, G. et al. (1998). Complementation cloning of NEMO, a component of the IκB kinase complex essential for NF-κB activation. Cell *93*(7), 1231–1240.

Yamazaki, S., Muta, T. et al. (2001). A novel IκB protein, IκB-ζ, induced by proinflammatory stimuli, negatively regulates nuclear factor-κB in the nuclei. J Biol Chem *276*(29), 27657–27662.

Yang, J., Lin, Y. et al. (2001). The essential role of MEKK3 in TNF-induced NF-κB activation. Nat Immunol *2*(7), 620–624.

Yang, L., Cohn, L. et al. (1998). Essential role of nuclear factor κB in the induction of eosinophilia in allergic airway inflammation. J Exp Med *188*(9), 1739–1750.

Yaron, A., Hatzubai, A. et al. (1998). Identification of the receptor component of the IκBα-ubiquitin ligase. Nature *396*(6711), 590–594.

Yin, L., Wu, L. et al. (2001). Defective lymphotoxin-β receptor-induced NF-κB transcriptional activity in NIK-deficient mice. Science *291*(5511), 2162–2165.

Zabel, U., and Baeuerle, P. A. (1990). Purified human IκB can rapidly dissociate the complex of the NF-κB transcription factor with its cognate DNA. Cell *61*(2), 255–265.

Zandi, E., Rothwarf, D. M. et al. (1997). The IκB kinase complex (IKK) contains two kinase subunits, IKKα and IKKβ, necessary for IκB phosphorylation and NF-κB activation. Cell *91*(2), 243–252.

Zhang, D. H., Cohn, L. et al. (1997). Transcription factor GATA-3 is differentially expressed in murine Th1 and Th2 cells and controls Th2-specific expression of the interleukin-5 gene. J Biol Chem *272*(34), 21597–21603.

Zhang, D. H., Yang, L. et al. (1999). Inhibition of allergic inflammation in a murine model of asthma by expression of a dominant-negative mutant of GATA-3. Immunity *11*(4), 473–482.

Zhang, G., and Ghosh, S. (2001). Toll-like receptor-mediated NF-κB activation: a phylogenetically conserved paradigm in innate immunity. J Clin Invest *107*(1), 13–19.

Zhang, S. Q., Kovalenko, A. et al. (2000). Recruitment of the IKK signalosome to the p55 TNF receptor: RIP and A20 bind to NEMO (IKKγ) upon receptor stimulation. Immunity *12*(3), 301–311.

Zhao, Q., and Lee, F. S. (1999). Mitogen-activated protein kinase/ERK kinase kinases 2 and 3 activate nuclear factor-κB through IκB kinase-α and IκB kinase-β. J Biol Chem *274*(13), 8355–8358.

Zheng, W., and Flavell, R. A. (1997). The transcription factor GATA-3 is necessary and sufficient for Th2 cytokine gene expression in CD4 T cells. Cell *89*(4), 587–596.

Zhong, H., Su Yang, H. et al. (1997). The transcriptional activity of NF-κB is regulated by the IκB-associated PKAc subunit through a cyclic AMP-independent mechanism. Cell *89*(3), 413–424.

Zhong, H., Voll, R. E. et al. (1998). Phosphorylation of NF-κB p65 by PKA stimulates transcriptional activity by promoting a novel bivalent interaction with the coactivator CBP/p300. Mol Cell *1*(5), 661–671.

Zhu, Z., Tang, W. et al. (1996). Rhinovirus stimulation of interleukin-6 in vivo and in vitro. Evidence for nuclear factor κB-dependent transcriptional activation. J Clin Invest *97*(2), 421–430.

CHAPTER 3

MOLECULAR MECHANISMS OF CANCER

AKRIT SODHI, SILVIA MONTANER, and J. SILVIO GUTKIND
Cell Growth Regulation Section, Oral and Pharyngeal Cancer Branch,
National Institute of Dental and Craniofacial Research, National
Institutes of Health, Bethesda, Maryland

INTRODUCTION

The 30 trillion cells that constitute the adult human body grow and differentiate to take on their many specialized functions in a tightly regulated fashion. They proliferate only when required, as a result of a delicate balance between growth-promoting and growth-inhibiting mechanisms that are controlled by an intricate network of intra- and extracellular molecules, including those mediating the communication with both neighboring and distal cells and tissues. In stark contrast, cancer cells override these controlling mechanisms and follow their own internal program for timing their reproduction. Indeed, cancer cells can grow in an unrestricted manner, and over time they can acquire the ability to migrate from their original site, invade nearby tissues, and form tumors (metastases) at distant organs. The primary tumor and its metastases become lethal when they invade and disrupt tissues whose function is vital for survival.

Historically, patients have suffered the devastating consequences of tumor growth and spread with limited therapeutic options. However, discoveries over the past two decades have dramatically increased our understanding of the most basic mechanisms controlling normal and malignant cell proliferation, thus providing golden opportunities for novel treatment modalities for neoplastic diseases. We now know that changes in cellular behavior occur progressively, from slightly deregulated proliferation to full malignancy, as a result of the accumulation of mutations in DNA caused by exposure to endogenous and exogenous genotoxic agents or altered expression of a limited set of genes. The process of genetic instability has been hypothesized to be a critical

Signal Transduction and Human Disease, Edited by Toren Finkel and
J. Silvio Gutkind
ISBN 0-471-02011-7   Copyright © 2003 John Wiley & Sons, Inc.

element in the accumulation of these mutations (Lengauer et al., 1998). Environmental and endogenous DNA-damaging agents and genetic instability drive tumor progression by generating mutations in two types of genes, oncogenes and tumor suppressor genes, providing cancer cells with a selective growth advantage and thereby leading to the clonal outgrowth of a tumor. In general, oncogenes (called protooncogenes in their normal, nonmutated form) promote cell proliferation and survival, whereas tumor suppressor genes inhibit cell growth. Normally, these genes control cell growth by integrating the information generated by extracellular stimuli and intracellular mediators. Recent advances in our understanding of the normal and aberrant function of these genes have helped us to understand—and may ultimately allow us to halt—the stepwise process leading to human cancer.

A classic example of an oncogene is the *ras* gene that is mutated in more than 40% of human cancers (Barbacid, 1987; Bos, 1989; Bos, 1997). As discussed in more detail below, the normal function of *ras* is to transduce signals specified by a wide variety of growth factors. When mutated, *ras* can persistently activate growth-stimulating pathways even in the absence of mitogenic stimuli, resulting in uncontrolled cell proliferation. In contrast, alterations in tumor suppressor genes frequently cause a functional loss of their protein product, thus depriving cells of key molecular brakes that normally forestall inappropriate cell growth. A typical example is the *p53* gene, which encodes for a tumor suppressor that is inactivated in the majority of human cancers (Vogelstein and Kinzler, 1992; Vousden and Lu, 2002). Of interest, the products of many oncogenes and tumor suppressor genes are often targeted by tumor viruses, such as human papillomaviruses (Scheffner et al., 1990; Vousden, 1995), and by epigenetic events leading to their increased (oncogenes) or reduced (tumor suppressor genes) expression, thus functionally activating or inactivating these proteins, respectively, in the absence of mutations in their encoding gene.

An emerging concept in molecular carcinogenesis is that activating and inactivating events must occur sequentially in several oncogenes and tumor suppressor genes for the initiation and progression of human cancer. Because a comprehensive review of all possible molecular mechanisms involved in cancerous growth is beyond the scope of this chapter, we will focus instead on key biochemical routes participating in inter- and intracellular communication whose deregulation contributes to the acquisition of the malignant phenotype in human cancer.

TYROSINE PHOSPHORYLATION

Retroviral Oncogenes and Growth Promotion

The discovery of retroviral oncogenes provided the foundation of our current understanding of the molecular mechanism controlling normal and aberrant cell growth. However, it took more than six decades after the discovery of the first transforming retrovirus, the Rous sarcoma virus (Rous, 1911), at the beginning of the last century for its oncogene, v-*src*, to be isolated (Duesberg and Vogt, 1970; Duesberg and Vogt, 1973; Schwartz et al., 1983) and found to represent an altered version of a cellular gene, c-*src* (Spector et al., 1978; Stehelin et al., 1976). This conceptual breakthrough led to the prediction that human cancers may result from the activation of otherwise normal cellular genes or protooncogenes. It only took a few more years to realize that the protein product of v-*src*, p60$^{\text{v-Src}}$, exhibits a protein kinase activity (Brugge and Erikson, 1977; Collett and Erikson, 1978; Levinson et al., 1978) and to find that this kinase exhibits an unusual specificity for tyrosine residues (Collett et al., 1980; Hunter and Sefton, 1980). Together, these findings provided compelling evidence of a key role for tyrosine phosphorylation in cell growth control.

Receptor Tyrosine Kinases and Cancer

These observations led to the discovery that polypeptide growth factors, such as platelet-derived growth factor (PDGF) and epidermal growth factor (EGF), can similarly promote the tyrosine phosphorylation of cellular proteins (Cooper et al., 1982; Ek et al., 1982; Hunter and Cooper, 1981; Ushiro and Cohen, 1980). The molecular cloning of the EGF receptor (EGFR) (Ullrich et al., 1984), and many other growth factor receptors soon after, provided the first evidence that mitogenic signaling by most polypeptide growth factors is mediated by receptors that act as ligand-dependent protein tyrosine kinases. These receptor tyrosine kinases (RTKs) can be structurally visualized as membrane-associated allosteric enzymes with an extracellular ligand-binding domain and an intracellular protein tyrosine kinase domain separated by the plasma membrane. Polypeptide growth factors interact with RTKs through their extracellular domain, thereby enhancing their integral cytoplasmic tyrosine kinase activity and promoting the transfer of the γ-phosphate of ATP to tyrosine residues of specific protein substrates. Tyrosine phosphorylation can thus be thought of as the language that these receptors use to communicate information from the extracellular environment (i.e., through bound growth factors) to the interior of the cell.

The Epidermal Growth Factor Receptor Family

As described below, many human malignancies express high levels of growth factors and their receptors and many malignant cells exhibit highly active RTKs because of their activation by an autocrine or paracrine mechanism or by activating mutations in their coding sequence. To date, more than 50 RTKs, which belong to 20 different receptor subfamilies, have been identified (Hunter, 2000). Among the best-studied group of these receptors is the EGFR family (also known as type I RTKs or ErbB tyrosine kinase receptors), and thus we will use this receptor family to illustrate the emerging concepts of how RTKs

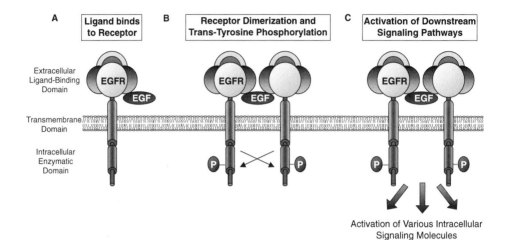

Figure 3.1. *Activation of receptor tyrosine kinases.* **A.** EGF binds to the extracellular ligand binding domain of its cognate receptor. **B.** The receptor becomes activated by ligand-induced dimerization and cross-phosphorylation in *trans*. For EGFRs, this dimerization can occur between two identical receptors (homodimerization) or between different receptors of the same family (heterodimerization). **C.** After receptor autophosphorylation, recruitment and phosphorylation of several intracellular substrates ultimately lead to the activation of a series of signal transduction pathways.

mediate growth promotion. In addition to being essential for several normal cellular processes, the aberrant activity of members of this receptor family has also been linked to the development and growth of numerous tumor types. Moreover, analysis of signal transduction through the EGFR family and the multiple processes that regulate their function, such as receptor heterodimerization and endocytosis, has been critical to understanding the underlying mechanisms by which deregulation of RTKs can lead to cancer.

The EGFR family is comprised of four highly related receptors: the epidermal growth factor receptor (EGFR/HER1/ErbB1), HER2 (ErbB2/neu), HER3 (ErbB3), and HER4 (ErbB4) (Hackel et al., 1999). As for other RTKs, these receptors exhibit an extracellular ligand-binding domain, a transmembrane lipophilic segment, and an intracellular protein tyrosine kinase domain with a carboxyl-terminal regulatory region (Fig. 3.1). In the case of HER3 (ErbB3), however, its tyrosine kinase domain lacks enzymatic activity.

Epidermal Growth Factor Receptor Signaling

The discovery that EGFRs become activated by ligand-induced dimerization and cross-phosphorylation in *trans* provided a general model by which RTKs can transmit signals across the plasma membrane upon ligand binding or spontaneously when largely overexpressed, as has been described in several human tumors (see below). For EGFRs, this dimerization can occur between two identical receptors (homodimerization) or between different receptors of the same family (heterodimerization). There is a multiplicity of EGFR ligands that drive the formation of homo- or heterodimeric complexes among the four EGFRs.

No ligand has been yet been identified for HER2 (ErbB2). However, HER2 is known to be the preferred coreceptor for EGFR (HER1), HER3, and HER4 (Olayioye et al., 2000). This preference for heterodimerization within the HER2 receptor family explains how HER2 can signal even in the absence of a cognate ligand. The heterodimers between HER2 and the other EGFRs have relatively high ligand affinity and potent signaling activity and are synergistic for cell transformation. These features may be related to the ability of HER2 to decelerate the rate of ligand dissociation, which could cause a prolonged signaling by all EGF family ligands. Thus HER2 has been proposed as a master coordinator of the signaling network that functions as a shared coreceptor for EGF-like ligands, rather than a receptor that mediates the action of a specific (yet to be identified) ligand (Olayioye et al., 2000). Within the EGFR family, the combinatorial possibilities are numerous and each homo- and heterodimer receptor complex could possibly activate divergent signaling pathways, thus eliciting specific cellular responses and resulting in an enormous signaling and biological diversity.

After receptor dimerization, activation of the intrinsic protein tyrosine kinase activity and tyrosine autophosphorylation occur. There is a strict requirement for RTK activity in receptor-mediated cellular signaling, as receptors lacking kinase function as a result of mutations of the ATP binding site do not display a full range of biochemical responses after ligand binding (Honegger et al., 1987a; Honegger et al., 1987b). After receptor autophosphorylation, recruitment and phosphorylation of several intracellular substrates ultimately lead to the activation of a series of signal transduction pathways (Fig. 3.2). Once a specific signaling molecule binds to the activated receptor, several modes of activation have been identified, as described below.

Activation by Tyrosine Phosphorylation. Intracellular molecules can bind to the cytoplasmic tail of the receptor by phosphotyrosine-mediated interaction and become substrates of the receptor kinase. It has been demonstrated that tyrosine phosphorylation of phospholipase C (PLC)-γ increases its catalytic activity and is essential for its activation by RTKs (Margolis et al., 1989; Wahl et al., 1989). Activation of PLC-γ results in hydrolysis of membrane-associated phosphatidylinositol 4,5-phosphate (PIP_2) to generate inositol triphosphate and diacylglycerol, two potent second messengers that activate a cascade of intracellular signaling events including elevation of the intracellular levels of Ca^{2+} and the activation of protein kinase C (PKC), respectively.

Signal transducers and activators of transcription molecules (STATs) are latent transcription factors that also can be activated as a result of tyrosine phosphorylation by RTKs. These transcription factors were first

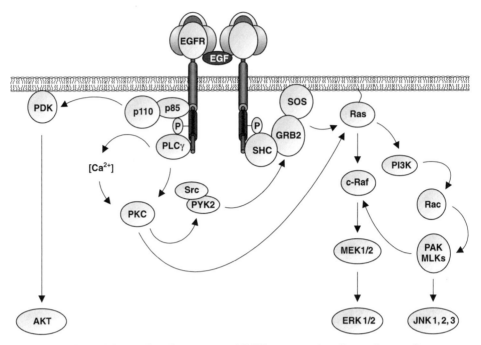

Figure 3.2. *Modes of activation of EGF receptor signaling pathways.* Once a specific signaling molecule has bound the receptor, several modes of activation have been identified: activation by tyrosine phosphorylation (e.g., PLC γ); activation by conformational change (e.g., PI3K); activation by membrane recruitment (e.g., Grb2/SOS/Ras); and recruitment of multiadapter proteins.

identified as components of interferon signaling, but a number of RTKs activate the STAT pathway as well (Darnell, 1997). For example, EGF, PDGF, and colony stimulating factor-1 all stimulate tyrosine phosphorylation and activation of STATs 1 and 3. EGF can also activate STAT5. The intrinsic tyrosine kinase functions of these RTKs are required for STAT phosphorylation and activation, independent of the activation of the Janus kinases (JAKs) that mediate the activation of STATs by cytokine receptors (see Chapter 9). The Src-homology 2 (SH2) domain of STATs first bind to tyrosine phosphorylated receptors, followed by the tyrosine phosphorylation of the receptor-associated STATs. This leads to the formation of homo- or heterodimers of STATS through intermolecular SH2 domain-phosphotyrosine-mediated interactions. They then translocate into the nucleus, where they bind to specific DNA elements, thereby stimulating the expression of a variety of downstream genes.

Activation by Conformational Change. Phosphatidylinositol (PI) 3-kinases constitute a family of enzymes composed of a heterodimeric complex of regulatory (p85) and catalytic (p110) subunits. RTKs can promote the activation of these enzymes, which phosphorylate the D3 position of PI (see Chapter 5 for additional details). Activation of enzymatic activity is initiated by binding of p85 to the receptor (Otsu et al., 1991; Skolnik et al., 1991). In particular, the activity of the p85/p110 heterodimer is increased when the p85 SH2 domains bind to the tyrosine phosphorylated receptor within a particular sequence motif. Because p85 remains bound to p110, it is believed that p85 binding to the receptor induces a conformational change that is then transmitted to p110. p85 appears to play a role in stabilizing and inhibiting the function of the p110 catalytic subunit. Thus it is possible that the conformational change induced by p85 binding to receptor phosphopeptide may eliminate its inhibition of the catalytic subunit and, therefore, increase PI 3-kinase activity.

Activation by Membrane Recruitment. This modality of signal transduction by RTKs is best exemplified by their mechanism of activation of Ras, a key downstream effector of the signal transduction pathway utilized by growth factors to initiate cell growth and differentiation. For EGFR, an adapter molecule known as Grb2 provides a link between RTK activation and the accumulation of GTP-bound Ras. Grb2 belongs to a group of proteins, composed virtually entirely of SH2 and SH3 domains (like crk, nck, and p85), that are thought to function as adapter molecules or regulatory components of specific catalytic subunits. These proteins differ from other SH2-containing molecules because they themselves lack a distinct enzymatic activity (Schlessinger, 1994).

Grb2 is a 23-kDa protein consisting of one SH2 domain intercalated between two SH3 domains. Like many signaling proteins, Grb2 uses its SH2 domain to bind to specific phosphorylated tyrosines in the cytoplasmic domain of ligand-activated RTKs, but Grb2 is not a direct substrate for RTKs, as it does not become phosphorylated. Instead, Grb2 uses its SH3 domains to bind a proline-rich region in the carboxyl-terminal tail of Sos (*Son of Sevenless*). Sos is a Ras guanine nucleotide exchange factor (GEF), which promotes the release of GDP from Ras and its exchange for GTP, thereby activating Ras (see below). Thus the accumulation of active Ras after growth factor stimulation results from the translocation of Sos to the plasma membrane upon binding and recruitment of Grb2 to the cytoplasmic tail of RTKs (Hunter, 2000). Activation of Ras in turn, results in the stimulation of a number of Ras pathways, including those regulating the activation of the MAP kinase (MAPK) cascade (see below).

Shc proteins are also adapter molecules that contain SH2 domains. Unlike Grb2, Shc binds activated receptors through their SH2 and/or PTB domains and becomes tyrosine phosphorylated, thus creating docking sites for additional molecules harboring phosphotyrosine recognition motifs, including Grb2. Thus Shc may provide an alternative docking site for recruiting the Grb2/Sos complex to the membrane. Indeed, the EGFR can use both adapters to activate Ras signaling (Hackel et al., 1999).

Recruitment of Multiadapter Proteins. RTK activation and autophosphorylation can lead to the recruitment of multiadapter molecules, which
in turn become phosphorylated and mediate recruitment of other signaling molecules. As discussed in further detail in Chapter 5, this modality in signal transduction expands the repertoire of pathways that a single
RTK can activate. Grb2-associated binder-1 (Gab1), the insulin-receptor
substrates 1 and 2 (IRS-1, -2), and the *Drosophila* protein DOS, are
members of a growing family of docking molecules. This family is characterized by a pleckstrin-homology (PH) domain at the amino terminus
of the protein along with additional binding sites for SH2 and SH3
domains. Gab1 was identified as a Grb2 binding protein as well as a
protein involved in EGF and insulin receptor signaling (Holgado-
Madruga et al., 1996). Tyrosine phosphorylation of Gab1 mediates its
interaction with several proteins that contain SH2 domains including
PLC-γ, PI 3-kinase, and Syp2, which, in turn, appear to mediate numerous biological effects.

Ligand-EGFR Complex Endocytosis

During signal termination, activated EGFR complexes are endocytosed
in clathrin-coated pits. Two distinct processes have been identified that
determine the fate of the internalized receptors. The first of these
processes involves an ubiquitin ligase known as Cbl (Clague and Urbe,
2001). Recruitment of Cbl to ligand-receptor complexes in early
endosomes targets receptors for lysosomal degradation by promoting
receptor ubiquitination. In the absence of Cbl or when v-Cbl, the viral,
oncogenic form of Cbl, is present, receptors are instead recycled to the
plasma membrane. A single tyrosine residue (tyrosine 1045) on the
EGFR is essential for Cbl-mediated downregulation of EGFR signaling.
Consistent with this model, mutation of Y1045 leads to a potentiation of
EGFR signaling. A second determinant of receptor fate is the stability
of the activated ligand-receptor complex in the mildly acidic endosomal
environment. Activated EGFR homodimers are relatively stable and
remain bound to Cbl. These relatively stable interactions result in endocytic sorting to lysosomes and receptor degradation. In contrast, EGFR-
HER2 heterodimers are less stable and uncouple in early endosomes,
causing Cbl to dissociate from the receptor complex. The receptors then
travel through the default recycling pathway, which returns them to the
cell surface.

Epidermal Growth Factor Receptor Family and Cancer

Since the first connection between a viral oncogene, a constitutively
active truncated mutant of Erb2, and human cancer was made in 1984,
it has been well known that aberrant signaling by growth factor receptors is critically involved in human neoplasias (Downward et al., 1984).
Constitutive activation of RTKs can occur by several mechanisms. In
most cases, gene amplification, overexpression, or mutations are respon-

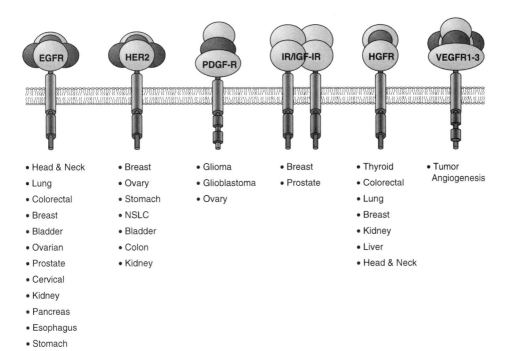

Figure 3.3. *Receptor tyrosine kinases frequently mutated or overexpressed in human cancer.* **Top**, Schematic representation of commonly mutated RTKs. **Bottom**, cancers in which RTK is commonly found mutated or overexpressed.

sible for the acquired transforming potential of oncogenic RTKs. Somatic and germline mutations, which are associated with distinct inherited and spontaneous human cancer syndromes, have been observed in at least 10 different RTK families. More recently, activation of autocrine growth factor loops has been recognized as a novel mechanism whereby RTKs acquire a constitutively-active phenotype. In many solid tumors it has been shown that elevated levels of both growth factor receptor and its ligand are expressed concomitantly. Indeed, investigation of members of the RTK family has revealed a clear association between aberrant RTK signaling and human cancer (Fig. 3.3).

EGFR (HER1, ErbB1). The EGFR is frequently overexpressed in bladder, cervical, ovarian, kidney, pancreatic, and non-small cell lung cancer and occurs with very high incidence in squamous cell carcinomas of the head and neck (Table 3.1). Tumorigenic changes in EGFR activity often occur via mutations that activate their kinase activity in the absence of ligand binding. A number of EGFR deletions have been also identified in human cancer, and most of these mutations alter the extracellular ligand-binding domain of the receptor and result in a truncated EGFR with constitutively active kinase function. The predominant mechanism leading to EGFR overexpression is EGFR gene amplification, with more than 15 copies per cell reported in certain tumors. In

TABLE 3.1. U.S. Incidence of Major Cancers with EGFR Overexpression

Primary Tumor Site	New Cases/Year	Frequency overexpressing EGFR (%)
Cervical/uterus	48,900	90
Head and neck	30,200	80–100
Prostate	180,400	65
Renal cell	31,200	50–90
Esophageal	12,300	43–89
Lung	164,000	40–80
Ovarian	23,100	35–70
Bladder	53,200	31–48
Pancreatic	28,300	30–50
Colon/rectum	130,200	25–77
Breast	184,200	14–91

Modified from Herbst, R.S. and Langer, C.J. (2002). Epidermal growth factor receptors as a target for cancer treatment. *Semin Oncol 29*, 27–36.

general, elevated levels of EGFR expression lead to the activation of their kinase activity by spontaneous dimerization, and are associated with late stages of disease progression and often correlate with an increased rate of tumor proliferation and metastasis.

A very potent mechanism of constitutive EGFR activation in a variety of human cancers is their autocrine stimulation via growth factor loops. The most prominent ligand involved in autocrine growth receptor activation is transforming growth factor (TGF)-α. Coexpression of TGF-α and EGFR is frequently observed in glioblastomas and squamous cell carcinomas of the head and neck, where it correlates with a poor prognosis. Interestingly, it was recently shown that G protein-coupled receptor (GPCR)-induced cleavage of EGF-like growth factors leads to EGFR transactivation and EGFR-related signaling in cancer cells, suggesting that GPCR-EGFR cross-communication may play a prominent role in the development and progression of human cancer (Gschwind et al., 2001).

HER2 (ErbB2, neu). The RTK HER2 was originally identified as both the transforming gene in a chemically transformed rat neuroblastoma cell line and as an EGFR-related cDNA clone (Schechter et al., 1985), and it has been the most frequently implicated RTK in human neoplasias. Despite the fact that no ligand is known to bind with high affinity to HER2, it is well known that this receptor acts as a coreceptor for the other EGFR family members (see above). These heteromolecular interactions are of pathophysiological relevance, because such receptor combinations show strong mitogenic signaling and tumorigenicity. Indeed, the increased availability of HER2 for heterodimer formation may lead to high-level autophosphorylation and constitutive signaling, resulting in prolonged and enhanced mitogenic signaling by the heterodimer receptors. Signaling pathways involving Ras, Src, PI 3-kinase, and the MAPKs extracellular signal-regulated protein kinase (ERK) and JNK (see below) have been shown to be activated by HER2-expressing cell lines and seem to be important for tumor management.

HER2 gene mutation and high-level overexpression are other potent mechanisms resulting in HER2 activation. A single point mutation within the transmembrane region of HER2 (at position 664), encoding a change from the hydrophobic amino acid valine to the negatively charged glutamic acid residue, renders the receptor constitutively active (Weiner et al., 1989). Although the mutation is not observed in human tumors, a polymorphism at codon 655, which results in a valine to isoleucine substitution, has been identified in the transmembrane coding region of the HER2 gene. Indeed, an increased risk of breast cancer, particularly among younger women, is associated with this mutation (Eccles, 2001).

The relatively low expression level of HER2 in normal epithelial cells is significantly enhanced in several types of human cancers, including breast, ovarian, gastric, lung, bladder, and kidney carcinomas. Overexpression and/or gene amplification of HER2 is especially prevalent in human breast carcinomas, where HER2 gene amplification has been identified with a frequency of 30%. Indeed, aberrantly elevated levels of HER2, either with or without gene amplification, correlate with a more aggressive progression of disease and a reduced patient survival time (Eccles, 2001). Clinical as well as laboratory data revealed that overexpression of HER2 increases the metastatic potential of human breast and lung cancer cells and correlates with the number of lymph node metastases in node-positive breast cancer patients. Moreover, in recent years, it became evident that HER2 is also a predictive marker for responses to various therapeutic agents used in cancer therapy. Of most important clinical relevance is the fact that HER2 overexpression correlates with a lack of response to antiestrogen hormonal therapy and confers resistance to tamoxifen, an antiestrogen that is administered as endocrine therapy in breast cancer patients. HER2 expression therefore represents a pivotal biological marker that can help to determine more accurately the prognosis for individual patients.

Other RTKs Implicated in Human Cancer

Platelet-Derived Growth Factor Receptor Family.
The PDGF receptor (PDGFR) and Kit are members of the PDGFR family of RTKs. These proteins are characterized by an extracellular domain with 5 Ig-like domains and an intracellular tyrosine kinase domain that is split by an insertion of around 100 amino acids. Two genes encoding PDGFR-α and PDGFR-β have been identified, and both receptors are activated by ligand dimers consisting of PDGF-A and/or PDGF-B. This in turn leads to receptor dimerization with three possible configurations: $\alpha\alpha$, $\beta\beta$, or $\alpha\beta$. Coexpression of PDGFR and its ligands has been identified in glioblastomas and other human astrocytotic brain tumors, whereas

normal brain tissue does not express these proteins (Fleming et al., 1992; Hermanson et al., 1992). These findings suggest an autocrine loop that stimulates the uncontrolled growth of human brain tumors. However, the PDGF receptor-ligand system may also exert its effects on tumor progression via the stimulation of neovascularization of solid tumors.

The RTK Kit is predominantly expressed in mast cells, melanocytes, and bone marrow, whereas its cognate ligand, stem cell factor (SCF), is found in stromal cells, fibroblasts, and endothelial cells. Kit has been shown to have seemingly opposing effects in human malignancy. On one hand, functional expression of Kit has a positive effect on the growth of small cell lung cancer cells and mutations in Kit have been identified in patients with familial and sporadic gastrointestinal tumors (Maeyama et al., 2001; Yamaguchi et al., 2000). On the other hand, loss of Kit is associated with progression of melanoma and thyroid carcinoma, suggesting a tumor suppressor function in this particular tumor type (Natali et al., 1995; Natali et al., 1992). An explanation of Kit function that reconciles its paradoxical effects in different tumors remains to be determined.

Insulin Growth Factor Receptor Family. The insulin receptor (IR) and the insulin-like growth factor (IGF) receptor (IGF-IR) are structurally quite different from other members of the EGFR family. IR and IGF-IR consist of two extracellular α subunits, which are responsible for ligand binding, and two membrane-spanning β subunits bearing the tyrosine kinase domain and the autophosphorylation sites. The ligands for the two receptors include insulin, IGF-I, and IGF-II. Whereas insulin is mostly a metabolic hormone, IGF-I and IGF-II are crucial for normal development and carcinogenesis (Furstenberger and Senn, 2002). In the circulation IGFs are found to be complexed to a number of different IGF-binding proteins (IGFBPs), which serve as transport vehicles for these ligands and modify the stability and the proliferative effect of the growth factors.

IGF-IR and its ligands are involved in the pathogenesis of a variety of human tumors, in particular breast and prostate cancer. In primary breast tumors, IGF-I and IGF-II are predominantly expressed in stromal fibroblasts surrounding the normal and malignant breast epithelium, whereas the IGF-IR is overexpressed by breast cancer cells and shows enhanced tyrosine kinase activity (Helle and Lonning, 1996). IGF-I has also been implicated in prostate cancer, where high plasma IGF-I levels correlate with an elevated risk for this malignancy (Yu and Rohan, 2000).

Hepatocyte Growth Factor Receptor. The protooncogene *met* encodes the hepatocyte growth factor receptor (HGFR), which was identified as a regulator of a variety of processes including cell migration, cell scattering, and invasion of extracellular matrices (Danilkovitch-Miagkova and Zbar, 2002; Furge et al., 2000). HGFR is a disulfide-linked heterodimer with a glycosylated extracellular α-chain and a β-chain, which consists of the transmembrane and cytoplasmic tyrosine kinase domains. Hepatocyte growth factor (HGF), or scatter factor (SF), the corresponding ligand, is expressed in mesenchyme-derived cells, where it is suggested to act in a paracrine manner on epithelial cells in close proximity and has been described as a growth modulator for hepatocytes, melanocytes, and keratinocytes in vitro.

HGF overexpression was demonstrated in a variety of human tumors, such as thyroid and colorectal carcinomas, and seems to have prognostic significance for non-small cell lung and breast cancer. HGFR mutations have been identified in patients with inherited predisposition to develop multiple papillary renal cell carcinomas (HPRCCs). Most of these mutations lie adjacent to the kinase domain, leading to enhanced enzymatic activity, transformation of fibroblasts, and invasive growth in vitro. Somatic mutations in the HGFR have been demonstrated in childhood hepatocellular as well as head and neck squamous cell carcinomas (Di Renzo et al., 2000). Because HGF/SF is an important motility factor, its contribution to human malignancies may involve tumor cell migration, invasive growth, and metastasis.

RET Receptor Tyrosine Kinase. The rearranged during transformation (RET) RTK is the gene responsible for multiple endocrine neoplasia type 2 (MEN2) (Jhiang, 2000). MEN2 is a dominant autosomal inherited cancer syndrome that exists in three different subtypes and is characterized by the development of medullary thyroid carcinoma. The RET protooncogene encodes an RTK that is characterized by a cadherin-like and a cysteine-rich domain in the extracellular part of the receptor. RET is normally expressed during embryogenesis in the peripheral nervous system and in the urogenital system and is involved in the development of the neural crest and the kidney (Schuchardt et al., 1994). MEN2 cancer syndromes are caused by dominant activating germline mutations in the RET protooncogene, often leading to constitutively activated receptors.

Vascular Endothelial Growth Factor Receptor Family. Angiogenesis, the formation of new blood vessels developing from preexisting ones, is essential in many physiological processes, including embryonic development, wound healing, and tissue regeneration, and also represents a critical pathogenic mechanism in a number of human diseases, including cancer. Vascular endothelial growth factor (VEGF), an endothelial cell-specific mitogen, is among the most potent angiogenic stimulators and plays a central role in the regulation of both physiological and pathologic neovascularization. The two RTKs that bind VEGFs, VEGF receptor (VEGFR)-1 and VEGFR-2, are expressed in endothelial cells during embryonic development and are the key regulators for angiogenesis. Blood vessel formation is usually quiescent in the adult organism except for wound repair and the female menstrual cycle. However, expansion of solid tumors beyond a diameter of 1–2 mm requires de novo formation of a vascular network that provides the growing tumor with oxygen and essential nutrients. Many studies have provided strong evidence for the role of the VEGF-VEGFR ligand-receptor system in tumor vascularization and metastasis (Carmeliet and Collen, 2000). Indeed, the hypoxic environment of solid tumors stimulates stromal expression of VEGF. Moreover, solid tumors further stimulate VEGF expression either directly, through multiple oncogenic signaling cascades, or indirectly by stimulating its secretion by adjacent stromal cells by still poorly understood mechanisms. Furthermore, investigation of endothelial cell-derived tumors, including hemangiomas, angiosarcomas, and Kaposi sarcoma, have demonstrated that the VEGF-VEGFR ligand-receptor system may also play an autocrine or paracrine role in tumor cell proliferation (Bais et al., 1998; Montaner et al., 2001; Sodhi et al., 2000). The VEGF-VEGFR ligand-receptor system thus appears to play an important role for tumor induction, progression, and metastasis.

Fibroblast Growth Factor Receptor Family. With more than 20 distinct members identified to date, the fibroblast growth factors (FGFs) represent the largest family of growth factor ligands (Goldfarb, 2001). Like the VEGF-VEGFR ligand-receptor system, the FGFs and their designated receptors (FGFRs) appear to play critical roles in both normal development and tumor formation and progression. There are two classes of FGFRs. The first class comprises the four high-affinity FGFRs, whereas the second class is defined by low-affinity FGF binding sites. Considerable evidence indicates that those low-affinity-binding sites represent heparan sulfate proteoglycan molecules (HSPG) located on the cell surface. These HSPG receptors may support the fine tuning of cell responses to the FGFs present and also regulate their availability and their transport within a tissue.

The first members of the large FGF family to be identified were FGF-1 (aFGF) and FGF-2 (bFGF). Purified on the basis of their mitogenic activity toward fibroblasts, both ligands are potent mitogens for a variety of other cells including those of mesodermal, ectodermal, and endodermal origin. In addition, they play a role as positive regulators for endothelial cell growth and angiogenesis. The family of high-affinity FGFRs comprises four members: FGFR1 (flg), FGFR2 (bek), FGFR3, and FGFR4. Ligand binding to FGFRs induces dimerization and phosphorylation of the cytoplasmic tyrosine residues, but full activation is only achieved in the presence of heparin. A common explanation is that heparin is able to bind a number of monovalent FGFs, allowing the formation of receptor oligomers, which bind to the clustered FGFs.

The fact that a remarkable number of FGFs were identified as genes isolated from tumors because of their ability to induce proliferation or transformation of fibroblasts is suggestive of their functional role in tumorigenesis. Indeed, a number of studies point out that changes in the expression patterns of FGFs may contribute to growth deregulation of human tumor cells. Furthermore, it has been demonstrated that expression of FGF-1 or FGF-8b in fibroblasts leads to transformation and tumor formation in nude mice (Jouanneau et al., 1995; MacArthur et al.,

1995). In addition, several studies suggest that endogenous FGFs are autocrine mediators of neoplastic cell growth.

Changes at the level of FGFRs, such as point mutations, elevated expression, or different splicing, result in dysregulated FGFR signaling and have been identified in a variety of human tumors. FGFR overexpression is observed in tumors arising from numerous tissues, including breast, prostate, melanoma, thyroid, and salivary gland. Somatic mutations in the FGFR-3 gene have been identified in bladder cancer and in multiple myeloma. Like the VEGFR, the FGFR signaling shows a physiological profile of action that includes both mitogenic and angiogenic activity and is frequently altered in human tumors. Together, these observations suggest a critical role for this RTK in cancer.

THE *RAS* ONCOGENE

Ras Proteins and Cancer

The many biological functions of Ras proteins have been the focus of intense investigation because of their direct relevance to the development of human cancer. The first appearance of this family of oncogenes was in the context of the genome of highly oncogenic retroviruses. While passaging the Moloney strain of murine leukemia virus (Harvey-MSV), Jennifer Harvey isolated the first sarcoma virus of mammalian origin. Since 1964, there have been four additional isolates of sarcoma viruses containing *ras* oncogenes. Improvements of in vitro culture techniques for mammalian cells, along with the development of continuous lines of mouse fibroblasts, made possible the biological and subsequent biochemical characterization of MSVs. Studies on fibroblasts transformed by ras-MSVs provided early evidence regarding the nature of the molecules involved in oncogenic growth, suggesting that the normal components of intracellular signaling pathways that control cell proliferation are probably involved in the processes leading to malignant transformation. Indeed, mutations in the three human *ras* genes, H-*ras*, K-*ras*, and N-*ras*, have been detected in many human cancers, with frequencies reported up to 90% in certain aggressive human carcinomas (Table 3.2). These tumors include those of lung, colon, urinary bladder, gallbladder, pancreas, breast, and ovary. Hematopoietic cell malignancies such as B- and T-cell lymphomas, lymphoid and acute myeloid leukemias, as well as neuroblastomas and sarcomas also present mutations in *ras* genes. K-*ras* is the most commonly mutated *ras* gene in these tumors (Ellis and Clark, 2000). H-*ras* and N-*ras* mutations are found less frequently.

The cellular *ras* genes encode proteins of 21 kDa that bind guanine nucleotides (Barbacid, 1987; Symons and Takai, 2001). The current biochemical model for Ras proteins suggest that these molecules are found in an inactive GDP-bound state, exchanging this nucleotide for GTP upon activation. In turn, GTP-bound forms of Ras proteins regulate the activity of downstream effector molecules until signaling

TABLE 3.2. Ras Mutations in Human Cancers

Primary Tumor Site	Ras Isoform	Frequency Mutated (%)
Pancreatic carcinoma	K	90
Thyroid		
Undifferentiated papillary	K, H, N	60
Follicular	K, H, N	53
Papillary	none	0
Colorectal	K	44
Seminoma	K, N	43
Myelodysplastic sutndrome	K, N	40
Non-small cell lung cancer (NSCLC)	K	33
Acute myelogenous leukemia (AML)	N	30
Liver cancer	N	30
Melanoma	N	13
Bladder cancer	H	10
Kidney cancer	H	10

Modified from Macaluso, M. et al., (2002). Ras family genes: An interesting link between cell cycle and cancer. *J Cell Physiol 192*, 125–130 and Adeji, A.A. (2001). Blocking oncogenic Ras signaling for cancer therapy. *J Nat Cancer Inst 93*, 1062–1074.

is terminated by hydrolysis of GTP to GDP. Ras proteins have served as a prototype for a large superfamily of small GTPases, whose mammalian members now number over 100. At least seven distinct branches have been described according to their sequence identity and protein function: Ras, Rab, Rho, Ran, Gem/Rad, Rit/Rin, Rheb, and Arf/Arl. The Ras family includes the four Ras proteins (H-Ras, N-Ras, and the two K-Ras splice variants, K-Ras4A and 4B), as well as the closely related (~50% amino acid identity) Rap (1A, 1B, 2A, and 2B), Ral (1A and 1B), and R-Ras (R-Ras, TC21/R-Ras2, and M-Ras/R-Ras3) proteins. This superfamily of GTPases is involved in multiple and diverse biological functions within the cell. Several members harbor transforming potential. However, we focus our attention on the three members of the Ras subfamily (H-Ras, N-Ras, and K-Ras4A/4B) as they are the only *ras* oncogenes detected in human cancer.

Ras Proteins: Molecular Regulators of Signal Transduction

In the past decade, a large body of information derived from genetic analyses of model organisms including *Drosophila, Saccharomyces cerevisiae,* and *Caenorhabditis elegans*, in addition to biochemical and biological studies in mammalian cells, converged on the discovery that Ras acts downstream of polypeptide growth factors receptors (see above) and upstream of a cascade of cytoplasmic kinases that culminate with the activation of the p42 and p44 MAPKs, also known as ERK2 and ERK1, respectively (Pearson et al., 2001; Whitmarsh and Davis, 1999).

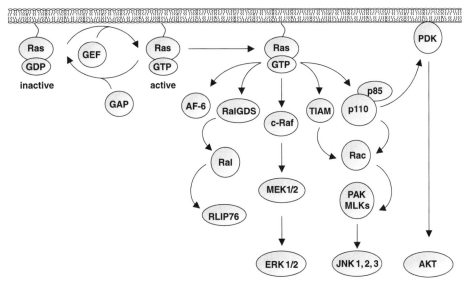

Figure 3.4. *Signaling pathways activated by the small GTPase protein Ras.* Schematic representation of Ras signaling molecules demonstrating the three common pathways induced through ras activation of the effectors Raf (**center**), Ral (**left**), and PI3K (**right**).

Activated MAPKs in turn direct the activities of nuclear transcription factors, thus regulating gene expression. This discovery represented a remarkable breakthrough in our ability to understand the molecular basis for the transduction of proliferative signals and further provided the first clue as to how the aberrant function of Ras proteins may contribute to the malignant conversion of cancer cells.

Ras is now known to transduce signals from a large variety of structurally dissimilar cell surface receptors, which converge on Ras activation to control a complex array of intracellular signaling pathways (Fig. 3.4). Ras function is dependent on whether it is associated with GDP (off state) or GTP (on state). Although the intrinsic GDP/GTP exchange and GTP hydrolyzing activity of Ras are very low, two types of regulatory proteins direct Ras GDP/GTP cycling. GEFs (RasGRF1, 2 and SOS1/2) catalyze the exchange of GDP for GTP, increasing the extent of active Ras-GTP. Conversely, GTPase activating proteins (GAPs; p120 and NF1) stimulate the intrinsic GTPase activity of the Ras protein, inducing the conversion of Ras-GTP to its inactive Ras-GDP form. This system allows Ras to function as a regulated molecular switch that is controlled by a coordinated stimulus-mediated regulation of GEFs and GAPs (Bar-Sagi and Hall, 2000). The single amino acid substitutions at residues 12, 13, or 61 that unmask Ras transforming activity render Ras resistant to the effects of GAPs (Bos, 1989). Thus they become trapped in the active, GTP-bound form, causing constitutive and inappropriate activation of signaling pathways. Indeed, activating mutations of the *ras* genes in human cancer most often occur within either codon 12 or codon 61.

Raf as an Effector for Ras

The best-characterized Ras signal transduction pathway is the Raf/MEK/ERK MAPK cascade. A critical discovery connecting this kinase cascade with Ras was the observation that active Ras-GTP, but not inactive Ras-GDP, forms a high-affinity complex with the serine-threonine protein kinase Raf-1, also known as c-Raf (Kerkhoff and Rapp, 2001; Vojtek et al., 1993). c-Raf is then recruited from the cytosol to the plasma membrane leading to its activation. Ras was first shown to bind to the amino terminus of c-Raf at a single site (residues 50–131), in an area that is essential for plasma membrane localization. However, it was later shown that after this initial event Ras then interacts with a second binding site that is localized in the cysteine-rich domain (Raf-CRD) of c-Raf. The Ras:Raf-CRD interaction appears to be dispensable for membrane localization but is instead essential for the full activation of Raf. Two other Raf isoforms, A-Raf and B-Raf, appear to play a similar role, although their tissue-specific distribution and kinetics of activation by Ras suggest that they may perform highly related but distinct biological functions. Animal models using knockout technology have shown that A-Raf is responsible for intestinal and neurological development, B-Raf regulates vascularization and suppresses programmed cell death, and c-Raf-1 is involved in growth control and skin, lung, and placental development (Hagemann and Rapp, 1999). Interestingly, activated c-Raf has been detected in primary human stomach cancer, breast and renal carcinoma, skin fibroblasts from a cancer-prone family (Li–Fraumeni syndrome), a radiation-resistant human laryngeal cancer, and a human glioblastoma cell line. Recently, a detailed genome-wide analysis of mutations in the Ras-ERK pathway has revealed that B-Raf may also be often mutated in human cancer, most notably in melanomas (Davies et al., 2002).

Other Ras Effectors

Although initial studies suggested that c-Raf might be the principal or even lone effector for the Ras protein, it is now clear that non-Raf effector-mediated pathways are equally important in mediating the biological effects of Ras GTPase (Marshall, 1995). For example, activated Ras, but not Raf, can cause transformation of a variety of epithelial cells and some Ras effector domain mutants that are impaired in c-Raf interaction retain transforming activity (Khosravi-Far et al., 1996; White et al., 1995).

Indeed, other Ras effectors were later identified and shown to contribute to the tumorigenic potential of this potent oncogene. Ras interacts with the PI 3-kinase complex, formed by the regulatory p85 subunit and the catalytic p110 subunit, in a GTP-dependent manner. The acute phosphorylation of phosphatidylinositol lipids at the D-3 position of the inositol ring sets in motion a coordinated set of events leading to cell growth, cell cycle entry, cell migration, and cell survival, through the activation of downstream molecules (see below). The Ral GEF (RalGDS) and two closely related proteins (RGL and RGL2/Rlf) also represent intriguing candidate effectors of Ras that may link this GTPase with other Ras-related proteins (Katz and McCormick, 1997). Members of this family have been identified repeatedly by yeast two-hybrid library screening searches for candidate effectors of Ras and Ras-related proteins (Rap, R-Ras, and TC21/R-Ras2), and coexpression of RalGDS cooperated synergistically with activated Raf-1 to induce transformation of NIH 3T3 cells. However, the implications of Ras-RalGDS interaction are still poorly understood. Another surprising and still not well-clarified finding is that two Ras GAPs, p120RasGAP and NF1/neurofibromin, may also act as downstream effectors for Ras.

Another molecule downstream from Ras is AF-6, which is expressed in a variety of tissues (Kuriyama et al., 1996). The sequence of this protein is highly similar to the *Drosophila* protein Canoe that functions in signaling pathways downstream from the Notch receptor and acts as a regulator of cellular differentiation. In vitro experiments have shown that the amino-terminal domains of AF-6 and Canoe interact specifically with Ras-GTP and this interaction interferes with the binding of Ras with Raf. Although the AF-6 function is still not clear, it has been shown to serve as one component of tight junctions in epithelial cells and in cell-cell adhesions in nonepithelial cells, suggesting that AF-6 may be involved in the regulation of cell-cell contacts. Interestingly, the AF-6 protein was also identified independently as a fusion partner of the MLL protein associated with chromosome translocation events in human leukemias. The MLL/AF-6 chimeric protein is the gene product of a reciprocal translocation t(6;11)(q27;q23) associated with a subset of human acute lymphoblastic leukemias (ALL), strongly implicating this pathway in the genesis of ALL.

THE MAP KINASE SIGNALING CASCADES

The MAPKs are a group of protein kinases that play an essential role in signal transduction pathways modulating gene transcription in the nucleus in response to changes in the cellular environment (Whitmarsh and Davis, 1999). MAPKs also play a key role in intracellular communication, and their activating pathways have been conserved throughout evolution, from plants, fungi, nematodes, insects, to mammals. The ERK group of MAPK includes two mammalian enzymes (ERK1 and ERK2). Functional studies indicate that the activation of ERK1 and ERK2 provides proliferative signals that may contribute to normal growth and to malignant transformation in some cells while promoting cellular differentiation processes in others. The induction of the kinase activity of ERK1 and ERK2 is mediated by dual phosphorylation on threonine and tyrosine residues included in a short sequence that forms the "activation loop," located close to the kinase active site. Phosphorylation is catalyzed by a group of dual-specificity protein kinases (MAPK kinases; MAPKKs), which include MEK1 and MEK2, and provokes conformational changes in the protein that lead to MAPK activation and nuclear localization. Termination of the MAPK activity requires dephosphorylation on either Thr or Tyr by a Ser/Thr or Tyr phosphatase. Dephosphorylation of both Thr and Tyr residues can be mediated also by a dual-specificity MAPK phosphatase. Similarly, the mammalian MAPKKs are activated by phosphorylation mediated by MAPKK kinases (MAPKKKs), which include the Ras effector Raf. Thus the Raf group of MAPKKK plays an important role in the coupling of ERK MAPK activation to the Ras signaling pathway.

There are at least 11 members of the MAPK superfamily, which can be divided into five groups: the ERKs (ERK1 and ERK2); c-Jun N-terminal kinases (JNK1, JNK2, and JNK3); p38s (p38α, p38β, p38γ, p38δ); ERK5; and ERK8 (Gutkind, 2000). Separate signal transduction pathways initiated by different extracellular stimuli activate each group of MAPKs (Fig. 3.5). The MAPKs, in turn, exhibit distinct substrate specificities. Thus specific extracellular stimuli lead to the differential activation of MAPKs and consequent phosphorylation of different groups of MAPK substrates in response to an individual stimulus. This complexity

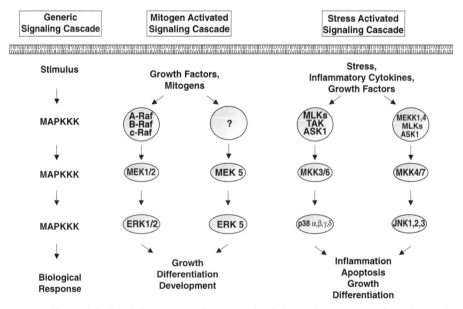

Figure 3.5. *MAP kinase signaling cascades.* Schematic representation of generic (**left**), mitogen-activated (**center**), and stress-activated (**right**) signaling cascades. Separate signal transduction pathways initiated by different extracellular stimuli activate each group of MAPKs that, in turn, exhibit distinct substrate specificities.

and specificity in the mechanism of signal transduction by MAPKs ensures that cells mount an appropriate response to extracellular stimulation, as the resulting pattern of gene expression will depend on the integration of the combinatorial signals provided by the temporal activation of each of these groups of MAPKs.

Whereas a detailed description of the Ras-ERK pathway has been presented above and in other chapters and the JNK pathway will be described in the context of the contribution to signaling by small GTPases of the Rho family (see below), we focus here on two less-known MAPK pathways, p38 and ERK5, and on some emerging common themes, such as the ability of MAPKs to select their substrates and regulate gene expression.

The p38 Pathway

The family of p38 kinases, included within the superfamily of MAPK, has grown rapidly over the past few years. To date, four p38 kinases have been described and named p38α(or CSBP-1), p38β, p38γ (also known as SAPK3 or ERK6). and p38δ (or SAPK4), some of which also have splice variants (Widmann et al., 1999). Other members are the *Drosophila* MAPK (p38α and p38β) and the yeast MAPK Hog1p. This family of proteins exhibits a common Thr-Gly-Tyr phosphorylation motif and shows a pattern of activation by environmental stress and cell surface receptors, very reminiscent of that displayed by JNKs, and by the treatment of cells with cytokines. These enzymes are also activated by exposure to endotoxins, including bacterial lipopolysaccharides. Genetic analysis of yeast indicates that the Hog1p MAPK is required for the response to osmotic stress. This response involves the transcriptional upregulation of genes that induce the biosynthesis of the osmotic stabilizer glycerol. In *Drosophila*, the activation of p38α and p38β by endotoxin is implicated in the downregulation of immunity gene expression. The role of p38 MAPK in regulating gene expression in mammalian cells has been established by studies of the transcription factors ATF2, CHOP, Elk-1, MEF-2C, and SAP-1 (Pearson et al., 2001).

Like the ERKs, the activation of p38s is also mediated by dual phosphorylation on Thr and Tyr residues. Although the upstream activators for these kinases are not yet well defined, potential p38 MAPKKKs include ASK1 and TAK1 (Gutkind, 2000; Ono and Han, 2000). MKK6 seems to be a very general stimulator for p38s, whereas MKK3 and -4 preferentially phosphorylate p38α. The specific transcription factors that can be regulated by p38s include cAMP-responsive element binding protein (CREB), ATF1 (activating transcription factor 1), ATF2, Max, CHOP (C/EBP homologous protein), and MEF2C. In addition, these MAPKs can also trigger the activation of other serine/threonine kinases such as MAPK-interacting kinase 1 (Mnk1) and Mnk2 and MAPK-activated protein kinases (MAPKAPKs) (Ono and Han, 2000; Pearson et al., 2001).

The BMK Group of MAP Kinases

A MAPK distantly related to ERK1 and ERK2 has recently been characterized and termed big mitogen-activated protein kinase 1 (BMK1) or ERK5 (Zhou et al., 1995). It contains a Thr-Glu-Tyr motif in its activation loop, similar to that of MAPKs. ERK5 is larger than any other known MAPK (~80 kDa) and is selectively activated by MEK5. This kinase can be stimulated by oxidative stress and may also play a role in early gene expression triggered by serum by directly phosphorylating the transcription factors MEF2C or c-Myc (English et al., 1998; Gutkind, 2000). It has been demonstrated that ERK5 is part of a distinct MAPK signaling pathway required for EGF-induced cell proliferation and progression through the cell cycle. Additional evidence further suggests that the ERK5 signaling pathway participates in the regulation of cell proliferation by neuregulin receptors. Moreover, it has been shown that MEK5 and ERK5 cooperate with the Raf effectors MEK1/2 and ERK1/2 to induce foci formation. The transcription factor NF-κB and p90 ribosomal S6 kinase seem to be involved in MEK5-ERK5-dependent focus formation and may serve as integration points for ERK5 and ERK1/2 signaling. Nevertheless, the physiological function of ERK5 is still unclear and further studies are required to establish the role of this MAPK signaling pathway.

Substrate Selection and Nuclear Signal Transmission by MAPK Signaling Pathways

Protein kinases typically interact with their substrates by recognizing primary sequence determinants that surround the phosphorylation site(s). In the case of MAPK, the minimum general primary sequence required to be a substrate for protein phosphorylation is the presence of a Ser or Thr residue followed by Pro (Whitmarsh and Davis, 1999). Additionally, most MAPKs also appear to interact with their substrates by binding to a second site. This "docking" interaction is required for substrate phosphorylation by MAPK in vitro and is likely to be a relevant mechanism for the selection of high-affinity substrates by these kinases in vivo (Sharrocks et al., 2000). For substrate recognition, MAPKs first dock to the substrate through a binding interaction at one site leading to phosphorylation of the substrate at a second site at one or more Ser/Thr-Pro motifs. Recent studies demonstrate that the selective interaction of MAPK with substrates may influence the physiological regulation of MAPK targets in vivo. Examples of such docking interactions include the interaction of ERK or JNK with the transcription factor Elk-1 (Sharrocks et al., 2000). Furthermore, there is substantial evidence that the normal function of MAPK signaling modules requires the physical organization of the MAPK pathway components within the cell. It is likely that scaffold proteins may be required for targeting of MAPK signaling to some, but not all, substrates, and that some MAPK pathways may normally function in the absence of a scaffold protein in vivo.

Although MAPK-mediated signaling pathways are generally initiated at the cell surface, transcription factors located in the nucleus are frequent targets for many of these pathways, thus suggesting the existence of mechanisms involved in the transmission of signals from the cytoplasm to the nucleus (Volmat and Pouyssegur, 2001). It has been showed that the activated form of c-Raf-1 (MAPKKK) is present at the plasma membrane, whereas activated MEK1 and MEK2 (MAPKK) are located in the cytoplasm, because of the presence of a nuclear export signal (NES) in the protein. The transmission of signals from the cytoplasm to the nucleus appears to be directly mediated by the ERK MAPK, because recent studies have shown that ERK is located in the cytoplasm of quiescent cells and in the nucleus of activated cells. It is still unclear which is the mechanism that accounts for this activation-induced translocation of ERK to the nucleus. One explanation is that ERK is sequestered in the cytoplasm by MEK1 and MEK2. ERK activation would lead to MAPKK dissociation from ERK1 and ERK2, which are then able to form homodimers and migrate to the nucleus. Several lines of evidence supporting this model have been reported. However, it remains to be established whether the release of ERK from the MAPKK is sufficient to allow nuclear accumulation or whether additional processes (including ERK phosphorylation) are involved. Further studies are required to clarify the molecular mechanism of nuclear redistribution of the ERK MAPK and to determine whether other MAPKs use similar modes to phosphorylate nuclear proteins.

Regulation of Gene Expression Through MAPK Signaling Pathways

Regulation of gene expression is a central function of MAPK signaling pathways. These intracellular routes modulate gene expression by multiple approaches, acting at different steps (pretranscriptional, transcriptional, and posttranscriptional). It is likely that this complexity may be important for the coordination of appropriate responses to a wide variety of extracellular stimuli (Whitmarsh and Davis, 1999).

The first approach that the MAPKs utilize to regulate gene expression is the phosphorylation of transcription factors that are then retained in the cytoplasm in an inactive form. Dephosphorylation of these factors is needed for migration to the nucleus to activate transcription. For example, the phosphorylation of the SMAD1 transcription factor by ERK or the phosphorylation of NFAT4 by JNK causes the cytoplasmic retention of these factors and therefore the inhibition of their transcriptional activity (Chow et al., 1997; Kretzschmar et al., 1997). Conversely, it is still unclear whether the phosphorylation by MAPKs of other cytoplasmic transcription factors may cause their nuclear translocation. Regulation of the DNA binding activity is also a mechanism employed by MAPK to control transcription factor activity. An example is the transcription factor Elk-1, which is phosphorylated by ERK, JNK, and p38 MAPK, leading to increased DNA binding (Yang et al., 1998).

MAPKs most often regulate the transcriptional activity of the nuclear factors that bind to DNA. For example, phosphorylation of the transcriptional activation domains of ATF2, CHOP, Elk-1, and c-Jun causes increased transcription. Detailed studies of phosphorylation-dependent transcriptional activation have been reported in the case of the transcription factor CREB. These studies indicate that transcription factor phosphorylation can cause changes in the interaction with coactivator molecules, including CBP, p300, and Rb family proteins, which function, in part, through the regulation of histone acetylation.

Regulation of translation represents a fourth mechanism of regulation of gene expression by the MAPKs. The best example is the translational regulation of cytokine expression observed in macrophages exposed to endotoxin. Translational regulation of the cytokine tumor necrosis factor-α (TNF-α) by p38 MAPK requires a *cis*-acting element in the mRNA. Modulation of mRNA processing and nuclear export has not been described, although these biosynthetic steps also represent potential sites for intervention by MAPKs.

The final mechanism employed by MAPKs to regulate gene expression is the regulation of protein degradation. This is exemplified by the short-lived transcription factor c-Jun, which is rapidly degraded by the ubiquitin-proteasome pathway. Phosphorylation of c-Jun by JNK inhibits the ubiquitination of c-Jun and, therefore, its rapid degradation (Fuchs et al., 1996; Musti et al., 1997). Consequently, JNK activation prolongs the half-life of c-Jun, leading to the accumulation of the c-Jun protein. The regulation of protein stability by MAPK therefore contributes to the regulation of gene expression mediated by these signaling pathways.

THE RHO FAMILY OF SMALL GTPASES

Rho, Rac, and Cdc42

The Ras-MAPK pathway provides a point of convergence of numerous signaling routes by which growth factor tyrosine kinase receptors control cell proliferation. However, emerging evidence suggests that small GTPases of the Rho family also play a central role in normal cell growth and malignant conversion (Jaffe and Hall, 2002). The Rho family forms a large subgroup within the Ras superfamily of GTP-binding proteins and regulates a wide spectrum of cellular functions, such as actin cytoskeleton reorganization, endocytosis and exocytosis, transcription activation, stimulation of DNA synthesis, and/or translational regulation. There are 24 predicted mammalian Rho GTPases, 18 of which have been described, that share between 50% and 90% amino acid sequence homology; RhoA, Rac1, and CDC42 are the most extensively studied (Bar-Sagi and Hall, 2000; Bishop and Hall, 2000) (Fig. 3.6). These proteins are ubiquitously expressed across the species, from yeast to humans. Unlike Ras, genetic analysis of human cancers has not revealed mutations in Rho-like GTPases, but changes in their expression levels or

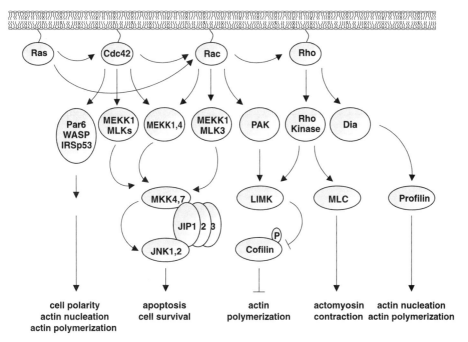

Figure 3.6. *Signaling pathways activated by the Rho family of small GTPase proteins.* Schematic representation of Rho, Rac, and Cdc42 signaling molecules demonstrating the interactions among these signaling molecules and the related small GTPase, Ras.

alterations in proteins that regulate their activity appear to be important factors in tumor development. Indeed, increased protein or mRNA levels for Rho proteins have been found in different tumor types such as breast, lung, pancreas, and colorectal cancer. A positive role for RhoA has been described in both migration and invasion of T-lymphoma cells, colon carcinoma cells, and hepatoma cells. Interestingly, a recent comparison of the gene expression profiles of highly metastatic versus weakly metastatic melanoma cell lines also revealed an increase in the expression of RhoC in the highly metastatic lines (Clark et al., 2000).

The involvement of the Rho family in cell growth control was previously suggested by the high oncogenic potential that the majority of their GEFs, such as Dbl, exhibit in vitro (Cerione and Zheng, 1996). Stimulation of growth factor receptors, cytokine receptors, cell-to-cell or extracellular matrix-to-cell adhesion receptors, or G protein-coupled receptors can all initiate intracellular signals that lead to Rho GTPase activation, a process that involves a variety of recently identified Dbl family members (Schmidt and Hall, 2002). Concomitantly, the incoming signals might also modulate Rho GTPase-activating proteins (RhoGAPs) and Rho GDP-dissociation inhibitors (RhoGDIs), both of which inhibit Rho signaling, resulting in increased levels of active

Rho-GTP. Rho-GTP then binds a wide variety of effector proteins that further convey the signals to downstream components.

Regulators of the Activity of Rho GTPases and Cancer

There are more than 50 predicted GEFs for the Rho family in the human genome. All members of the Dbl family contain a pleckstrin-homology domain (PH) immediately carboxyl-terminal to a Dbl-homology (DH) domain (Zheng, 2001). These domains are the structural elements responsible for the GDP/GTP exchange activity. Whereas the DH domain is accountable for GEF catalytic activity per se, the PH domain is involved in the intracellular targeting of the DH domain, by association with the plasma membrane, actin, or actin-binding components. In addition, a general finding among Rho GEFs is that the DH and PH module represents the minimum structural unit bearing transforming function. Demonstration of substrate-binding and catalytic activities in reconstituted GEF reactions and examination of Rho GTPase activation states induced by overexpression of GEFs have provided valuable information on the biochemical functions of Dbl members, although much of this is currently limited to the three most prominent Rho GTPases, RhoA, Rac1, and Cdc42. Some of the GEFs, for example, Cdc24, FGD1, Tiam1, and Lbc, are specific for a single Rho protein, whereas others, including Dbl, Ost, Ect2, and Bcr, are more promiscuous, targeting multiple Rho GTPases.

Many of the Rho GEFs were initially identified by their ability to induce cell transformation in in vitro experiments and were only later found to be exchange factors for Rho GTPases (Van Aelst and D'Souza-Schorey, 1997). For example, the *dbl* oncogene product was originally isolated from a diffuse B cell lymphoma, whereas transfection of NIH3T3 cells with an osteosarcoma expression cDNA library led to the appearance of cell foci that were found to harbor the *ost* oncogene. Vav protein has been also implicated in lymphocytic proliferation and lymphopenia. Tiam1 was first identified as an invasion-inducing gene using proviral tagging in combination with in vitro selection for invasiveness. This Rac-specific GEF was found to have a point mutation in the PH domain (amino acid 441) of one of the two alleles, in around 10% (4 of 35) of the renal cell carcinoma (RCC) samples examined (Engers et al., 2000). Expression of this mutant in fibroblasts induced focus formation to a similar degree as constitutively active (amino-terminally deleted) Tiam1, suggesting that the A441G mutation may act in a dominant gain-of-function manner in vivo. Transformation by these exchange factors has been shown to induce cellular phenotypes that closely resemble those induced by expressing constitutively activated forms of Rho GTPases and to be inhibited by dominant-negative forms of Rho GTPases. These data indicate that Rho GEFs may mediate cell transformation by deregulation of endogenous Rho activity. Members of the Dbl family also have been shown to regulate cell migration and tumor metastasis.

The DH domain is also present in the Bcr protein and in the p210 isoform of Bcr-Abl fusions in chronic myelogenous and acute lymphocytic leukemias (Cilloni et al., 2002). Another Rho GEF, leukemia-associated Rho GEF (LARG), has been isolated as a fusion partner of the mixed-lineage leukemia (MLL) gene in a patient with primary acute myeloid leukemia (AML) (Kourlas et al., 2000). This fusion protein contained the amino-terminal portion of MLL in frame with the carboxyl-terminal 80% of LARG, which includes the DH and PH domains. It remains to be established whether the MLL-LARG fusion protein is sufficient to induce transformation in myeloid cells and how frequently this locus is affected in human cancer.

Several studies have also described deletions or mutations in Rho GAPs in different cancers, showing that deregulated activity of Rho GTPases can also be reached by a loss-of-function mutation or decreased expression of this type of regulator. One of these GAPs, the human ortholog of the chicken GRAF (GTPase regulator associated with the focal adhesion kinase pp125FAK) gene, was isolated as an MLL fusion partner from a patient with juvenile myelomonocytic leukemia. The fusion protein consisted of the amino-terminal portion of MLL fused to the carboxyl-terminal region of GRAF, which lacks the GAP domain. Of interest, human GRAF maps to chromosomal region 5q31, which is commonly deleted in myelodysplastic syndromes (MDSs) and AMLs. Sequence analysis revealed that 3 of 13 patients in whom one allele of GRAF was deleted had mutations in the other GRAF allele that either inactivate or remove the GAP domain (Borkhardt et al., 2000). These results suggest that GRAF is a tumor suppressor gene that may contribute to the development of some hematological tumors. Two other GAPs, p190-A and *deleted in liver cancer* (DLC-1), are also localized to regions that are deleted in gliomas/astrocytomas and hepatocellular carcinoma (HCC), respectively.

Further experimental data implicate Rho GTPase signaling in the progression of the malignant phenotype. This family of GTPases has been shown to be transforming in vitro (Perona et al., 1993) and might potentially influence the invasiveness and metastatic potential of tumor cells in different manners (Jaffe and Hall, 2002). Expression of constitutively activated forms of Rho, Rac, or CDC42 in fibroblasts results in focus formation. These foci are morphologically distinct from the swirling and spread out foci induced by activated Ras and instead are highly compact, resembling foci induced by overexpression of members of the Rho GEF family of oncogenes (Jaffe and Hall, 2002; Zohar et al., 1998). Although constitutively active forms of Rho, Rac, and Cdc42 have a weak transforming activity when expressed alone, they each enhance the focus formation induced by Ras (Qiu et al., 1995). Interestingly, the morphologic changes induced by coexpression of activated Rho or Rac and activated Raf are somewhat different from those seen when the factors are expressed alone, supporting the idea that the contribution of Rho GTPases to transformation is mediated, at least in part, by signaling events that are different from those induced by Ras.

Rho GTPases were initially described as essential molecules in the regulation of various aspects of actin cytoskeleton dynamics. RhoA, Rac1, and CDC42 activation lead to the formation of actin stress fibers, membrane ruffles/lamellipodia, and filopodia, respectively (Ridley, 2001). These structures are commonly observed in migrating cells in culture. Additionally, RhoA, Rac1, and CDC42 promote the formation of integrin-containing adhesion complexes, which mediate attachment to the extracellular matrix. They may also modify the strength of cadherin-mediated cell-cell adhesions, modulating the ability of cells to detach from the tumor mass. The ability of Rho proteins to control actin structures and adhesion complexes is likely to be an important factor in their ability to influence the invasiveness of tumor cells in vivo. In addition, Rho proteins also regulate expression of metalloproteinases and phospholipid metabolism and have been implicated in various vesicular transport events, all of which may have an impact on tumor cell invasiveness. Nonetheless, more investigation must be completed to clarify which of the many effector molecules and downstream activated signaling pathways of the Rho family GTPases unambiguously relay an oncogenic potential in vivo.

Rho GTPases and Gene Expression Regulation

Most efforts addressing the function of Rho GTPases have focused on the molecular dissection of the mechanism(s) involved in their ability to control the dynamic assembly and remodeling of actin-containing cytostructures. A detailed analysis of the downstream targets for these GTPases participating in this function is provided in Chapter 7. In an interesting turn, following the mechanism by which growth factor receptors promote the expression of the transcription factor c-*jun*, it was observed that this response did not correlate with ERK activation, suggesting the involvement of a distinct biochemical route regulating the expression of this gene (Coso et al., 1995a). Instead, the transcriptional activity of c-Jun and its ability to promote its own expression were shown to be regulated by a family of enzymes structurally related to ERKs, termed JNKs or stress-activated protein kinases (SAPKs), that phosphorylate c-Jun on its amino-terminal domain (Derijard et al., 1994; Kyriakis et al., 1994). An unexpected prediction from these studies was that a distinct set of upstream signaling molecules might regulate JNK and MAPK. Indeed, Rac1 and Cdc42 were found to initiate an independent kinase cascade regulating JNK activity (Coso et al., 1995b; Minden et al., 1995). Other components of this pathway include several MAPKKKs: MAPK or ERK kinase kinases (MEKK1, MEKK2, MEKK3, and MEKK4), ASK1 (apoptosis-stimulated kinase 1), MLK3 (mixed-lineage kinase 3), TPL2 (tumor progression locus 2, also known as Cot), TAK (TGF-ß-activated kinase), MUK (MAPK upstream kinase), GCK (germinal center kinase), and PAK, which can all contribute to activation of JNKs, apparently through two MAPKKs, MKK4 (also called Sek or JNKK1) and MKK7. So far, however, there is limited information regarding the precise architecture of the signaling pathways in which each of these kinases act. Substrates for JNK include transcription factors of the c-Jun family, as well as other transcription factors, including Elk-1 and Elk-2, Sap-1 (serum response factor accessory protein-1), NFAT4 (nuclear factor of activated T cells 4), and ATF2 (activating transcription factor 2) (Gutkind, 2000).

Three genes that encode mammalian JNK protein kinases have been molecularly cloned. The JNK1 and JNK2 genes are expressed ubiquitously, and a total of eight JNK isoforms are translated by alternative splicing (Weston, 2002). The JNK3 gene has a more restricted pattern of expression with highest levels in brain and testis. Two forms of JNK3 are translated by alternative splicing of transcripts derived from the JNK3 gene. Studies of mammalian cells demonstrate that JNK is required for some forms of stress-induced neuronal apoptosis. JNK is also required for inflammatory responses and for the survival and malignant transformation of certain types of cells (e.g. B cells). Rho, Rac, and CDC42 have each been reported also to activate serum response factor (SRF)-dependent transcription (Hill et al., 1995), the transcription factor NF-κB (Perona et al., 1997), and c-myc expression (Chiariello et al., 2001). In turn, how these events lead to cellular transformation is still under intense investigation. A necessary requirement for a cell to transit from a normal to a transformed state is the dysregulation of its cell cycle machinery. As described below, several components of the cell cycle machinery, such as cyclins and cyclin-dependent kinase inhibitors, also represent central points of convergence of growth-promoting pathways emanating from Ras and Rho GTPases.

THE WNT SIGNAL TRANSDUCTION PATHWAY

Wnt: A New Paradigm in Signaling

The first *Wnt* gene (*int-1/Wnt-1*) was cloned in 1982 as a gene whose expression was upregulated in mouse mammary tumor virus (MMTV)-induced breast tumors. Further analysis showed that the locus encoded a cysteine-rich secreted glycoprotein that binds to the cell surface and initiates signal transduction. Its discovery in murine breast tumors prompted investigation of Wnt expression in human cancer (van Leeuwen and Nusse, 1995). Recent data have since shown that Wnt is overexpressed in several tumors. However, it is the downstream components of the canonical Wnt signaling pathway that have been most commonly reported to be altered in human malignancies. Among them, mutations and deletions in the adenomatous polyposis coli (*APC*) tumor suppressor gene, which encodes a large, 310-kDa protein that interferes with the Wnt signaling pathway, are present in 85% of all colorectal tumors. Moreover, approximately half of the remaining 15% contain mutations in β-catenin, a downstream target of Wnt. β-Catenin mutations are never observed in *APC*-deficient tumors, which is consistent with the idea that APC and β-catenin function in the same pathway involved in suppressing colorectal tumor growth (Bienz and Clevers, 2000). Although of great interest because of its role in carcinogenesis, the study of the Wnt signaling route is not restricted to cancer biology. In fact, the function of Wnt proteins and their receptors, regulatory molecules, and intracellular targets has recently yielded some surprising observations, which are also shaping our understanding of the process of normal development, and how dysregulation of developmentally coordinated programs can result in cancer.

Signaling Pathways of Wnt

Wnt signaling plays a central role in the development of many phylogenetically diverse organisms, from *Drosophila* and *C. elegans* to *Xenopus* and higher vertebrates. Genetic alterations of components in the Wnt pathway are associated with tumorigenesis both in animal models and in humans. These mutations are most common in regenerating epithelial tissues, in which this pathway normally operates. In these tissues, oncogenic mutations result in constitutive (ligand independent) activation of the Wnt pathway and consequent cellular proliferation. Indeed, mutations in the Wnt pathway are commonly associated with melanomas and carcinomas of the breast and colon (Moon et al., 2002).

Intracellular signaling of the Wnt pathway diversifies into at least four branches: the β-catenin pathway (the canonical Wnt signaling pathway), which activates target genes in the nucleus; the planar cell polarity pathway, which involves JNK and cytoskeletal rearrangements; the Wnt/Ca^{2+} pathway, which involves activation of PLC and PKC; and a pathway that regulates spindle orientation and asymmetric cell division (Fig. 3.7). In brief, in the canonical Wnt signaling pathway, glycogen synthase kinase 3 (GSK-3)-mediated β-catenin ubiquitination and degradation are inhibited by the binding of Wnt to the receptor-coreceptor complex, Frizzled/LRP (see below). In the planar cell polarity pathway, Wnt signaling activates JNK and directs asymmetric cytoskeletal organization and coordinated polarization of cell morphology within the plane of epithelial sheets (Mlodzik, 1999). This pathway branches from the canonical Wnt pathway downstream of Frizzled, at the level of Dishevelled, and involves downstream components like the small GTPase Rho and a kinase cascade including Misshapen (a Ste20 homolog), JNK kinase, and JNK. Several members of the canonical Wnt-signaling pathway, including GSK-3 and APC, have also been implicated in spindle orientation and asymmetric cell division of *C. elegans* and *Drosophila*. Wnt has been further shown to play a role in the release of intracellular calcium, possibly mediated via G proteins (Kuhl et al., 2000). This pathway involves activation of PLC, PKC,, and calmodulin-dependent kinase II and is implicated in *Xenopus* ventralization and in the regulation of convergent extension movements.

Despite the multiple signaling pathways activated by Wnt, the canonical Wnt signaling pathway remains the most frequently implicated in

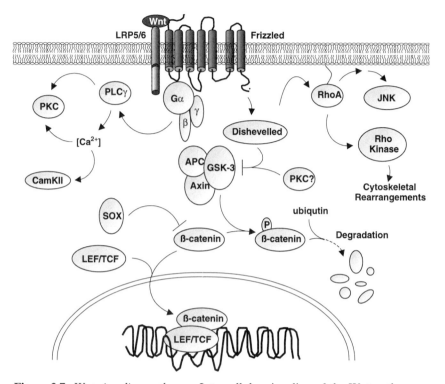

Figure 3.7. *Wnt signaling pathways.* Intracellular signaling of the Wnt pathway diversifies into at least four branches: the β-catenin pathway (the canonical Wnt signaling pathway), which activates target genes in the nucleus (**center**); the Wnt/Ca²⁺ pathway, which involves activation of PLC and PKC (**left**); the planar cell polarity pathway, which involves JNK and cytoskeletal rearrangements (**right**); and a pathway that regulates spindle orientation and asymmetric cell division (not shown).

human malignancies (Bienz and Clevers, 2000). Thus we will describe in detail each component of this still not fully understood signal transduction system.

The Canonical Wnt Signaling Pathway

Wnt. Since its initial discovery, multiple Wnt genes have been found in mice (16 genes), humans (13), *Xenopus* (15), zebra fish (9), *Drosophila* (4), and *C. elegans* (5) (Cadigan and Nusse, 1997; Seidensticker and Behrens, 2000). These cysteine-rich glycoproteins are associated with the Golgi network, whereas the small amount of secreted Wnt is tightly associated with components of the extracellular matrix. The presence of multiple Wnt genes suggests a great deal of specificity in expression and function. Wnt genes are divided into two classes, the Wnt-1 class (including Wnt-1, Wnt-2, Wnt-3A, and Wnt-8) and the Wnt-5A class (including Wnt-5A, Wnt-4, and Wnt-11). Only the Wnt-1 class has been shown to have the ability to induce morphological transformation, although this distinction remains unclear.

Frizzled. The reception of Wnt signaling is mediated, at least in part, by a member of the Frizzled family of receptors. Frizzleds are seven transmembrane spanning proteins that share extensive homology throughout their coding regions (Cadigan and Nusse, 1997). Frizzled appears to comply with the canonical signaling pathway linking seven transmembrane spanning proteins to the activation of heterotrimeric G proteins (see Chapter 6). Whereas the preference of individual Wnts for binding to different Frizzleds remains unclear, a third molecule, the LDL receptor-related proteins (LRPs), may now help explain the specificity of Wnt signaling.

LDL Receptor-Related Proteins. LRPs (LRP5 and LRP6 in vertebrates) are single transmembrane proteins (Pandur and Kuhl, 2001). LRPs act upstream of Dishevelled and downstream of the Wnt ligand. Given its transmembrane structure LRP6 is likely to function as a coreceptor for Frizzleds. In fact, it has been shown that the extracellular domain of LRP6 binds Wnt-1 and forms a complex with Frizzled in a Wnt-dependent manner. It is speculated that a heterodimeric receptor complex of Frizzled and LRP brings intracellular proteins of the Wnt signaling cascade into close proximity to initiate signal transduction. Thus heterodimer formation might be required for signal transduction through β-catenin. However, LRPs do not appear to be required for all Wnt signaling events acting through Fizzled. It is believed that it is the combination of distinct Wnts, Frizzleds, and LRPs that is crucial for the activation of each particular pathway.

Dishevelled. Although the mechanisms underlying Wnt-induced activation of Frizzled are unclear, they ultimately lead to the activation of Dishevelled, a cytosolic phosphoprotein of unknown function (Novak and Dedhar, 1999). The mechanisms underlying Dishevelled's activation and subsequent signaling are unknown. Phosphorylation of Dishevelled correlates with Wnt activity, but it is not sufficient to transduce Wnt signals. The PDZ domain in the central part of the protein is the best characterized and is required for Dishevelled function. Recently available information suggests that this domain may participate in the recruitment and activation of a novel formin homology protein termed Daam1, which is highly related to the Rho downstream target Diaphanus that is involved in promoting actin reorganization and transcriptional responses (Habas et al., 2001). A second domain in Dishevelled is the DIX (Dishevelled and axin) domain, which may be responsible for subcellular localization. The third shared region of homology in this molecule is the DEP domain, found in a number of proteins associated with G protein signaling. This region is dispensable for Wnt signaling but is required for tissue polarity.

GSK-3. Once activated through Frizzled and LRP, Dishevelled initiates poorly characterized cellular events that result in the downregulation of GSK-3 kinase activity (Manoukian and Woodgett, 2002). GSK-3 is a serine/threonine protein kinase first identified on the basis of its ability to phosphorylate and inactivate glycogen synthase. Subsequent analysis revealed that highly homologous isoforms of this kinase, GSK-3α (51 kDa) and GSK-3β (47 kDa), are encoded by two genes in vertebrates.

Wnt acts by directly inhibiting GSK-3 kinase activity. Inhibitors of PKC can block the Wnt-induced inhibition of GSK-3, suggesting that PKC mediates the Wnt-induced inhibition of GSK-3 activity. Furthermore, some isoforms of PKC can phosphorylate GSK-3 and inhibit its activity in vitro. It is not known whether all Wnts will induce this reduction in GSK-3 activity. Signaling pathways that respond to insulin, insulin-like growth factor (IGF-1), and EGF also inhibit GSK-3 kinase activity (Harwood, 2001). However, regulation of GSK-3 by those factors is pharmacologically distinct from that induced by Wnt and mediated by protein kinase B/Akt or Rsk-90. GSK-3 has many potential protein substrates including c-Jun, c-Myb, c-Myc, Tau, the eukaryotic initiation factor 2B, CREB, the regulatory subunit of cAMP-dependent kinase (RII), NFAT, and armadillo/β-catenin (Woodgett, 2001). All of these proteins contain at least one consensus site for GSK-3 phosphorylation in which the target serine or threonine residue is followed by any three amino acids followed by another serine or threonine that is usually prephosphorylated. Both protein kinase A and casein kinase II are candidates to mediate this priming phosphorylation reaction.

β-Catenin and Plakoglobin. Two targets of GSK-3 implicated in the Wnt signaling pathway are β-catenin and plakoglobin (also referred to as γ-catenin) (Novak and Dedhar, 1999). Both were initially identified as components of the cadherin-based cell-cell adherence junctions, where they play critical roles in cell-cell contact and cytoskeletal anchorage. The amino terminus of β-catenin and plakoglobin contain a GSK-3 consensus phosphorylation site. The central region of these proteins consists of 12–13 "armadillo repeats," which function as nonhomotypic protein interaction domains. In the case of β-catenin, these mediate interactions with cadherin, axin, conductin, APC, and lymphoid enhancer factor (LEF)/T-cell factors (TCFs) and with itself (Polakis, 1999).

The nonmembranous fractions of β-catenin and plakoglobin are both involved in Wnt signaling. Although research has focused on β-catenin, plakoglobin is thought to signal analogously. In the absence of a Wnt stimulus, GSK-3 constitutively phosphorylates serine and threonine residues in β-catenin's amino terminus. Phosphorylated β-catenin is ubiquitinated and proteolytically degraded through a process mediated by Slimb and possibly APC (see below). With Wnt stimulation, GSK-3 activity is downmodulated. Unphosphorylated β-catenin then accumulates and translocates to the nucleus, to signal as a heterodimer with LEF/TCF proteins. The strength of the transduced signal correlates with β-catenin's cytosolic levels (Polakis, 1999).

Mutant forms of β-catenin have been recently identified in human colorectal cancers and melanomas (Moon et al., 2002; Polakis, 1999). These mutants have altered or deleted GSK phosphorylation sites and consequently are able to resist degradation. They are thought to drive oncogenesis by increasing β-catenin levels, thereby activating Wnt target genes. Deregulation of β-catenin signaling is also an important event in the genesis of a number of other malignancies, including hepatocellular carcinoma, ovarian cancer, endometrial cancer, medulloblastoma pilomatrixomas, and prostate cancer (Morin, 1999). β-Catenin mutations appear to be a crucial step in the progression of a subset of these cancers, suggesting an important role in the control of cellular proliferation or cell death. In contrast, no plakoglobin mutations have been identified in neoplasms.

LEF and TCF. β-Catenin signaling occurs in the nucleus. However, it lacks a classic nuclear localization signal. One explanation contends that β-catenin binds to LEF/TCFs in the cytoplasm and then, guided by the nuclear localization signal of LEF/TCF, translocates to the nucleus. LEF and TCFs belong to a group of proteins, which bind to DNA via an HMG box domain, a region homologous to high-mobility-group I proteins (Novak and Dedhar, 1999). LEF-1 and the TCFs were initially identified as lymphoid-specific transcription factors, on the basis of their affinities for the enhancer regions of T-cell receptor α chain and CD3ε, respectively. It has subsequently become clear that these factors play far more global roles and are expressed in both lymphoid and nonhematopoietic tissues.

To date, four vertebrate homologs have been identified: LEF-1, TCF-1, TCF-3, and TCF-4. LEF-1 and the TCFs bind to and dimerize with β-catenin and plakoglobin (Sharpe et al., 2001). These LEF/TCF-β-catenin/plakoglobin heterodimeric complexes transduce the Wnt signal to the nucleus, where they are thought to alter transcription of Wnt-responsive genes by altering DNA bending and, accordingly, are described as architectural transcription factors.

Regulation of the Canonical Wnt Signaling Pathway

Axin and Conductin. The complexity of the Wnt signaling pathway is magnified by the existence of numerous cellular regulatory mechanisms that act at multiple levels of the pathway. Axin and conductin appear to regulate Wnt signaling downstream of *Dishevelled* and upstream of ß-catenin (Novak and Dedhar, 1999; Seidensticker and Behrens, 2000). The mechanism underlying the inhibitory effect of axin and conductin on Wnt signaling is beginning to be understood. Axin and conductin bind several proteins including APC, GSK-3, and β-catenin. Thus axin and conductin may mediate the formation of multiprotein complexes that spatially approximate GSK-3 and β-catenin. This could promote GSK-3-directed phosphorylation of β-catenin, leading to its degradation. Axin and conductin may also be phosphorylated by GSK-3, but the significance of this is uncertain, because binding affinities for GSK-3 and β-catenin are unaffected by such modification. The significance of APC interaction with axin and conductin is also unclear, because the expression of conductin fragments that cannot bind APC still downregulate β-catenin protein levels in wild-type and mutant APC-expressing cell lines.

APC. Of particular interest is the regulation of the Wnt pathway by the *APC* gene product (Bienz, 2002). APC was originally identified as the tumor suppressor gene that is inactivated in humans with familial adenomatous polyposis (FAP) syndrome (Kinzler et al., 1991). Such individuals develop hundreds to thousands of benign colorectal polyps, of which a small number progress to colorectal adenocarcinomas. FAP patients inherit one defective copy of the *APC* gene and subsequently lose their wild-type copy before polyp development (Powell et al., 1992).

As described above, approximately 85% of all colorectal tumors have mutations in *APC*, and nearly half of the remaining 15% contain mutations in β-catenin that result in the stabilization of β-catenin protein (Srivastava et al., 2001). These genetic alterations do not overlap, which suggests that APC and β-catenin function in the same pathway suppressing colorectal tumor growth. Functional support for this came from studies of *APC*-deficient colon tumor cell lines, which contain elevated levels of β-catenin protein and have high activity from LEF/TCF-responsive reporter genes, both of which are dramatically reduced on introduction of a wild-type *APC* gene (Morin et al., 1997). Although the function of APC and its relationship to Wnt signaling have been genetically examined in several systems, the role of APC in the Wnt signaling pathway still remains unclear. APC can positively or negatively regulate signaling of β-catenin dependent pathways depending on the cellular context. The reasons underlying these differences are still not understood.

Hedgehog. Several other genes not directly involved in Wnt signaling have nonetheless been suggested to have overlapping functions with the canonical Wnt signaling pathway. Many of these genes are components of another signaling pathway, the hedgehog pathway. There are several indications that the two pathways may transduce signals in a similar manner. Furthermore, there is some evidence that the two pathways may affect one another during development.

The role of the hedgehog pathway in tumorigenesis was established with the discovery that inactivating mutations in the *patched* gene, which encodes one component of the hedgehog receptor, are responsible for the inherited cancer predisposition disorder known as Gorlin or nevoid basal cell carcinoma syndrome (NBCCS) (Ruiz i Altaba et al., 2002).

Inactivation of *patched* has been shown to be a major factor in basal cell carcinoma (BCC) formation, with mutations detected in between 12% and 40% of sporadic BCCs. In addition, loss of heterozygosity for markers encompassing the *patched* locus has been detected in more than half of BCCs, suggesting that in many tumors one of the *patched* alleles is inactivated by deletion (Gailani and Bale, 1997). In addition to BCCs, the *patched* gene has also been implicated in the etiology of a range of other tumors including medulloblastoma, squamous cell carcinomas of the esophagus, transitional cell carcinomas of the bladder, and the benign skin lesions trichoepitheliomas (Bale, 2002). Experiments in animals further suggest that hedgehog signaling may play a role in the pathogenesis of human rhabdomyosarcoma.

Several other members of the hedgehog pathway have also been shown to have a role in tumor formation. Most notably, activating mutations in the Smoothened seven transmembrane protein, which is required for hedgehog signaling and is highly homologous to Frizzleds in several regions, have been detected in 10–20% of BCCs (Lam et al., 1999). These discoveries have highlighted the potential role of developmentally important genes in controlling cell growth and differentiation and added to the increasing list of such genes whose aberrant function contributes to the process of tumorigenesis.

Downstream Targets of Wnt Signaling

Although genetic analysis has identified numerous genes required to carry out Wnt-induced functions, in most cases, it is not clear whether Wnt signaling has primary or secondary effects on the transcriptional activation of a target gene. However, with the identification of the LEF/TCFs as mediators of Wnt signaling, it is now possible to examine putative target gene promoters for LEF/TCF binding sites and directly test their responsiveness to Wnt signaling in cell culture or in the animal. Indeed, LEF/TCFs have been shown to promote the expression of several genes that have important roles in the development and progression of tumors, namely: c-myc, cyclin D1, gastrin, cyclooxygenase (COX)-2, matrix metalloproteinase (MMP)-7, urokinase-type plasminogen activator receptor (aPAR), CD44 proteins, and P-glycoprotein (Novak and Dedhar, 1999). These observations may help provide an explanation as to how constitutive activation of the Wnt signaling pathway results in tumorigenesis in humans (Polakis, 2000) and guide in the search for molecules interfering with the aberrant function of this signaling pathway in cancer.

TGF-β SIGNALING AND CANCER

The Transforming Growth Factor Superfamily

Another prominent signal transduction pathway important for a variety of cellular processes involves the multifunctional polypeptide TGF-β.

Like the versatile Wnt signaling pathway, the TGF-β superfamily comprises critical regulators of cell proliferation and differentiation, developmental patterning and morphogenesis, and disease pathogenesis (Massague, 2000). Since the discovery of the first member, TGF-β1, approximately 20 years ago (Anzano et al., 1982), the family has expanded significantly and now includes over 30 members in vertebrates and around a dozen related proteins in invertebrates.

Abnormal activation or inhibition of these TGF-β-regulated processes is implicated in many diseases including pulmonary, renal, hepatic, and neurodegenerative disorders (Blobe et al., 2000). However, most investigations have focused on the role of TGF-β in oncogenesis (Hata et al., 1998). The TGF-β pathway regulates many of the cellular processes exploited by most cancers, including an ability to grow independently of exogenous growth factors and to divide indefinitely, tumor neovascularization, and tissue invasion (Akhurst and Derynck, 2001). Indeed, TGF-β plays an important and intriguing role in cancer progression, functioning first as an antiproliferative factor and later as a tumor promoter (Pasche, 2001). Several components of its signal transduction pathway are functionally altered in cancer. Understanding the components of the TGF-β signaling pathway is essential to appreciating its complex and seemingly contradictory roles in human cancer.

The Transforming Growth Factor Receptor Complex

The TGF-β family of growth factors can be divided into two general branches: the TGF/activin/nodal and BMP/GDF branches (Miyazono et al., 2001). Members of both subclasses have different, although often complementary, effects in both normal and pathological processes. TGF-β is a growth factor that is secreted from cells in a latent, inactive form that is unable to bind its cognate receptor because of noncovalent interactions of two copies of its prosegment with the processed mature TGF-β dimer. To exert its biological activity, mature TGF-β must therefore be released, through proteolytic cleavage of the prosegment (Gentry et al., 1988). On cleavage, TGF-β is able to bind its cell surface receptor, a complex of single-pass transmembrane receptors that contain an intracellular kinase domain that phosphorylates serine and threonine residues. This serine/threonine kinase receptor complex consists of two different transmembrane proteins, named type I and type II receptors (Fig. 3.8). Ligand binding to the type II receptor induces its association with the type I receptor. This event leads to a unidirectional phosphorylation of the type I receptor by the type II receptor, thereby activating its kinase domain (Massague, 1998). The type I receptors propagate the signal by phosphorylating the Smad family of intracellular mediators at their carboxyl-terminal end (see below). Five type II receptors and eight type I receptors (also called ALKs, or activin-like kinases) are currently known. The complexity of TGF ligand-receptor complexes may help explain the diversity of biological phenomena defined by this unique family of signaling molecules.

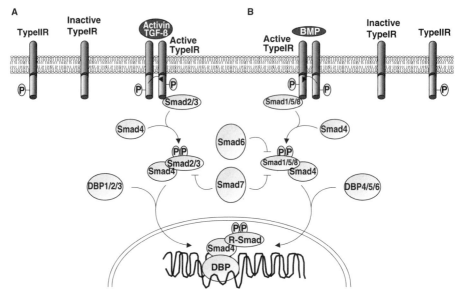

Figure 3.8. *TGF-β Signaling Pathways*. The TGF-β family of growth factors can be divided into two general branches: **A.** the TGF/activin/nodal and **B.** BMP/GDF branches. In general, ligand binding to type II receptor induces its association with the type I receptor, leading to a unidirectional phosphorylation and activation of the type I receptor. Activated type I receptor then phosphorylates an R-Smad, which is released from the receptors and accumulates in the nucleus as heterodimeric complexes with Smad4. In the nucleus, this activated Smad complex associates with other molecules that ultimately determine the transcriptional effect on their target genes.

Smads

Smad proteins are a family of eight mammalian proteins that comprise the basic components of the TGF-β intracellular signaling cascade (Wotton and Massague, 2001). Smads can be divided into three distinct classes. The receptor-regulated, or R-Smads (Smads1, -2, -3, -5, and -8), are directly phosphorylated by activated type I receptors on two conserved serines at the carboxyl-terminus. Phosphorylation of R-Smads induces their release from the receptor complex as well as from SMAD anchor for receptor activation (SARA), a protein that recruits SMADs to the membrane (Tsukazaki et al., 1998). Phosphorylation stimulates R-Smads to heterodimerize (Kawabata et al., 1998), and accumulate in the nucleus as heteromeric complexes with a second class of SMADs, the co-Smads, of which Smad4 is the only known member (Lagna et al., 1996). Receptor-mediated phosphorylation of an R-Smad decreases its affinity for SARA and increases its affinity for Smad4. In the nucleus, the activated Smad complex associates with two classes of proteins: DNA binding cofactors that will help select target genes and coactivators or corepressors that will determine the transcriptional effect on the target genes. The third class of SMADs, the inhibitory Smad6 and Smad7, counteract the effects of the R-Smads and thus antagonize TGF-β signaling (Wotton and Massague, 2001).

Despite this apparent complexity, the biological consequences of these signaling pathways appear to be entirely determined by the type I receptor activated, which, in turn, is defined by the ligand family (TGF-β-like or BMP-like) binding to the receptor (Chen et al., 1998b). In general, the signal emanating from the type I receptor is directed into one of two intracellular pathways, at the level of the plasma membrane. The TGFs, activins, and nodals engage receptors that phosphorylate Smads 2 and -3. The BMPs and related GDFs, as well as AMH/MIS, engage receptors that signal through Smads -1, -5, and -8 (although Activins and BMPs may share some of their type II receptors). Each of the R-Smads can then interact with a wide variety of specific DNA binding proteins to regulate transcriptional responses.

Structure-based investigations have revealed important determinants that mediate the interaction of SMADs with specific receptors, transcriptional partners, and other associating proteins (Kretzschmar and Massague, 1998). The Smad proteins contain two conserved globular domains, known as the Mad homology 1 (MH1) and MH2 domains, coupled by a more divergent linker region. The MH1 and MH2 domains are conserved in all R-Smads and co-Smads. The MH1 domain recognizes the DNA consensus sequence CAGAC, whereas the MH2 domain binds the transcriptional coactivators p300 and CBP, in competition with the corepressors TGIF, Ski, and SnoN (Liu et al., 2001).

Localization of these signaling mediators plays an important role in the TGF-β signaling pathway (Chen et al., 1998b). Most Smads are located in the cytoplasm, in the basal state, for their prompt exposure to activated receptors. However, Smads 2 and -3 (and possibly other Smads) also have intrinsic nuclear import activity within the MH2 domain. Smads are maintained in the cytoplasm, in part, by binding to the protein SARA. This anchor protein serves to tether Smads in the cytoplasm, occlude their nuclear import signal on the MH2 domain, and facilitate Smad presentation to the activated receptors. Smad4 contains a nuclear export signal (NES) that keeps it out of the nucleus in the absence of agonist stimulation.

Smad-Independent TGF Signaling

Smad function is involved in most actions of the TGF family. However, accumulating data suggests that SMAD-independent pathways also exist (Attisano and Wrana, 2002). Several Smad4-defective cell lines from human or mouse retain some level of responsiveness to TGF, suggesting that, if R-Smads are involved in these responses, they can do so without Smad4. MAPK activity (JNK, p38, and ERK) can also be induced by TGF-β, although this activation seems to be dependent on the cell type and experimental conditions. The biochemical link between the TGF-β receptors and MAPK pathways has been hard to determine, although evidence suggests that the MAPKKK family member TAK1 and Rho proteins could be involved in this link (Yamaguchi et al., 1995). At least one TGF response, fibronectin induction, has been partly ascribed to JNK activation (Hocevar et al., 1999). Smads can also interact in vitro with the JNK and p38 substrates c-Jun and ATF2 (Sano et al., 1999; Zhang et al., 1998), respectively, raising the possibility that TGF may simultaneously activate Smad and MAPK pathways that then physically converge on target genes. Nevertheless, the physiological role of MAPKs in TGF signaling still remains unclear.

TGF-β Signaling and Cancer

The complexity of TGF-β signaling is further exacerbated by the seemingly contradictory role that TGF-β plays in cancer (Akhurst and Derynck, 2001). Early in the course of cancer development, TGF-β acts as a tumor suppressor by inhibiting growth or inducing apoptosis. Cancers may therefore select for disruptions in the tumor suppressive effects of TGF-β. Conversely, during later stages of carcinogenesis, TGF-β appears to act as a tumor enhancer. Consequently, cancers often develop selective alterations of the TGF-β signaling cascade that ultimately favor tumor growth (including immunosuppression, angiogenesis, and extracellular matrix secretion). Dissection of the TGF-β signal transduction cascade helps provide an explanation for this multifunctional role of TGF-β in tumorigenesis. Indeed, nearly every component of the TGF-β pathway can be altered in cancer. However, the three main targets appear to be TGF-β receptors, Smads, and their target genes.

Receptor Mutations in Cancer. Mutations in the TGF-β receptor, particularly the type II receptor, have been reported in a wide variety of cancers (Blobe et al., 2000). Gastrointestinal cancers, including esophageal, gastric, colorectal, and hepatocellular cancers, frequently contain mutations in the type II TGF-β receptor (Markowitz et al., 1995; Park et al., 1994). Mutations in type I receptor mutations have also been reported in several cancers (including cervical carcinoma, lymphoma, and breast cancer metastases) but are less common (Anbazhagan et al., 1999; Chen et al., 1998a; Schiemann et al., 1999). Additionally, tumor endothelial cells often overexpress the TGF-β coreceptor endoglin, with a higher incidence in aggressive tumors (Miller et al., 1999), suggestive of an important role for stromal TGF-β signaling in tumor progression.

Mutations of Smads in Cancer. Deletions or mutations in the Smads may also disrupt signaling downstream of the TGF-β receptor. Indeed, the co-Smad Smad4 was originally identified as a tumor suppressor gene lost in pancreatic cancers, Deleted in Pancreatic Cancer 4 (DPC4) (Lagna et al., 1996). Mutations in Smad4 were later identified in colorectal cancers with a higher incidence of mutation in metastatic colon carcinomas (Riggins et al., 1997). Mutations in other Smads are only rarely found in human cancer. Smad2 is occasionally altered in human colorectal tumors (Takagi et al., 1998), whereas mice lacking Smad3 uniformly develop metastatic colon cancer (Zhu et al., 1998).

Alterations of Coactivators and Corepressors in Cancer. The loss of coactivators involved in Smad gene transactivation and increased expression of corepressors are both associated with the loss of TGF-β-mediated antiproliferative effects. Not surprisingly, mutations in both have also been identified in human cancer. CBP and p300, two coactivators of the TGF-β signaling pathway, are altered in several types of cancers, including leukemia and gastric and colon carcinomas (Muraoka et al., 1996). Conversely, the Smad3 inhibitors Evi-1 and Ski are overexpressed in hematologic malignancies and melanomas, respectively (Jolkowska and Witt, 2000; Luo et al., 1999; Xu et al., 2000).

Mutations in Complementary Signaling Pathways. Alterations in the TGF-β signal transduction pathway can also cooperate with other genetic and epigenetic alterations frequently found in cancer to change cellular behavior. Indeed, disruption of the TGF-β signaling pathway, combined with oncogenic *ras*, is associated with the development of aneuploidy and malignant transformation (Glick et al., 1999). Furthermore, in the context of Smad4 loss-of-function mutations, oncogenic *ras* can also cause complementary loss of the antiproliferative response to TGF-β (Calonge and Massague, 1999; Dai et al., 1999).

Loss of TGF-β Target Genes in Cancer. The growth inhibitory effects of TGF-β are frequently mediated through the induction of cyclin-dependent kinase inhibitors (CKIs) p15, p21 and p27 or downregulation of the cyclin-dependent kinases (CDKs) CDK4 and CDK2, which are critical for the regulated progression through the cell cycle. The retinoblastoma protein, Rb, is therefore maintained in the hypophosphorylated state, which prevents the transcription factor E2F from inducing expression of genes required for G_1-S phase transition (see **CELL CYCLE IN CANCER**, below).

CELL CYCLE IN CANCER

Introduction to the Cell Cycle

Despite the complexity of the myriad of intracellular signaling cascades, the effect of growth-promoting signals is ultimately appreciated by their impact on cell proliferation. Cell proliferation progresses in a cyclical fashion through a series of well-recognized steps (Fig. 3.9). After dividing, cells first enter a gap phase (G_1) during which cells grow in preparation for DNA replication. This is followed by the synthesis (S) phase when DNA, packaged into chromosomes, is replicated. Subsequent to this is a second gap phase (G_2), after which the cells enter mitosis (M),

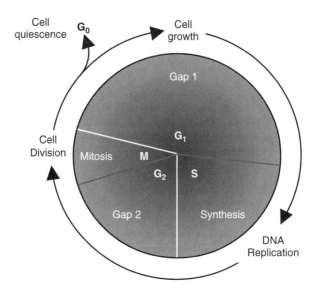

Figure 3.9. *The cell cycle.* Cell proliferation progresses in a cyclical fashion through a series of well-recognized steps, known as the cell cycle. After dividing, cells first enter a gap phase (G_1) during which cells grow in preparation for DNA replication. This is followed by the synthesis (S) phase when the DNA is replicated. Subsequent to this is a second gap phase (G_2), after which the cells enter mitosis (M), when the cell prepares for and completes cell division. The G_1/S and G_2/M transitions comprise the major restriction points (in red) of cell cycle progression. Cells that are not actively proliferating exit the cycle into quiescence (G_0).

when the cell prepares for and completes cell division. Cells that are not actively proliferating exit the cycle into quiescence (G_0). The G_1 and G_2 phases serve as checkpoints for the cell to make sure that it is ready to proceed in the cell cycle. If it is not, the cell will use this time to make proper adjustments that can include cell growth, correction or completion of DNA synthesis, and duplication of intracellular components.

The assembly and disassembly of a series of protein kinase complexes, which phosphorylate their substrates on specific serine and/or threonine residues, drive progression through the cell cycle. These enzyme complexes consist of a catalytic subunit, CDK; a regulatory cyclin subunit; and proliferating cell nuclear antigen (PCNA), a component of the DNA polymerase δ enzyme complex (Miller and Cross, 2001; Obaya and Sedivy, 2002). An abnormality in any component of this machinery could lead to deregulation of proper cell cycle progression, to abnormal cell proliferation, and ultimately to the occurrence of disease.

In cancer, neoplastic cells possess intrinsic derangements in the progression of the normal cell cycle (Ho and Dowdy, 2002). In contrast to normal cells, tumor cells are unable to stop at one or both of the predetermined checkpoints of the cell cycle (G_1/S and G_2/M). Absence of these pauses in the cell cycle, necessary to verify the integrity of the genome

before cells advance to the next phase, results in deregulated cell proliferation. With the discovery of the function of oncogenes and tumor suppressor genes, it quickly became evident that tumor cells frequently
acquire either mutations or deletions in those genes important to cell
cycle regulation. Indeed, almost all tumors have an abnormality in some
component of the cell cycle pathway. As a consequence of these cell cycle
alterations, tumor cells are able to transverse cell cycle checkpoints in
a way that ignores normal growth or regulatory factors, resulting in
unchecked cellular proliferation.

Regulation of CDK Activity

CDK activity is regulated by a number of mechanisms, reflecting the
importance of maintaining precise control of cell proliferation in
response to a continuously changing environment. An active complex
requires the association of the CDK with its cognate cyclin (Miller and
Cross, 2001). Complex formation may be regulated by the availability of
the cyclin molecule within the cell. For example, transcription and translation of cyclin genes are temporally regulated, such that cyclin proteins
are synthesized only at specific times during the cell cycle. Once
expressed, cyclins are subject to ubiquitin-mediated regulated proteolysis. Additionally, temporal regulation of the subcellular compartmentalization of CDKs regulates the availability of the CDKs to complex with
cyclins.

CDK activity is also regulated by the phosphorylation status of the
kinase itself. CDKs are subject to both activating and inhibitory phosphorylations on conserved threonine and tyrosine residues, which must
either be added [by CDK-activating kinase (CAK) and CDK7-cyclin H-
mat1] or removed (by members of the cdc25 phosphatase family),
respectively, to enable full activation of the kinase.

CDK activity is subject to further regulation by association with
members of two classes of proteins—the Cip/Kip family and the Ink4
family of CKIs (Lee and Yang, 2001) (see below). The Cip/Kip family
consists of p21 (also known as WAF1), p27 (Kip1), and p57 (Kip2). These
molecules are capable of inhibiting the function of G_1/S-phase (CDK4/6-
cyclin D and CDK2-cyclin E) and S-phase (CDK2-cyclin A) complexes,
thereby blocking progression through G_1 into S phase (Nakayama, 1998).
The Ink4 CKIs show considerable specificity for CDK4 and CDK6,
inhibiting their kinase activity, and hence negatively influence progression through G_1 (Roussel, 1999) (Fig. 3.10).

Function of CDK Complexes

In response to mitogenic signaling, primarily via the Ras-Raf-MEK-
ERK cascade, transcription of D-type cyclins is induced and assembly of
CDK4/6-cyclin D complexes is facilitated in an ERK-dependent manner.
Further mitogenic signaling through phosphatidylinositol 3-OH kinase
(PI-3K) to protein kinase B (PKB/AKT) maintains cyclin D in the

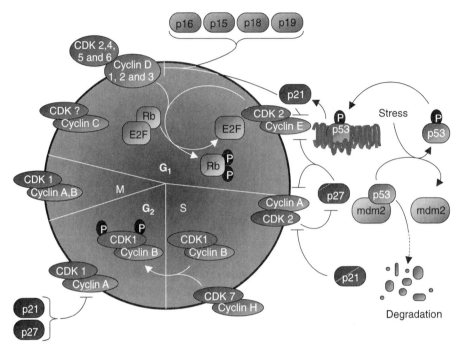

Figure 3.10. *Regulation of the cell cycle.* Progression through the cell cycle is driven by the CDK/cyclin complexes, whose activity, in turn, is regulated by the Cip/Kip family [p21 (also known as WAF1), p27 (Kip1), and p57 (Kip2)] and the Ink4 family (p16, p15, p18, and p19) of CDK Inhibitors (CKIs). The Cip/Kip family inhibits the function of G_1/S (CDK4/6-cyclin D and CDK2-cyclin E) and S phase (CDK2-cyclin A) complexes, thereby blocking progression through G_1 into S phase. The Ink4 CKIs inhibit the kinase activity of CDK4 and CDK6, thereby negatively influencing progression through G_1.

nucleus by preventing its phosphorylation by GSK-3β and thus its subsequent export and proteolysis.

The major substrate of G_1 CDK-cyclin D complexes is the product of the retinoblastoma gene, Rb (see below), which becomes phosphorylated on serine and threonine residues during mid-G_1 (Hickman et al., 2002). Rb, together with its related proteins p107 and p130, plays a central role in regulating cell cycle progression (Tonini et al., 2002) (Fig. 3.11). In its unphosphorylated state, Rb is able to bind to and sequester members of the E2F family of transcription factors. However, phosphorylation by CDKs results in a conformational change in Rb, which probably opens it up to subsequent phosphorylation in late G_1 by CDK2-cyclin E. The net result of this is the release of E2F, which then coordinates transcription of genes whose products are required for DNA synthesis (including cyclins E and A) and, thus, progression through S phase. This results in increased levels and availability of A and E cyclins, which are necessary for activation of CDK2. CDK2 activation is further enhanced by the sequestration of Cip/Kip CKIs by cyclin D-CDK complexes, thereby

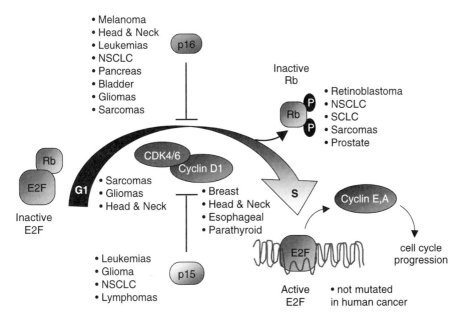

Figure 3.11. *Mutations in the Rb/E2F pathway.* In its unphosphorylated state, Rb is able to bind to and sequester members of the E2F family of transcription factors. However, phosphorylation of Rb by CDKs 4 and 6 (complexed to cyclin D) results in a conformational change, which inhibits Rb from binding to and sequestering E2F, releasing E2F, which then coordinates transcription of genes whose products are required for DNA synthesis (including cyclins E and A) and, thus, progression through S phase. CDK4/6/Cyclin D, in turn, is inhibited by the CDK Inhibitor (CKI), p16 (and, to a lesser extent, p15). Human cancers in which mutations are commonly found in signaling molecules involved in this pathway are listed next to the respective molecule.

relieving their inhibitory effects on CDK2. Indeed, Cip/Kip CKIs are also thought to act as assembly factors for cyclin D-CDK complexes, because they bind to both the kinase and cyclin subunits and enhance their association.

Further mechanisms that enable the cell to progress into and through S phase include phosphorylation of p27/Kip1 by CDK2, and its subsequent targeting for degradation by the ubiquitin-proteasome machinery, and maintenance of Rb in a hyperphosphorylated state. These mechanisms relieve the dependence of the cell on mitogens. Increased proteasome-mediated degradation of p27 has been reported in aggressive colorectal carcinomas and may be another mechanism to abrogate growth repression in a subset of cancers (Loda et al., 1997).

On cessation of mitogenic stimulation, D cyclins are rapidly degraded, releasing Cip/Kip proteins, which are then available to bind and inhibit CDK2 complexes and mediate cell cycle arrest in G_1. In the case of p27/Kip1, this occurs as a result of the CKI disrupting the ATP binding site of the CDK subunit. In contrast, complexes containing equimolar amounts of cyclin D, CDK4/6, and p21/p27 remain catalytically active.

Cell Cycle Mediators

Cip/Kip Family. As mentioned above, p21/Cip/Kip CKIs act as assembly factors for G_1 CDK-cyclin D complexes, while also inhibiting CDK2-cyclin E/A. The fact that the expression of p21 is regulated by the tumor suppressor protein p53 (see below) underscores the importance of p21 as a functional brake for unregulated growth (Dotto, 2000). It is now clear that p21 is responsible to a large extent for the G_1 arrest that occurs in response to DNA damaging agents, including ultraviolet or γ-irradiation, and commonly used cancer chemotherapeutics such as adri-amycin, 5-fluorouracil, bleomycin, and cisplatin (Dotto, 2000). Thus, as p53 is inactivated at high frequency in a broad spectrum of human tumors, p21-dependent G_1 cell cycle arrest in response to DNA damage is often abrogated in cancer (Lee and Yang, 2001). Consequently, cancer cells continue to proliferate and duplicate their DNA content without repairing the damage. Furthermore, overexpression of molecules that induce transcription of S-phase genes, such as E2F or B-myb, may enable tumor cells to overcome p21-mediated growth arrest. p21 appears to play an important role in the regulation of programmed cell death. Indeed, p21 inhibits the kinases JNK and p38, both of which have been implicated in cell death in response to stress signals.

Although mutation of p53 is common in human cancer, some tumors retain one or more wild-type alleles. Because p21 is a critical mediator of p53-dependent growth arrest, it might therefore be predicted that tumors expressing a functional p53 would subvert this mechanism by expressing aberrant forms of p21. However, it appears that intragenic mutation of p21 is relatively rare in human cancers, although polymorphisms have been identified, some of which are associated with increased risk of neoplasia. This may indicate a fundamental requirement for p21 in cellular growth control and development.

Because p21 plays a central role in regulating cell cycle progression, it is perhaps not surprising that factors other than wild-type p53 are capable of inducing p21 expression. Growth factors, including PDGF and FGF, can induce p21 (Kivinen and Laiho, 1999). This elevation of p21 expression in response to mitogenic agents is entirely consistent with its role as an assembly factor for cyclin D-CDK4/6 complexes. A number of other effectors of p21 expression have been identified. These include TGF-α, TGFβ and activin A, Smad proteins (which signal downstream of TGF-β/activin), IL-1, TNF-α, interferon γ (IFN-γ), EGF, and steroids such as progesterone. Interestingly, members of the *jun* family of transcription factors (c-Jun, junB, junD, ATF-2) cooperate with the transcription factor Sp1 to transactivate transcription from the p21 promoter. This likely provides an important mechanism for integrating signals from a number of biochemical pathways to regulate cell cycle progression and/or cell death.

A second member of the p21-related proteins is p27/Kip1 (Olashaw and Pledger, 2002). The gene encoding p27 shares 44% identity with p21 in the amino-terminal region of the protein and functions in a manner similar to p21. Deregulation of p27 activity is likely to play a role in malignant disease. Several human tumor types contain reduced levels of p27, including breast and colon cancers, and this is further indicative of more aggressive disease. Diminished levels of p27 are likely to contribute to the malignant phenotype of other human tumors as well. The BCR-ABL oncoprotein, which is expressed in many leukemic cell lines as a result of chromosomal translocation, results in a decrease in p27 expression through a PI3-K/AKT-dependent mechanism (Gesbert et al., 2000). And increased activity of AKT appears to enhance prostate cancer progression by blocking p27 expression (Graff et al., 2000). This is likely to be a direct effect of AKT to inhibit the activity of AFX/Forkhead transcription factors, which transcriptionally transactivates p27 (Nakamura et al., 2000). In addition, it is probable that neoplasms in which function of the PTEN tumor suppressor protein has been lost will also contain low levels of p27, because PTEN, a phosphatidylinositol 3,4,5-trisphosphate (PIP_3) phosphatase, negatively regulates AKT activity. Furthermore, oncogenic signaling from HER-2/neu results in decreased cellular levels of p27 by promoting its export into the cytoplasm and thus enhancing ubiquitin-mediated degradation. Similarly, overexpression of Myc, a frequent alteration found in cancer, may contribute to increased cellular proliferation through induction of cyclins D1 and D2, leading to sequestration of p27 and release of the block on cyclin A/E-CDK2 activities. Downregulation of p27 may prove to be a critical step in the progression of many human cancers.

A third member of the Cip/Kip family of CKIs is p57/Kip2 (Lee and Yang, 2001). The gene encoding human p57, *KIP2*, is located on chromosome 11p15.5, which may be altered in a number of tumor types, and in a familial syndrome—Beckwith–Wiedemann syndrome (BWS)—which is characterized by a number of developmental abnormalities related to increased growth and which carries an elevated risk of childhood tumors. *KIP2* is paternally imprinted, and this is consistent with the suggestion that BWS is carried maternally, thus suggesting that p57^{KIP2} may play a central role in this syndrome. However, the precise contribution of p57^{KIP2} to the development of human cancer remains unclear.

Ink4 Family. The Ink4 family of CKIs consists of four members: p16/Ink4A, p15/Ink4B, p18/Ink4C, and p19/Ink4D (Roussel, 1999). These molecules show specificity for G_1 CDKs, namely CDK4 and CDK6. Binding of Ink4 CKIs to the CDK results in inhibition of kinase activity, in contrast to the effects of Cip/Kip proteins. INK4 proteins probably compete with Cip/Kip CKIs for CDK4 subunits and block binding of Cip/Kips and cyclins. This will have the effect of making more Cip/Kip proteins available to bind cyclin E-CDK2, thus blocking exit from G_1 into S phase.

The gene encoding p16 is located on the short arm of chromosome 9 in humans, at 9p21 (Serrano et al., 1993). This site is an important tumor suppressor locus in a wide range of human cancers. As a negative regulator of cell growth, p16 acts as a tumor suppressor. Homozygous deletion of the p16 locus has been identified in cell lines from a wide range of tumor cell lines, including lung, breast, bone, epidermis, ovary, lymphoreticular malignancies, and melanoma (Kamb et al., 1994). Furthermore, in samples in which one allele was retained, the remaining allele frequently contained missense or nonsense mutations. p16 expression may also be silenced in tumors, primarily by methylation of its promoter (Rocco and Sidransky, 2001). Thus inactivation of p16 by deletion, mutation, or transcriptional silencing is likely to play a major role in tumorigenesis by deregulating CDK activity during G_1 (Okamoto et al., 1994). Loss of p16 is functionally equivalent to inactivation of Rb as, in such cases, Rb remains in a hyperphosphorylated (nonsuppressive) state. Indeed, analysis of human lung cancers reveals an inverse correlation between these two molecules, such that 89% of small cell lung tumors retained expression of p16, but either lost Rb or expressed an aberrant form, whereas the remainder lacking p16 retained a normal Rb protein. Conversely, in non-small cell lung cancers where Rb is generally normal, p16 was found to be absent or otherwise inactivated, except in the minority of lesions where Rb was deleted or mutated (Otterson et al., 1994).

p16 may also play a role in the development of some inherited cancers. For example, the genetic locus for familial melanoma (MLM) maps to chromosome 9p21, and it is likely that p16 is a candidate melanoma susceptibility gene, as germline alterations have been found in a high proportion of cases (Ranade et al., 1995). Pancreatic adenocarcinomas, which also show 9p losses, exhibit p16 alterations at high frequency. This type of malignancy is also very common in familial melanoma cohorts.

CDK4, cyclin D, p16, and Rb are components of the same growth-regulatory pathway that functions to control progression through G_1 into S phase. Although it was initially thought that p16-mediated growth arrest was entirely dependent on the presence of a functional Rb protein, Rb alone is insufficient for p16-mediated growth arrest, suggesting that, although these molecules may share some functions, nonredundant activities are required for G_1 arrest by p16.

The INK4A locus, in addition to encoding the p16 protein, is also capable of utilizing an upstream exon 1b together with an alternative reading frame within exon 2, thereby directing synthesis of the protein p19/ARF (alternative reading frame) in mouse and p14/ARF in humans (Sharpless and DePinho, 1999). Like p16, ARF is also a tumor suppressor. However, its mode of action is distinct from that of p16 or any of the other INK4 proteins. Whereas p16 acts on the pRb pathway by inhibiting CDK4/6, ARF acts on the p53 pathway by interacting with MDM2, thereby blocking degradation of p53 and resulting in p53 stabilization and activation. Furthermore, binding of ARF to MDM2 results in sequestration of MDM2 in the nucleolus. This prevents shuttling of MDM2 between the nucleus and cytoplasm (to where it must export p53 for degradation), thereby leaving p53 free within the nucleus to transactivate transcription of responsive genes. Interestingly, ARF is required for p53 activation by proliferative signals such as Myc but not for activation in response to ultraviolet or ionizing radiation. In addition, ARF is under the transcriptional control of E2F, thus providing an additional mechanism whereby cells that have deregulated G_1 progression (through loss of p16, pRb, or elevation of CDK activity, for instance), and an elevated pool of free E2F, are still subject to G_1 arrest. Additionally, ARF can target certain E2F species for degradation. Notably, germline deletion of *CDKN2A* exon 1β sequences in a melanoma-astrocytoma syndrome family has been reported, which results in loss of ARF function but not p16 or p15 and implicates ARF inactivation as a predisposing factor for this disease (Randerson-Moor et al., 2001). Taken together, it is clear that deletion of all or part of the *INK4A/ARF* locus has considerable impact on two of the major growth-suppressive pathways that are inactivated in human cancers.

The other members of the Ink4 family of CDK inhibitors are $p15^{INK4b}$, $p18^{INK4c}$, and $p19^{INK4d}$. p15 is encoded by a gene also situated on chromosome 9p21, proximal to that which encodes p16, and therefore represents an additional candidate for a tumor suppressor at this locus. p15 function is lost in some human tumors and cell lines, including non-small cell lung cancer, leukemias and lymphomas, and glioblastomas (Tsihlias et al., 1999). Although a few samples harbored intragenic mutations within the p15 coding region, homozygous deletion was found to be more common. Thus loss of 9p may represent an efficient method through which cells deregulate their growth control mechanisms by deleting both p16 and p15 genes in one event.

$p18^{INK4c}$ maps to human chromosome 1p32, a locus that shows structural abnormalities in a range of different malignancies. p18 associates strongly with cyclin D-CDK6 complexes and inhibits phosphorylation of pRb. The fourth member of the Ink4 family is $p19^{INK4d}$. In humans, the gene encoding $p19^{INK4d}$ is located on chromosome 19. Thus far, mutations in $p19^{INK4d}$ have not been identified in human cancer.

The Small GTPase Family. The transforming potential of the small GTPase family is, in part, defined by their effect as regulators of the cell cycle. An important function of the Ras/Raf/MEK/ERK pathway is the transcriptional upregulation of cyclin D1, a critical element in cell cycle progression (Cheng et al., 1998; Lavoie et al., 1996). Similarly, Rac can enhance the expression of cyclin D1 by stimulating the activity of its promoter (Westwick et al., 1997). As cyclin D1 levels are upregulated inside the cell, this protein is able to bind its CDK partners, CDK4 and CDK6, forming active CyclinD1-CDK4 and CyclinD1-CDK6 complexes, which can phosphorylate and inactivate Rb. Indeed, cyclin D1 can be considered an oncogene on its own. For example, a survey of human breast cancer etiology in patients has revealed the cyclin D1 gene to be amplified in 20% of the cases examined and cyclin D1 protein levels also appeared to be elevated in more than 50% of mammary carcinomas (Sutherland and Musgrove, 2002). Furthermore, cyclin D1-deficient mice are able to resist cancer development induced by the Ras and Neu onco-genes (Yu et al., 2001).

Rho GTPases may affect cyclin D1 levels in different manners. It is possible that Rac could contribute to the full oncogenic upregulation of cyclin D1 transcription by Ras. However, Rac can also activate the cyclin D1 gene independently of Ras GTPase. Surprisingly, activation of inte-grin signaling has been also shown to regulate translationally the levels of cyclin D1 protein in a Rac-dependent manner (Mettouchi et al., 2001). PDGF-induced c-*myc* expression has been also demonstrated to rely on Rac activity, in a Ras-independent fashion (Chiariello et al., 2001). In addition, Rho activity may ensure sustained cyclin D1 expression throughout G_1 in an ERK-dependent manner, while Rac1 and Cdc42 induce cyclin D1 early in G_1 phase of the cell cycle. Rho also regulates the function of the cyclin-CDK complex inhibitors, p21Cip1 and p27Kip1. It has been demonstrated that activation of Rho both sup-presses p21Cip1 induction and triggers p27Kip1 degradation (Pruitt and Der, 2001). These Rho-mediated events lead to a stabilization of cyclin/CDK activity, which consequently promotes progression through the cell cycle.

One of the Rac effectors molecules involved in the signaling to the cell cycle machinery to induce cell transformation is the Ste20-like p21 PAK kinase (Manser et al., 1994). It has been shown that PAK activity correlates with cyclin D1 promoter induction. A catalytically inactive form of PAK can effectively suppress Ras-induced transformation in some cell lines. Similarly, the function of Rac3 in the proliferation of breast tumor cells appears to correlate with increased PAK activity (Mira et al., 2000). Nonetheless, other data indicate that Rac may act on the cell cycle through PAK-dependent or PAK-independent pathways.

Another protein involved in the signaling of the family of Rho GTPases is Merlin, which is the product of the tumor suppressor gene NF2 (Shaw et al., 2001). Mutations in this gene predispose humans and mice to the development of neurofibromatosis type II (NF2). Functional state of Merlin seems to be controlled by Rac GTPase (Sherman and Gutmann, 2001). Expression of constitutively active Rac, Dbl, and Tiam induces the phosphorylation of Merlin at the Ser-518 residue (Xiao et al., 2002). This modification inhibits the "closed" con-formation of the molecule that relies on the head-to-tail interaction between amino- and carboxyl-terminal residues within the Merlin protein and promotes its "open" conformation. Because it is the closed state that exhibits a negative growth regulatory function in vitro and in vivo, the release of its inhibitory functions by activated Rac may con-tribute to Rac tumorigenic potential. Indeed, Rac-induced transforma-tion is drastically reduced by Merlin expression. NF2-deficient fibroblasts also display features that strongly resemble those elicited by Rac, such as membrane ruffling and an increase in the number of intracellular vesi-cles (Shaw et al., 2001). Thus Merlin may link events at the plasma mem-brane with the cell cycle machinery and in particular cell cycle-promoting cyclin D1 levels.

TUMOR SUPPRESSOR GENES

Although an extensive review of the current knowledge of the normal and aberrant function of tumor suppressor proteins is beyond the scope of this chapter, we focus our attention on a few key examples to illustrate how the functional activity of this diverse group of proteins can be regulated by growth-promoting signaling pathways or control signaling events leading to cell proliferation or death.

The Retinoblastoma (Rb) Pathway

Retinoblastoma (Rb) is a well-characterized tumor suppressor that plays a critical role in controlling cell growth through its interaction with the transcription factor E2F (Hickman et al., 2002) (Fig. 3.11). The Rb/E2F pathway is critical in regulating the initiation of DNA replication and S phase, and this key pathway is disrupted in virtually all human cancers. Analysis of human tumors has revealed a wide spectrum of mutations that alter the Rb/E2F pathway. However, mutations within the Rb/E2F pathway are usually not duplicative: Tumors carrying an Rb mutation do not generally exhibit a mutation in a second gene within this pathway. Common mutations disrupting this pathway are the Rb loss of function mutations (Friend et al., 1986). Although originally observed in inherited retinoblastoma, mutations of the Rb gene of sporadic (somatic) origin occur at considerably higher frequency than the inherited eye tumors. These mutations have been identified in a wide spectrum of tumors including osteosarcomas, small cell lung carcinomas, and breast carcinomas.

p16INK4a cyclin kinase inhibitor, which controls CDK4-cyclin D kinase activity, is also a frequent mutation disrupting the regulation of the Rb/E2F pathway. In the absence of p16, CDK4-cyclin D activity is elevated, leading to Rb phosphorylation and the resultant accumulation of E2F. Thus loss of function mutation of p16 is functionally equivalent to Rb loss. Inherited mutation of p16INK4a and subsequent loss of the wild-type allele in tumors is observed in melanoma, but, like Rb, loss of p16 function is much more prevalent in sporadic cancers of a variety of types.

Deregulation of the Rb/E2F pathway can also result from dysregulated and increased expression of D-type cyclins or CDK4. Both amplification and translocation of the D1 cyclin gene has been observed in a variety of human cancers. For example, amplification of the CDK4 gene has been seen in sarcomas and gliomas.

Interestingly, genetic alterations involving the Rb/E2F pathway do not involve mutations of the E2F genes themselves, the activity of which is the ultimate event in the activation of the pathway. The E2F genes encode proteins with varied functions. E2F1 functions as a signal for apoptosis, E2F2 and E2F3 appear to play a positive role in cell proliferation, and E2F4 and E2F5 function together with Rb family members to repress transcription in quiescent cells. Although loss of E2F1 in the mouse leads to tumor formation, no examples of E2F mutations have proven to be causative of the cancer phenotype, suggesting that the E2F proteins must serve a function that is essential for cell survival (Trimarchi and Lees, 2002).

p53

p53 is a tumor-suppressor gene that encodes for a tetrameric protein that is a sensor of multiple forms of genotoxic, oncogenic, and nongenotoxic stress. *p53* is involved in several diverse and critical cellular processes and is one of the most extensively studied genes in cancer research (Vousden and Lu, 2002). Mutations of this single gene during tumorigenesis can have extensive consequences including inappropriate cell growth, increased cell survival, and genetic instability, and as such are the focal point of selection pressures in tissues exposed to carcinogens or to oncogenic changes. Thus the clonal expansion of cells with mutations in *p53* may be seen as the result of a selection process intrinsic to the natural history of cancer. Indeed, *p53* gene mutations occur in more than half of all malignancies from a wide range of human tumors and are often associated with poor prognosis and treatment failure. Consequently, a full appreciation of the cellular processes involved in tumorigenesis will require a thorough understanding of the signaling pathways that control p53 function (Sharpless and DePinho, 2002).

Regulation of p53 activity is critical to enable both normal cell division and the suppression of tumorigenesis. As such, there are many mechanisms by which p53 is regulated. Initial attention focused on the control of transcription and translation of *p53*. However, it now appears that the *p53* function is controlled by a number of molecules governing its protein levels, localization, and transactivating activity. Indeed, p53 is subject to several posttranslational modifications and interacts with numerous other cell proteins. The contribution of these modifications and protein-protein interactions to the regulation of p53 stability, location, and activity has only recently come to light and is still an active area of investigation.

p53 Protein Stability. The levels of p53 protein are usually quite low, as its rapid degradation prevents the accumulation of p53 in normal, proliferating cells. The stability of p53 is highly regulated, primarily by its ability to bind the cellular protooncogene MDM2, which functions as an ubiquitin ligase for p53, mediating its ubiquitination and targeting this tumor suppressor for proteasome degradation (Michael and Oren, 2002) (Fig. 3.12). The expression of MDM2, in turn, is regulated by p53, producing a feedback loop that tightly regulates p53 function. The contribution of MDM2 to p53 regulation is underscored by the finding that the *mdm2* gene is often amplified in tumors that retain wild-type *p53*, suggesting that enhanced expression of MDM2 prevents the activation of *p53* as a tumor suppressor and can substitute for mutational loss of *p53* during tumorigenesis (Momand et al., 1998). Degradation of p53 by

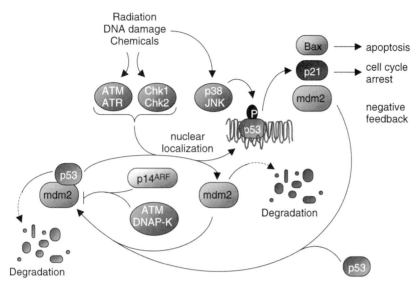

Figure 3.12. *Regulation of the tumor suppressor p53.* The stability of p53 is primarily regulated by its ability to bind the ubiquitin ligase, MDM2. The expression of MDM2, in turn, is regulated by p53, producing a feedback loop tightly regulating p53 function. The kinases ATM, ATR, Chk1, Chk2, and JNK phosphorylate p53, thereby inhibiting MDM2-mediated degradation of p53. p53 transactivation function is also directly modulated, independent of protein stability, by regulation of the stress-activated kinase p38.

MDM2 depends on the direct interaction of a small domain in the amino terminus of p53 with a deep hydrophobic cleft within the amino terminus of MDM2. This interaction can be impaired by phosphorylation of p53 within its MDM2 binding region, resulting in an impaired ability of MDM2 to target p53 for degradation. Phosphorylation of these residues can occur in response to stress such as DNA damage. Several stress-induced signaling pathways leading to the inhibition of MDM2-mediated degradation of p53 have been identified. Kinases activated by genotoxic damage, including ATM and ATR, Chk1 and Chk2, have been shown to phosphorylate and thus stabilize p53 (Abraham, 2001). Other stress-induced kinases have similarly been shown to phosphorylate p53, thereby promoting its stability. JNK, for example, has been shown to phosphorylate p53 at threonine 81 in response to UV irradiation, whereas the Polo-like kinase-3 has been implicated in the phosphorylation of p53 at serine 20 in response to reactive oxygen species (ROS), suggesting that different stress signals give rise to unique patterns of p53 phosphorylation. Furthermore, the response to p53 phosphorylation on a particular residue may depend on the specific stress signal initiating the phosphorylation event, underscoring the complexity of p53 phosphorylation and stabilization. Moreover, MDM2 can itself be phosphorylated by stress-induced kinases (e.g., ATM and DNA-PK), resulting in an impaired ability of MDM2 to target p53 for degradation.

Indeed, multiple phosphorylation sites in MDM2 are likely to play a role in the regulation of MDM2 and p53 function. Recently, phosphorylation of MDM2 by cyclin A-CDK2 at threonine 216 has been shown to inhibit its ability to bind p53. However, the CDK-dependent induction of MDM2 expression may counteract the ability of CDKs to promote p53 stability.

Independent of MDM2, additional mechanisms also exist to control p53 degradation, reflecting the importance of maintaining precise control of this critical tumor suppressor. For example, JNK has been shown to target p53 for degradation independent of MDM2. However, the contribution of these pathways to the full activation of a p53 response remains poorly understood. Moreover, in some situations phosphorylation has been shown to be unnecessary for p53 stabilization. Indeed, many other mechanisms that may promote p53 stabilization have been described, including downregulation of MDM2 expression, downregulation of the levels of free ubiquitin, or competition between MDM2 and transcriptional coactivators that bind to the same region of p53. The cellular protooncogene c-Abl forms a complex with p53 and is believed to be necessary for efficient stabilization of p53 in response to DNA damage. However, rather than inhibiting p53 from interacting with MDM2, c-Abl appears to prevent p53 ubiquitination by MDM2 (Sionov et al., 2001). Similarly, an interaction between p53 and the transcription factor hypoxia-inducible factor 1α (HIF-1α)—a critical player in tumor angiogenesis—stabilizes and activates p53 in response to hypoxia. Additionally, ARF plays an important role in p53 stabilization in response to many oncogenes that contribute to abnormal proliferation. Although loss of ARF does not prevent stabilization of p53 in response to all stress signals, loss of ARF can, to some extent, substitute for loss of p53. In fact, mice lacking *Arf* are highly susceptible to tumor formation (Weber et al., 2000). Although ARF has activities independent of p53, activation of p53 by ARF may protect an organism from cancer development by selectively eliminating cells that develop proliferative abnormalities.

p53 Nuclear Localization. Regulation of p53 nuclear localization also plays an important role in controlling p53 function. Localization of p53 to the nucleus enables p53 to transactivate p53-responsive genes, one of its key functions as a tumor suppressor. p53 contains two nuclear export sequences, one in the carboxyl terminal oligomerization domain and the other in the amino terminal MDM2-binding region. Although not absolutely dependent on its presence, the ability of p53 to be exported from the nucleus is certainly enhanced by MDM2 ubiquitin ligase activity (Michael and Oren, 2002). Stress-induced phosphorylation of the amino terminal nuclear export sequence (which is within the MDM2 binding domain) inhibits nuclear export through this amino terminal signal.

The p53 carboxyl terminal nuclear export sequence is not normally accessible to the export pathway when p53 is tetramerized. The p53-stabilizing protein ZBP-89 inhibits nuclear export by binding the central and carboxyl terminal domains of p53, thereby blocking ubiquitination and subsequent unmasking of the carboxyl terminal nuclear export sequence (Bai and Merchant, 2001). Together, these observations suggest that amino terminal phosphorylation of p53 may augment nuclear retention of p53 both directly (by inhibiting the amino terminal nuclear export sequence) and indirectly (by inhibiting MDM2 binding and subsequent ubiquitination and activation of the carboxyl terminal nuclear export sequence). The complexity of p53 subcellular localization is further regulated by poorly understood regulatory mechanisms that control the export of p53 back out to the cytoplasm.

p53 Transactivation. p53 transactivation function is also directly modulated, independent of nuclear localization, by regulation of the ability of p53 to bind DNA or interact with other components of the transcriptional machinery. There are numerous covalent and noncovalent posttranslational modifications within the carboxyl terminus of p53 that could modulate DNA binding activity. Particularly interesting are modifications such as phosphorylation, sumoylation, or glycosylation. However, at the moment there are no firm conclusions to be made about how, or even if, p53 activity is regulated through modification of the carboxyl terminal region.

Better understood are modifications in the amino terminus of p53, which contains the p53 transcriptional activation domain. Phosphorylation within this domain can stimulate p53 transcriptional activity by stabilizing the interaction between p53 and DNA. Phosphorylation of either serine 15 or serine 33 leads to enhanced binding between p53 and histone acetyl transferases such as p300/CBP or pCAF. The interactions between p53 and CBP/p300 have been associated with subsequent acetylation at the carboxyl terminus of p53.

p38 MAPK(see above) plays a key role in the activation of p53 by genotoxic stress (Sanchez-Prieto et al., 2000). However, p38 does not affect the accumulation of p53 in response to DNA damage or its nuclear localization. In contrast, p38 directly associates with p53 and phosphorylates serine 33 and serine 46, located within its amino-terminal transactivation domain, thereby stimulating its functional activity. Mutation of these sites decreases p53-mediated and UV-induced apoptosis. Inhibition of p38 activation after UV irradiation or treatment with chemotherapeutic agents decreases phosphorylation of serine 33, serine 37, and serine 15 and also markedly reduces UV-induced apoptosis in a p53-dependent manner. These results suggest that p38 activation plays a prominent role in an integrated regulation of amino-terminal phosphorylation events that regulate p53-mediated apoptosis after UV radiation.

Adenomatous Polyposis Coli (APC)

As described above, FAP is an autosomal dominant inherited disease characterized by the presence of adenomatous polyps in the colon and rectum, with inevitable development of colorectal cancer if left untreated. FAP is associated with mutations in the APC gene, an important tumor suppressor implicated in the majority of sporadic colorectal cancers (Bienz, 2002). Loss of APC function initiates a chain of molecular and histological changes—known as the "adenoma-carcinoma" sequence—each accompanied by a genetic alteration in a specific oncogene or tumor suppressor gene. Inactivation of APC promotes the two essential requirements for the development of cancer, namely, a selective advantage to allow for the initial clonal expansion and genetic instability to allow for additional hits at other genes responsible for tumor progression. The signaling events defining the role of APC in colon cancer are described in greater detail in the sections on Wnt signaling (see above).

PTEN

The *pten* gene (phosphatase and tensin homolog deleted on chromosome 10) was identified in 1997 as a tumor suppressor gene at chromosome 10q23 that is mutated in a large variety of sporadic cancers and in two autosomal dominant hamartoma syndromes known as Cowden disease and Bannayan–Riley–Ruvalcaba syndrome (Simpson and Parsons, 2001). To date, more than three hundred somatic point mutations of PTEN have been reported in primary tumors and metastasis, particularly in endometrial carcinomas and glioblastomas, as well as in brain, prostate, and breast cancers and melanoma (Teng et al., 1997). Rather than controlling cell cycle progression or genomic stability as most tumor suppressor proteins, PTEN functions primarily as a lipid phosphatase. Its key target is PIP_3, dephosphorylating the same 3′ site in the inositol ring of membrane phosphatidylinositols that is phosphorylated by the action of PI3K, a central downstream target for RTKs and Ras (see above). By doing so, PTEN functions in the regulation of many normal cellular processes, including growth, adhesion, migration, invasion and apoptosis (Yamada and Araki, 2001). This prominent role in regulating cell growth, migration, and survival makes PTEN a critical player in cancer suppression. Thus, not surprisingly, defects in PTEN, either by inherited or somatic mutations or epigenetic events, can cooperate with the loss of other tumor suppressors and/or activation of oncogenes in tumor progression.

At the structural level, PTEN resembles the large family of protein tyrosine phosphatases, but its enlarged active site together with the presence of a phospholipid-binding domain, known as C2, accounts for its ability to bind PIP_3 (Simpson and Parsons, 2001). This C2 domain may function not in substrate recognition but in targeting PTEN to the plasma membrane and providing the right orientation to the catalytic domain of PTEN to bind PIP_3 and other substrates. The function of the tensinlike domain of PTEN is still not fully defined. PTEN also exhibits a PDZ-binding motif in its carboxyl-terminal tail, which may play a regulatory function or participate in the formation of multiprotein complexes. Similarly, PTEN can be regulated by phosphorylation of certain serine and threonine residues, which can modulate the enzymatic activity and the stability of PTEN. One candidate to regulate PTEN by phosphorylation is the protein kinase CK2, also known as casein kinase II.

However, the likely complexity of the molecular mechanisms controlling PTEN are still poorly understood and are under intensive current investigation.

PTEN in Growth, Apoptosis, and Anoikis. As expected for a tumor suppressor protein, transient expression of PTEN suppresses proliferation. However, results are quite variable and often dependent on a variety of factors, including culturing conditions. Indeed, when cells are maintained in low-serum-containing medium, PTEN causes cell growth arrest in G_1 phase of the cell cycle and induces an increase in the level of cell cycle inhibitors such as p27 KIP1 and a decrease in the levels of phospho-Rb (Mamillapalli et al., 2001; Paramio et al., 1999). This cell cycle arrest in G_1 is caused by the lipid phosphatase activity of PTEN against PIP_3 and can be mimicked by expression of SHIP-2, an enzyme that hydrolyzes another phosphate group on PIP_3 (Taylor et al., 2000).

Much better understood is the role of PTEN in apoptosis (Di Cristofano and Pandolfi, 2000). When PTEN is reexpressed in many carcinoma cell lines, it can cause the initiation of cell death programs directly or sensitize cells to the apoptotic effect of other death-inducing agents, including chemotherapeutic drugs. A particularly important role of PTEN is in cell death after loss of contact with the extracellular matrix, also known as anoikis (Yamada and Araki, 2001). This process is likely to be of biological significance in normal epithelial cell function, as it prevents growth in suspension and in abnormal sites. This anchorage-dependent cell survival, however, is often defective in malignant cells and can be restored by reexpression of PTEN in cells lacking a functional PTEN. The mechanism by which PTEN modulates apoptosis is by reducing levels of PIP_3, thereby diminishing the activity of Akt, a central regulator of apoptosis (See Chapters 4 and 5 for Akt function and its targets and Chapter 4 for apoptosis and its regulation in cancer).

These observations may have important therapeutic implications. For example, amplification or overactivity of the PI3K-Akt pathway is frequent in many neoplasias and a hallmark of serous epithelial ovarian cancers. The resultant activation of the PI3K pathway in ovarian cancers contributes to cell cycle progression, decreased apoptosis, and increased metastatic potential. Of interest, both ovarian and breast cancer cells that also have overactive Akt are highly sensitive to the pharmacologic and genetic inhibition of the PI3K-Akt pathway, thus suggesting that this pathway may represent an excellent target for pharmacologic intervention in these and other cancers.

INTRACELLULAR SIGNALING MOLECULES

Today's Research Defining Tomorrow's Drug Targets

Over the past decade tremendous progress has been made in the elucidation of cellular signaling mechanisms and the establishment of molec-

TABLE 3.3. Signaling Molecules as Targets for Anticancer Drugs

Target	Drug	Company	Status
RTKs			
EGFR	ZD 1839 (Iressa)	AstraZeneca	Phase III
	OSI774	OSI/Roche/Genentech	Phase III
	Cetuximab C225	ImClone Systems	Phase III
	EGF fusion protein	Seragen	Phase II
EGFR/HER2	CI1033	Pfizer	Phase I
	EKB569	Wyeth-Ayerst	Phase I
	GW2016	GlaxoSmithKline	Phase I
	PKI1666	Novartis	Phase I
HER2	Trastuzumab (Herceptin)	Genentech	FDA Approved
VEGFR	SU5416	SUGEN	Phase II
	SU6668	SUGEN	Phase I
BCR-AbI	STI571 (Gleevec)	Novartis	FDA Approved
Ser/Thr kinases			
PKC	PKC412	Novartis	Phase I
MKK1	PD184352	Pfizer	Phase I
Raf	BAY43-906	ONYX/Bayer	Phase I
ChkI, PKC, PDK1	UCN-01	Kyowa-Hakko	Phase II
mTOR	CCI778	Wyeth-Ayerst	Phase II
	RAD001	Novartis	Phase I
Cell cycle			
CDKs	Flavopiridol	Aventis	Phase II
CDK2	CYC202	Cyclacel	Phase I

Modified from Cohen, P. (2002). Protein kinases—the major drug targets of the twenty-first century? *Nat Rev Drug Disc 1*, 309–315.

ular pathways that define the cancer phenotype. Empirical observations and experimental studies have led to an emerging consensus that specific signaling pathways play a critical role in cancer development and progression. Understanding how these pathways communicate information from the extracellular environment to the interior of the cell may help elucidate fundamental molecular mechanisms triggering cell transformation and further identify novel therapeutic targets in the treatment of human cancer. Of note, the recent success of Gleevec, the first gene product targeted drug for a specific protein kinase, Abl, in the treatment of chronic myelogenous leukemia (CML) suggests that mechanism-based therapies may be more efficacious than conventional chemotherapeutic approaches to cancer. Indeed, some of the most promising anticancer drugs in development are inhibitors of protein kinases (Table 3.3). As protein kinases comprise the largest enzyme family, it is not surprising that serine-threonine and tyrosine protein kinases have now emerged as one of the most important group of drug targets, accounting for 20%–30% of the drug discovery programs of many pharmaceutical companies. We can conclude that the recent elucidation of the molecular mechanisms by which normal cells transduce proliferative signals and grow or make life or death decisions, together with the dramatic expansion of our knowledge on how altered function of key signaling and cell cycle-regulating molecules contributes to neoplastic conversion, have now provided an unprecedented understanding of the most basic mechanisms controlling normal and aberrant cell growth, as well as a golden opportunity in the search of novel molecular targets for pharmacologic intervention in cancer.

REFERENCES

Abraham, R. T. (2001). Cell cycle checkpoint signaling through the ATM and ATR kinases. Genes Dev *15*, 2177–2196.

Akhurst, R. J., and Derynck, R. (2001). TGF-β signaling in cancer—a double-edged sword. Trends Cell Biol *11*, S44–S51.

Anbazhagan, R., Bornman, D. M., Johnston, J. C., Westra, W. H., and Gabrielson, E. (1999). The S387Y mutations of the transforming growth factor-β receptor type I gene is uncommon in metastases of breast cancer and other common types of adenocarcinoma. Cancer Res *59*, 3363–3364.

Anzano, M. A., Roberts, A. B., Meyers, C. A., Komoriya, A., Lamb, L. C., Smith, J. M., and Sporn, M. B. (1982). Synergistic interaction of two classes of transforming growth factors from murine sarcoma cells. Cancer Res *42*, 4776–4778.

Attisano, L., and Wrana, J. L. (2002). Signal transduction by the TGF-β super-family. Science *296*, 1646–1647.

Bai, L., and Merchant, J. L. (2001). ZBP-89 promotes growth arrest through stabilization of p53. Mol Cell Biol *21*, 4670–4683.

Bais, C., Santomasso, B., Coso, O., Arvanitakis, L., Raaka, E. G., Gutkind, J. S., Asch, A. S., Cesarman, E., Gershengorn, M. C., Mesri, E. A., and Gerhengorn, M. C. (1998). G-protein-coupled receptor of Kaposi's sarcoma-associated herpesvirus is a viral oncogene and angiogenesis activator. Nature *391*, 86–89.

Bale, A. E. (2002). Hedgehog signaling and human disease. Annu Rev Genomics Hum Genet *3*, 47–65.

Barbacid, M. (1987). ras genes. Annu Rev Biochem *56*, 779–827.

Bar-Sagi, D., and Hall, A. (2000). Ras and Rho GTPases: a family reunion. Cell *103*, 227–238.

Bienz, M. (2002). The subcellular destinations of APC proteins. Nat Rev Mol Cell Biol *3*, 328–338.

Bienz, M., and Clevers, H. (2000). Linking colorectal cancer to Wnt signaling. Cell *103*, 311–320.

Bishop, A. L., and Hall, A. (2000). Rho GTPases and their effector proteins. Biochem J *348*, 241–255.

Blobe, G. C., Schiemann, W. P., and Lodish, H. F. (2000). Role of transforming growth factor β in human disease. N Engl J Med *342*, 1350–1358.

Borkhardt, A., Bojesen, S., Haas, O. A., Fuchs, U., Bartelheimer, D., Loncarevic, I. F., Bohle, R. M., Harbott, J., Repp, R., Jaeger, U., et al. (2000). The human GRAF gene is fused to MLL in a unique t(5;11)(q31;q23) and both alleles are disrupted in three cases of myelodysplastic syndrome/acute myeloid leukemia with a deletion 5q. Proc Natl Acad Sci USA *97*, 9168–9173.

Bos, J. L. (1989). ras oncogenes in human cancer: a review. Cancer Res *49*, 4682–4689.

Bos, J. L. (1997). Ras-like GTPases. Biochim Biophys Acta *1333*, M19–M31.

Brugge, J. S., and Erikson, R. L. (1977). Identification of a transformation-specific antigen induced by an avian sarcoma virus. Nature *269*, 346–348.

Cadigan, K. M., and Nusse, R. (1997). Wnt signaling: a common theme in animal development. Genes Dev *11*, 3286–3305.

Calonge, M. J., and Massague, J. (1999). Smad4/DPC4 silencing and hyperactive Ras jointly disrupt transforming growth factor-β antiproliferative responses in colon cancer cells. J Biol Chem *274*, 33637–33643.

Carmeliet, P., and Collen, D. (2000). Molecular basis of angiogenesis. Role of VEGF and VE-cadherin. Ann NY Acad Sci *902*, 249–262; discussion 262–244.

Cerione, R. A., and Zheng, Y. (1996). The Dbl family of oncogenes. Curr Opin Cell Biol *8*, 216–222.

Chen, T., Carter, D., Garrigue-Antar, L., and Reiss, M. (1998a). Transforming growth factor β type I receptor kinase mutant associated with metastatic breast cancer. Cancer Res *58*, 4805–4810.

Chen, Y. G., Hata, A., Lo, R. S., Wotton, D., Shi, Y., Pavletich, N., and Massague, J. (1998b). Determinants of specificity in TGF-β signal transduction. Genes Dev *12*, 2144–2152.

Cheng, M., Sexl, V., Sherr, C. J., and Roussel, M. F. (1998). Assembly of cyclin D-dependent kinase and titration of p27Kip1 regulated by mitogen-activated protein kinase kinase (MEK1). Proc Natl Acad Sci USA *95*, 1091–1096.

Chiariello, M., Marinissen, M. J., and Gutkind, J. S. (2001). Regulation of c-myc expression by PDGF through Rho GTPases. Nat Cell Biol *3*, 580–586.

Chow, C. W., Rincon, M., Cavanagh, J., Dickens, M., and Davis, R. J. (1997). Nuclear accumulation of NFAT4 opposed by the JNK signal transduction pathway. Science *278*, 1638–1641.

Cilloni, D., Guerrasio, A., Giugliano, E., Scaravaglio, P., Volpe, G., Rege-Cambrin, G., and Saglio, G. (2002). From genes to therapy: the case of Philadelphia chromosome-positive leukemias. Ann NY Acad Sci *963*, 306–312.

Clague, M. J., and Urbe, S. (2001). The interface of receptor trafficking and signalling. J Cell Sci *114*, 3075–3081.

Clark, E. A., Golub, T. R., Lander, E. S., and Hynes, R. O. (2000). Genomic analysis of metastasis reveals an essential role for RhoC. Nature *406*, 532–535.

Collett, M. S., and Erikson, R. L. (1978). Protein kinase activity associated with the avian sarcoma virus src gene product. Proc Natl Acad Sci USA *75*, 2021–2024.

Collett, M. S., Purchio, A. F., and Erikson, R. L. (1980). Avian sarcoma virus-transforming protein, pp60src shows protein kinase activity specific for tyrosine. Nature *285*, 167–169.

Cooper, J. A., Bowen-Pope, D. F., Raines, E., Ross, R., and Hunter, T. (1982). Similar effects of platelet-derived growth factor and epidermal growth factor on the phosphorylation of tyrosine in cellular proteins. Cell *31*, 263–273.

Coso, O. A., Chiariello, M., Kalinec, G., Kyriakis, J. M., Woodgett, J., and Gutkind, J. S. (1995a). Transforming G protein-coupled receptors potently activate JNK (SAPK). Evidence for a divergence from the tyrosine kinase signaling pathway. J Biol Chem *270*, 5620–5624.

Coso, O. A., Chiariello, M., Yu, J. C., Teramoto, H., Crespo, P., Xu, N., Miki, T., and Gutkind, J. S. (1995b). The small GTP-binding proteins Rac1 and Cdc42 regulate the activity of the JNK/SAPK signaling pathway. Cell *81*, 1137–1146.

Dai, J. L., Schutte, M., Bansal, R. K., Wilentz, R. E., Sugar, A. Y., and Kern, S. E. (1999). Transforming growth factor-β responsiveness in DPC4/SMAD4-null cancer cells. Mol Carcinog *26*, 37–43.

Danilkovitch-Miagkova, A., and Zbar, B. (2002). Dysregulation of Met receptor tyrosine kinase activity in invasive tumors. J Clin Invest *109*, 863–867.

Darnell, J. E., Jr. (1997). STATs and gene regulation. Science *277*, 1630–1635.

Davies, H., Bignell, G. R., Cox, C., Stephens, P., Edkins, S., Clegg, S., Teague, J., Woffendin, H., Garnett, M. J., Bottomley, W., et al. (2002). Mutations of the BRAF gene in human cancer. Nature *417*, 949–954.

Derijard, B., Hibi, M., Wu, I. H., Barrett, T., Su, B., Deng, T., Karin, M., and Davis, R. J. (1994). JNK1: a protein kinase stimulated by UV light and Ha-Ras that binds and phosphorylates the c-Jun activation domain. Cell *76*, 1025–1037.

Di Cristofano, A., and Pandolfi, P. P. (2000). The multiple roles of PTEN in tumor suppression. Cell *100*, 387–390.

Di Renzo, M. F., Olivero, M., Martone, T., Maffe, A., Maggiora, P., Stefani, A. D., Valente, G., Giordano, S., Cortesina, G., and Comoglio, P. M. (2000). Somatic mutations of the MET oncogene are selected during metastatic spread of human HNSC carcinomas. Oncogene *19*, 1547–1555.

Dotto, G. P. (2000). p21(WAF1/Cip1): more than a break to the cell cycle? Biochim Biophys Acta *1471*, M43–M56.

Downward, J., Yarden, Y., Mayes, E., Scrace, G., Totty, N., Stockwell, P., Ullrich, A., Schlessinger, J., and Waterfield, M. D. (1984). Close similarity of epidermal growth factor receptor and v-erb-B oncogene protein sequences. Nature *307*, 521–527.

Duesberg, P. H., and Vogt, P. K. (1970). Differences between the ribonucleic acids of transforming and nontransforming avian tumor viruses. Proc Natl Acad Sci USA *67*, 1673–1680.

Duesberg, P. H., and Vogt, P. K. (1973). RNA species obtained from clonal lines of avian sarcoma and from avian leukosis virus. Virology *54*, 207–219.

Eccles, S. A. (2001). The role of c-erbB-2/HER2/neu in breast cancer progression and metastasis. J Mammary Gland Biol Neoplasia *6*, 393–406.

Ek, B., Westermark, B., Wasteson, A., and Heldin, C. H. (1982). Stimulation of tyrosine-specific phosphorylation by platelet-derived growth factor. Nature *295*, 419–420.

Ellis, C. A., and Clark, G. (2000). The importance of being K-Ras. Cell Signal *12*, 425–434.

Engers, R., Zwaka, T. P., Gohr, L., Weber, A., Gerharz, C. D., and Gabbert, H. E. (2000). Tiam1 mutations in human renal-cell carcinomas. Int J Cancer *88*, 369–376.

English, J. M., Pearson, G., Baer, R., and Cobb, M. H. (1998). Identification of substrates and regulators of the mitogen-activated protein kinase ERK5 using chimeric protein kinases. J Biol Chem *273*, 3854–3860.

Fleming, T. P., Saxena, A., Clark, W. C., Robertson, J. T., Oldfield, E. H., Aaronson, S. A., and Ali, I. U. (1992). Amplification and/or overexpression of platelet-derived growth factor receptors and epidermal growth factor receptor in human glial tumors. Cancer Res *52*, 4550–4553.

Friend, S. H., Bernards, R., Rogelj, S., Weinberg, R. A., Rapaport, J. M., Albert, D. M., and Dryja, T. P. (1986). A human DNA segment with properties of the gene that predisposes to retinoblastoma and osteosarcoma. Nature 323, 643–646.

Fuchs, S. Y., Dolan, L., Davis, R. J., and Ronai, Z. (1996). Phosphorylation-dependent targeting of c-Jun ubiquitination by Jun N-kinase. Oncogene 13, 1531–1535.

Furge, K. A., Zhang, Y. W., and Vande Woude, G. F. (2000). Met receptor tyrosine kinase: enhanced signaling through adapter proteins. Oncogene 19, 5582–5589.

Furstenberger, G., and Senn, H. J. (2002). Insulin-like growth factors and cancer. Lancet Oncol 3, 298–302.

Gailani, M. R., and Bale, A. E. (1997). Developmental genes and cancer: role of patched in basal cell carcinoma of the skin. J Natl Cancer Inst 89, 1103–1109.

Gentry, L. E., Lioubin, M. N., Purchio, A. F., and Marquardt, H. (1988). Molecular events in the processing of recombinant type 1 pre-protransforming growth factor β to the mature polypeptide. Mol Cell Biol 8, 4162–4168.

Gesbert, F., Sellers, W. R., Signoretti, S., Loda, M., and Griffin, J. D. (2000). BCR/ABL regulates expression of the cyclin-dependent kinase inhibitor p27Kip1 through the phosphatidylinositol 3-kinase/AKT pathway. J Biol Chem 275, 39223–39230.

Glick, A., Popescu, N., Alexander, V., Ueno, H., Bottinger, E., and Yuspa, S. H. (1999). Defects in transforming growth factor-β signaling cooperate with a Ras oncogene to cause rapid aneuploidy and malignant transformation of mouse keratinocytes. Proc Natl Acad Sci USA 96, 14949–14954.

Goldfarb, M. (2001). Signaling by fibroblast growth factors: the inside story. Sci STKE 2001, PE37.

Graff, J. R., Konicek, B. W., McNulty, A. M., Wang, Z., Houck, K., Allen, S., Paul, J. D., Hbaiu, A., Goode, R. G., Sandusky, G. E., et al. (2000). Increased AKT activity contributes to prostate cancer progression by dramatically accelerating prostate tumor growth and diminishing p27Kip1 expression. J Biol Chem 275, 24500–24505.

Gschwind, A., Zwick, E., Prenzel, N., Leserer, M., and Ullrich, A. (2001). Cell communication networks: epidermal growth factor receptor transactivation as the paradigm for interreceptor signal transmission. Oncogene 20, 1594–1600.

Gutkind, J. S. (2000). Regulation of mitogen-activated protein kinase signaling networks by G protein-coupled receptors. Sci STKE 2000, RE1.

Habas, R., Kato, Y., and He, X. (2001). Wnt/Frizzled activation of Rho regulates vertebrate gastrulation and requires a novel Formin homology protein Daam1. Cell 107, 843–854.

Hackel, P. O., Zwick, E., Prenzel, N., and Ullrich, A. (1999). Epidermal growth factor receptors: critical mediators of multiple receptor pathways. Curr Opin Cell Biol 11, 184–189.

Hagemann, C., and Rapp, U. R. (1999). Isotype-specific functions of Raf kinases. Exp Cell Res 253, 34–46.

Harwood, A. J. (2001). Regulation of GSK-3: a cellular multiprocessor. Cell 105, 821–824.

Hata, A., Shi, Y., and Massague, J. (1998). TGF-β signaling and cancer: structural and functional consequences of mutations in Smads. Mol Med Today *4*, 257–262.

Helle, S. I., and Lonning, P. E. (1996). Insulin-like growth factors in breast cancer. Acta Oncol *35*, 19–22.

Hermanson, M., Funa, K., Hartman, M., Claesson-Welsh, L., Heldin, C. H., Westermark, B., and Nister, M. (1992). Platelet-derived growth factor and its receptors in human glioma tissue: expression of messenger RNA and protein suggests the presence of autocrine and paracrine loops. Cancer Res *52*, 3213–3219.

Hickman, E. S., Moroni, M. C., and Helin, K. (2002). The role of p53 and pRB in apoptosis and cancer. Curr Opin Genet Dev *12*, 60–66.

Hill, C. S., Wynne, J., and Treisman, R. (1995). The Rho family GTPases RhoA, Rac1, and CDC42Hs regulate transcriptional activation by SRF. Cell *81*, 1159–1170.

Ho, A., and Dowdy, S. F. (2002). Regulation of G_1 cell-cycle progression by oncogenes and tumor suppressor genes. Curr Opin Genet Dev *12*, 47–52.

Hocevar, B. A., Brown, T. L., and Howe, P. H. (1999). TGF-β induces fibronectin synthesis through a c-Jun N-terminal kinase-dependent, Smad4-independent pathway. EMBO J *18*, 1345–1356.

Holgado-Madruga, M., Emlet, D. R., Moscatello, D. K., Godwin, A. K., and Wong, A. J. (1996). A Grb2-associated docking protein in EGF- and insulin-receptor signalling. Nature *379*, 560–564.

Honegger, A. M., Dull, T. J., Felder, S., Van Obberghen, E., Bellot, F., Szapary, D., Schmidt, A., Ullrich, A., and Schlessinger, J. (1987a). Point mutation at the ATP binding site of EGF receptor abolishes protein-tyrosine kinase activity and alters cellular routing. Cell *51*, 199–209.

Honegger, A. M., Szapary, D., Schmidt, A., Lyall, R., Van Obberghen, E., Dull, T. J., Ullrich, A., and Schlessinger, J. (1987b). A mutant epidermal growth factor receptor with defective protein tyrosine kinase is unable to stimulate proto-oncogene expression and DNA synthesis. Mol Cell Biol 7, 4568–4571.

Hunter, T. (2000). Signaling—2000 and beyond. Cell *100*, 113–127.

Hunter, T., and Cooper, J. A. (1981). Epidermal growth factor induces rapid tyrosine phosphorylation of proteins in A431 human tumor cells. Cell *24*, 741–752.

Hunter, T., and Sefton, B. M. (1980). Transforming gene product of Rous sarcoma virus phosphorylates tyrosine. Proc Natl Acad Sci USA *77*, 1311–1315.

Jaffe, A. B., and Hall, A. (2002). Rho GTPases in transformation and metastasis. Adv Cancer Res *84*, 57–80.

Jhiang, S. M. (2000). The RET proto-oncogene in human cancers. Oncogene *19*, 5590–5597.

Jolkowska, J., and Witt, M. (2000). The EVI-1 gene—its role in pathogenesis of human leukemias. Leukoc Res *24*, 553–558.

Jouanneau, J., Moens, G., Montesano, R., and Thiery, J. P. (1995). FGF-1 but not FGF-4 secreted by carcinoma cells promotes in vitro and in vivo angiogenesis and rapid tumor proliferation. Growth Factors *12*, 37–47.

Kamb, A., Gruis, N. A., Weaver-Feldhaus, J., Liu, Q., Harshman, K., Tavtigian, S. V., Stockert, E., Day, R. S. 3rd, Johnson, B. E., and Skolnick, M. H. (1994). A cell cycle regulator potentially involved in genesis of many tumor types. Science *264*, 436–440.

Katz, M. E., and McCormick, F. (1997). Signal transduction from multiple Ras effectors. Curr Opin Genet Dev 7, 75–79.

Kawabata, M., Inoue, H., Hanyu, A., Imamura, T., and Miyazono, K. (1998). Smad proteins exist as monomers in vivo and undergo homo- and hetero-oligomerization upon activation by serine/threonine kinase receptors. EMBO J 17, 4056–4065.

Kerkhoff, E., and Rapp, U. R. (2001). The Ras-Raf relationship: an unfinished puzzle. Adv Enzyme Regul 41, 261–267.

Khosravi-Far, R., White, M. A., Westwick, J. K., Solski, P. A., Chrzanowska-Wodnicka, M., Van Aelst, L., Wigler, M. H., and Der, C. J. (1996). Oncogenic Ras activation of Raf/mitogen-activated protein kinase-independent pathways is sufficient to cause tumorigenic transformation. Mol Cell Biol 16, 3923–3933.

Kinzler, K. W., Nilbert, M. C., Su, L. K., Vogelstein, B., Bryan, T. M., Levy, D. B., Smith, K. J., Preisinger, A. C., Hedge, P., McKechnie, D., et al. (1991). Identification of FAP locus genes from chromosome 5q21. Science 253, 661–665.

Kivinen, L., and Laiho, M. (1999). Ras- and mitogen-activated protein kinase kinase-dependent and - independent pathways in p21Cip1/Waf1 induction by fibroblast growth factor-2, platelet-derived growth factor, and transforming growth factor-β1. Cell Growth Differ 10, 621–628.

Kourlas, P. J., Strout, M. P., Becknell, B., Veronese, M. L., Croce, C. M., Theil, K. S., Krahe, R., Ruutu, T., Knuutila, S., Bloomfield, C. D., and Caligiuri, M. A. (2000). Identification of a gene at 11q23 encoding a guanine nucleotide exchange factor: evidence for its fusion with MLL in acute myeloid leukemia. Proc Natl Acad Sci USA 97, 2145–2150.

Kretzschmar, M., Doody, J., and Massague, J. (1997). Opposing BMP and EGF signalling pathways converge on the TGF-β family mediator Smad1. Nature 389, 618–622.

Kretzschmar, M., and Massague, J. (1998). SMADs: mediators and regulators of TGF-β signaling. Curr Opin Genet Dev 8, 103–111.

Kuhl, M., Sheldahl, L. C., Park, M., Miller, J. R., and Moon, R. T. (2000). The Wnt/Ca^{2+} pathway: a new vertebrate Wnt signaling pathway takes shape. Trends Genet 16, 279–283.

Kuriyama, M., Harada, N., Kuroda, S., Yamamoto, T., Nakafuku, M., Iwamatsu, A., Yamamoto, D., Prasad, R., Croce, C., Canaani, E., and Kaibuchi, K. (1996).Identification of AF-6 and canoe as putative targets for Ras. J Biol Chem 271, 607–610.

Kyriakis, J. M., Banerjee, P., Nikolakaki, E., Dai, T., Rubie, E. A., Ahmad, M. F., Avruch, J., and Woodgett, J. R. (1994). The stress-activated protein kinase subfamily of c-Jun kinases. Nature 369, 156–160.

Lagna, G., Hata, A., Hemmati-Brivanlou, A., and Massague, J. (1996). Partnership between DPC4 and SMAD proteins in TGF-βsignalling pathways. Nature 383, 832–836.

Lam, C. W., Xie, J., To, K. F., Ng, H. K., Lee, K. C., Yuen, N. W., Lim, P. L., Chan, L. Y., Tong, S. F., and McCormick, F. (1999). A frequent activated smoothened mutation in sporadic basal cell carcinomas. Oncogene 18, 833–836.

Lavoie, J. N., L'Allemain, G., Brunet, A., Muller, R., and Pouyssegur, J. (1996). Cyclin D1 expression is regulated positively by the p42/p44MAPK and negatively by the p38/HOGMAPK pathway. J Biol Chem 271, 20608–20616.

Lee, M. H., and Yang, H. Y. (2001). Negative regulators of cyclin-dependent kinases and their roles in cancers. Cell Mol Life Sci 58, 1907–1922.

Lengauer, C., Kinzler, K. W., and Vogelstein, B. (1998). Genetic instabilities in human cancers. Nature *396*, 643–649.

Levinson, A. D., Oppermann, H., Levintow, L., Varmus, H. E., and Bishop, J. M. (1978). Evidence that the transforming gene of avian sarcoma virus encodes a protein kinase associated with a phosphoprotein. Cell *15*, 561–572.

Liu, X., Sun, Y., Weinberg, R. A., and Lodish, H. F. (2001). Ski/Sno and TGF-β signaling. Cytokine Growth Factor Rev *12*, 1–8.

Loda, M., Cukor, B., Tam, S. W., Lavin, P., Fiorentino, M., Draetta, G. F., Jessup, J. M., and Pagano, M. (1997). Increased proteasome-dependent degradation of the cyclin-dependent kinase inhibitor p27 in aggressive colorectal carcinomas. Nat Med *3*, 231–234.

Luo, K., Stroschein, S. L., Wang, W., Chen, D., Martens, E., Zhou, S., and Zhou, Q. (1999). The Ski oncoprotein interacts with the Smad proteins to repress TGFβ signaling. Genes Dev *13*, 2196–2206.

MacArthur, C. A., Lawshe, A., Shankar, D. B., Heikinheimo, M., and Shackleford, G. M. (1995). FGF-8 isoforms differ in NIH3T3 cell transforming potential. Cell Growth Differ *6*, 817–825.

Maeyama, H., Hidaka, E., Ota, H., Minami, S., Kajiyama, M., Kuraishi, A., Mori, H., Matsuda, Y., Wada, S., Sodeyama, H., et al. (2001). Familial gastrointestinal stromal tumor with hyperpigmentation: association with a germline mutation of the c-kit gene. Gastroenterology *120*, 210–215.

Mamillapalli, R., Gavrilova, N., Mihaylova, V. T., Tsvetkov, L. M., Wu, H., Zhang, H., and Sun, H. (2001). PTEN regulates the ubiquitin-dependent degradation of the CDK inhibitor p27(KIP1) through the ubiquitin E3 ligase SCF(SKP2). Curr Biol *11*, 263–267.

Manoukian, A. S., and Woodgett, J. R. (2002). Role of glycogen synthase kinase-3 in cancer: regulation by Wnts and other signaling pathways. Adv Cancer Res *84*, 203–229.

Manser, E., Leung, T., Salihuddin, H., Zhao, Z. S., and Lim, L. (1994). A brain serine/threonine protein kinase activated by Cdc42 and Rac1. Nature *367*, 40–46.

Margolis, B., Rhee, S. G., Felder, S., Mervic, M., Lyall, R., Levitzki, A., Ullrich, A., Zilberstein, A., and Schlessinger, J. (1989). EGF induces tyrosine phosphorylation of phospholipase C-II: a potential mechanism for EGF receptor signaling. Cell *57*, 1101–1107.

Markowitz, S., Wang, J., Myeroff, L., Parsons, R., Sun, L., Lutterbaugh, J., Fan, R. S., Zborowska, E., Kinzler, K. W., Vogelstein, B., et al. (1995). Inactivation of the type II TGF-β receptor in colon cancer cells with microsatellite instability. Science *268*, 1336–1338.

Marshall, M. S. (1995). Ras target proteins in eukaryotic cells. FASEB J *9*, 1311–1318.

Massague, J. (1998). TGF-β signal transduction. Annu Rev Biochem *67*, 753–791.

Massague, J. (2000). How cells read TGF-β signals. Nat Rev Mol Cell Biol *1*, 169–178.

Mettouchi, A., Klein, S., Guo, W., Lopez-Lago, M., Lemichez, E., Westwick, J. K., and Giancotti, F. G. (2001). Integrin-specific activation of Rac controls progression through the G_1 phase of the cell cycle. Mol Cell *8*, 115–127.

Michael, D., and Oren, M. (2002). The p53 and Mdm2 families in cancer. Curr Opin Genet Dev *12*, 53–59.

Miller, D. W., Graulich, W., Karges, B., Stahl, S., Ernst, M., Ramaswamy, A., Sedlacek, H. H., Muller, R., and Adamkiewicz, J. (1999). Elevated expression of endoglin, a component of the TGF-β-receptor complex, correlates with proliferation of tumor endothelial cells. Int J Cancer *81*, 568–572.

Miller, M. E., and Cross, F. R. (2001). Cyclin specificity: how many wheels do you need on a unicycle? J Cell Sci *114*, 1811–1820.

Minden, A., Lin, A., Claret, F. X., Abo, A., and Karin, M. (1995). Selective activation of the JNK signaling cascade and c-Jun transcriptional activity by the small GTPases Rac and Cdc42Hs. Cell *81*, 1147–1157.

Mira, J. P., Benard, V., Groffen, J., Sanders, L. C., and Knaus, U. G. (2000). Endogenous, hyperactive Rac3 controls proliferation of breast cancer cells by a p21-activated kinase-dependent pathway. Proc Natl Acad Sci USA *97*, 185–189.

Miyazono, K., Kusanagi, K., and Inoue, H. (2001). Divergence and convergence of TGF-β/BMP signaling. J Cell Physiol *187*, 265–276.

Mlodzik, M. (1999). Planar polarity in the *Drosophila* eye: a multifaceted view of signaling specificity and cross-talk. EMBO J *18*, 6873–6879.

Momand, J., Jung, D., Wilczynski, S., and Niland, J. (1998). The MDM2 gene amplification database. Nucleic Acids Res *26*, 3453–3459.

Montaner, S., Sodhi, A., Pece, S., Mesri, E. A., and Gutkind, J. S. (2001). The Kaposi's sarcoma-associated herpesvirus G protein-coupled receptor promotes endothelial cell survival through the activation of Akt/protein kinase B. Cancer Res *61*, 2641–2648.

Moon, R. T., Bowerman, B., Boutros, M., and Perrimon, N. (2002). The promise and perils of Wnt signaling through β-catenin. Science *296*, 1644–1646.

Morin, P. J. (1999). β-Catenin signaling and cancer. Bioessays *21*, 1021–1030.

Morin, P. J., Sparks, A. B., Korinek, V., Barker, N., Clevers, H., Vogelstein, B., and Kinzler, K. W. (1997). Activation of β-catenin-Tcf signaling in colon cancer by mutations in β-catenin or APC. Science *275*, 1787–1790.

Muraoka, M., Konishi, M., Kikuchi-Yanoshita, R., Tanaka, K., Shitara, N., Chong, J. M., Iwama, T., and Miyaki, M. (1996). p300 gene alterations in colorectal and gastric carcinomas. Oncogene *12*, 1565–1569.

Musti, A. M., Treier, M., and Bohmann, D. (1997). Reduced ubiquitin-dependent degradation of c-Jun after phosphorylation by MAP kinases. Science *275*, 400–402.

Nakamura, N., Ramaswamy, S., Vazquez, F., Signoretti, S., Loda, M., and Sellers, W. R. (2000). Forkhead transcription factors are critical effectors of cell death and cell cycle arrest downstream of PTEN. Mol Cell Biol *20*, 8969–8982.

Nakayama, K. (1998). Cip/Kip cyclin-dependent kinase inhibitors: brakes of the cell cycle engine during development. Bioessays *20*, 1020–1029.

Natali, P. G., Berlingieri, M. T., Nicotra, M. R., Fusco, A., Santoro, E., Bigotti, A., and Vecchio, G. (1995). Transformation of thyroid epithelium is associated with loss of c-kit receptor. Cancer Res *55*, 1787–1791.

Natali, P. G., Nicotra, M. R., Winkler, A. B., Cavaliere, R., Bigotti, A., and Ullrich, A. (1992). Progression of human cutaneous melanoma is associated with loss of expression of c-kit proto-oncogene receptor. Int J Cancer *52*, 197–201.

Novak, A., and Dedhar, S. (1999). Signaling through β-catenin and Lef/Tcf. Cell Mol Life Sci *56*, 523–537.

Obaya, A. J., and Sedivy, J. M. (2002). Regulation of cyclin-Cdk activity in mammalian cells. Cell Mol Life Sci *59*, 126–142.

Okamoto, A., Demetrick, D. J., Spillare, E. A., Hagiwara, K., Hussain, S. P., Bennett, W. P., Forrester, K., Gerwin, B., Serrano, M., Beach, D. H., et al. (1994). Mutations and altered expression of p16INK4 in human cancer. Proc Natl Acad Sci USA *91*, 11045–11049.

Olashaw, N., and Pledger, W. J. (2002). Paradigms of growth control: relation to Cdk activation. Sci STKE *2002*, RE7.

Olayioye, M. A., Neve, R. M., Lane, H. A., and Hynes, N. E. (2000). The ErbB signaling network: receptor heterodimerization in development and cancer. EMBO J *19*, 3159–3167.

Ono, K., and Han, J. (2000). The p38 signal transduction pathway: activation and function. Cell Signal *12*, 1–13.

Otsu, M., Hiles, I., Gout, I., Fry, M. J., Ruiz-Larrea, F., Panayotou, G., Thompson, A., Dhand, R., Hsuan, J., Totty, N., et al. (1991). Characterization of two 85 kd proteins that associate with receptor tyrosine kinases, middle-T/pp60c-src complexes, and PI3-kinase. Cell *65*, 91–104.

Otterson, G. A., Kratzke, R. A., Coxon, A., Kim, Y. W., and Kaye, F. J. (1994). Absence of p16INK4 protein is restricted to the subset of lung cancer lines that retains wildtype RB. Oncogene *9*, 3375–3378.

Pandur, P., and Kuhl, M. (2001). An arrow for wingless to take-off. Bioessays *23*, 207–210.

Paramio, J. M., Navarro, M., Segrelles, C., Gomez-Casero, E., and Jorcano, J. L. (1999). PTEN tumour suppressor is linked to the cell cycle control through the retinoblastoma protein. Oncogene *18*, 7462–7468.

Park, K., Kim, S. J., Bang, Y. J., Park, J. G., Kim, N. K., Roberts, A. B., and Sporn, M. B. (1994). Genetic changes in the transforming growth factor β (TGF-β) type II receptor gene in human gastric cancer cells: correlation with sensitivity to growth inhibition by TGF-β. Proc Natl Acad Sci USA *91*, 8772–8776.

Pasche, B. (2001). Role of transforming growth factor β in cancer. J Cell Physiol *186*, 153–168.

Pearson, G., Robinson, F., Beers Gibson, T., Xu, B. E., Karandikar, M., Berman, K., and Cobb, M. H. (2001). Mitogen-activated protein (MAP) kinase pathways: regulation and physiological functions. Endocr Rev *22*, 153–183.

Perona, R., Esteve, P., Jimenez, B., Ballestero, R. P., Ramon y Cajal, S., and Lacal, J. C. (1993). Tumorigenic activity of rho genes from *Aplysia californica*. Oncogene *8*, 1285–1292.

Perona, R., Montaner, S., Saniger, L., Sanchez-Perez, I., Bravo, R., and Lacal, J. C. (1997). Activation of the nuclear factor-κB by Rho, CDC42, and Rac-1 proteins. Genes Dev *11*, 463–475.

Polakis, P. (1999). The oncogenic activation of β-catenin. Curr Opin Genet Dev *9*, 15–21.

Polakis, P. (2000). Wnt signaling and cancer. Genes Dev *14*, 1837–1851.

Powell, S. M., Zilz, N., Beazer-Barclay, Y., Bryan, T. M., Hamilton, S. R., Thibodeau, S. N., Vogelstein, B., and Kinzler, K. W. (1992). APC mutations occur early during colorectal tumorigenesis. Nature *359*, 235–237.

Pruitt, K., and Der, C. J. (2001). Ras and Rho regulation of the cell cycle and oncogenesis. Cancer Lett *171*, 1–10.

Qiu, R. G., Chen, J., Kirn, D., McCormick, F., and Symons, M. (1995). An essential role for Rac in Ras transformation. Nature *374*, 457–459.

Ranade, K., Hussussian, C. J., Sikorski, R. S., Varmus, H. E., Goldstein, A. M., Tucker, M. A., Serrano, M., Hannon, G. J., Beach, D., and Dracopoli, N. C. (1995). Mutations associated with familial melanoma impair p16INK4 function. Nat Genet *10*, 114–116.

Randerson-Moor, J. A., Harland, M., Williams, S., Cuthbert-Heavens, D., Sheridan, E., Aveyard, J., Sibley, K., Whitaker, L., Knowles, M., Bishop, J. N., and Bishop, D. T. (2001). A germline deletion of p14(ARF) but not CDKN2A in a melanoma-neural system tumour syndrome family. Hum Mol Genet *10*, 55–62.

Ridley, A. J. (2001). Rho GTPases and cell migration. J Cell Sci *114*, 2713–2722.

Riggins, G. J., Kinzler, K. W., Vogelstein, B., and Thiagalingam, S. (1997). Frequency of Smad gene mutations in human cancers. Cancer Res *57*, 2578–2580.

Rocco, J. W., and Sidransky, D. (2001). p16(MTS-1/CDKN2/INK4a) in cancer progression. Exp Cell Res *264*, 42–55.

Rous, P. (1911). A sarcoma of the fowl transmissible by an agent separable from the tumor cells. J Exp Med *13*, 397–411.

Roussel, M. F. (1999). The INK4 family of cell cycle inhibitors in cancer. Oncogene *18*, 5311–5317.

Ruiz i Altaba, A., Sanchez, P., and Dahmane, N. (2002). Gli and hedgehog in cancer: tumours, embryos and stem cells. Nat Rev Cancer *2*, 361–372.

Sanchez-Prieto, R., Rojas, J. M., Taya, Y., and Gutkind, J. S. (2000). A role for the p38 mitogen-activated protein kinase pathway in the transcriptional activation of p53 on genotoxic stress by chemotherapeutic agents. Cancer Res *60*, 2464–2472.

Sano, Y., Harada, J., Tashiro, S., Gotoh-Mandeville, R., Maekawa, T., and Ishii, S. (1999). ATF-2 is a common nuclear target of Smad and TAK1 pathways in transforming growth factor-β signaling. J Biol Chem *274*, 8949–8957.

Schechter, A. L., Hung, M. C., Vaidyanathan, L., Weinberg, R. A., Yang-Feng, T. L., Francke, U., Ullrich, A., and Coussens, L. (1985). The neu gene: an erbB-homologous gene distinct from and unlinked to the gene encoding the EGF receptor. Science *229*, 976–978.

Scheffner, M., Werness, B. A., Huibregtse, J. M., Levine, A. J., and Howley, P. M. (1990). The E6 oncoprotein encoded by human papillomavirus types 16 and 18 promotes the degradation of p53. Cell *63*, 1129–1136.

Schiemann, W. P., Pfeifer, W. M., Levi, E., Kadin, M. E., and Lodish, H. F. (1999). A deletion in the gene for transforming growth factor β type I receptor abolishes growth regulation by transforming growth factor β in a cutaneous T-cell lymphoma. Blood *94*, 2854–2861.

Schlessinger, J. (1994). SH2/SH3 signaling proteins. Curr Opin Genet Dev *4*, 25–30.

Schmidt, A., and Hall, A. (2002). Guanine nucleotide exchange factors for Rho GTPases: turning on the switch. Genes Dev *16*, 1587–1609.

Schuchardt, A., D'Agati, V., Larsson-Blomberg, L., Costantini, F., and Pachnis, V. (1994). Defects in the kidney and enteric nervous system of mice lacking the tyrosine kinase receptor Ret. Nature *367*, 380–383.

Schwartz, D. E., Tizard, R., and Gilbert, W. (1983). Nucleotide sequence of Rous sarcoma virus. Cell *32*, 853–869.

Seidensticker, M. J., and Behrens, J. (2000). Biochemical interactions in the wnt pathway. Biochim Biophys Acta *1495*, 168–182.

Serrano, M., Hannon, G. J., and Beach, D. (1993). A new regulatory motif in cell-cycle control causing specific inhibition of cyclin D/CDK4. Nature *366*, 704–707.

Sharpe, C., Lawrence, N., and Martinez Arias, A. (2001). Wnt signalling: a theme with nuclear variations. Bioessays *23*, 311–318.

Sharpless, N. E., and DePinho, R. A. (1999). The INK4A/ARF locus and its two gene products. Curr Opin Genet Dev *9*, 22–30.

Sharpless, N. E., and DePinho, R. A. (2002). p53: good cop/bad cop. Cell *110*, 9–12.

Sharrocks, A. D., Yang, S. H., and Galanis, A. (2000). Docking domains and substrate-specificity determination for MAP kinases. Trends Biochem Sci *25*, 448–453.

Shaw, R. J., Paez, J. G., Curto, M., Yaktine, A., Pruitt, W. M., Saotome, I., O'Bryan, J. P., Gupta, V., Ratner, N., Der, C. J., et al. (2001). The Nf2 tumor suppressor, merlin, functions in Rac-dependent signaling. Dev Cell *1*, 63–72.

Sherman, L. S., and Gutmann, D. H. (2001). Merlin: hanging tumor suppression on the Rac. Trends Cell Biol *11*, 442–444.

Simpson, L., and Parsons, R. (2001). PTEN: life as a tumor suppressor. Exp Cell Res *264*, 29–41.

Sionov, R. V., Coen, S., Goldberg, Z., Berger, M., Bercovich, B., Ben-Neriah, Y., Ciechanover, A., and Haupt, Y. (2001). c-Abl regulates p53 levels under normal and stress conditions by preventing its nuclear export and ubiquitination. Mol Cell Biol *21*, 5869–5878.

Skolnik, E. Y., Margolis, B., Mohammadi, M., Lowenstein, E., Fischer, R., Drepps, A., Ullrich, A., and Schlessinger, J. (1991). Cloning of PI3 kinase-associated p85 utilizing a novel method for expression/cloning of target proteins for receptor tyrosine kinases. Cell *65*, 83–90.

Sodhi, A., Montaner, S., Patel, V., Zohar, M., Bais, C., Mesri, E. A., and Gutkind, J. S. (2000). The Kaposi's sarcoma-associated herpes virus G protein-coupled receptor up-regulates vascular endothelial growth factor expression and secretion through mitogen-activated protein kinase and p38 pathways acting on hypoxia-inducible factor 1α. Cancer Res *60*, 4873–4880.

Spector, D. H., Varmus, H. E., and Bishop, J. M. (1978). Nucleotide sequences related to the transforming gene of avian sarcoma virus are present in DNA of uninfected vertebrates. Proc Natl Acad Sci USA *75*, 4102–4106.

Srivastava, S., Verma, M., and Henson, D. E. (2001). Biomarkers for early detection of colon cancer. Clin Cancer Res *7*, 1118–1126.

Stehelin, D., Varmus, H. E., Bishop, J. M., and Vogt, P. K. (1976). DNA related to the transforming gene(s) of avian sarcoma viruses is present in normal avian DNA. Nature *260*, 170–173.

Sutherland, R. L., and Musgrove, E. A. (2002). Cyclin D1 and mammary carcinoma: new insights from transgenic mouse models. Breast Cancer Res *4*, 14–17.

Symons, M., and Takai, Y. (2001). Ras GTPases: singing in tune. Sci STKE *2001*, PE1.

Takagi, Y., Koumura, H., Futamura, M., Aoki, S., Ymaguchi, K., Kida, H., Tanemura, H., Shimokawa, K., and Saji, S. (1998). Somatic alterations of the SMAD-2 gene in human colorectal cancers. Br J Cancer *78*, 1152–1155.

Taylor, V., Wong, M., Brandts, C., Reilly, L., Dean, N. M., Cowsert, L. M., Moodie, S., and Stokoe, D. (2000). 5′ phospholipid phosphatase SHIP-2 causes protein kinase B inactivation and cell cycle arrest in glioblastoma cells. Mol Cell Biol *20*, 6860–6871.

Teng, D. H., Hu, R., Lin, H., Davis, T., Iliev, D., Frye, C., Swedlund, B., Hansen, K. L., Vinson, V. L., Gumpper, K. L., et al. (1997). MMAC1/PTEN mutations in primary tumor specimens and tumor cell lines. Cancer Res *57*, 5221–5225.

Tonini, T., Hillson, C., and Claudio, P. P. (2002). Interview with the retinoblastoma family members: do they help each other? J Cell Physiol *192*, 138–150.

Trimarchi, J. M., and Lees, J. A. (2002). Sibling rivalry in the E2F family. Nat Rev Mol Cell Biol *3*, 11–20.

Tsihlias, J., Kapusta, L., and Slingerland, J. (1999). The prognostic significance of altered cyclin-dependent kinase inhibitors in human cancer. Annu Rev Med *50*, 401–423.

Tsukazaki, T., Chiang, T. A., Davison, A. F., Attisano, L., and Wrana, J. L. (1998). SARA, a FYVE domain protein that recruits Smad2 to the TGFβa receptor. Cell *95*, 779–791.

Ullrich, A., Coussens, L., Hayflick, J. S., Dull, T. J., Gray, A., Tam, A. W., Lee, J., Yarden, Y., Libermann, T. A., Schlessinger, J., et al. (1984). Human epidermal growth factor receptor cDNA sequence and aberrant expression of the amplified gene in A431 epidermoid carcinoma cells. Nature *309*, 418–425.

Ushiro, H., and Cohen, S. (1980). Identification of phosphotyrosine as a product of epidermal growth factor-activated protein kinase in A-431 cell membranes. J Biol Chem *255*, 8363–8365.

Van Aelst, L., and D'Souza-Schorey, C. (1997). Rho GTPases and signaling networks. Genes Dev *11*, 2295–2322.

van Leeuwen, F., and Nusse, R. (1995). Oncogene activation and oncogene cooperation in MMTV-induced mouse mammary cancer. Semin Cancer Biol *6*, 127–133.

Vogelstein, B., and Kinzler, K. W. (1992). p53 function and dysfunction. Cell *70*, 523–526.

Vojtek, A. B., Hollenberg, S. M., and Cooper, J. A. (1993). Mammalian Ras interacts directly with the serine/threonine kinase Raf. Cell *74*, 205–214.

Volmat, V., and Pouyssegur, J. (2001). Spatiotemporal regulation of the p42/p44 MAPK pathway. Biol Cell *93*, 71–79.

Vousden, K. H. (1995). Regulation of the cell cycle by viral oncoproteins. Semin Cancer Biol *6*, 109–116.

Vousden, K. H., and Lu, X. (2002). Live or let die: the cell's response to p53. Nat Rev Cancer *2*, 594–604.

Wahl, M. I., Nishibe, S., Suh, P. G., Rhee, S. G., and Carpenter, G. (1989). Epidermal growth factor stimulates tyrosine phosphorylation of phospholipase C-II independently of receptor internalization and extracellular calcium. Proc Natl Acad Sci USA *86*, 1568–1572.

Weber, J. D., Jeffers, J. R., Rehg, J. E., Randle, D. H., Lozano, G., Roussel, M. F., Sherr, C. J., and Zambetti, G. P. (2000). p53-independent functions of the p19(ARF) tumor suppressor. Genes Dev *14*, 2358–2365.

Weiner, D. B., Kokai, Y., Wada, T., Cohen, J. A., Williams, W. V., and Greene, M. I. (1989). Linkage of tyrosine kinase activity with transforming ability of the p185neu oncoprotein. Oncogene *4*, 1175–1183.

Weston, C. R. and Davis, R. J. (2002). The JNK signal transduction pathway. Curr Opin Genet Dev *12*, 14–21.

Westwick, J. K., Lambert, Q. T., Clark, G. J., Symons, M., Van Aelst, L., Pestell, R. G., and Der, C. J. (1997). Rac regulation of transformation, gene expression, and actin organization by multiple, PAK-independent pathways. Mol Cell Biol *17*, 1324–1335.

White, M. A., Nicolette, C., Minden, A., Polverino, A., Van Aelst, L., Karin, M., and Wigler, M. H. (1995). Multiple Ras functions can contribute to mammalian cell transformation. Cell *80*, 533–541.

Whitmarsh, A. J., and Davis, R. J. (1999). Signal transduction by MAP kinases: regulation by phosphorylation- dependent switches. Sci STKE *1999*, PE1.

Widmann, C., Gibson, S., Jarpe, M. B., and Johnson, G. L. (1999). Mitogen-activated protein kinase: conservation of a three-kinase module from yeast to human. Physiol Rev *79*, 143–180.

Woodgett, J. R. (2001). Judging a protein by more than its name: GSK-3. Sci STKE *2001*, RE12.

Wotton, D., and Massague, J. (2001). Smad transcriptional corepressors in TGF β family signaling. Curr Top Microbiol Immunol *254*, 145–164.

Xiao, G. H., Beeser, A., Chernoff, J., and Testa, J. R. (2002). p21-activated kinase links Rac/Cdc42 signaling to merlin. J Biol Chem *277*, 883–886.

Xu, W., Angelis, K., Danielpour, D., Haddad, M. M., Bischof, O., Campisi, J., Stavnezer, E., and Medrano, E. E. (2000). Ski acts as a co-repressor with Smad2 and Smad3 to regulate the response to type β transforming growth factor. Proc Natl Acad Sci USA *97*, 5924–5929.

Yamada, K. M., and Araki, M. (2001). Tumor suppressor PTEN: modulator of cell signaling, growth, migration and apoptosis. J Cell Sci *114*, 2375–2382.

Yamaguchi, K., Shirakabe, K., Shibuya, H., Irie, K., Oishi, I., Ueno, N., Taniguchi, T., Nishida, E., and Matsumoto, K. (1995). Identification of a member of the MAPKKK family as a potential mediator of TGF-β signal transduction. Science *270*, 2008–2011.

Yamaguchi, M., Miyaki, M., Iijima, T., Matsumoto, T., Kuzume, M., Matsumiya, A., Endo, Y., Sanada, Y., and Kumada, K. (2000). Specific mutation in exon 11 of c-kit proto-oncogene in a malignant gastrointestinal stromal tumor of the rectum. J Gastroenterol *35*, 779–783.

Yang, S. H., Whitmarsh, A. J., Davis, R. J., and Sharrocks, A. D. (1998). Differential targeting of MAP kinases to the ETS-domain transcription factor Elk-1. EMBO J *17*, 1740–1749.

Yu, H., and Rohan, T. (2000). Role of the insulin-like growth factor family in cancer development and progression. J Natl Cancer Inst *92*, 1472–1489.

Yu, Q., Geng, Y., and Sicinski, P. (2001). Specific protection against breast cancers by cyclin D1 ablation. Nature *411*, 1017–1021.

Zhang, Y., Feng, X. H., and Derynck, R. (1998). Smad3 and Smad4 cooperate with c-Jun/c-Fos to mediate TGF-β-induced transcription. Nature *394*, 909–913.

Zheng, Y. (2001). Dbl family guanine nucleotide exchange factors. Trends Biochem Sci *26*, 724–732.

Zhou, G., Bao, Z. Q., and Dixon, J. E. (1995). Components of a new human protein kinase signal transduction pathway. J Biol Chem *270*, 12665–12669.

Zhu, Y., Richardson, J. A., Parada, L. F., and Graff, J. M. (1998). Smad3 mutant mice develop metastatic colorectal cancer. Cell *94*, 703–714.

Zohar, M., Teramoto, H., Katz, B. Z., Yamada, K. M., and Gutkind, J. S. (1998). Effector domain mutants of Rho dissociate cytoskeletal changes from nuclear signaling and cellular transformation. Oncogene *17*, 991–998.

CHAPTER 4

APOPTOTIC PATHWAYS IN CANCER PROGRESSION AND TREATMENT

JOYA CHANDRA and SCOTT H. KAUFMANN

Division of Oncology Research, Mayo Clinic, and Department of Molecular Pharmacology, Mayo Graduate School, Rochester, Minnesota

INTRODUCTION

The term apoptosis was coined in 1972 by Currie, Kerr, and Wyllie to denote a form of cell death that involved cell shrinkage and chromatin condensation (Wyllie et al., 1980). On the basis of its morphological and biochemical features, apoptosis was shown to be distinct from necrosis, a form of cell death characterized by ATP and NAD^+ depletion, cell swelling, membrane rupture, and disorganized digestion of the cellular contents. Studies over the ensuing two decades demonstrated that a variety of cellular stresses, including withdrawal of trophic factors and treatment with anticancer drugs, provoked this type of cell death (Arends and Wyllie, 1991). Indeed, it has been suggested that the induction of apoptosis might be the default pathway whenever eukaryotic cells encounter unfavorable growth conditions (Raff, 1992). Because of the widespread occurrence of apoptosis, as well as the recognition that it is a regulated process, there has been considerable interest in defining the biochemical pathways involved.

Over the past 10–15 years, there has also been a growing recognition that apoptotic pathways are dysregulated in cancer cells. Colon cancer explants, for example, exhibit less apoptosis than normal colonic epithelium (Bedi et al., 1995). Studies in model systems have provided a potential rationale for this observation. The oncoprotein myc, which induces proliferation when cells are maintained under favorable nutrient conditions, induces apoptosis when cells encounter unfavorable growth conditions (Cory et al., 1999). As a consequence, the ability of *myc* to

Signal Transduction and Human Disease, Edited by Toren Finkel and J. Silvio Gutkind
ISBN 0-471-02011-7 Copyright © 2003 John Wiley & Sons, Inc.

transform cells is enhanced by the cooperation of an antiapoptotic gene such as *Bcl-2* (Cory et al., 1999; Evan and Vousden, 2001). Other dominant, growth-promoting oncoproteins (e.g., mutant ras isoforms) also require cooperating antiapoptotic changes to facilitate transformation.

To devise rational therapeutic approaches to deal with the inhibition of apoptosis that accompanies the process of carcinogenesis, it is necessary to first understand the biochemical basis for this phenomenon. In the following sections we describe the components of the core apoptotic machinery, review what is known about their regulation, illustrate some of the ways that this machinery is altered in cancer cells, and then discuss therapeutic approaches that are suggested by current understanding of apoptotic pathways.

APOPTOTIC PROTEASE ACTIVATION

A Central Role for Caspases in Apoptosis

Cell shrinkage, chromatin condensation, internucleosomal DNA fragmentation, and formation of apoptotic bodies are features that distinguish apoptosis from necrosis (Arends and Wyllie, 1991). These and other classic features of apoptosis are the end result of protease activation. Although a number of different proteases now appear to be activated during apoptosis (see below), the most extensively studied apoptotic proteases are caspases, a family of cysteine proteases that cleave substrates at aspartate residues. The role of this enzyme family in apoptosis has been extensively studied over the past decade and has been comprehensively reviewed (Budihardjo et al., 1999; Earnshaw et al., 1999; Hengartner, 2000; Thornberry and Lazebnik, 1998). Of the 11 human caspases, 7 have definite (caspases 3, 6, 8, and 9) or probable (caspases 2, 7, and 10) roles in apoptosis, whereas the remainder are thought to be involved in cytokine processing and the regulation of inflammation.

The cleavage specificity of each caspase is determined by the last four or five amino acids preceding the scissile bond. For example, caspase 3, the enzyme responsible for many of the substrate cleavages that occur during apoptosis, preferentially recognizes the sequences asp-glu-val-asp↓X and asp-met-gln-asp↓X. In contrast, caspase 6, which appears to be responsible for more limited apoptotic cleavages, prefers val-glu-ile-asp↓X with the more lipophilic valine in the −4 position (Earnshaw et al., 1999). These sequence preferences have provided the opportunity to develop somewhat selective synthetic substrates and inhibitors that can be utilized to study the roles of these enzymes during apoptosis.

Active caspases are tetrameric enzymes consisting of two large 17- to 37-kDa and two small 10- to 12-kDa subunits. Each caspase is synthesized as a single-chain zymogen that contains an N-terminal prodomain followed by one large and one small subunit. As described below, the prodomains of some of the procaspases participate in protein-protein interactions that help transduce certain signals into protease activation.

Maturation of the procaspases involves cleavage at critical aspartate residues, that is, at the same type of bonds that caspases themselves cleave. This feature of caspase activation provides the opportunity for autoactivation as well as the possibility of caspase cascades.

Two Pathways of Caspase Activation

According to current understanding, there are at least two major pathways of apoptotic caspase activation (Budihardjo et al., 1999; Hengartner, 2000). One, called the "death receptor" or "extrinsic" pathway, utilizes caspase 8 and/or 10 as the initiator caspase in a protease cascade. The other, called the "mitochondrial" or "intrinsic" pathway, involves caspase 9. As indicated in Figure 4.1, each of these signal transduction pathways leads to activation of caspases 3 and 6, which are the two widely studied effector caspases.

Receptor-Mediated Caspase Activation. Certain receptors, generically termed "death receptors," play important roles in immune cell-mediated killing. Ligation of these receptors can, under certain conditions, lead to apoptotic signaling (Ashkenazi, 2002). Current models indicate that binding of Fas ligand to Fas (a receptor also known as CD95 or Apo-1), tumor necrosis factor-α to tumor necrosis factor-α receptor 1 (TNF-R1), Apo-3 ligand/TWEAK to death receptor 3, or tumor necrosis factor-α-related apoptosis-inducing ligand (TRAIL) to death receptor 4 or 5 promotes the assembly of a protein complex called the DISC (death-inducing signaling complex). For example, binding of Fas ligand to the extracellular domain of Fas results in an alteration in a cytoplasmic domain of the receptor (the so-called "death domain") that allows its interaction with a homologous domain on the adaptor molecule FADD. As a consequence, the other domain of FADD ("death effector domain") acquires the ability to bind homologous motifs present in the prodomains of procaspases 8 and 10. The net result is the recruitment of multiple procaspase 8 or procaspase 10 zymogens, each of which has limited but detectable intrinsic enzyme activity, to the vicinity of the ligated receptor (Kischkel et al., 2001; Wang et al., 2001). It is currently thought that this receptor-mediated juxtaposition of the zymogens allows them to begin to digest one another, thereby generating active caspase 8 and/or 10 (Salvesen and Dixit, 1999) and transducing a series of protein-protein interactions into a proteolytic signal.

Although the events surrounding Fas ligation have been most intensively studied, a similar DISC involving FADD is assembled after ligation of DR4 or DR5 (Peter, 2000). In the case of TNF-R1, the death domain-containing adaptor molecule TRADD is interposed between the receptor and FADD.

If activation of caspase 8 and/or 10 at these DISCs is extensive, caspase 3 can be directly activated. Alternatively, if initiator caspase activation is weak, cleavage of the cytoplasmic substrate Bid generates a fragment that is capable of amplifying the signal by activating the mito-

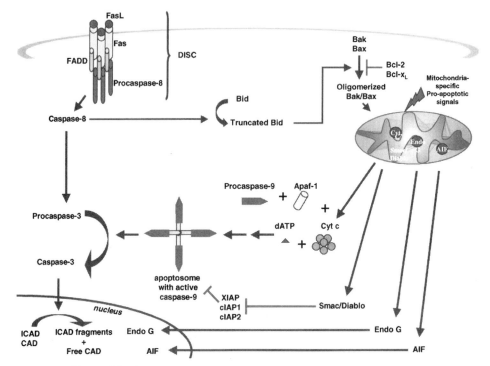

Figure 4.1. *Apoptotic pathways in a cellular context.* Caspases are activated as a consequence of signaling through two major pathways. The "extrinsic" or "death receptor" pathway begins with ligation of specific cell surface receptors. In the case of Fas, binding of its cognate ligand (Fas ligand) results in an alteration of the so-called "death domain" in the cytoplasmic tail of the receptor to permit binding of the death domain-containing adaptor molecule FADD (Fas-associated death domain). As a consequence of this binding, the "death effector domain" of FADD acquires the ability to bind homologous domains in procaspase 8, thereby drawing this zymogen to the site. Molecular interactions within this assembly, which is called a DISC (death-inducing signaling complex), result in cleavage of procaspase 8 and release of mature caspase 8 to the cytosol, where it cleaves procaspase 3 and the small "BH3 only" Bcl-2 family member Bid. Depending on the ratio of these cleavages, apoptosis might be triggered directly or might depend on activation of the mitochondrial pathway.

The "intrinsic" or "mitochondrial" pathway involves release of proapoptotic polypeptides from mitochondria. One of these is cytochrome *c*, which facilitates the dATP- or ATP-dependent binding of procaspase 9 to apoptotic protease activating factor 1 (Apaf-1) through protein-protein interaction domains known as caspase recruitment domains (CARDs). Formation of this complex, which is known as the apoptosome, results in enhanced activity of caspase 9, which then catalyzes caspase 3 activation.

Once activated, caspase 3 cleaves a number of substrates. As described in the text, these cleavages contribute to the apoptotic phenotype. For example, caspase 3-mediated cleavage of inhibitor of caspase-activated deoxyribonuclease (ICAD) releases CAD, which then cleaves DNA.

Other polypeptides released from mitochondria also contribute to apoptotic changes. Endonuclease G and apoptosis-inducing factor (AIF) promote DNA chondrial pathway (Budihardjo et al., 1999) as described in greater detail below.

Mitochondria-Dependent Caspase Activation. Induction of apoptosis by a variety of toxic insults, including radiation and chemotherapeutic agents, is accompanied by changes in mitochondrial function (Budihardjo et al., 1999). These stimuli cause several mitochondrial changes, including decreases in mitochondrial membrane potential and the release of polypeptides that normally reside in the mitochondrial intermembrane space (Green and Reed, 1998). Among the polypeptides released is cytochrome c. Once in the cytoplasm, this polypeptide binds to a scaffolding protein called apoptotic protease-activating factor-1 (Apaf-1). In the presence of cytochrome c and dATP, Apaf-1 undergoes a conformational change that facilitates binding of a protein interaction domain called a caspase recruitment domain (CARD) on Apaf-1 to a similar motif present in the prodomain of procaspase 9. The result is the formation of a M_r ~700,000 complex called an apoptosome that has enhanced ability to cleave procaspases 3 and 7 to active enzymes. These caspases then contribute to the apoptotic phenotype as described below.

The nucleotide requirements for apoptosome formation have been actively investigated. Although dATP is the preferred endogenous nucleotide, ATP will substitute at higher (but still physiological) concentrations. Interestingly, F-Ara-ATP (the activated intracellular metabolite of the antineoplastic agent fludarabine), Ara-CTP (the activated product of cytarabine), and 2', 2'-difluoro-2'-deoxycytidine-5'-triphosphase (the active metabolite of gemcitabine) are more potent than dATP in facilitating caspase 9 activation. It has been postulated that this effect might contribute to the anticancer activity of these agents (Genini et al., 2000).

Whether caspase 9 is truly the initiating caspase in the intrinsic cell death pathway is somewhat controversial. Although cells containing targeted deletions of caspase 9 or Apaf-1 were initially reported to be resistant to induction of apoptosis by a variety of cellular stresses, a recent study has called these observations into question (Marsden et al., 2002). Lymphocytes lacking Apaf-1 or procaspase 9 remain sensitive to the cytotoxic effects of dexamethasone, ionizing radiation, and growth factor

---

fragmentation in a caspase-independent manner. Smac/DIABLO binds to the caspase inhibitors XIAP, cIAP1, and cIAP2, limiting their ability to inhibit active caspases.

As indicated in the text, recent studies have raised the possibility that additional Bcl-2-regulated caspase activation events occur upstream of mitochondrial cytochrome c release (Lassus et al., 2002; Marsden et al., 2002; Robertson et al., 2002). The nature of these steps and their regulation is currently a subject of intense investigation.

withdrawal. Additional studies have suggested that procaspase 2 down-regulation inhibits DNA damage-induced cytochrome c release and subsequent caspase 9 activation in some but not all cell lines (Lassus et al., 2002; Robertson et al., 2002). Curiously, however, mice with targeted *Caspase-2* gene deletions have a very subtle phenotype (Earnshaw et al., 1999). One possibility is that procaspase 2 and another currently unidentified procaspase serve redundant roles as initiator caspases upstream of mitochondria. Although further study is required to test this model, there is general agreement that caspase 9 is the first caspase activated downstream of cytochrome c release.

The Consequences of Caspase Activation

Once caspase 8, 9, or 10 activates caspase 3, biochemical changes that characterize the apoptotic phenotype begin to occur. First, caspase 3 directly participates in endonuclease activation. The nuclease caspase-activated deoxyribonuclease (CAD/DFF40) is a constitutively expressed nuclear protein that is ordinarily complexed with its inhibitor ICAD (inhibitor of CAD/DFF45). Caspase 3-mediated cleavage of ICAD results in the liberation of CAD, which begins to digest the chromatin. Because chromatin in the linker regions is more accessible than chromatin wound around histones (Hewish and Burgoyne, 1973), the net result is a characteristic internucleosomal pattern of DNA degradation. This internucleosomal degradation is accompanied by chromatin condensation, another hallmark of apoptosis. Caspase 3 also cleaves procaspase 6, liberating the active form of this enzyme. Together, caspases 3 and 6 cleave structural proteins of the nucleus (lamins and NuMA), facilitating nuclear fragmentation. In addition, these two proteases (primarily caspase 3) cleave >200 other cellular substrates, thereby inhibiting DNA repair and cell cycle progression, inactivating signal transduction pathways that are critical for survival, and activating a series of enzymes that are thought to participate in cellular disassembly (Earnshaw et al., 1999).

Caspase-Independent Pathways

Even though caspases appear to be critical for cleavage of key substrates during apoptosis, other effector molecules might also contribute to the apoptotic phenotype. Apoptosis-inducing factor (AIF) is an oxidoreductase that is released from mitochondria into the cytosol during apoptosis (Daugas et al., 2000). It subsequently localizes to nuclei, where it is capable of generating large (>50 kb) DNA fragments and inducing chromatin condensation by unknown mechanisms. Targeted disruption of the *AIF* gene results in death early during embryonic development, suggesting that AIF might be required for events in the early embryo. Further studies have demonstrated a failure of central cavitation in blastula-like embryoid bodies formed from AIF-deficient embryonic stem cells. Because AIF-deficient cells are resistant to the proapoptotic effects of menadione but remain sensitive to the induction of apoptosis by a variety of other triggers, including staurosporine, etoposide, hydrogen peroxide, and Fas ligand (Joza et al., 2001), the importance of AIF to cancer treatment remains an open question.

Endonuclease G, another polypeptide released from mitochondria, has also been implicated in apoptosis. This endonuclease, which ordinarily resides in the mitochondrial intermembrane space, reportedly enters nuclei and cleaves chromatin after its release to the cytosol (Li et al., 2001a). It is possible that loss of the normal compartmentalization of endonuclease G explains the chromatin degradation observed in some cell types that cannot properly activate CAD, for example, MCF-7 cells, which lack caspase-3.

In a number of model systems, ranging from tissue culture cell lines in vitro to models of limb development in vivo, it has been observed that caspase inhibition does not prevent cell death. Although caspase inhibitors prevent substrate cleavages, cells ultimately take up vital dyes, indicating rupture of the plasma membrane. In some models, translocation of phosphatidylserine from the inner leaflet of the plasma membrane (where it normally resides) to the extracellular surface also proceeds despite inhibition of caspase activity, providing a recognition signal that allows phagocytes to engulf and clear the cell. The observation that molecules such as AIF and endonuclease G as well as the protease HtRA2 (Suzuki et al., 2001) can be released from mitochondria and exert their effects independent of caspase activity provides a potential explanation for the ability of caspase inhibitors to alter some of the biochemical and morphological features of apoptosis without preventing cell death.

It is also important to recognize that some caspase-independent cell deaths appear to be necrotic, that is, to involve ATP depletion and loss of membrane integrity. When caspases are inhibited, the primary lesions induced by various treatments (e.g., microtubule disruption or DNA damage) and the cytochrome c release that follows still occur. These changes might be sufficiently disruptive to cellular metabolism that cells will ultimately die even if they cannot activate caspases.

REGULATION OF APOPTOTIC PATHWAYS

Regulation of the Extrinsic Pathway

Because of the importance of death receptor signaling for regulation of the immune system, factors that affect signaling through the extrinsic pathway have been extensively studied. It was recognized relatively early that certain death receptors, particularly DR4 and DR5, are expressed in a tissue-restricted pattern in adult mammals (Schulze-Osthoff et al., 1998). In addition, binding of ligand to death receptors is modulated by the tissue specific expression of so-called decoy receptors, that is, cell surface or secreted polypeptides that are capable of binding ligands but not signaling to intracellular signal transduction components (Ashkenazi, 2002). Decoy receptors capable of binding Fas ligand and TRAIL have been well characterized.

Molecules that can modulate death receptor signaling after receptor ligation have also been described. In the case of TNF-R1, DISC assembly cannot proceed until an inhibitor called silencer of death domains (SODD) dissociates from the cytoplasmic domain of the ligated receptor (Jiang et al., 1999). Whether other death receptors are regulated in a similar fashion remains to be determined.

Death domains of at least some of the death receptors also appear to be subject to inhibitory phosphorylation. Although the kinases responsible for this phosphorylation remain to be identified, a phosphatase that enhances death receptor signaling has been described (Sato et al., 1995).

Finally, procaspase 8 activation is regulated. Mammalian cells express endogenous polypeptides called cellular Flice-like inhibitory polypeptides (c-FLIPs) that contain sequences homologous to the procaspase 8 prodomain but lack a caspase active site. Initial studies suggested that c-FLIP overexpression inhibited binding of procaspase 8 to FADD. More recent studies, however, have suggested that at least some splice variants of c-FLIP inhibit procaspase 8 activation after the initial cleavage between the large and small subunits occurs (Krueger et al., 2001).

Bcl-2 Family Members and the Regulation of Mitochondrial Protein Release

Bcl-2 Family Members. The Bcl-2 family consists of a growing list of both pro- and antiapoptotic polypeptides (reviewed in Cory and Adams, 2002; Gross et al., 1999; Kaufmann and Hengartner, 2001). Bcl-2, the founding member of this family, was identified as a polypeptide that is overexpressed when the corresponding gene becomes juxtaposed to the immunoglobulin heavy chain promoter as a consequence of the t(14;18) chromosomal translocation that frequently occurs in indolent B cell lymphomas. Subsequent analysis has identified at least 17 family members, which can be divided into three groups based on the presence or absence of four conserved Bcl-2 homology (BH) domains. The N-terminal BH4 domain is present in all antiapoptotic family members, including Bcl-2, Bcl-w, Bcl-x_L, Mcl-1, A-1, Boo/Diva, and Nrf3. Proapoptotic family members, all of which lack the BH4 domain, fall into one of two categories. Either they are small family members containing only the BH3 domain (e.g., Bid, Bad, Bik, Bim, Blk, Bmf, Hrk, Bnip3, Nix, Noxa, and PUMA) or are larger and resemble Bcl-2 more closely but merely lack the BH4 domain (e.g., Bax, Bak, and Bok).

A growing number of "BH3 only" family members have been recognized, many within the past few years. Examples include Bmf and Bim, which are released on cytoskeletal disruption. Bmf is normally associated with dynein light chain 2 but is released to the cytoplasm and translocated to mitochondria when cell adhesion is disrupted, thereby driving the process of detachment-induced apoptosis or anoikis (Puthalakath et al., 2001). Bim is ordinarily associated with microtubules through its binding to dynein light chain 8 but translocates to mitochondria on treatment of cells with microtubule poisons such as paclitaxel. In each case, the "BH3 only" protein is thought to facilitate release of cytochrome c and subsequent caspase activation as described below.

Other "BH3 only" proteins become active after different processes. Noxa and PUMA are synthesized in a p53-dependent manner after DNA damage. In contrast, Bad is constitutively expressed but is regulated by phosphorylation. Under favorable growth conditions, Bad is phosphorylated by Akt and/or protein kinase A and sequestered in the cytoplasm by 14-3-3 proteins that bind the phosphorylated epitopes. After growth factor deprivation, Akt activity decreases and dephosphorylated Bad translocates to mitochondria, where it is thought to facilitate subsequent apoptosis.

How the proapoptotic Bcl-2 family members regulate apoptosis once they bind to mitochondria has remained an enigma. Some of the models that have been advanced have been based on studies performed in lower eukaryotes such as the nematode *Caenorhabditis elegans*.

Lessons from Worms. Examination of *C. elegans* development has provided considerable insight into the process of programmed cell death (Ellis et al., 1991). Of the 1090 somatic cells that are generated during ontogeny, 131 undergo developmentally programmed cell death. Genetic analysis has demonstrated that a variety of *ced* or "cell death" genes regulate this process. Two genes, *Ced-3* and *Ced-4*, are absolutely required for all developmental cell deaths in *C. elegans*. *Ced-3* encodes a caspase, and *Ced-4* encodes an Apaf-1 ortholog. Additional studies have demonstrated that *Ced-9* loss-of-function mutations result in excess cell deaths; and *Egl-1* loss-of-function mutations result in diminished deaths. The polypeptides encoded by *Ced-9* and *Egl-1* are, respectively, homologous to antiapoptotic and "BH3 only" Bcl-2 family members (Conradt and Horvitz, 1998).

It appears that *Ced-9*, *Ced-4*, and *Ced-3* are constitutively expressed in *C. elegans*. Binding of the CED-9 protein to CED-4 ordinarily blocks oligomerization of CED-4 and concomitant activation of CED-3. When Egl-1 is induced in a cell-specific manner, it binds to CED-9, releasing CED-4 to the cytoplasm, where it oligomerizes, activates CED-3, and kills the cell (Hengartner, 2000). This relationship between anti- and proapoptotic signaling proteins provides a convenient mechanism for keeping the ubiquitously expressed apoptotic machinery inactive until the organism needs to activate it in a highly selective fashion. Although the exact functions of human Bcl-2 family members are slightly different [e.g., the antiapoptotic family members do not appear to interact with human Apaf-1 (Moriishi et al., 1999)], the paradigm of using protein-protein interactions between pro- and antiapoptotic Bcl-2 family members to regulate apoptosis appears to be conserved from worms to humans.

Mitochondria as a Focal Point for Bcl-2 Protein Function. As indicated above, the proapoptotic "BH3 only" proteins present in higher eukaryotes (e.g., Bid, Bim, Bmf) translocate from cytoplasmic sites to the outer surfaces of mitochondria in response to apoptotic stimulii. It is important to emphasize, however, that other Bcl-2 family members also localize to mitochondria. In particular, the proapoptotic protein Bax translocates to the mitochondrial surface after treatment with various stimuli; and Bak appears to constitutively reside on mitochondria. In addition, the antiapoptotic family members also localize to mitochondria, raising the possibility that they might regulate apoptosis by inhibiting cytochrome c release. Consistent with this hypothesis, overexpression of Bcl-2 (Kluck et al., 1997; Yang et al., 1997) or Bcl-x_L (Kim et al., 1997) diminishes release of cytochrome c from mitochondria after treatment of cells with a variety of stimuli.

Overexpression of Bcl-2 or Bcl-x_L also inhibits apoptosis when cytochrome c is directly injected into the cytosol of some cells (Brustugun et al., 1998; Li et al., 1997). This observation raises the possibility that the antiapoptotic effects of Bcl-2 homologs might not be completely explained by their effects on mitochondrial polypeptide release. Consistent with this possibility, it has also been observed that Bcl-2 can inhibit apoptosis in cells that lack Apaf-1 or procaspase 9, suggesting that Bcl-2 might regulate processes that are independent of the cytochrome c-initiated portion of the intrinsic pathway (Marsden et al., 2002). Additional studies have shown that Bcl-2 is also found on the cytoplasmic aspect of nuclear pore complexes, where it might play a role in regulating glutathione flux or nuclear/cytoplasmic transport, and on the cytoplasmic surface of the endoplasmic reticulum, where it is postulated to regulate Ca^{2+} transport (Reed, 1997). Alternatively, cells that fail to activate caspases after cytoplasmic injection of cytochrome c might also require mitochondrial release of other proapoptotic polypeptides, notably Smac/DIABLO (see below), to facilitate caspase activation. Further study is required to distinguish between these possibilities.

Mitochondrial Release of Apoptogenic Polypeptides. Part of the current uncertainty regarding the mechanism by which Bcl-2 family members inhibit apoptosis arises because the process of cytochrome c release from mitochondria remains poorly understood (Cory and Adams, 2002; Desagher and Martinou, 2000). Existing models for cytochrome c release can roughly be divided into those that are mitochondrial permeability transition (MPT) dependent and those that are not. MPT is the term applied to the opening of a pore that allows solutes, ions, and possibly small polypeptides to traverse the mitochondrial membrane. Increases in cytosolic Ca^{2+}, oxidants, and inorganic phosphate are all classical triggers of MPT in isolated mitochondria. Although some possible pore constituents have been identified, including the adenine nucleotide translocase (ANT), voltage-dependent anion channel (VDAC), and cyclophilin D, the exact composition of the pore that is opened during MPT is still under investigation.

Some current models of cytochrome *c* release suggest that components of the MPT pore, perhaps in association with other polypeptides, serve as a conduit for the transit of cytochrome *c* to the cytosol. According to this model, antiapoptotic Bcl-2 family members might interact with components of this pore and inhibit its function. Proapoptotic Bcl-2 family members such as Bax and Bak are postulated to bind antiapoptotic family members, inhibiting their binding to pore components and facilitating the release of cytochrome *c*. These types of models are particularly effective for explaining the proapoptotic effects of agents that directly cause MPT (e.g., calcium ionophore and atractylocide) as well as the proapoptotic effects of polypeptides such as Bad and Bmf, whose only demonstrated function is the ability to bind antiapoptotic Bcl-2 family members. These models have difficulty, however, explaining the observation that cytochrome *c* release in some apoptotic models occurs under conditions in which mitochondria remain fully polarized (Bossy-Wetzel et al., 1998).

Other models suggest an MPT-independent role for proapoptotic Bcl-2 family members. As indicated above, caspase 8-mediated cleavage of Bid leads to a truncated fragment that is capable of facilitating cytochrome *c* release from mitochondria. One set of experiments has suggested that truncated Bid is *N*-myristoylated, which enables it to localize to mitochondria, where it binds cardiolipin, a phospholipid present exclusively on mitochondrial membranes. How the binding of truncated Bid to cardiolipin would result in cytochrome *c* release is not entirely clear. Another model suggests that Bid, which (like other Bcl-2 family members) has pore-forming abilities, oligomerizes or binds to other polypeptides to form a pore that shuttles proteins from the mitochondrial intermembrane space to the cytoplasm. A third model proposes that Bid binds to Bax and Bak, causing a conformational change that facilitates their insertion into the outer mitochondrial membrane, where they then cause release of cytochrome *c*.

A number of recent observations are consistent with this last model (Kaufmann and Hengartner, 2001). First, studies in knockout mice indicate that Bax and Bak both participate in cytochrome *c* release. When the gene for either of these polypeptides is disrupted, cytochrome *c* release and caspase activation are slightly diminished. In contrast, cells lacking both Bax and Bak exhibit a profound defect in cytochrome *c* release after treatment with a variety of different stimuli. Second, truncated Bid has been shown to induce a conformational change in Bax and Bak. Third, truncated Bid has also been shown under cell-free conditions to facilitate binding of Bax and Bak to mitochondria, where they release cytochrome *c*. Fourth, cells lacking both Bax and Bak have been shown to be resistant to the proapoptotic effects of a number of "BH3 only" polypeptides, including Bid, Bad, Bim, and Noxa, suggesting that all "BH3 only" polypeptides might utilize a similar mechanism (Cheng et al., 2001). Thus the suggestion that Bid-induced conformational changes in Bax and Bak are responsible for membrane insertion and pore formation by the latter proteins is consistent with many recent observations.

Despite recent evidence in support of this model, a number of issues remain unresolved. First, it remains unclear how Bax and Bak facilitate cytochrome c release. One possibility is that these polypeptides form pores upon insertion into the mitochondrial outer membrane (Pavlov et al., 2001; Kuwana et al., 2000). Another possibility is that Bax and Bak can interact with components of the MPT pore (Tsujimoto and Shimizu, 2000). Second, the binding of Bid to Bax and Bak has proven difficult to detect in many laboratories. Although it is possible that specific conditions are required or the interaction is transient, it is also possible that the postulated interaction does not exist. The binding of other "BH3 only" proteins to Bax and Bak has likewise proven elusive. Third, the role of the antiapoptotic Bcl-2 family members requires further clarification. Some observations suggest that Bcl-2 and Bcl-x_L bind the active conformation of Bax and Bak, for example, the Bid-induced conformation, preventing their insertion into the outer mitochondrial membrane (Kaufmann and Hengartner, 2001). Other observations raise the possibility that Bcl-2 and Bcl-x_L bind the "BH3 only" family members, abrogating their ability to activate Bax and Bak (Cheng et al., 2001). More recent observations raise the possibility that antiapoptotic Bcl-2 family members might even regulate caspase activation upstream of cytochrome c release (Marsden et al., 2002). Ongoing studies will undoubtedly resolve these issues and provide a clearer picture of how Bcl-2 family members regulate the intrinsic cell death pathway.

Regulation of Apoptosis by IAP Family Members

The IAP Family. In addition to Bcl-2 family members, a family of polypeptides called IAP (inhibitor of apoptosis) proteins also regulates apoptotic processes (Deveraux and Reed, 1999; Miller, 1999; Salvesen and Duckett, 2002). Human members of this family include cIAP1, cIAP2, XIAP, NAIP, and livin. Common structural features include one or more baculovirus inhibitor repeat (BIR) motifs and RING domains. The zinc finger-like BIR domains bind the surfaces of caspases, allowing sequences between the BIRs to block the caspase active site. In this manner, XIAP, cIAP1, and cIAP2 are able to block caspase 8- and 9-initiated events in vitro and to inhibit the activities of purified caspases 3, 7, and 9. The RING domain acts as a ubiquitin ligase and presumably facilitates proteasome-mediated degradation of whatever these IAP proteins bind (Kaufmann and Hengartner, 2001).

Although forced overexpression of any of these polypeptides can inhibit apoptosis, the physiological roles of the IAP proteins remain somewhat uncertain. Deletion of the *XIAP* gene, for example, fails to yield a phenotype (Harlin et al., 2001). cIAP1 and cIAP2 have lower affinity for caspases than XIAP and are postulated to instead regulate tumor necrosis factor-α receptor-associated factors (Li et al., 2002). Thus the ability of IAPs to inhibit caspases in vitro does not establish that this is their function in situ.

It is also important to emphasize that not all BIR-containing poly-peptides necessarily function as caspase inhibitors (Kaufmann and Hengartner, 2001). Survivin, for example, is a small polypeptide that contains a single BIR and no RING domain. This polypeptide is specifically expressed during late G_2 and M phases of the cell cycle. Immunocyto-chemical studies have demonstrated that survivin localizes to centro-meres during early mitosis and at the midbody during telophase (Uren et al., 1999; Wheatley et al., 2001). Interestingly, survivin[-/-] cells display defects in mitosis rather than excessive apoptosis, consistent with a postulated role of small BIR proteins in regulating mitosis (Miller, 1999; Uren et al., 1999).

Regulating the Regulators: Proapoptotic IAP Protein Inhibitors. In contrast to survivin, cIAP1, cIAP2, and XIAP appear to be bona fide regulators of apoptosis. XIAP, for example, is capable of binding and inhibiting partially cleaved caspase 9 but not procaspase 9 (Datta et al., 2000; Srinivasula et al., 2001), providing a convenient way for cells to counter the effects of inadvertent caspase activation. Interestingly, the antiapoptotic activity of XIAP, cIAP1, and cIAP2 is itself regulated. Smac (second mitochondria derived activator of caspase)/DIABLO (direct IAP binding protein with low pI) has been identified as a 19-kDa polypeptide that is capable of binding these antiapoptotic IAP proteins, displacing cleaved caspase 9 and facilitating caspase activation (Srinivasula et al., 2001; Verhagen and Vaux, 2002). More recently, the serine protease HtrA2/Omi has been shown to affect IAP protein action in a similar manner. Interestingly, both Smac/DIABLO and HtrA2 are generally released from mitochondria under conditions similar to those that release cytochrome *c*. Thus a variety of polypeptides that facilitate caspase activation are released from mitochondria downstream of the regulatory effects of Bcl-2 family members.

The Akt Pathway

Although there are undoubtedly a variety of ways in which signal transduction pathways impinge on the core apoptotic machinery, including the effects of the NF-κB and raf/MEK/erk pathways on expression of IAP proteins and antiapoptotic Bcl-2 family members, respectively (Boucher et al., 2000; Earnshaw et al., 1999), one of the most widely studied signal transduction pathways that modulates apoptotic events is the phosphatidylinositol-3 (PI3) kinase/Akt pathway (Datta et al., 1999). In brief, binding of a variety of hormones and cytokines to their cognate receptor tyrosine kinases results in activation of PI3 kinase, which catalyzes the production of 3-phosphoinositides. These lipid mediators then activate phosphoinositide-dependent protein kinase 1, which phosphorylates and activates the protein kinase Akt. Akt, in turn, phosphorylates a number of substrates that directly affect the core apoptotic machinery, including the transcription factor FKHRL1, which loses the ability to transactivate the *Fas ligand* and *Bim* genes when phosphorylated, and the "BH3 only" protein Bad, which displays diminished affinity for the antiapoptotic protein Bcl-x_L when phosphorylated. In addition, Akt-mediated phosphorylation of other unidentified mitochondrial proteins reportedly inhibits mitochondrial release of cytochrome *c*. Thus Akt appears to affect apoptotic events on several levels.

ALTERATIONS IN APOPTOTIC PATHWAYS IN CANCER CELLS

The possibility that apoptotic pathways might be deranged in cancer cells has been extensively investigated. Alterations are commonly found in the death receptor pathway, the mitochondrial pathway, and in regulators that affect both pathways. Because this literature has been recently reviewed (Kaufmann and Gores, 2000; Kaufmann and Vaux, 2003), primary references are provided for only the most recent findings.

Alterations in Death Receptor Pathways

Although death receptor signaling can be interrupted at a variety of steps, a number of alterations that occur in cancer cells interrupt signaling at the level of the receptors themselves. Decreased Fas expression, for example, has been observed in hepatomas compared to normal hepatocytes. Other tumor cells overexpress decoy receptors that bind Fas ligand but cannot induce DISC assembly. The gene for decoy receptor 3, for example, is amplified in ~50% of lung and colon cancers. In addition, this gene is overexpressed without amplification in carcinomas of the stomach, esophagus, and rectum.

Other mechanisms for downregulating the death receptor pathway impinge on steps distal to DISC assembly. Cell lines derived from a variety of neoplasms, including small cell lung cancer, neuroblastoma, and primitive neuroectodermal tumors, fail to express the *Caspase 8* gene as a consequence of DNA methylation or, more rarely, mutational inactivation. The *Caspase 10* gene is mutated in a fraction of gastric cancers, non-small cell lung cancers, and non-Hodgkin lymphomas. In other neoplasms, notably melanoma, pancreatic cancer, and colon cancer, elevated levels of c-FLIP appear to prevent caspase activation.

How does inactivation of the death receptor pathway facilitate carcinogenesis? Interruption of death receptor signaling renders cancer cells less susceptible to killing by lymphocytes that express Fas ligand or TRAIL. In addition, it might also convey resistance to unfavorable environmental conditions. Recent studies have demonstrated that cytotoxic events precipitated by withdrawal of soluble growth factors or adhesion-mediated survival signals can involve activation of the Fas/Fas ligand pathway (Aoudjit and Vuori, 2001; Le-Niculescu et al., 1999; Thangaraju et al., 2000). Accordingly, disruption of this pathway might facilitate the survival of cells during periods of nutritional stress or loss of adhesion, for example, during the process of metastasis.

Alterations in the Mitochondrial Pathway

The mitochondrial pathway is also extensively targeted during the process of carcinogenesis. In principle, this pathway could be inhibited by diminished expression of proapoptotic Bcl-2 family members, over-expression of antiapoptotic Bcl-2 family members, or diminished expression of apoptosome components. Many of these mechanisms have been identified in tumor cell lines or clinical cancer specimens.

Downregulation of Proapoptotic Bcl-2 Family Members. The *Bax* gene contains a nucleotide repeat that is frequently mutated to give a frameshift mutation during development of mismatch repair deficiency-associated gastric and colon cancer (Rampino et al., 1997). Colon cancer cell lines that are heterozygous for this mutation rapidly inactivate the normal allele when grown as xenografts (Ionov et al., 2000), suggesting that loss of the second *Bax* allele conveys a survival advantage to the tumor cells. Consistent with this result, targeted deletion of *Bax* genes also renders HCT116 colon cancer cells resistant to variety of agents (Zhang et al., 2000). Interestingly, patients with colon cancer containing *Bax* mutations have a poorer prognosis than patients with mismatch repair-deficient colon cancer without *Bax* mutations.

The preceding results, although internally consistent, are difficult to reconcile with other data suggesting that Bax and Bak have parallel, redundant roles in apoptosis (Wei et al., 2001). Studies in murine fibroblasts have revealed that targeted deletion of either *Bax* or *Bak* has minimal effect on DNA damage-induced apoptosis, whereas deletion of both genes results in a profound defect. One possible explanation is that Bax and Bak are differentially expressed in various cell types. If Bak levels are particularly low in colon cells, inhibition of Bax expression might have a more important effect on ability to activate the mitochondrial pathway. If this model is correct, then Bak might play a corresponding role in other cell types. It remains to be determined whether the *Bak* gene is mutated or methylated in various cancers.

Inactivating p53 Mutations. The tumor suppressor gene *p53* is mutated or deleted in ~50% of all cancers. These tumors undoubtedly have a variety of defects, including the failure of certain cell cycle checkpoints. Particularly germane to the present discussion, however, is the observation that p53 transcriptionally activates a number of proapoptotic genes, including *Noxa*, *PUMA*, and *p53AIP1*. Two of these three encode "BH3 only" Bcl-2 family members; and all three of the resulting polypeptides localize to mitochondria, where they appear (at least in overexpression studies) to facilitate cytochrome *c* release. Accordingly, deletion of *p53* results in diminished DNA damage-induced cytochrome *c* release.

In addition to inactivating mutations in the *p53* gene, there also appear to be other mechanisms for blunting the p53-induced proapoptotic response. Recent studies have demonstrated that p53-mediated transcription of proapoptotic genes (but not cell cycle arrest genes) requires transcriptional coactivation by apoptosis stimulating protein for p53 no. 1 (ASPP1) or ASPP2 (Roth et al., 2001). Interestingly, downregulation of these transcriptional coactivators is frequently observed in breast cancers that contain wild-type p53 (Roth et al., 2001). The result of this downregulation would be a decrease in DNA damage-induced cytochrome *c* release. In addition, expression of a naturally occurring truncated form of the p53 homolog p73 can suppress p53-mediated transcriptional activation of proapoptotic genes (Melino et al., 2002). Whether this truncated p73 isoform is differentially expressed in various tumors remains to be established.

Overexpression of Antiapoptotic Bcl-2 Family Members. Inhibition of the mitochondrial pathway can also result from overexpression of antiapoptotic Bcl-2 family members. Bcl-2 overexpression is observed in up to 70% of human cancers (Reed, 1999) and is associated with a poor prognosis in patients with acute myelogenous leukemia, intermediate-grade lymphomas, and carcinomas of the prostate, ovary, and upper aerodigestive tract. Other antiapoptotic Bcl-2 family members have more recently been shown to play similar roles in other neoplasms. The gene encoding Bcl-x_L, for example, is transcriptionally activated by signal transducer and activator of transcription-3 (STAT3) in interleukin-6-dependent multiple myeloma cells (Catlett-Falcone et al., 1999). STAT3-dependent signaling likewise appears to upregulate Mcl-1 in large granular lymphocytic leukemia (Epling-Burnette et al., 2001), although the signaling pathway responsible for STAT3 activation remains to be elucidated in the latter case. Finally, STAT5-dependent transcriptional activation of the *Bcl-x_L* and *Mcl-1* genes appears to account, at least in part, for the antiapoptotic effects of bcr/abl, the transforming kinase in chronic myelogenous leukemia and some cases of acute lymphocytic leukemia (Amarante-Mendes et al., 1998; Horita et al., 2000; Mow et al., 2002). In short, antiapoptotic Bcl-2 family members are frequently upregulated in various neoplasms.

Silencing of the Apaf-1 Gene. Diminished expression of cytochrome *c*, Apaf-1, or procaspase 9 would also be expected to dampen the mitochondrial pathway. Although this prediction has been confirmed in fibroblasts from cytochrome $c^{-/-}$ mice, the associated electron transport defect makes it difficult to envision that cytochrome *c*-deficient tumor cells would have a survival advantage. On the other hand, ~50% of metastatic melanomas and corresponding cell lines were recently shown to lack Apaf-1 as a consequence of gene methylation (Soengas et al., 2001). It remains to be determined whether Apaf-1 is similarly downregulated in other neoplasms and whether procaspase 9 suffers a similar fate.

Inactivation of Both Pathways

From the outline of apoptotic pathways presented above, it should be clear that certain alterations can affect signaling through both pathways.

Extensive Alteration of the PI3 Kinase/Akt Pathway in Cancer Cells. A variety of alterations result in activation of the PI3 kinase/Akt pathway in cancer cells (Kaufmann and Gores, 2000; Vivanco and Sawyers, 2002). The gene for Akt1 is occasionally amplified in gastric cancer; and the gene for Akt2 is sometimes amplified in pancreatic and ovarian cancer. More frequently, however, this pathway is activated by changes that occur upstream. A variety of growth factors that signal through PI3 kinase are expressed by tumor cells in an autocrine or paracrine fashion. Mutated growth factor receptors, for example, mutated epidermal growth factor receptors in gliomas, can also activate this pathway. Alternatively, the catalytic subunit of PI3 kinase is overexpressed in some ovarian cancers; and the inhibitory regulatory subunit of PI3 kinase is mutated in colon cancer (Bates and Edwards, 2001).

Perhaps the most frequent alterations affecting this pathway involve decreased expression of the lipid phosphatase PTEN, which removes the 3′ phosphate group from phosphatidylinositol-(3,4,5)-trisphosphate. Diminished PTEN expression results in elevated 3-phosphoinositides and persistent elevation of Akt activity. Interestingly, the *PTEN* gene is inactivated by loss of one allele and mutation of the other in a variety of neoplasms, especially those arising in the brain, prostate, ovary, or endometrium. Current estimates indicate that the PI3 kinase/PTEN/Akt axis is altered in 50% of cancers, making this the second most frequently mutated/altered pathway in cancer after p53. In prostate cancer, on the other hand, PTEN protein is downregulated without mutation, indicating that tumors with mutations might represent only a fraction of the neoplasms in which this pathway is deranged. Additional analysis has revealed that *PTEN* is sometimes mutated even in cells displaying atypical hyperplasia (a so-called preneoplastic phenotype), raising the possibility that PTEN alterations might occur early during the process of carcinogenesis.

IAP Proteins and Cancer. In contrast to alterations in the PI3 kinase pathway, which occur commonly in a wide variety of malignancies, the link between elevated IAP proteins and the development of cancer is less clear. Forced XIAP overexpression in lymphoid cells does not result in an excess of lymphomas (Conte et al., 2001). In addition, elevated tumor cell XIAP expression has been associated with a poorer prognosis in patients with acute myelogenous leukemia (Tamm et al., 2000) but a better prognosis in non-small cell lung cancer (Ferreira et al., 2001). Whether these associations reflect causal relationships or correlations between XIAP expression and unidentified prognostic factors remains to be determined.

Perhaps the clearest link between IAP proteins and malignancy is observed in chronic neutrophilic leukemia (CNL), a rare disorder characterized by clonal accumulation of mature neutrophils in the absence of an identifiable cause (Elliott et al., 2001). Recent work has demonstrated that calpain-mediated turnover of XIAP in normal neutrophils gradually lowers the apoptotic threshold, facilitating apoptosis in this short-lived cell type (Kobayashi et al., 2002). A deficiency of μ-calpain has been linked to persistence of XIAP and inhibition of spontaneous as well as cytokine-induced apoptosis in CNL neutrophils (Kobayashi et al., 2002). Data from other CNL patients, however, have implicated elevated cIAP1 or cIAP2 levels in resistance of CNL neutrophils to apoptosis (Hasegawa et al., 2002). Although it is possible that these results reflect heterogeneity in the pathogenesis of CNL, both of these results remain to be confirmed in samples from more patients.

Apoptotic Dysregulation Is Common in Cancer

In summary, the majority of cancers appear to have one or more alterations in their apoptotic machinery or its regulation. These changes presumably give transformed subclones a survival advantage by inhibiting the apoptosis that would occur in normal cells if they sustained the same degree of aneuploidy and aberrant activation of proliferation-associated genes in the face of sometimes hostile conditions.

INDUCTION OF APOPTOSIS BY CANCER CHEMOTHERAPY

Conventional Anticancer Drugs Induce Apoptosis

Virtually all of the currently utilized anticancer drugs induce apoptosis in susceptible cell types (Herr and Debatin, 2001; Kaufmann and Earnshaw, 2000). Most of the data supporting this view have come from experiments in which tissue culture cells were treated with various agents and examined for apoptotic morphological or biochemical changes. In a few instances, notably acute leukemia patients receiving antileukemic therapy, apoptosis has been shown to occur in the clinical setting as well (Li et al., 1994; Stahnke et al., 2001). Because of its potential relevance to mechanisms of drug resistance, the question of which apoptotic pathways are activated by various agents has received considerable attention (Herr and Debatin, 2001; Kaufmann and Earnshaw, 2000; Villunger and Strasser, 1998). Current evidence suggests that most anticancer drugs, particularly DNA-damaging agents and microtubule poisons, activate the mitochondrial pathway. It is important to emphasize, however, that a growing body of evidence has implicated caspase 8-initiated processes in the action of some drugs as well. The antimetabolite 5-fluorouracil, for example, induces p53-dependent upregulation of Fas ligand followed by autocrine activation of Fas/Fas ligand signaling in some colon cancer lines (Kaufmann and Earnshaw, 2000). Caspase 8 has also been identified as the initiator caspase in cells treated with the DNA-damaging agent camptothecin (Shao et al., 2001) and the chemopreventative agents 2-cyano-3,12-dioxoolean-1,9-dien-28-oic acid (Ito et al., 2000), sulindac sulfide (Huang and He, 2001), and fenretinide (Kalli et al., 2001).

With all of the alterations that inhibit apoptotic pathways, it is natural to wonder how any of the anticancer agents can ever kill neoplastic cells. It is important to realize that other changes in cancer cells sensitize them to these agents. Cells that lack p53, for example, have enhanced sensitivity to DNA-damaging agents because of an inability to arrest in G_1 and repair DNA damage before replication (Hartwell and Kastan, 1994). In a similar fashion, alterations in mitotic checkpoint function that contribute to chromosomal instability and aneuploidy abrogate the ability of cells to arrest in the face of spindle damage (Amon, 1999), thereby enhancing sensitivity to microtubule-directed agents. Finally, as indicated above, alterations that contribute to proliferation (e.g., myc overexpression or ras mutation) appear to render cells more susceptible to apoptosis under unfavorable growth conditions. Thus multiple alterations counterbalance the antiapoptotic changes in cancer cells.

Novel Approaches to Enhance Apoptosis

The growing understanding of apoptotic pathways makes it possible to ask whether neoplastic cells can be killed or sensitized to the effects of anticancer treatments by manipulations that reverse the underlying apoptotic dysregulation. Over the past few years, several approaches for doing this have been examined in preclinical or early clinical studies.

Interruption of Bcl-2 Function. On the basis of estimates that Bcl-2 might be upregulated in as many as 70% of all tumors, there has been considerable interest in inhibiting Bcl-2 function. G3139, a phosphorothioate antisense oligonucleotide that is complementary to the first 18 nucleotides of the Bcl-2 coding sequence, has been developed as a molecule that specifically downregulates Bcl-2 mRNA (Cotter, 1999). Phase I clinical testing has demonstrated the successful downregulation of tumor cell Bcl-2 protein levels in some patients (Jansen et al., 2000; Marcucci et al., 2002; Waters et al., 2000). Because Bcl-2 downregulation would be expected to be most effective if combined with agents that trigger the mitochondrial pathway, G3139 is currently undergoing extensive clinical testing in combination with other anticancer agents. An antisense oligonucleotide that downregulates Bcl-x_L function is likewise being tested clinically.

Potential problems with antisense oligonucleotides include their short serum half-lives, which necessitate continuous intravenous administration for up to 2 or 3 weeks, and the occurrence of sequence independent toxicities. An alternative approach would be the development of a small molecule that inhibits the antiapoptotic function of Bcl-2. HA14-1, a small amphipathic organic molecule that was predicted to bind to a surface pocket formed by domains BH1 to BH3 on Bcl-2, has recently been shown to induce Apaf-1-dependent apoptosis in Bcl-2-expressing cells (Wang et al., 2000). Further studies are required to demonstrate that this agent actually targets Bcl-2 in situ and to assess the feasibility of developing this molecule into a drug. Nonetheless, the identification of HA14-1 raises the possibility that inhibition of antiapoptotic Bcl-2 family members by small molecules might be feasible.

Restoration of p53 Function. As indicated above, loss or mutation of *p53* would be expected to diminish DNA damage-induced expression of proapoptotic p53 transcriptional targets, including PUMA, Noxa, and p53AIP1. Genes encoding Bax, Bid, Fas ligand, the TRAIL receptor DR5, procaspase 6, and Apaf-1 are also transcriptionally activated by p53 in some cellular contexts. Collectively, these observations have provided a rationale for reintroducing wild-type p53 into cells. One approach is to infect cells with adenovirus encoding p53 (Roth et al., 2001). Although this strategy is currently being tested in a number of tumor models, the ability of mutant p53 to bind and inactivate wild-type p53 creates a potential problem. To circumvent this pitfall, adenoviral delivery of Bax is also being investigated (Li et al., 2001b). Alternatively, recent studies have identified a class of small molecules that facilitate refolding of mutant p53 into the wild-type conformation (Foster et al., 1999). Although this type of molecule would be ineffective in cells harboring p53 deletions, further studies will determine whether this approach is effective in tumors with mutant p53.

Forced Caspase Expression. A totally different approach would be to bypass Bcl-2 family members and introduce active caspases directly into cells. Adenoviral vectors expressing caspases have recently been introduced into two different models. In one model, adenovirus-mediated introduction of caspase 8 facilitated anoikis and inhibited peritoneal dissemination of gastric carcinoma cells in a murine xenograft model (Nishimura et al., 2001). Whether this approach will have sufficient selectivity to permit clinical development is unclear. Moreover, the propensity of tumor cells to disseminate before neoplasms become clinically manifest limits enthusiasm for this approach.

In an attempt to activate apoptosis without requiring a second stimulus like loss of adhesion-mediated signaling, another study examined adenovirus encoding procaspase 9 under the control of a prostate-specific promoter (Xie et al., 2001). Because the procaspase 9 was fused to an FK506 binding protein domain, selective dimerization and activation of caspase 9 could be triggered by treatment with a dimerized FK506 homolog. Likewise, it has been proposed that the telomerase promoter might selectively drive caspase transgene expression in tumor cells (Komata et al., 2002). Further studies utilizing these approaches are awaited with interest.

Smac Peptides. A potential problem with introducing active caspases into tumor cells is the possibility that elevated levels of IAP proteins, notably XIAP, might inhibit the apoptosis. Under physiological conditions, this problem is overcome by the mitochondrial release of Smac/DIABLO, which binds IAP proteins and inhibits their antiapoptotic function. Structural analyses of XIAP bound to Smac (reviewed in Kaufmann and Hengartner, 2001; Verhagen and Vaux, 2002) have demonstrated that the four N-terminal amino acids of Smac are critical for this binding. On the basis of additional studies demonstrating the ability of synthetic peptides with this IAP binding sequence to facilitate caspase activation under cell-free conditions, cell-permeant peptides and small molecule peptidomimetics are being developed by a number of groups in hopes of overcoming drug resistance or sensitizing cells to chemotherapeutic agents that exert their effects via mitochondria (Arnt et al., 2002; Fulda et al., 2002).

Inhibitors of PI3 Kinase/Akt Signaling. A number of experimental approaches also have the potential to inhibit antiapoptotic signaling through the PI3 kinase/Akt pathway. Several of these approaches target signaling upstream of PI3 kinase. For example, inhibitors that target one or more epidermal growth factor receptor family members are currently in development (Fry, 1999). Likewise, STI571, the recently approved bcr/abl inhibitor, inhibits bcr/abl-mediated activation of the PI3 kinase/Akt pathway (Neshat et al., 2000) as well as STAT5-induced transcription of Bcl-x_L and Mcl-1 (Horita et al., 2000; Mow et al., 2002).

Because receptor tyrosine kinase inhibitors and STI571 inhibit upstream signaling that is tumor type specific, these molecules are applicable to a limited range of neoplasms. The demonstration that PI3 kinase and Akt are components of a pathway downstream of diverse antiapoptotic signals has prompted renewed interest in PI3 kinase inhibitors (e.g., LY294004) as well as the development of Akt inhibitors. Reports of the antineoplastic effects of these compounds, alone and in combination with other proapoptotic treatments, are awaited with interest.

SUMMARY

Although initial descriptions of cancer highlighted the role of uncontrolled proliferation, more recent studies have demonstrated altered regulation of apoptosis in neoplastic cells. It now appears that the vast majority of cancer cells have changes that inhibit activation of the mitochondrial pathway and/or the death receptor pathway. On the basis of these results, studies are beginning to examine strategies to potentiate apoptosis, including inhibition of antiapoptotic Bcl-2 family members, reintroduction of wild-type p53, adenoviral delivery of caspase cDNA, inhibition of IAP function, and interruption of PI3 kinase/Akt survival signaling. We predict that these rationally developed proapoptotic strategies will assume increasingly important roles in anticancer therapy over the next decade.

REFERENCES

Amarante-Mendes, G. P., McGahon, A. J., Nishioka, W. K., Afar, D. E., Witte, O. N., and Green, D. R. (1998). Bcl-2-independent Bcr-Abl-mediated resistance to apoptosis: protection is correlated with up regulation of Bcl-xL. Oncogene *16*, 1383–1390.

Amon, A. (1999). The spindle checkpoint. Curr Opin Genet Dev *9*, 69–75.

Aoudjit, F., and Vuori, K. (2001). Matrix attachment regulates Fas-induced apoptosis in endothelial cells: A ROLE for c-Flip and implications for anoikis. J Cell Biol *152*, 633–643.

Arends, M. J., and Wyllie, A. H. (1991). Apoptosis: Mechanisms and roles in pathology. Intl Rev Exp Pathol *32*, 223–254.

Arnt, C. R., Chiorean, M. V., Heldebrant, M. P., Gores, G. J., and Kaufmann, S. H. (2002). Synthetic Smac/DIABLO peptides enhance the effects of chemotherapeutic agents by binding XIAP and cIAP1 in situ. J Biol Chem *277*, 44236–44243.

Ashkenazi, A. (2002). Targeting death and decoy receptors of the tumour-necrosis factor superfamily. Nat Rev Cancer *2*, 420–430.

Bates, R. C., and Edwards, N. S. (2001). A CD44 survival pathway triggers chemoresistance vs. Lyn kinase and phosphoinositide 3-kinase/Akt in colon carcinoma cells. Cancer Res *61*, 5275–5283.

Bedi, A., Basricha, P. J., Akhtar, A. J., Barber, J. P., Bedi, G. C., Giardiello, F. M., Zehnbauer, B. A., Hamilton, S. R., and Jones, R. J. (1995). Inhibition of apoptosis during development of colorectal cancer. Cancer Res *55*, 1811–1816.

Bossy-Wetzel, E., Newmeyer, D. D., and Green, D. R. (1998). Mitochondrial cytochrome *c* release in apoptosis occurs upstream of DEVD-specific caspase activation and independently of mitochondrial transmembrane depolarization. EMBO J *17*, 37–49.

Boucher, M. J., Morisset, J., Vachon, P. H., Reed, J. C., Laine, J., and Rivard, N. (2000). MEK/ERK signaling pathway regulates the expression of Bcl-2, Bcl-X(L), and Mcl-1 and promotes survival of human pancreatic cancer cells. J Cell Biochemi *79*, 355–369.

Brustugun, O. T., Fladmark, K. E., Doskeland, S. O., Orrenius, S., and Zhivotovsky, B. (1998). Apoptosis Induced by microinjection of cytochrome *c* is caspase-dependent and is inhibited by Bcl-2. Cell Death Differ *5*, 660–668.

Budihardjo, I., Oliver, H., Lutter, M., Luo, X., and Wang, X. (1999). Biochemical pathways of caspase activation during apoptosis. Annu Rev Cell Devel Biol *15*, 269–290.

Catlett-Falcone, R., Landowski, T. H., Oshiro, M. M., Turkson, J., Levitzki, A., Savino, R., Ciliberto, G., Moscinski, L., Fernández-Luna, J. L., Nuñez, G., Dalton, W. C., and Jove, R. (1999). Constitutive activation of Stat3 signaling confers resistance to apoptosis in human U266 myeloma cells. Immunity *10*, 105–115.

Cheng, E. H., Wei, M. C., Weiler, S., Flavell, R. A., Mak, T. W., Lindsten, T., and Korsmeyer, S. J. (2001). BCL-2, BCL-X(L) Sequester BH3 domain-only molecules preventing BAX- and BAK-mediated mitochondrial apoptosis. Mol Cell *8*, 705–711.

Conradt, B., and Horvitz, H. R. (1998). The *C. elegans* protein EGL-1 is required for programmed cell death and interacts with the Bcl-2-like protein CED-9. Cell *93*, 519–529.

Conte, D., Liston, P., Wong, J. W., Wright, K. E., and Korneluk, R. G. (2001). Thymocyte-targeted overexpression of XIAP transgene disrupts T lymphoid apoptosis and maturation. Proc Natl Acad Sci USA *98*, 5049–5054.

Cory, S., and Adams, J. M. (2002). The Bcl2 family: regulators of the cellular life-or-death switch. Nat Rev Cancer *2*, 647–656.

Cory, S., Vaux, D. L., Strasser, A., Harris, A. W., and Adams, J. M. (1999). Insights from Bcl-2 and Myc: Malignancy involves abrogation of apoptosis as well as sustained proliferation. Cancer Res *59*, 1685s–1692s.

Cotter, F. E. (1999). Antisense therapy of hematologic malignancies. Semin Hematol *36*, 9–14.

Datta, R., Oki, E., Endo, K., Biedermann, V., Ren, J., and Kufe. D. (2000). XIAP regulates DNA damage-induced apoptosis downstream of caspase-9 cleavage. J Biol Chem *275*, 31733–31738.

Datta, S. R., Brunet, A., and Greenberg, M. E. (1999). Cellular survival: a play in three Akts. Genes Dev *13*, 2905–2927.

Daugas, E., Nochy, D., Ravagnan, L., Loeffler, M., Susin, S. A., Zamzami, N., and Kroemer, G. (2000). Apoptosis-inducing factor (AIF): A ubiquitous mitochondrial oxidoreductase involved in apoptosis. FEBS Lett *476*, 118–123.

Desagher, S., and Martinou, J. C. (2000). Mitochondria as the central control point of apoptosis. Trends Cell Biol *10*, 369–377.

Deveraux, Q. L., and Reed, J. C. (1999). IAP family proteins—suppressors of apoptosis. Genes Dev *13*, 239–252.

Earnshaw, W. C., Martins, L. M., and Kaufmann, S. H. (1999). Mammalian caspases: Structure, activation, substrates and functions during apoptosis. Annu Rev Biochem *68*, 383–424.

Elliott, M. A., Dewald, G. W., Tefferi, A., and Hanson, C. A. (2001). Chronic neutrophilic leukemia (CNL): A clinical, pathologic, and cytogenetic study. Leukemia *15*, 35–40.

Ellis, R. E., Yuan, J. Y., and Horvitz, H. R. (1991). Mechanisms and functions of cell death. Annu Rev Cell Biol *7*, 663–698.

Epling-Burnette, P. K., Liu, J. H., Catlett-Falcone, R., Turkson, J., Oshiro, M., Kothapalli, R., Li, Y., Wang, J.-M., Yang-Yen, H.-F., Karras, J., Jove, R., and Loughran, T. P. (2001). Inhibition of STAT3 signaling leads to apoptosis of leukemic large granular lymphocytes and decreased Mcl-1 expression. J Clin Invest *107*, 351–362.

Evan, G. I., and Vousden, K. H. (2001). Proliferation, cell cycle and apoptosis in cancer. Nature *411*, 342–348.

Ferreira, C. G., van der Valk, P., Span, S. W., Ludwig, I., Smit, E. G., Kruyt, F. A. E., Pinedo, H. M., van Tinteren, H., and Giaccone, G. (2001). Expression of X-linked inhibitor of apoptosis as a novel prognostic marker in radically resected non-small cell lung cancer patients. Clin Cancer Res *7*, 2468–2474.

Foster, B. A., Coffey, H. A., Morin, M. J., and Rastinejad, F. (1999). Pharmacological rescue of mutant p53 conformation and function. Science *286*, 2507–2510.

Fry, D. W. (1999). Inhibition of the epidermal growth factor receptor family of transferase kinases as an approach to cancer chemotherapy: progression from reversible to irreversible inhibitors. Pharmacol Therapeut *82*, 207–218.

Fulda, S., Wick, W., Weller, M., and Debatin, K.-M. (2002). Smac agonists sensitize for Apo2L/TRAIL- or anticancer drug-induced apoptosis and induce regression of malignant glioma in vivo. Nat Med *8*, 808–815.

Genini, D., Budihardjo, I., Plunkett, W., Wang, X., Carrera, C. J., Cottam, H. B., Carson, D. A., and Leoni, L. M. (2000). Nucleotide requirements for the in vitro activation of the apoptosis protein-activating factor-1-mediated caspase pathway. J Biol Chem *275*, 29–34.

Green, D. R., and Reed, J. C. (1998). Mitochondria and apoptosis. Science *281*, 1309–1312.

Gross, A., McDonnell, J. M., and Korsmeyer, S. J. (1999). BCL-2 family members and the mitochondria in apoptosis. Genes Dev *13*, 1899–1911.

Harlin, H., Reffey, S. B., Duckett, C.S., Lindsten, T., and Thompson, C. B. (2001). Characterization of XIAP-deficient mice. Mol Cell Biol *21*, 3604–3608.

Hartwell, L. H., and Kastan, M. B. (1994). Cell cycle control and cancer. Science *266*, 1821–1828.

Hasegawa, T., Suzuki, K., Sakamoto, C., Ohta, K., Nishiki, S., Hino, M., Tatsumi, N., and Kitagawa, S. (2003). Expression of the inhibitor of apoptosis (IAP) family members in human neutrophils: up-regulation of cIAP2 by granulocyte colony-stimulating factor and overexpression of cIAP2 in chronic neutrophilic leukemia. Blood *101*, 1164–1171.

Hengartner, M. O. (2000). The biochemistry of apoptosis. Nature *407*, 770–776.

Herr, I., and Debatin, K.-M. (2001). Cellular stress response and apoptosis in cancer therapy. Blood *98*, 2603–2614.

Hewish, D. R., and Burgoyne, L. A. (1973). Chromatin sub-structure. The digestion of chromatin DNA at regularly spaced sites by a nuclear deoxyribonuclease. Biochem Biophys Res Commun *52*, 504–510.

Horita, M., Andreu, E. J., Benito, A., Arbona, C., Sanz, C., Benet, I., Prosper, F., and Fernandez-Luna, J. L. (2000). Blockade of the Bcr-Abl kinase activity induces apoptosis of chronic myelogenous leukemia cells by suppressing signal transducer and activator of transcription 5-dependent expression of Bcl-xL. J Exp Med *191*, 977–984.

Huang, Y., and He, Q. (2001). Sulindac sulfide-induced apoptosis involves death receptor 5 and the caspase 8-dependent pathway in human colon and prostate cancer cells. Cancer Res *61*, 6918–6924.

Ionov, Y., Yamamoto, H., Krajewski, S., Reed, J. C., and Perucho, M. (2000). Mutational inactivation of the proapoptotic gene BAX confers selective advantage during tumor clonal evolution. Proc Natl Acad Sc USA *97*, 10872–10877.

Ito, Y., Pandey, P., Place, A., Sporn, M. B., Gribble, G. W., Honda, T. K. S., and Kufe, D. (2000). The novel triterpenoid 2-cyano-3,12-dioxoolean-1,9-dien-28-oic acid induces apoptosis of human myeloid leukemia cells by a caspase-8-dependent mechanism. Cell Growth Differ *11*, 261–267.

Jansen, B., Wacheck, V., Heere-Ress, E., Schlagbauer-Wadl, H., Hoeller, C., Lucas, T., Hoermann, M., Hollenstein, U., Wolff, K., and Pehamberger, H. (2000). Chemosensitisation of malignant melanoma by BCL2 antisense therapy. Lancet *356*, 1728–1733.

Jiang, Y., Woronicz, J. D., Liu, W., and Goeddel, D. V. (1999). Prevention of constitutive TNF receptor 1 signaling by silencer of death domains. Science *283*, 543–546.

Joza, N., Susin, S. A., Daugas, E., Stanford, W. L., Cho, S. K., Li, C. Y., Sasaki, T., Elia, A. J., Cheng, H. Y., Ravagnan, L., Ferri, K. F., Zamzami, N., Wakeham, A., Hakem, R., Yoshida, H., Kong, Y. Y., Mak, T. W., Zuniga-Pflucker, J. C., Kroemer, G., and Penninger, J. M. (2001). Essential role of the mitochondrial apoptosis-inducing factor in programmed cell death. Nature *410*, 549–554.

Kalli, K. R., Devine, K.E., Erlichman, C., Hartmann, L. C., Jenkins, R. B., Conover, C. A., and Kaufmann, S. H. (2001). Fenretinide-induced apoptosis is initiated by procaspase-8 in an epithelial ovarian carcinoma cell line. Proc Am Assn Cancer Res *42*, 639–640.

Kaufmann, S. H., and Earnshaw, W. C. (2000). Induction of apoptosis by cancer chemotherapy. Exp Cell Res *256*, 42–49.

Kaufmann, S. H., and Gores, G. J. (2000). Apoptosis in cancer: cause and cure. Bioessays *22*, 1007–1017.

Kaufmann, S. H., and Hengartner, M. (2001). Programmed cell death: alive and well in the new millenium. Trends Cell Biol *11*, 526–534.

Kaufmann, S. H., and Vaux. (2003). Oncogene (in press).

Kim, C. N., Wang, X., Huang, Y., Ibrado, A. M., Liu, L., Frang, G., and Bhalla, K. (1997). Overexpression of Bcl-X(L) Inhibits Ara-C-induced mitochondrial loss of cytochrome *c* and other perturbations that activate the molecular cascade of apoptosis. Cancer Res *57*, 3115–3120.

Kischkel, F. C., Lawrence, D. A., Tinel, A., LeBlanc, H., Virmani, A., Schow, P., Gazdar, A., Blenis, J., Arnott, D., and Ashkenazi, A. (2001). Death receptor recruitment of endogenous caspase-10 and apoptosis initiation in the absence of caspase-8. J Biol Chem *276*, 46639–46646.

Kluck, R. M., Bossy-Wetzel, E., Green, D. R., and Newmeyer, D. D. (1997). The release of cytochrome *c* from mitochondria: a primary site for Bcl-2 regulation of apoptosis. Science *275*, 1132–1136.

Kobayashi, S., Takahashi, A., Yamashita, K., Takeoka, T., Ohtsuki, T., Suzuki, Y., Takahashi, R., Yamamoto, K., Kaufmann, S. H., Uchiyama, T., and Sasada, M. (2002). Calpain-mediated XIAP degradation in neutrophil apoptosis and its impairment in chronic neutrophilic leukemia. J Biol Chem *277*, 33968–33977.

Komata, T., Kondo, Y., Kanzawa, T., Ito, H., Hirohata, S., Koga, S., Sumiyoshi, H., Takakura, M., Inoue, M., Barna, B.P., Germano, I. M., Kyo, S., and Kondo, S. (2002). Caspase-8 gene therapy using the human telomerase reverse transcriptase promoter for malignant glioma cells. Hum Gene Therapy *13*, 1015–1025.

Krueger, A., Schmitz, I., Baumann, S., Krammer, P. H., and Kirchhoff, S. (2001). Cellular FLICE-inhibitory protein splice variants inhibit different steps of caspase-8 activation at the CD95 death-inducing signaling complex. J Biol Chem *276*, 20633–20640.

Kuwana, T., Mackey, M. R., Perkins, G., Ellisman, M. H., Latterich, M., Schneiter, R., Green, D. R., and Newmeyer, D. D. (2002). Bid, Bax, and lipids cooperate to form supramolecular openings in the outer mitochondrial membrane. Cell *111*, 331–342.

Lassus, P., Opitz-Araya, X., and Lazebnik, Y. (2002). Requirement for caspase-2 in stress-induced apoptosis before mitochondrial permeabilization. Science *297*, 1352–1354.

Le-Niculescu, H., Bonfoco, E., Kasuya, Y., Claret, F. X., Green, D. R., and Karin, M. (1999). Withdrawal of survival factors results in activation of the JNK pathway in neuronal cells leading to Fas ligand induction and cell death. Mol Cell Biol *19*, 751–763.

Li, F. L., Srinivasan, A., Wang, Y., Armstrong, R. C., Tomaselli, K. J., and Fritz, L. C. (1997). Cell-specific induction of apoptosis by microinjection of cytochrome *c*. Bcl-xL has activity independent of cytochrome *c* release. J Biol Chem *272*, 30299–30305.

Li, L. Y., Luo, X., and Wang, X. (2001a). Endonuclease G is an apoptotic DNase when released from mitochondria. Nature *412*, 95–99.

Li, X., Gong, J., Feldman, E., Seiter, K., Traganos, F., and Darzynkiewicz, Z. (1994). Apoptotic cell death during treatment of leukemias. Leukemia Lymphoma *13*, 65–70.

Li, X., Marani, M., Yu, J., Nan, B., Roth, J.A., Kagawa, S., Fang, B., Denner, L., and Marcelli, M. (2001b). Adenovirus-mediated Bax overexpression for the induction of therapeutic apoptosis in prostate cancer. Cancer Res *61*, 186–191.

Li, X., Yang, Y., and Ashwell, J. D. (2002). TNF-RII and c-IAP1 mediate ubiqui-tination and degradation of TRAF2. Nature *416*, 345–347.

Marcucci, G., Byrd, J. C., Dai, G., Klisovic, M. I., Young, D. C., Cataland, S. R., Fisher, D. B., Lucas, D., Chan, K. K., Porcu, P., Lin, Z. P., Farag, S. F., Frankel, S. R., Zweibel, J. A., Kraut, E. H., Balcerzak, S. P., Bloomfield, C. D., Grever, M. R., and Caligiuri, M. A. (2003). Phase I and pharmacodynamic studies of G3139, a Bcl-2 antisense oligonucleotide, in combination with chemotherapy in refractory or relapsed acute leukemia. Blood *101*, 425–432.

Marsden, V. S., O'Connor, L., O'Reilly, L. A., Silke, J., Metcalf, D., Ekert, P. G., Huang, D. C. S., Cecconi, F., Kuida, K., Tomaselli, K. J., Roy, S., Nicholson, D. W., Vaux, D. L., Bouillet, P., Adams, J. M., and Strasser, A. (2002). Apoptosis initiated by Bcl-2-regulated caspase activation independently of the cytochrome *c*/Apaf-1/caspase-9 apoptosome. Nature *419*, 634–637.

Melino, G., De Laurenzi, V., and Vousden, K. H. (2002). p73: Friend or foe in tumorigenesis. Nat Rev Cancer *2*, 605–615.

Miller, L. K. (1999). An EXEGESIS of IAPs: Salvation and surprises from BIR motifs. Trends Cell Biol *9*, 323–328.

Moriishi, K., Huang, D.C., Cory, S., and Adams, J. M. (1999). Bcl-2 family members do not inhibit apoptosis by binding the caspase activator Apaf-1. Proc Natl Acad Sci USA *96*, 9683–9688.

Mow, B. M. F., Chandra, J., Svingen, P.A., Hallgren, C. G., Weisberg, E., Kottke, T. J., Narayanan, V. L., Litzow, M .R., Friggen, J. D., Sausville, E. A., Tefferi, A., and Kaufmann, S. H. (2002). Effects of the Bcr/Abl kinase inhibitors STI571 and adaphostin (NSC 680410) on chronic myelogenous leukemia cells in vitro. Blood *99*, 664–671.

Neshat, M. S., Raitano, A. B., Wang, H. G., Reed, J. C., and Sawyers, C. L. (2000). The survival function of the Bcr-Abl oncogene is mediated by Bad-dependent and -independent pathways: roles for phosphatidylinositol 3-kinase and Raf. Mol Cell Biol *20*, 1179–1186.

Nishimura, S., Adachi, M., Ishida, T., Matsunaga, T., Uchida, H., Hamada, H., and Imai, K. (2001). Adenovirus-mediated transfection of caspase-8 augments anoikis and inhibits peritoneal dissemination of human gastric carcinoma cells. Cancer Res *61*, 7009–7014.

Pavlov, E. V., Priault, M., Pietkiewicz, D., Cheng, E. H., Antonsson, B., Manon, S., Korsmeyer, S. J., Mannella, C. A., and Kinnally, K. W. (2001). A novel, high conductance channel of mitochondria linked to apoptosis in mammalian cells and Bax expression in yeast. J Cell Biol *155*, 725–732.

Peter, M. E. (2000). The TRAIL Discussion: It is FADD and caspase-8! Cell Death Differ *7*, 759–760.

Puthalakath, H., Villunger, A., O'Reilly, L. A., Beaumont, J. G., Coultas, L., Cheney, R. E., Huang, D. C., and Strasser, A. (2001). Bmf: A proapoptotic BH3-only protein regulated by interaction with the myosin V actin motor complex, activated by anoikis. Science *293*, 1829–1832.

Raff, M. C. (1992). Social controls on cell survival and cell death. Nature *356*, 397–400.

Rampino, N., Yamamoto, H., Ionov, Y., Li, Y., Sawai, H., Reed, J. C., and Perucho, M. (1997). Somatic frameshift mutations in the BAX gene in colon cancers of the microsatellite mutator phenotype. Science *275*, 967–969.

Reed, J. C. (1997). Bcl-2 family proteins: Strategies for overcoming chemoresistance in cancer. Adv Pharmacol *41*, 501–532.

Reed, J. C. (1999). Dysregulation of apoptosis in cancer. J Clin Oncol *17*, 2941–2953.

Robertson, J. D., Enoksson, M., Suomela, M., Zhivotovsky, B., and Orrenius, S. (2002). Caspase-2 acts upstream of mitochondria to promote cytochrome *c* release during etoposide-induced apoptosis. J Biol Chem *277*, 29803–29809.

Roth, J. A., Grammer, S. F., Swisher, S. G., Komaki, R., Nemunaitis, J., Merritt, J., Fujiwara, T., and Meyn, R. E., Jr. (2001). Gene therapy approaches for the management of non-small cell lung cancer. Semin Oncol *28*, 50–56.

Salvesen, G. S., and Dixit, V. M. (1999). Caspase activation: The induced-proximity model. Proc Natl Acad Sci USA *96*, 10964–10967.

Salvesen, G. S., and Duckett, C. S. (2002). IAP proteins: Blocking the road to death's door. Mol Cell Biol *3*, 401–410.

Sato, T., Irie, S., Kitada, S., and Reed, J. C. (1995). FAP-1: A protein tyrosine phosphatase that associates with Fas. Science *268*, 411–415.

Schulze-Osthoff, K., Ferrari, D., Los, M., Wesselborg, S., and Peter, M. E. (1998). Apoptosis signaling by death receptors. Eur J Biochem *254*, 439–459.

Shao, R. G., Cao, C. X., Nieves-Neira, W., Dimanche-Boitrel, M. T., Solary, E., and Pommier, Y. (2001). Activation of the Fas pathway independently of Fas ligand during apoptosis induced by camptothecin in p53 mutant human colon carcinoma cells. Oncogene *20*, 1852–1859.

Soengas, M. S., Capodieci, P., Polsky, D., Mora, J., Esteller, M., Opitz-Araya, X., McCombie, R., Herman, J. G., Gerald, W. L., Lazebnik, Y. A., Cordon-Cardo, C., and Lowe, S. W. (2001). Inactivation of the apoptosis effector Apaf-1 in malignant melanoma. Nature *409*, 207–211.

Srinivasula, S. M., Hegde, R., Saleh, A., Datta, P., Shiozaki, E., Chai, J., Lee, R. A., Robbins, P. D., Fernandes-Alnemri, T., Shi, Y., and Alnemri, E. S. (2001). A conserved XIAP-interaction motif in caspase-9 and Smac/DIABLO regulates caspase activity and apoptosis. Nature *410*, 112–116.

Stahnke, K., Fulda, S., Friesen, C., Staub, G., and Debatin, K.-M. (2001). Activation of apoptosis pathways in peripheral blood lymphocytes by in vivo chemotherapy. Blood *98*, 3066–3073.

Suzuki, Y., Imai, Y., Nakayama, H., Takahashi, K., Takio, K., and Takahashi, R. (2001). A serine protease, HtrA2, is released from the mitochondria and interacts with XIAP, inducing cell death. Mol Cell *8*, 613–621.

Tamm, I., Kornblau, S. M., Segall, H., Krajewski, S., Welsh, K., Kitada, S., Scudiero, D. A., Tudor, G., Qui, Y. H., Monks, A., Andreeff, M., and Reed, J. C. (2000). Expression and prognostic significance of IAP-family genes in human cancers and myeloid leukemias. Clin Cancer Res 6, 1796–1803.

Thangaraju, M., Kaufmann, S. H., and Couch, F. (2000). BRCA1 facilitates stress-induced apoptosis in breast and ovarian cancer. J Biol Chem 275, 33487–33496.

Thornberry, N. A., and Lazebnik, Y. (1998). Caspases: Enemies within. Science 281, 1312–1316.

Tsujimoto, Y., and Shimizu, S. (2000). VDAC regulation by the Bcl-2 family of proteins. Cell Death Differ 7, 1174–1181.

Uren, A. G., Beilharz, T., O'Connell, M. J., Bugg, S. J., van Driel, R., Vaux, D. L., and Lithgow, T. (1999). Role for yeast inhibitor of apoptosis (IAP)-like proteins in cell division. Proc Natl Acad Sci USA 96, 10170–10175.

Verhagen, A. M., and Vaux, D. L. (2002). Cell death regulation by the mammalian IAP antagonist Diablo/Smac. Apoptosis 7, 163–166.

Villunger, A., and Strasser, A. (1998). Does "death receptor" signaling play a role in tumorigenesis and cancer therapy? Oncol Res 10, 541–550.

Vivanco, I., and Sawyers, C. L. (2002). The phosphatidylinositol 3-kinase AKT pathway in human cancer. Nat Rev Cancer 2, 489–501.

Wang, J., Chun, H. J., Wong, W., Spencer, D. M., and Lenardo, M. J. (2001). Caspase-10 is an initiator caspase in death receptor signaling. Proc Natl Acad Sci USA 98, 13884–13888.

Wang, J. L., Liu, D., Zhang, Z. J., Shan, S., Han, X., Srinivasula, S. M., Croce, C. M., Alnemri, E. S., and Huang, Z. (2000). Structure-based discovery of an organic compound that binds Bcl-2 protein and induces apoptosis of tumor cells. Proc Natl Acad Sci USA 97, 7124–7129.

Waters, J. S., Webb, A., Cunningham, D., Clarke, P. A., Raynaud, F., di Stefano, F., and Cotter, F. E. (2000). Phase I clinical and pharmacokinetic study of Bcl-2 antisense oligonucleotide therapy in patients with non-Hodgkin's lymphoma. J Clin Oncol 18, 1812–1823.

Wei, M. C., Zong, W. X., Cheng, E .H., Lindsten, T., Panoutsakopoulou, V., Ross, A. J., Roth, K. A., MacGregor, G. R., Thompson, C. B., and Korsmeyer, S. J. (2001). Proapoptotic BAX and BAK: A requisite gateway to mitochondrial dysfunction and death. Science 292, 727–730.

Wheatley, S. P., Carvalho, A., Vagnarelli, P., and Earnshaw, W. C. (2001). INCENP is required for proper targeting of survivin to the centromeres and the anaphase spindle during mitosis. Curr Biol 11, 886–890.

Wyllie, A. H., Kerr, J. F. R., and Currie, A. R. (1980). Cell death: The significance of apoptosis. Intl Rev Cytol 68, 251–306.

Xie, X., Zhao, X., Liu, Y., Zhang, J., Matusik, R. J., Slawin, K. M., and Spencer, D. M. (2001). Adenovirus-mediated tissue-targeted expression of a caspase-9-based artificial death switch for the treatment of prostate cancer. Cancer Res 61, 6795–6804.

Yang, J., Liu, X., Bhalla, K., Kim, C. N., Ibrado, A. M., Cai, J., Peng, T.-I., Jones, D. P., and Wang, X. (1997). Prevention of apoptosis by Bcl-2: Release of cytochrome c from mitochondria blocked. Science 275, 1129–1132.

Zhang, L., Yu, J., Park, B. H., Kinzler, K. W., and Vogelstein, B. (2000). Role of BAX in the apoptotic response to anticancer agents. Science 290, 989–992.

CHAPTER 5

MOLECULAR AND CELLULAR ASPECTS OF INSULIN RESISTANCE: IMPLICATIONS FOR DIABETES

DEREK LE ROITH, MICHAEL J. QUON, and YEHIEL ZICK
Diabetes Branch, NIH, Bethesda, Maryland; Diabetes Unit, NCCAM, NIH, Bethesda, Maryland; Department of Molecular Cell Biology, The Weizmann Institute of Science, Rehovot 76100, ISRAEL

INTRODUCTION

Type 1 diabetes is an autoimmune disease in which autoantibodies to the beta cells of the islets progressively destroy the beta cells, leading to an absolute deficiency of insulin secretion and a lifelong dependence on daily insulin injections (Eisenbarth, 1986). Insulin produced and secreted by the beta cell is necessary to control whole body homeostasis of glucose and lipids as well as to maintain normal protein metabolism. In the absence of insulin, hepatic glucose production will continue unabated, leading to hyperglycemia. Fat cells will undergo lipolysis, leading to breakdown of lipid stores with release of free fatty acids (FFAs) and glycerol. The FFAs will in turn increase glucose and VLDL triglyceride production by the liver. Skeletal muscle will undergo protein catabolism (DeFronzo, 1988). Thus insulin action at the target tissues is vital for normal bodily functions. This section outlines the normal aspects of cellular responses to insulin. As shown in Figure 5.1, insulin is a multipurpose hormone. The specific signaling pathways involved in insulin action are dealt with in a subsequent section.

Type 2 diabetes is a genetic disorder with strong environmental influences such as obesity and aging (DeFronzo, 1988). Although the genes involved in this disease are as yet undefined, the effects of obesity and aging on the disease are currently being determined. The disorder is defined as a "dual defect," with insulin resistance being detected early

Signal Transduction and Human Disease, Edited by Toren Finkel and J. Silvio Gutkind
ISBN 0-471-02011-7   Copyright © 2003 John Wiley & Sons, Inc.

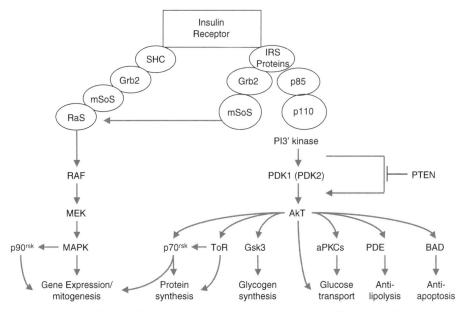

Figure 5.1. Insulin has numerous pleiotropic effects on cells.

followed by dysfunction of the pancreatic beta cells. Both obesity and aging enhance the degree of insulin resistance, and this places excessive pressure on the beta cells, which are unable to cope with extra demand. Clinical examples of this process are seen in patients with impaired glucose tolerance who progress to true type 2 diabetes, such as gestational diabetes, Pima Indians, and the growing epidemic of type 2 diabetes in minority groups at the age of adolescence; the insulin resistance influences the pancreas in genetically predisposed individuals.

The mechanisms involved in this pathophysiological process are numerous, but "gluco-lipotoxicity" stands out above all (Boden, 1997; Yki-Jarvinen, 1992). Gluco-lipotoxicity is a process whereby high glucose and lipid levels interfere with the normal function of the major organs normally required to maintain normal glucose homeostasis. Dysfunction of the pancreatic beta cells, liver, fat cells, and muscle may all be due to gluco-lipotoxicity, and although this process does not necessarily explain the basic pathogenesis of the disease it certainly goes a long way to identify mechanisms involved in worsening of the disease and impaired responses to therapeutic agents used to treat the disease.

In this chapter we approach the issues involved in insulin signaling and insulin resistance by initially describing the normal molecular mechanisms involved in insulin signaling pathways and then describing those aspects that represent known defects in these pathways causing insulin resistance. Finally, we present examples of mouse models that represent these defects to give the reader an idea of the disease state from a pathophysiological viewpoint.

MOLECULAR MECHANISMS OF METABOLIC INSULIN SIGNALING PATHWAYS

To understand how abnormalities in insulin action contribute to the pathogenesis of diabetes it is useful to first review the normal molecular mechanisms of metabolic insulin signaling pathways. Insulin initiates its biological actions by binding to specific cell surface receptors. The insulin receptor (IR) belongs to a large family of ligand-activated receptor tyrosine kinases (RTKs). Signaling pathways from the IR regulate a diverse array of functions, including essential metabolic functions such as glucose transport into skeletal muscle and adipose tissue. Below, we explain how IR signaling follows a general paradigm for RTK signal transduction and highlight the pathways involved with insulin-stimulated translocation of the insulin-responsive glucose transporter GLUT4. In addition, we explore some recently discovered complexities in insulin signaling networks and discuss how metabolic insulin signaling pathways contribute not only to glucose utilization but also to regulation of other important physiological processes such as hemodynamic homeostasis and insulin secretion.

Ligand Binding

RTKs are activated when ligand binds to the extracellular portion of the cell surface receptor. Monomeric EGF and PDGF receptors dimerize in response to ligand binding, and this receptor dimerization is a necessary first step in RTK activation (Heldin, 1995; van der Geer et al., 1994). The IR exists in a dimerized form even in the absence of ligand. Insulin binding to the extracellular α-subunit of the receptor results in a rapid conformational change and activation of the tyrosine kinase domain in the β-subunit of the receptor (Lee et al., 1997).

Receptor Autophosphorylation

The kinase region of RTKs contains a conserved catalytic domain and ATP binding site (van der Geer et al., 1994). When ligand binds to the receptor dimer, the kinase domain in one half of the dimer phosphorylates tyrosine residues in the activation loop of the kinase domain of the other half. In the absence of ligand, the activation loop of the IR occludes the catalytic site so that access to ATP and substrates is blocked. On trans-autophosphorylation, the phosphorylated activation loop is stabilized in an open conformation that gives unrestricted access to ATP and substrates, resulting in a large increase in kinase activity of the receptor (Hubbard, 1997).

Proximal Signals

The autophosphorylated receptor can bind to other signaling molecules through interactions between phosphotyrosine motifs on the receptor and src homology-2 (SH2) domains on downstream molecules. SH2 domains are protein domains that share homology with a noncatalytic region of the src protooncogene product. Many molecules that mediate RTK signaling contain SH2 domains, including phosphoinositol 3-kinase (PI3K), growth factor receptor bound protein 2 (GRB-2), SH2-containing phosphatase-2 (SHP-2), GTPase activating protein (GAP), and phospholipase C-γ (PLC-γ), among others. Specific SH2 binding motifs are defined by the three amino acid residues on the COOH-terminal side of the phosphotyrosine residue. These motifs provide specificity for interaction with particular SH2 domains (Songyang et al., 1993). The phosphotyrosine sites that engage specific SH2 domains of various signaling molecules have been well mapped for the EGF and PDGF receptors. Although the autophosphorylated IR can also directly interact with SH2 domains in molecules such as PI3K, SHP-2, and GAP (Staubs et al., 1994), these interactions are not critical for insulin signaling. Instead, a number of IR substrate proteins such as IRS-1, IRS-2, IRS-3, IRS-4, Shc, and GAB-1 are essential to functionally couple the IR with downstream signaling molecules (White, 1998). The various members of the IRS family share several common features, including NH_2-terminal pleckstrin homology (PH) domains and protein tyrosine binding (PTB) domains that are important for mediating interactions with the IR (Myers and White, 1996). In particular, the PTB domain binds to the phosphorylated tyrosine 972 in the NPEY motif of the juxtamembrane region of the IR β-subunit (Myers and White, 1996). The COOH-terminal portion of IRS proteins contain multiple tyrosine-containing motifs that undergo phosphorylation by the IR and serve as docking sites for SH2-domain containing proteins. These phosphotyrosine motifs are highly conserved between the various IRS proteins. For example, multiple YXXM motifs that bind to the SH2 domains in the p85 regulatory subunit of PI3K are present in all IRS family members. Thus phosphorylated IRS proteins can form signaling complexes consisting of several SH2 domain-containing molecules. However, differences between various IRS family members exist and suggest that these molecules do not have completely redundant functions. For example, IRS-3 is approximately 50% shorter than IRS-1 and IRS-2 and, unlike IRS-1 and IRS-2, IRS-3 does not contain a phosphotyrosine motif predicted to bind to the SH2 domain of GRB-2 (Lavan et al., 1997). Furthermore, the phenotype of transgenic mice homozygous for null alleles of IRS-1 is distinct from the phenotype of IRS-2 knockout mice. IRS-1 knockout mice are mildly insulin resistant but do not develop diabetes, whereas IRS-2 knockout mice have both insulin resistance and severe pancreatic beta-cell defects leading to abnormal insulin secretion and the development of diabetes. Thus various IRS family members may have some overlapping functions but are not completely interchangeable.

SH2 and SH3 Domain-Containing Signaling Molecules

Signaling molecules immediately downstream from RTKs and their substrates often contain SH2 and/or SH3 domains. As mentioned above,

SH2 domains interact specifically with phosphorylated tyrosine motifs. SH3 domains bind with high affinity to particular proline-rich sequences. Some SH2 domain-containing proteins (e.g., SHP-2 and PLC-γ) are effectors that possess intrinsic catalytic activity regulated by interactions of the SH2 domain with phosphotyrosine motifs on other proteins (e.g., IRS-1). Other SH2/SH3 domain-containing proteins (e.g., GRB-2, Nck, and the p85 regulatory subunit of PI3K) have no intrinsic catalytic activity and function strictly as adaptor proteins. These help to form specific signaling complexes via simultaneous interactions of multiple SH2/SH3 domains on the adaptor with both upstream and downstream signaling molecules. Two major effectors of pathways that are activated by a number of RTKs, including the IR, are Ras and PI3K. For example, GRB-2 is prebound to the guanine nucleotide exchange factor SOS (two SH3 domains of GRB-2 bind proline-rich regions of SOS). When phosphotyrosine motifs on IRS-1 or Shc bind to the SH2 domain of GRB-2, the prebound SOS catalyzes the exchange of GTP for GDP on Ras, leading to its activation. Similarly, the p85 regulatory subunit of PI3K is preassociated with the p110 catalytic subunit. On insulin stimulation, phosphorylated YXXM motifs on IRS proteins engage the SH2 domains of p85, leading to activation of p110 (for review see Nystrom and Quon, 1999).

Distal Phosphorylation Cascades

Distal RTK signaling pathways are difficult to dissect because branching pathways emerge from single effectors and multiple upstream inputs converge on single branch points. However, some distal signaling mechanisms such as phosphorylation cascades are shared by many growth factors including insulin. For example, Ras directly activates Raf, a serine/threonine kinase that phosphorylates and activates MEK, which in turn phosphorylates MAP kinase, which can then phosphorylate transcription factors such as Elk-1 leading to induction of early-immediate genes such as the protooncogenes c-jun and c-fos. Insulin signaling downstream from PI3K also involves serine/threonine phosphorylation cascades. Phospholipid products generated by PI3K activate phosphoinositide-dependent kinase-1 (PDK1) by binding to its PH domain. The constitutive catalytic activity of PDK-1 can be further enhanced by a phosphorylation-dependent mechanism in response to insulin stimulation (Chen et al., 2001). PDK-1 phosphorylates T308 in the regulatory region of Akt (another serine/threonine kinase), contributing to activation of Akt, which in turn phosphorylates and inactivates glycogen synthase kinase-3 (GSK-3), leading to activation of glycogen synthase (Chan et al., 1999).

Protein Tyrosine Phosphatases

Another aspect of regulation common to RTK signaling is the dephosphorylation of RTKs and their substrates by protein tyrosine phosphatases (PTPases). The number and diversity of PTPases rivals that of the RTKs (Walton and Dixon, 1993). PTPases are classified into two broad categories: cytoplasmic proteins containing a single catalytic PTPase domain and transmembrane "receptor-like" PTPases that typically contain tandem PTPase domains. Particular PTPases show selectivity for specific RTKs (Lammers et al., 1993). The transmembrane PTPases, PTP-α, PTP-ε, and LAR have all been implicated as modulators of insulin action. In particular, LAR interacts with and dephosphorylates the IR in intact cells. Among the nontransmembrane PTPases, PTP1B and SHP-2 both modulate insulin signaling. PTP1B dephosphorylates the IR both in vitro and in intact cells (Lammers et al., 1993) and negatively regulates both mitogenic and metabolic actions of insulin (Chen et al., 1997; Tonks, 1990). PTP1B knockout mice are more sensitive to insulin and resistant to becoming obese (Elchebly et al., 1999). Binding of the SH2 domains of SHP-2 to phosphotyrosine motifs either on the IR or on IRS-1 results in activation of SHP-2 phosphatase activity. A number of studies have shown that SHP-2 participates in Ras- and MAP kinase-dependent pathways as a positive mediator of mitogenic actions of insulin (Milarski and Saltiel, 1994; Yamauchi et al., 1995). Studies showing that expression of dominant-negative SHP-2 in transgenic mice causes insulin resistance are also consistent with a role for SHP-2 as a positive mediator of metabolic actions of insulin (Maegawa et al., 1999).

Lipid Phosphatases

Lipid phosphatases whose substrates include phosphoinositol products of PI3K also participate in the regulation of insulin signaling. PI $(3,4,5)P_3$ is a lipid product of PI3K that binds to the PH domain of various signaling molecules including PDK-1 and Akt. PI $(3,4,5)P_3$ localizes these molecules to signaling complexes and may also directly participate in stimulation of their kinase activity (Chen et al., 2001). 5'-Lipid phosphatases such as SHIP2 convert PI $(3,4,5)P_3$ to PI $(3,4)P_2$, and 3'-lipid phosphatases such as PTEN convert PI $(3,4,5)P_3$ to PI $(4,5)P_2$. Both of these phosphatases tend to oppose PI3K-dependent functions, but they seem to have distinct biological effects. For example, PTEN is a tumor suppressor and mutations in PTEN result in a number of neoplastic syndromes without metabolic phenotypes (Eng, 1999). In contrast, SHIP2 knockout mice have enhanced insulin sensitivity but no evidence of increased tumorigenesis (Clement et al., 2001). Thus SHIP2 plays a role to negatively modulate metabolic actions of insulin.

Feedback Pathways

Recently, the presence of feedback mechanisms in insulin signaling pathways have been identified that adds to the complexity of signal transduction. Some downstream kinases in insulin signaling pathways (e.g., Akt, GSK-3, and PKCζ) can also phosphorylate upstream signaling components such as IRS-1 on serine residues (Eldar-Finkelman and Krebs,

1997; Paz et al., 1999; Ravichandran et al., 2001b). Serine phosphorylation of IRS-1 is generally associated with impairment of its function (Hotamisligil et al., 1996). For example, tyrosine phosphorylation of IRS-1 as well as the ability of IRS-1 to bind and activate PI3K in response to insulin stimulation are both diminished when IRS-1 is phosphorylated on serine residues by PKCζ or Akt (Li et al., 1999; Ravichandran et al., 2001b). These events represent negative feedback in insulin signaling. However, phosphorylation of IRS-1 on certain residues by Akt may also enhance signaling by IRS-1 representing a positive feedback loop (Paz et al., 1999). Another example of feedback in insulin signaling is the ability of Akt to phosphorylate PTP1B at Ser^{50}, resulting in impairment of the ability of PTP1B to dephosphorylate the IR (Ravichandran et al., 2001a). Because PTP1B is a negative regulator of insulin signaling, the impairment of PTP1B function by Akt represents a positive feedback mechanism. The presence of these multiple positive and negative feedback mechanisms may help to confer specificity in insulin signaling and provides a potential mechanism for distinguishing signaling by the IR from signaling by other RTKs.

Insulin Signaling Related to Translocation of GLUT4 in Adipose Cells

The classic metabolic action of insulin to promote glucose uptake into adipose tissue and skeletal muscle is mediated by translocation of the insulin-responsive glucose transporter GLUT4 from an intracellular pool to the cell surface. The chain of signaling molecules regulating translocation of GLUT4 includes the IR (Quon et al., 1994b) phosphorylating IRS family members (Quon et al., 1994a; Zhou et al., 1997; Zhou et al., 1999) that then bind and activate PI3K (Quon et al., 1995) leading to activation of PDK-1 (Chen et al., 2001) that then phosphorylates and activates Akt (Cong et al., 1997) and PKCζ (Bandyopadhyay et al., 1999). Of note, Akt-2 knockout mice are diabetic because of defects in insulin-stimulated glucose uptake whereas Akt-1 knockout mice have normal glucose homeostasis(Cho et al., 2001a, Cho et al., 2001b). PTP1B and SHIP2 are both negative modulators of GLUT4 translocation that oppose insulin-stimulated glucose transport by dephosphorylating the IR and lipid products of PI3K, respectively (Chen et al., 1997; Clement et al., 2001; Elchebly et al., 1999). Even though PI3K is a key signaling molecule for metabolic actions of insulin and it is necessary for insulin-stimulated translocation of GLUT4, activation of PI3K per se is not sufficient for translocation of GLUT4. A PI3K-independent insulin signaling pathway involving the adaptor protein CAP recruits Cbl to the IR, resulting in tyrosine phosphorylation of Cbl (Baumann et al., 2000). The CAP-Cbl complex then interacts with flotillin in membrane lipid rafts and recruits a CrkII-C3G complex. C3G then activates the small GTP binding protein TC10 that leads to translocation of GLUT4 (Chiang et al., 2001). Interestingly, insulin-stimulated translocation of GLUT4 requires distinct compartmentalization of both PI3K and TC10 signaling pathways (Watson et al., 2001). The insulin signaling events described above couple to machinery controlling GLUT4 trafficking, resulting in enhanced translocation of GLUT4 to the cell surface (Pessin et al., 1999). However, the link between the distal signaling molecules Akt, PKCζ, and TC10 and the GLUT4 trafficking machinery is unknown and is currently an area of intensive investigation.

Insulin Signaling Related to Production of NO in Endothelial Cells

Recently, it has been recognized that insulin may play important physiological roles in tissues that are not classic targets for metabolic actions of insulin. For example, physiological concentrations of insulin cause vasodilation and increased blood flow that is mediated by production of nitric oxide (NO) in the vascular endothelium. Moreover, insulin resistance manifested in metabolic actions correlates highly with insulin resistance with respect to vasodilator actions of insulin. One reason for this may be that many key insulin signaling molecules important for mediating translocation of GLUT4 in adipose tissue and skeletal muscle also participate in insulin-stimulated activation of endothelial nitric oxide synthase (eNOS) in vascular endothelial cells. The IR, IRS-1, PI3K, and Akt are all necessary for insulin-stimulated production of NO in endothelium (Montagnani et al., 2001; Zeng et al., 2000; Zeng and Quon, 1996). Akt can directly phosphorylate eNOS, resulting in increased catalytic activity of eNOS (Dimmeler et al., 2000). Interestingly, the phosphorylation of eNOS at Ser^{1179} by Akt in response to insulin stimulation is necessary for insulin-mediated production of NO. Moreover, this phosphorylation-dependent mechanism is independent and separable from the classic calcium-dependent activation of eNOS (Montagnani et al., 2001). Thus one physiological action of insulin in the vasculature may be to couple regulation of hemodynamic homeostasis with metabolic homeostasis through production of NO (Montagnani and Quon, 2000). These observations provide a potential molecular explanation for how insulin resistance may contribute to the frequent associations among diabetes, obesity, and hypertension.

Insulin Signaling Related to Insulin Secretion in Beta Cells

Another nonclassic target for insulin action is the beta cell of the pancreas. Intriguingly, metabolic insulin signaling pathways may also play important roles in beta cell growth and development as well as in insulin secretion by the beta cell. Mice with targeted disruption of the IR exclusively in beta cells develop defects in glucose-sensitive insulin secretion, suggesting that insulin signaling in the beta cell is necessary for normal insulin secretory function (Kulkarni et al., 1999). Furthermore, IRS-1 and PI3K are both important for the ability of insulin to regulate insulin secretion by the beta cell (Aspinwall et al., 2000). IRS-2 knockout mice have significant defects in beta cell growth and development that lead to diabetes (Withers et al., 1998), whereas overexpression of constitutively active Akt-1 in mouse beta cells enhances beta cell mass and insulin secretory capability (Tuttle et al., 2001). Thus many of the key signaling molecules important for insulin-mediated glucose disposal may also be important for normal insulin secretion by the beta cell. This suggests an attractive unifying hypothesis for how insulin resistance may simultaneously impair both metabolic actions of insulin and insulin secretion leading to the development of diabetes.

THE MOLECULAR BASIS OF INSULIN RESISTANCE

Insulin resistance is a common pathological state in which target cells fail to respond to ordinary levels of circulating insulin (Kahn and Flier, 2000; Matthaei et al., 2000). Individuals with insulin resistance are predisposed to developing type 2 diabetes, and insulin resistance is frequently associated with a number of other health disorders including obesity, hypertension, chronic infection, and cardiovascular diseases (Saltiel, 2001; Taylor, 1999). Insulin resistance is manifested by decreased insulin-stimulated glucose transport and metabolism in adipocytes and skeletal muscle and by impaired suppression of hepatic glucose output (Reaven, 1995). The ensuing deregulation of carbohydrate and lipid metabolism that occurs as a consequence of insulin resistance further exacerbates its propagation. When the pancreatic β cells fail to compensate for the increasing demand for insulin, it results in the deterioration of glucose homeostasis and the development of glucose intolerance. This leads to the development of frank diabetes (Kahn, 1998), which is an upcoming epidemic of the twenty-first century. At the molecular level, insulin resistance is the consequence of impaired insulin signaling that may result from mutations or posttranslation modifications of the IR itself or any of its downstream effector molecules (see Kahn, 1998; Le Roith and Zick, 2001; Zick, 2001 for recent reviews).

Inducers of Insulin Resistance and Insulin Sensitivity

Several agents and metabolic conditions were implicated as inducers of insulin resistance. Most common are FFAs and their metabolites; TNF-α and other cytokines; catabolic hormones such as epinephrine, and fat-derived hormones such as resistin. It therefore appears that this syndrome is the consequence of action of a multitude of different inducers. On the other side, we find agents that act as insulin-sensitizers, with leptin being the best-studied candidate within this group.

Free Fatty Acids. Increased concentrations of plasma FFA are associated with many insulin-resistant states, including obesity and type 2 diabetes (Reaven et al., 1988). In humans, the triglyceride content of muscle correlates directly with insulin resistance and the fatty acid composition of muscle phospholipids influences insulin sensitivity. Indeed, increases in plasma fatty acid concentrations initially induce insulin resistance by inhibiting insulin-stimulated glucose transport, which is followed by a reduction in muscle glycogen synthesis and glucose oxidation (Shulman, 2000). In β cells, long-chain fatty acids may induce apoptosis through overproduction of ceramide, a known inducer of insulin resistance (Shimabukuro et al., 1998). It has further been proposed that the two mechanisms whereby TNF-α causes insulin resistance, and whereby the thiazolidine desines (TZDs) improve insulin sensitivity, may be triggered indirectly, via a reduction in the levels of FFAs.

Tumor Necrosis Factor. TNF-α is an endogenous cytokine produced by macrophages and lymphocytes after inflammatory stimulation, which has been implicated as a cause and a link to obesity-induced insulin resistance (Kanety et al., 1996; Spiegelman and Flier, 1996). TNF-α is secreted by adipocytes and enlarged adipocytes from obese animals, and humans overexpress this factor (Peraldi and Spiegelman, 1998). TNF-α presumably acts in a paracrine rather than endocrine fashion, and this could account for the failure to consistently detect elevated serum levels of TNF-α in obesity (Kahn and Flier, 2000). Further support for the role of TNF-α as an inducer of insulin resistance derives from the beneficial effect of knockout of TNF-α or TNF receptor genes on insulin resistance in animal models of obesity-associated insulin resistance (Peraldi and Spiegelman, 1998). Complete lack of TNF-α signaling in mice with a targeted mutation of both TNF receptor isoforms, p55 and p75, results in improved insulin sensitivity (Uysal et al., 1998), with the p55 receptor isoform having a stronger impact. Interestingly, activation of PPAR-γ by TZDs reduces the expression of TNF-α and hinders TNF-α's inhibition of insulin action (Jiang et al., 1998). Still, improvement of insulin resistance in response to loss of TNF signaling is at best partial, and the effect of TNF-α neutralization has not been seen in all experimental models. Thus TNF may be a partial contributor to insulin resistance, but other factors must exist.

TNF-α receptors are transmembrane glycoproteins, devoid of any enzymatic activity, that associate with different intracellular effectors (Locksley et al., 2001). Activation of TNFR1 is sufficient to mediate most biological responses of TNF, including the induction of insulin-resistant states (Skolnik and Marcusohn, 1996). TNF receptors have been shown to utilize distinct mechanisms to couple to proximal cytoplasmic signaling molecules. Recruitment of the signal transducer FADD (also known as MORT1) to the TNFR1 complex mediates apoptosis through activation of an intricate protease cascade (Locksley et al., 2001), whereas two other signal transducers, RIP and TRAF2, mediate both Jun NH$_2$-terminal kinase (JNK) and NF-κB activation (Van Antwerp et al., 1998). Similarly, FAN, a WD-repeat protein, couples TNFR1 to neutral sphingomyelinase (SMase) (Adam-Klages et al., 1996), whose stimulation results in the production of ceramide, which activates several kinases (Segui et al., 2001). As discussed below, TNF-α diminishes insulin-dependent Tyr-phosphorylation, while it induces Ser/Thr phosphorylation of IRS proteins (Hotamisligil et al., 1996; Kanety et al., 1995; Paz et al., 1997). Such Ser phosphorylation uncouples the IRS proteins from insulin signaling elements and induces insulin resistance (Zick, 2001).

Resistin. Resistin, a novel adipokine, is released from white adipocytes during differentiation. Resistin 3 mRNA is elevated in rodent models of obesity, with intravenous administration of resistin causing glucose intolerance and insulin resistance. Conversely, resistin-neutralizing antibodies reduce insulin resistance and hyperglycemia in mice with dietary obesity and type 2 diabetes (Steppan et al., 2001). PPAR-γ agonists have been proposed to enhance insulin sensitivity by decreasing resistin expression, implicating resistin as a link between obesity and diabetes, with elevated levels of resistin promoting insulin resistance (Steppan et al., 2001). In contrast, insulin resistance in several common rodent genetic models such as ob/ob and db/db mice was found to be associated with a decrease (rather than an increase) in resistin expression (Way et al., 2001). In addition, different PPAR-γ agonists were shown to stimulate, rather then inhibit, resistin expression in two standard rodent models of type 2 diabetes (Way et al., 2001). Hence, the role of resistin as an effector of insulin resistance awaits further clarification.

Insulin Sensitizers

Several adipose-derived hormones like TNF and resistin induce insulin resistance; however, other adipokines, like leptin and adiponectin, act as insulin sensitizers.

Leptin. Leptin, a product of the ob gene, is an adipocyte-derived hormone that exerts profound effects on satiety, energy expenditure, and neuroendocrine function (Kahn and Flier, 2000). Circulating leptin concentrations in humans correlate closely with fasting insulin concentrations and the percentage of body fat, making leptin a marker of obesity and the insulin resistance syndrome (Lonnqvist et al., 1999). Severe insulin resistance is a well-known feature of deficiency of leptin or its receptor in the ob/ob or db/db mouse strains, and these models were among the first to be investigated for the pathogenesis of insulin resistance. Leptin's major site of action is the hypothalamus, where neurons that are directly regulated by leptin reside (Elmquist et al., 1998). Leptin has a clear insulin-sensitizing effect; still, the molecular basis for the insulin-sensitizing effect of leptin remains a topic of great interest. Major unanswered questions include which signaling mechanism(s) and cellular targets in the periphery respond to the autonomic nerve output by which leptin affects metabolic pathways in relevant tissues, such as muscle, liver, and fat. Substantial data also support the notion that leptin may have important effects through direct action on peripheral target cells. The leptin receptor is a member of the cytokine family of receptors and occurs in five isoforms. Only the long form of the leptin receptor (OB-RL) has been shown to possess significant signaling capacity (Tartaglia, 1997). This involves activation of the STAT and MAPK pathways (Kim et al., 2000b); activation of PI3K (Cohen et al., 1996; Kellerer et al., 1997); promotion of lipid oxidation; and inhibition of lipid synthesis, which would promote insulin 4 sensitivity in muscle and fat (Muoio et al., 1997). Still, the relative importance of central versus peripheral actions of leptin in the metabolic actions of the hormone remains a matter of debate.

Adiponectin. Adiponectin is an adipocyte-derived hormone that functions as an insulin sensitizer (Yamauchi et al., 2001). Adiponectin decreases insulin resistance by decreasing triglyceride content in muscle and liver in obese mice. Insulin resistance in lipoatrophic mice is completely reversed by the combination of physiological doses of adiponectin and leptin, implicating adiponectin deficiency as an underlying cause for the development of insulin resistance. Adiponectin increases expression of the genes encoding CD36, acyl CoA oxidase, and uncoupling protein-2 (UCP2), which might enhance fatty acid transport, fat combustion, and dissipation, respectively. These data indicate that the insulin-sensitizing effects of adiponectin on insulin signaling is secondary to its ability to burn fat, presumably due to alterations in gene expression (Saltiel, 2001).

Signaling Pathways in Insulin Action as Molecular Targets of Insulin Resistance

Functional defects in insulin action and the induction of insulin resistance may result, in part, from impaired insulin signaling in insulin target tissues. At the molecular level, this could be the consequence of alterations in content, mutations, or posttranslational modifications of the IR itself or any of its downstream effector molecules (see Le Roith and Zick, 2001 for a recent review}. In some cases insulin resistance could be accounted for by a defect in insulin binding to its receptor (Roach et al., 1994); however, most often insulin resistance is attributed to a postbinding defect in insulin action. Several mechanisms could account for the impaired insulin signaling observed under conditions of insulin resistance. Reduced expression of IR, IRS proteins, and PI3K can contribute to the development of insulin resistance (Goodyear et al., 1995). Similarly, increased expression and activity of PTPases may enhance the dephosphorylation of IR or its downstream effectors and thus terminate insulin signaling propagated through Tyr phosphorylation events (Kahn and Flier, 2000; Matthaei et al., 2000). At least three PTPases, including PTP1B, leukocyte antigen–related phosphatase (LAR), and SHP-2, are increased in expression and/or activity in muscle and adipose tissue of obese humans and rodents (Goldstein et al., 1998). The activity of some of these PTPases is inhibited by insulin, whereas insulin counterregulatory hormones enhance PTPase activity. Accordingly, insulin stimulates Tyr phosphorylation and inactivation of PTP1B in vivo (Tao et al., 2001b), whereas the heterotrimeric G protein Giα2 prevents insulin resistance and enhances insulin signaling via suppression of PTP1B activity (Tao et al., 2001a). Similarly, PTP1B-null mice manifest increased insulin 5 sensitivity and resistance to diet-induced obesity (Elchebly et al., 1999). In contrast, cAMP-elevating agents activate the cAMP-dependent protein kinase (PKA), which inhibits insulin signaling through the phosphorylation and activation of PTP1B (Tao et al., 2001b). These results indicate that alterations in PTPase activity contribute to the development of insulin resistance.

Still, as discussed below, Ser/Thr phosphorylation of the IR itself or its downstream effectors, the IRS proteins and Shc, is presumably the prevailing pathological pathway leading to the induction of insulin resistance. In fact, agents that induce insulin resistance exploit phosphorylation-based negative feedback control mechanisms, otherwise utilized by insulin itself, to uncouple the IR from its downstream effectors and thereby terminate insulin signal transduction. Here we present some recent viewpoints on the molecular basis of insulin resistance, focusing on the cardinal role of Ser/Thr protein kinases as emerging key players in this arena (Le Roith and Zick, 2001; Tao et al., 2001b; Zick, 2001).

The Insulin Receptor. Some forms of insulin resistance may involve the receptor itself. Alterations in insulin receptor expression, binding, phosphorylation state, trafficking, and/or kinase activity have been identified in rare cases of severe insulin resistance, such as the type A syndrome, leprechaunism, and Rabon–Mendenhall syndrome. These arise from point mutations in the receptor, many of which produce a dominant-negative phenotype (Taylor and Arioglu, 1998). Nonetheless, impairment of the receptor function, due to extrinsic inhibition of its Tyr kinase activity, is a more common manifestation of insulin resistance. Such inhibited kinase activity can be attributed to the action of PTPases that dephosphorylate the Tyr-phosphorylated, active conformation of IR (Kusari et al., 1994), or it can be the result of Ser/Thr phosphorylation of the IR itself (Dunaif et al., 1995), observed, for example, in insulin-resistant patients with polycystic ovary syndrome (Venkatesan et al., 2001).

Insulin counterregulatory hormones and cytokines can activate serine kinases, particularly PKA and members of the protein kinase C (PKC) family, which function as IR kinases and were implicated in the development of peripheral insulin resistance (Zick, 2001). Elevation of intracellular cAMP levels increases Ser phosphorylation of the receptor in intact cells (Stadtmauer and Rosen, 1986) and decreases its ability to function as an insulin-stimulated Tyr kinase. Phosphorylation of IR by PKA results in incorporation of 1 mole of phosphate per mole of receptor, and this occurs concomitantly with a 25% decrease in the receptor kinase activity (Roth and Beaudoin, 1987).

Tumor-promoting phorbol esters like phorbol 12-myristate 13 acetate (TPA) elicit their antagonizing effects on insulin action by activating different PKC isoforms. TPA-mediated activation of PKCs stimulates Ser/Thr phosphorylation of insulin receptors in intact cells; inhibits insulin-stimulated receptor autophosphorylation, and increases the IR internalization rate (Hachiya et al., 1987; Takayama et al., 1984). Several

PKC isoforms are chronically activated under conditions of insulin resistance (Avignon et al., 1996; Considine et al., 1995) and can catalyze Ser/Thr phosphorylation of the insulin receptor or its substrates. Accordingly, the isolated IR serves as a direct substrate for PKC, which leads to inhibition of the intrinsic receptor Tyr kinase activity (Bollag et al., 1986), whereas pharmacological inhibition of PKC activity or reduction in PKC expression enhances insulin sensitivity and IRK activity (Donnelly and Qu, 1998).

The chronic elevation in insulin levels that occurs as a result of insulin resistance might stimulate serine kinases that phosphorylate the IR and inhibit its kinase activity. This could provide a mechanism for a vicious cycle of insulin-induced insulin resistance. The nature of these kinases still remains elusive. Insulin-stimulated Ser/Thr kinases can form complexes with the insulin receptor (Zick et al., 1983) and can phosphorylate the receptor with a high stoichiometry (Carter et al., 1996). A potential candidate is IKKβ, a downstream target of PKC-ζ, which is an effector of PI3K along the insulin signaling pathway. IKKβ inhibition by high doses of salicylates (Yin et al., 1998) or by a reduction in IKKβ gene dose (Yuan et al., 2001) reverses obesity- and diet-induced insulin resistance (Yuan et al., 2001). At the molecular level, activation or overexpression of IKKβ attenuates IRK activity and insulin signaling, whereas inhibition of IKKβ improves insulin-stimulated Tyr phosphorylation of IR in liver and muscle of insulin-resistant Zucker fatty rats, as well as in 3T3l-1 adipocytes treated with TNF-α or phosphatase inhibitors, thus implicating IKKβ or its downstream effectors as potential IR kinases.

Insulin Receptor Substrates (IRS Proteins)

Recent studies have focused on Ser/Thr phosphorylation of IRS proteins as a key negative feedback control mechanism that uncouples the IRS proteins from their upstream and downstream effectors and terminates signal transduction in response to insulin under physiological conditions (Le Roith and Zick, 2001; Zick, 2001). Emerging data further suggest that agents such as TNF-α, FFAs, and cellular stress, which inhibit insulin signaling and induce insulin resistance, take advantage of this mechanism by activating Ser/Thr kinases known as IRS kinases that phosphorylate the IRS proteins and inhibit their function. Thus, although the underlying molecular pathophysiology of insulin resistance is still not well understood, Ser phosphorylation of IRS proteins represents a new and possibly unifying mechanistic theme. As already mentioned, Ser/Thr phosphorylation of IRS proteins has a dual function in serving either as a positive or a negative modulator of insulin signaling. Phosphorylation of Ser residues within the P-Tyr-binding (PTB) domain of IRS-1 by insulin-stimulated PKB protects IRS proteins from the rapid action of PTPases and enables them to maintain their Tyr-phosphorylated active conformation, thus implicating PKB as a positive regulator of IRS-1 functions (Paz et al., 1999). By contrast, Ser/Thr phosphorylation of IRS proteins by other insulin-stimulated Ser/Thr kinases such as PKCζ or mTOR (vide supra) serves as a negative feedback control mechanism that inhibits further Tyr phosphorylation of IRS proteins.

PKB, mTOR. and PKCζ are downstream effectors of PI3K in the insulin signaling pathway. This suggests that their action should be orchestrated to allow phosphorylation by PKB and sustained activation of IRS-1 before the activation of mTOR or PKCζ, the actions of which are expected to terminate insulin signal transduction. Of note, the negative feedback control mechanism induced by PKCζ (or mTOR) includes a self-attenuation mode, whereby PI3K-mediated activation of PKCζ inhibits IRS-1 function; reduces complex formation between IRS-1 and PI3K, and thereby inhibits further activation of PKCζ itself.

Other aspects of insulin signaling are also subjected to homologous desensitization. Chronic stimulation with insulin results in persistent phosphorylation of the GDP/GTP exchange factor mSOS, which keeps it dissociated from the adaptor Grb2 and allows the GTPase Ras to return to its GDP-bound, inactive phase (Langlois et al., 1995). This process is apparently mediated by a MAPK that phosphorylates mSOS (Fucini et al., 1999). Hence, two major insulin signaling pathways, mediated by IRS proteins and Shc, are subjected to homologous desensitization in the form of insulin-induced Ser/Thr phosphorylation. Ser/Thr phosphorylation can induce the dissociation of IRS proteins from the IR (Liu et al., 2001; Paz et al., 1997); hinder Tyr phosphorylation sites (Mothe and Van Obberghen, 1996); release the IRS proteins from intracellular complexes that maintain them in close proximity to the receptor (Clark et al., 2000; Tirosh et al., 1999); induce IRS proteins degradation (Haruta et al., 2000; Pederson et al., 2001); or turn IRS proteins into inhibitors of the IRK (Hotamisligil et al., 1996). Because Ser/Thr phosphorylation of IRS proteins is stimulated by insulin treatment and by inducers of insulin resistance, the question is raised whether the same kinases and signaling pathways are being activated under both physiological and pathological conditions.

Role of PKCζ and IKK

A central role in the induction of insulin resistance is attributed to agents such as FFA, phorbol esters, and TNF-α (Moller, 2000) whose common feature is their ability to enhance Ser/Thr phosphorylation and inhibit insulin-stimulated Tyr phosphorylation of IRS proteins (Feinstein et al., 1993). The idea that insulin might stimulate the same IRS kinases emerged when it was realized that TNF-α activates PKCζ and its downstream target IKKb (Lallena et al., 1999; Martin et al., 2001). Potential mechanisms could involve TNF-α-mediated activation of sphingomyelinase (Adam-Klages et al., 1996) and production of ceramide, which stimulates PKCζ activity (Muller et al., 1995). Indeed, the effects of TNF-α on Ser/Thr phosphorylation of IRS proteins are mimicked by sphingomyelinase and ceramide analogs (Kanety et al., 1996; Paz et al., 1997), suggesting that TNF-α triggers a ceramide-activated kinase such as PKCζ. Alternatively, TNF-α can induce complex formation between

PKCζ, p62 ,and RIP proteins that serve as adaptors of the TNF receptor and link PKCζ to TNF-α signaling.

IKKβ. IKKβ, a downstream target of PKCζ and a potent inducer of insulin resistance (Kim et al., 2001a; Kim et al., 2001b; Yuan et al., 2001), is another potential mediator of IRS phosphorylation. Activation of IKKβ (e.g., by TNF-α) attenuated insulin signaling, and inhibited insulin-stimulated Tyr phosphorylation of IRS proteins, whereas IKKβ inhibition by high doses of salicylates (Yuan et al., 2001) prevented Ser/Thr phosphorylation of IRS proteins induced by a high-fat diet, TNF-α, or phosphatase inhibitors. This improved insulin-stimulated Tyr phosphorylation of IRS proteins, indicating that IKKβ or its downstream effectors serve as IRS kinases. As mentioned above, the effects of salicylates on IRS protein function were in part secondary to the enhanced IRK activity induced by salicylate treatment of insulin-resistant animals (Yuan et al., 2001), suggesting that IKKβ can negatively regulate the activity of both IR and IRS proteins (Yuan et al., 2001). The fact that inducers of insulin resistance activate IKKβ, whereas salicylates, which selectively inhibit IKKβ activity, prevent Ser/Thr phosphorylation of IRS proteins and insulin resistance, implicate IKKβ as a potential IRS kinase. Still, there is no direct evidence to indicate that IKKβ indeed phosphorylates either IR or the IRS proteins, and it might well be that downstream effectors of IKKβ play this role. The rapid phosphorylation of IRS proteins, which occurs on activation of IKKβ, argues against the possibility that IKKβ-mediated activation of NF-κB leads to de novo synthesis of inducers of insulin resistance. Nevertheless, further studies are required to address this possibility.

c-Jun NH$_2$-Terminal Kinase. JNK promotes insulin resistance by associating with IRS-1 and phosphorylating Ser[307], which inhibits insulin-stimulated Tyr-phosphorylation of IRS-1 (Aguirre et al., 2000). Because Ser[307] is adjacent to the PTB domain of IRS-1, its phosphorylation might disrupt the interaction between the juxtamembrane domain of the IR and the PTB domain of IRS-1. Interestingly, insulin and TNF-α stimulate phosphorylation of IRS-1 at Ser[307] via distinct pathways. Whereas insulin stimulates Ser/Thr kinases downstream of PI3K, TNF-α effects are mediated by members of the MAPK pathway (Rui et al., 2001). Of note, JNK itself is unlikely to serve as the sole insulin- or TNF-α-stimulated IRS kinase (at Ser[307]), because its activity is insensitive to inhibitors that block phosphorylation of Ser[307] in response to these stimuli in preadipocytes (Rui et al., 2001). Ser[307], phosphorylated by yet unknown kinases, in addition to JNK, might therefore integrate feedback and heterologous signals to attenuate IRS-1-mediated signals and contribute to insulin resistance.

Conventional PKCs and MAP Kinases. The kinases described above, PKCζ, IKKβ, and JNK, join a respected list of Ser/Thr kinases already implicated in phosphorylating IRS proteins, when triggered by agents that induce insulin resistance. These include "conventional" members of the PKC family, such as PKCα, activated by phorbol esters or endothelin-1 (Li et al., 1999), the activity of which is mediated, at least partially, by members of the MAPK pathway. These kinases phosphorylated IRS-1 at Ser^{612} {located in a consensus MAPK phosphorylation site} and at additional sites in its COOH tail. Such phosphorylation prevents the association of IRS-1 with the juxtamembrane domain of IR, impairs the ability of IRS-1 to undergo insulin-stimulated Tyr phosphorylation, and inhibits recruitment of downstream effectors such as PI3K (Li et al., 1999).

Indeed, conventional PKC isoforms were implicated in the induction of insulin resistance. Elevations in plasma FFA abolish insulin-stimulated IRS-1-associated PI3K activity (Dresner et al., 1999), and this could be accounted for by a reduction of insulin-stimulated Tyr phosphorylation of IRS-1, which was associated with activation of diacylglycerol-dependent PKCθ (Griffin et al., 1999). Similarly, nutritionally induced insulin resistance in skeletal muscle of *Psammomys obesus* is associated with overexpression of PKCε, which precedes the onset of hyperinsulinemia and hyperglycemia (Ikeda et al., 2001).

Obesity, Insulin Resistance, and IRS Phosphorylation

Elevated levels of FFA are characteristics of obesity, insulin resistance, and type 2 diabetes, and increasing evidence supports the contention that FFAs inhibit insulin action at peripheral target tissues (Shulman, 2000; Spiegelman and Flier, 2001). A recent study (Kim et al., 2001a) combined an observation made 120 years ago, indicating that high doses of salicylates lower blood glucose concentrations in diabetic patients (Ebstein, 1876), with contemporary knowledge regarding obesity and high-fat diet-induced activation of IKKb to show that salicylates prevent fat-induced muscle insulin resistance by inhibiting the activity of IKKb and its ability to mediate phosphorylation and inactivation of IRS-1 function (Kim et al., 2001a). Lipid infusion failed to alter insulin signaling in skeletal muscle of IKKb knockout mice, further implicating a protective role for IKKb inactivation in fat-induced development of insulin resistance (Kim et al., 2001a). Although the key data in this study is correlational, it positions IKKb as a potential mediator of Ser phosphorylation of IRS proteins. The mechanism by which lipids might activate IKKb presumably involves an increase in FFA-derived metabolites, such as diacylglycerol and ceramide, which are potent activators of PKCθ and PKCζ (Muller et al., 1995), both known to activate IKKb. Accordingly, defective insulin signaling in skeletal muscle of high fat-fed rats is associated with increased basal activity of PKCζ/λ (Tremblay et al., 2001). Obesity-induced insulin resistance is not limited to the effects of increased levels of FFA or TNF-α. Other "adipokines" secreted by fat cells, such as resistin (Steppan et al., 2001), might also contribute to the development of insulin resistance through Ser/Thr phosphorylation of IRS proteins, but further studies are required to address this possibility. Hence, it appears that increasing intracellular fatty acid metabolites, such as diacylglycerol, fatty acyl CoAs, or ceramides activates a Ser/Thr kinase cascade, leading to the phosphorylation of IRS proteins and their uncoupling from the insulin signaling elements, thus resulting in decreased activation of glucose transport. Further evidence supporting this hypothesis is provided by studies in transgenic mice that are almost totally devoid of fat because their adipocytes express the A-ZIP/F-1 protein, which blocks the function of several classes of transcription factors (Gavrilova et al., 2000). This lipodystrophy is associated with a twofold increase in muscle and liver triglyceride content. Interestingly, these mice are severely insulin resistant, because of defects in insulin action, particularly IRS-1/IRS-2-dependent activation of PI3K in muscle and liver, whereas transplantation of fat tissue into these mice returned the triglyceride content in muscle and liver to normal, as did insulin signaling and action (Kim et al., 2000a). These findings are consistent with the hypothesis that insulin resistance develops in obesity, type 2 diabetes, and lipodystrophy because of alterations in the partitioning of fat between the adipocyte and muscle or liver. This change leads to the intracellular accumulation of triglycerides and, probably more importantly, of intracellular fatty acid metabolites (e.g., fatty acyl CoAs, diacylglycerol, and ceramides) in these insulin-responsive tissues, which activates IRS kinases and leads to acquired insulin signaling defects and insulin resistance. This hypothesis might also explain how thiazolidinediones improve insulin sensitivity in muscle and liver tissue. By activating PPAR-γ receptors in adipocytes and promoting adipocyte differentiation, these agents might promote a redistribution of fat from liver and muscle into the adipocytes, much as fat transplantation does in fat-deficient mice (Kim et al., 2000a).

In summary, our understanding at the molecular level of insulin signal transduction, insulin resistance, and the connection between the two is evolving extremely rapidly. Current findings implicate the insulin receptor and IRS proteins as major targets for insulin-induced, phosphorylation-based negative feedback control mechanisms that uncouple the IR from its downstream effectors and terminate insulin signaling under physiological conditions. Recent studies further strengthen the concept that the varied agents and conditions that induce insulin resistance such as TNF-α, FFA, and obesity, activate IR and IRS kinases, with IKKβ and its downstream effectors being key candidates. Other inducers of insulin resistance such as endothelin-1 presumably utilize additional kinases such as MAPK to phosphorylate IRS proteins. Because each of the potential IRS kinases has a unique substrate specificity, the question remains as to which Ser sites are being modified by each kinase and what the consequences of such phosphorylation are. Given the large number of stimuli, pathways, kinases, and potential sites involved, it appears that Ser/Thr phosphorylation of IR and IRS proteins represents a combinatorial consequence of several kinases, activated by different pathways, acting in concert to phosphorylate multiple sites. Although many questions await answers, the new paradigms and emerging target kinases described here give a novel viewpoint of the molecular basis for insulin resistance. This should enable rational drug design to selectively inhibit the activity of the relevant enzymes and generate a novel class of therapeutic agents for insulin resistance and type 2 diabetes.

MOUSE MODELS DEVELOPED TO STUDY THE PATHOPHYSIOLOGY OF TYPE DIABETES

Under this section we discuss the signaling processes of type 2 diabetes by describing mouse models (both gene deletion and transgenic) that have been created in an attempt to understand the human disease as well as discussing the normal insulin signaling process at the target tissues with emphasis on the process of insulin resistance that is so commonly seen in type 2 diabetes.

Mouse models using gene targeting or transgenic approaches have been utilized to attempt to mimic the human disorder. Because mutations in the insulin receptor were shown to be the cause (albeit uncommon) of severe insulin resistance in humans, a mouse carrying a null mutation for the insulin receptor (IR) was created and homozygous mice died (not surprisingly) in the early postnatal stages from diabetic ketoacidosis (Accili et al., 1996). The IRS family of substrates apparently play important roles in insulin receptor signaling and are therefore obvious targets for gene deletion experiments. Interestingly, IRS-1 inactivation led to growth retardation without diabetes, initially suggesting that IRS-1 may be more significant in the IGF-I receptor signaling pathway than for the IR signaling pathway. However, its role in IR signaling was demonstrated by the impaired insulin secretion from b cells as well as the development of diabetes in combined heterozygous IR and IRS-1 knockout mice.

Mice with a homozygous deletion for the IRS-2 gene, on the other hand, readily develop diabetes. This is primarily secondary to a major defect in β cell growth, in addition to insulin resistance at the level of the liver. Gene deletion experiments with IRS-3 and IRS-4 did not result in any significant phenotype, suggesting that IRS-1 and IRS-2 are the major players in insulin signaling, although they may function differently in different tissues.

To determine the role of the IR and signaling molecules in various tissues and how this may impact on the pathogenic mechanisms involved in type 2 diabetes, various models have been created using the transgenic or cre-loxP systems. The transgenic approach has primarily utilized the overexpression of dominant-negative mutants driven by tissue-specific promoters or the cre-LoxP system that deletes the gene of interest in a tissue specific manner by expressing the cre recombinase in a specific tissue utilizing a specific promoter. Overexpression of a dominant-negative IR in muscle or IR gene deletion specifically in muscle using the cre-loxP system both resulted in insulin resistance, but neither mouse model developed marked hyperinsulinemia or diabetes (Bruning et al., 1998; Chang et al., 1994). Although the results of these studies were interpreted as suggesting that other tissues were also important in insulin-mediated glucose disposal, it remains a possibility that the IGF-IR was able to compensate for the loss of IR in muscle (vide infra). Alternatively, glucose uptake in muscle may have been mediated by another mechanism, namely, exercise-induced contraction. Similarly, pancreatic or liver-specific IR deletion mouse models failed to produce severe diabetes, again supporting the hypothesis that type 2 diabetes is a multiorgan disorder (Kulkarni et al., 1999; Lauro et al., 1998). When glucose transport protein-4 (GLUT-4) was deleted in muscle, the mice developed insulin resistance and diabetes (Kim et al., 2001b). This result could be explained by the fact that GLUT-4 mediates the downstream effects of both IR and IGF-IR on glucose uptake into muscle. These data are supported by a recent report on the overexpression of the dominant-negative mutant IGF-IR in muscle. In a separate study, GLUT 4 was deleted in adipocytes with the Cre-Lox/P system and this resulted in the expected insulin resistance in fat cells. Unexpectedly, the mice also demonstrated insulin resistance in both muscle and liver. These findings strongly suggested that fat cells were secreting factors that interfered with the normal function of insulin in these organs. These may include TNF-α, leptin, FFAs, triglycerides, and adiponectin. Some of these are known to cause insulin resistance (FFAs, resistin), whereas others are insulin sensitizers (leptin, adiponectin).

Using a mutant IGF-IR previously shown to be dominant negative in cultured cells, with a lysine residue in the ATP binding site of the tyrosine kinase domain, Fernandez et al. expressed this mutant in muscle with a muscle-specific creatine kinase promoter (Fernandez et al., 2001). The markedly overexpressed dominant-negative IGF-IR formed hybrids with the endogenous IGF-IRs as well as with the endogenous IR, leading to severe insulin resistance (Frattali et al., 1992). This was followed rapidly by insulin resistance at the level of the liver and fat cell, eventually leading to b cell dysfunction and type 2 diabetes. The progression of insulin resistance from muscle to liver and fat cell and eventual b cell dysfunction could be explained by the significant increase in circulating FFAs and triglycerides and triglyceride accumulation in various organs, although other explanations are also feasible. For example, the insulin resistance in muscle may lead to fat cell resistance secondary to hyperinsulinemia and/or glucose intolerance followed by release of FFAs, resistin, TNF-α, and other as yet unidentified factors leading to liver insulin resistance and beta cell dysfunction.

Akt 2(PKB β) gene deletion in mice resulted in insulin resistance and diabetes. This was associated with insulin resistance at the level of the liver and muscle, suggesting that both tissues played a role in the development of diabetes in these mice (Cho et al., 2001).

CONCLUSION

From these and other studies it seems clear that insulin resistance in one organ can lead to secondary insulin resistance in a second major organ and progressively lead to the full-blown picture of type 2 diabetes. However, it is also clear that multiple organ defects are more commonly associated with the development of diabetes. Although this does not necessarily replicate the human situation in which patients present with the full-blown picture, these models have become the cornerstone of research attempting to understand the human disease.

Understanding the molecular mechanisms involved in insulin action and the development of insulin resistance is important for the development of new tools for treating these common disorders. Pathophysiological studies in mouse models will enable investigators to test these new therapeutic agents.

REFERENCES

Accili, D., Drago, J., Lee, E. J., Johnson, M. D., Cool, M. H., Salvatore, P., Asico, L. D., Jose, P. A., Taylor, S. I., and Westphal, H. (1996). Early neonatal death in mice homozygous for a null allele of the insulin receptor gene. Nat Genet 12, 106–109.

Adam-Klages, S., Adam, D., Wiegmann, K., Struve, S., Kolanus, W., Schneider-Mergener, J., and Kronke, M. (1996). FAN, a novel WD-repeat protein, couples the p55 TNF-receptor to neutral sphingomyelinase. Cell 86, 937–947.

Aguirre, V., Uchida, T., Yenush, L., Davis, R., and White, M. F. (2000). The c-Jun NH$_2$-terminal kinase promotes insulin resistance during association with insulin receptor substrate-1 and phosphorylation of Ser[307]. J Biol Chem 275, 9047–9054.

Aspinwall, C. A., Qian, W. J., Roper, M. G., Kulkarni, R. N., Kahn, C. R., and Kennedy, R. T. (2000). Roles of insulin receptor substrate-1, phosphatidylinositol 3-kinase, and release of intracellular Ca^{2+} stores in insulin-stimulated insulin secretion in β-cells. J Biol Chem 275, 22331–22338.

Avignon, A., Yamada, K., Zhou, X., Spencer, B., Cardona, O., Saba-Siddique, S., Galloway, L., Standaert, M. L., and Farese, R. V. (1996). Chronic activation of protein kinase C in soleus muscles and other tissues of insulin-resistant type II diabetic Goto-Kakizaki (GK), obese/aged, and obese/Zucker rats. A mechanism for inhibiting glycogen synthesis. Diabetes 45, 1396–1404.

Bandyopadhyay, G., Standaert, M. L., Sajan, M. P., Karnitz, L. M., Cong, L., Quon, M. J., and Farese, R. V. (1999). Dependence of insulin-stimulated glucose transporter 4 translocation on 3-phosphoinositide-dependent protein kinase-1 and its target threonine- 410 in the activation loop of protein kinase C-ζ. Mol Endocrinol 13, 1766–1772.

Baumann, C. A., Ribon, V., Kanzaki, M., Thurmond, D. C., Mora, S., Shigematsu, S., Bickel, P. E., Pessin, J. E., and Saltiel, A. R. (2000). CAP defines a second signalling pathway required for insulin-stimulated glucose transport. Nature 407, 202–207.

Boden, G. (1997). Role of fatty acids in the pathogenesis of insulin resistance and NIDDM. Diabetes 46, 3–10.

Bollag, G. E., Roth, R. A., Beaudoin, J., Mochly-Rosen, D., and Koshland, D. E., Jr. (1986). Protein kinase C directly phosphorylates the insulin receptor in vitro and reduces its protein-tyrosine kinase activity. Proc Natl Acad Sci USA 83, 5822–5824.

Bruning, J. C., Michael, M. D., Winnay, J. N., Hayashi, T., Horsch, D., Accili, D., Goodyear, L. J., and Kahn, C. R. (1998). A muscle-specific insulin receptor knockout exhibits features of the metabolic syndrome of NIDDM without altering glucose tolerance. Mol Cell 2, 559–569.

Carter, W. G., Sullivan, A. C., Asamoah, K. A., and Sale, G. J. (1996). Purification and characterization of an insulin-stimulated insulin receptor serine kinase. Biochemistry 35, 14340–14351.

Chan, T. O., Rittenhouse, S. E., and Tsichlis, P. N. (1999). AKT/PKB and other D3 phosphoinositide-regulated kinases: kinase activation by phosphoinositide-dependent phosphorylation., Annu Rev Biochem 68, 965–1014.

Chang, P. Y., Benecke, H., Le Marchand-Brustel, Y., Lawitts, J., and Moller, D. E. (1994). Expression of a dominant-negative mutant human insulin receptor in the muscle of transgenic mice. J Biol Chem 269, 16034–16040.

Chen, H., Nystrom, F. H., Dong, L. Q., Li, Y., Song, S., Liu, F., and Quon, M. J. (2001). Insulin stimulates increased catalytic activity of phosphoinositide-dependent kinase-1 by a phosphorylation-dependent mechanism, Biochemistry 40, 11851–11859.

Chen, H., Wertheimer, S. J., Lin, C. H., Katz, S. L., Amrein, K. E., Burn, P., and Quon, M. J. (1997). Protein-tyrosine phosphatases PTP1B and syp are modulators of insulin-stimulated translocation of GLUT4 in transfected rat adipose cells. J Biol Chem 272, 8026–8031.

Chiang, S. H., Baumann, C. A., Kanzaki, M., Thurmond, D. C., Watson, R. T., Neudauer, C. L., Macara, I. G., Pessin, J. E., and Saltiel, A. R. (2001). Insulin-stimulated GLUT4 translocation requires the CAP-dependent activation of TC10. Nature 410, 944–948.

Cho, H., Mu, J., Kim, J. K., Thorvaldsen, J. L., Chu, Q., Crenshaw, E. B., 3rd, Kaestner, K. H., Bartolomei, M. S., Shulman, G. I., and Birnbaum, M. J. (2001a). Insulin resistance and a diabetes mellitus-like syndrome in mice lacking the protein kinase Akt2 (PKB β). Science 292, 1728–1731.

Cho, H.,Thorvaldsen, J. L., Chu, Q., Feng, F.,and Birnbaum, M. J. (2001b) Akt1/pkbα is required for normal growth but dispensable for maintenance of glucose homeostasis in mice. J Biol Chem 276, 38349–38352.

Clark, S. F., Molero, J. C., and James, D. E. (2000). Release of insulin receptor sub-strate proteins from an intracellular complex coincides with the development of insulin resistance. J Biol Chem 275, 3819–3826.

Clement, S., Krause, U., Desmedt, F., Tanti, J.-F., Behrends, J., Pesesse, X., Sasaki, T., Penninger, J., Doherty, M., Malaisse, W., et al. (2001). The lipid phosphatase SHIP2 controls insulin sensitivity. Nature 409, 92–97.

Cohen, B., Novick, D., and Rubinstein, M. (1996). Modulation of insulin activi-ties by leptin. Science 274, 1185–1188.

Cong, L. N., Chen, H., Li, Y., Zhou, L., McGibbon, M. A., Taylor, S. I., and Quon, M. J. (1997). Physiological role of Akt in insulin-stimulated translocation of GLUT4 in transfected rat adipose cells. Mol Endocrinol 11, 1881–1890.

Considine, R. V., Nyce, M. R., Allen, L. E., Morales, L. M., Triester, S., Serrano, J., Colberg, J., Lanza-Jacoby, S., and Caro, J. F. (1995). Protein kinase C is increased in the liver of humans and rats with non-insulin-dependent diabetes mellitus: an alteration not due to hyperglycemia. J Clin Invest 95, 2938–2944.

DeFronzo, R. A. (1988). Lilly lecture 1987. The triumvirate: β-cell, muscle, liver. A collusion responsible for NIDDM. Diabetes 37, 667–687.

Dimmeler, S., Dernbach, E., and Zeiher, A. M. (2000). Phosphorylation of the endothelial nitric oxide synthase at ser-1177 is required for VEGF-induced endothelial cell migration. FEBS Lett *477*, 258–262.

Donnelly, R., and Qu, X. (1998). Mechanisms of insulin resistance and new pharmacological approaches to metabolism and diabetic complications. Clin Exp Pharmacol Physiol *25*, 79–87.

Dresner, A., Laurent, D., Marcucci, M., Griffin, M. E., Dufour, S., Cline, G. W., Slezak, L. A., Andersen, D. K., Hundal, R. S., Rothman, D. L., et al. (1999). Effects of free fatty acids on glucose transport and IRS-1-associated phosphatidylinositol 3-kinase activity. J Clin Invest *103*, 253–259.

Dunaif, A., Xia, J., Book, C. B., Schenker, E., and Tang, Z. (1995). Excessive insulin receptor serine phosphorylation in cultured fibroblasts and in skeletal muscle. A potential mechanism for insulin resistance in the polycystic ovary syndrome. J Clin Invest *96*, 801–810.

Ebstein, W. (1876) Berliner Klinische Wochnschrift *13*, 337–340.

Eisenbarth, G. S. (1986). Type I diabetes mellitus. A chronic autoimmune disease. N Engl J Med *314*, 1360–1368.

Elchebly, M., Payette, P., Michaliszyn, E., Cromlish, W., Collins, S., Loy, A. L., Normandin, D., Cheng, A., Himms-Hagen, J., Chan, C. C., et al. (1999). Increased insulin sensitivity and obesity resistance in mice lacking the protein tyrosine phosphatase-1B gene. Science *283*, 1544–1548.

Eldar-Finkelman, H., and Krebs, E. G. (1997). Phosphorylation of insulin receptor substrate 1 by glycogen synthase kinase 3 impairs insulin action. Proc Natl Acad Sci USA *94*, 9660–9664.

Elmquist, J. K., Maratos-Flier, E., Saper, C. B., and Flier, J. S. (1998). Unraveling the central nervous system pathways underlying responses to leptin. Nat Neurosci *1*, 445–450.

Eng, C. (1999). The role of PTEN, a phosphatase gene, in inherited and sporadic nonmedullary thyroid tumors. Recent Prog Horm Res *54*, 441–452.

Feinstein, R., Kanety, H., Papa, M. Z., Lunenfeld, B., and Karasik, A. (1993). Tumor necrosis factor-α suppresses insulin-induced tyrosine phosphorylation of insulin receptor and its substrates. J Biol Chem *268*, 26055–26058.

Fernandez, A. M., Kim, J. K., Yakar, S., Dupont, J., Hernandez-Sanchez, C., Castle, A. L., Filmore, J., Shulman, G. I., and Le Roith, D. (2001). Functional inactivation of the IGF-I and insulin receptors in skeletal muscle causes type 2 diabetes. Genes Dev *15*, 1926–1934.

Frattali, A. L., Treadway, J. L., and Pessin, J. E. (1992). Insulin/IGF-1 hybrid receptors: implications for the dominant-negative phenotype in syndromes of insulin resistance. J Cell Biochem *48*, 43–50.

Fucini, R. V., Okada, S., and Pessin, J. E. (1999). Insulin-induced desensitization of extracellular signal-regulated kinase activation results from an inhibition of Raf activity independent of Ras activation and dissociation of the Grb2-SOS complex. J Biol Chem *274*, 18651–18658.

Gavrilova, O., Marcus-Samuels, B., Graham, D., Kim, J. K., Shulman, G. I., Castle, A. L., Vinson, C., Eckhaus, M., and Reitman, M. L. (2000). Surgical implantation of adipose tissue reverses diabetes in lipoatrophic mice. J Clin Invest *105*, 271–278.

Goldstein, B. J., Ahmad, F., Ding, W., Li, P. M., and Zhang, W. R. (1998). Regulation of the insulin signalling pathway by cellular protein-tyrosine phosphatases. Mol Cell Biochem *182*, 91–99.

Goodyear, L. J., Giorgino, F., Sherman, L. A., Carey, J., Smith, R. J., and Dohm, G. L. (1995). Insulin receptor phosphorylation, insulin receptor substrate-1 phosphorylation, and phosphatidylinositol 3-kinase activity are decreased in intact skeletal muscle strips from obese subjects. J Clin Invest 95, 2195–2204.

Griffin, M. E., Marcucci, M. J., Cline, G. W., Bell, K., Barucci, N., Lee, D., Goodyear, L. J., Kraegen, E. W., White, M. F., and Shulman, G. I. (1999). Free fatty acid-induced insulin resistance is associated with activation of protein kinase C θ and alterations in the insulin signaling cascade. Diabetes 48, 1270–1274.

Hachiya, H. L., Takayama, S., White, M. F., and King, G. L. (1987). Regulation of insulin receptor internalization in vascular endothelial cells by insulin and phorbol ester. J Biol Chem 262, 6417–6424.

Haruta, T., Uno, T., Kawahara, J., Takano, A., Egawa, K., Sharma, P. M., Olefsky, J. M., and Kobayashi, M. (2000). A rapamycin-sensitive pathway down-regulates insulin signaling via phosphorylation and proteasomal degradation of insulin receptor substrate-1. Mol Endocrinol 14, 783–794.

Heldin, C. H. (1995). Dimerization of cell surface receptors in signal transduction. Cell 80, 213–223.

Hotamisligil, G. S., Peraldi, P., Budavari, A., Ellis, R., White, M. F., and Spiegelman, B. M. (1996). IRS-1-mediated inhibition of insulin receptor tyrosine kinase activity in TNF-α- and obesity-induced insulin resistance. Science 271, 665–668.

Hubbard, S. R. (1997). Crystal structure of the activated insulin receptor tyrosine kinase in complex with peptide substrate and ATP analog. EMBO J 16, 5572–5581.

Ikeda, Y., Olsen, G. S., Ziv, E., Hansen, L. L., Busch, A. K., Hansen, B. F., Shafrir, E., and Mosthaf-Seedorf, L. (2001). Cellular mechanism of nutritionally induced insulin resistance in Psammomys obesus: overexpression of protein kinase Cε in skeletal muscle precedes the onset of hyperinsulinemia and hyperglycemia. Diabetes 50, 584–592.

Jiang, C., Ting, A. T., and Seed, B. (1998). PPAR-γ agonists inhibit production of monocyte inflammatory cytokines. Nature 391, 82–86.

Kahn, B. B. (1998). Type 2 diabetes: when insulin secretion fails to compensate for insulin resistance. Cell 92, 593–596.

Kahn, B. B., and Flier, J. S. (2000). Obesity and insulin resistance. J Clin Invest 106, 473–481.

Kanety, H., Feinstein, R., Papa, M. Z., Hemi, R., and Karasik, A. (1995). Tumor necrosis factor α-induced phosphorylation of insulin receptor substrate-1 (IRS-1). Possible mechanism for suppression of insulin-stimulated tyrosine phosphorylation of IRS-1. J Biol Chem 270, 23780–23784.

Kanety, H., Hemi, R., Papa, M. Z., and Karasik, A. (1996). Sphingomyelinase and ceramide suppress insulin-induced tyrosine phosphorylation of the insulin receptor substrate-1. J Biol Chem 271, 9895–9897.

Kellerer, M., Koch, M., Metzinger, E., Mushack, J., Capp, E., and Haring, H. U. (1997). Leptin activates PI-3 kinase in C2C12 myotubes via janus kinase-2 (JAK- 2) and insulin receptor substrate-2 (IRS-2) dependent pathways. Diabetologia 40, 1358–1362.

Kim, J. K., Gavrilova, O., Chen, Y., Reitman, M. L., and Shulman, G. I. (2000a). Mechanism of insulin resistance in A-ZIP/F-1 fatless mice. J Biol Chem 275, 8456–8460.

Kim, J. K., Kim, Y. J., Fillmore, J. J., Chen, Y., Moore, I., Lee, J., Yuan, M., Li, Z. W., Karin, M., Perret, P., et al. (2001a). Prevention of fat-induced insulin resistance by salicylate, J Clin Invest *108*, 437–446.

Kim, J. K., Zisman, A., Fillmore, J. J., Peroni, O. D., Kotani, K., Perret, P., Zong, H., Dong, J., Kahn, C. R., Kahn, B. B., and Shulman, G. I. (2001b). Glucose toxicity and the development of diabetes in mice with muscle-specific inactivation of GLUT4. J Clin Invest *108*, 153–160.

Kim, Y. B., Uotani, S., Pierroz, D. D., Flier, J. S., and Kahn, B. B. (2000b). In vivo administration of leptin activates signal transduction directly in insulin-sensitive tissues: overlapping but distinct pathways from insulin. Endocrinology *141*, 2328–2339.

Kulkarni, R. N., Bruning, J. C., Winnay, J. N., Postic, C., Magnuson, M. A., and Kahn, C. R. (1999). Tissue-specific knockout of the insulin receptor in pancreatic β cells creates an insulin secretory defect similar to that in type 2 diabetes. Cell *96*, 329–339.

Kusari, J., Kenner, K. A., Suh, K. I., Hill, D. E., and Henry, R. R. (1994). Skeletal muscle protein tyrosine phosphatase activity and tyrosine phosphatase 1B protein content are associated with insulin action and resistance. J Clin Invest *93*, 1156–1162.

Lallena, M. J., Diaz-Meco, M. T., Bren, G., Paya, C. V., and Moscat, J. (1999). Activation of IκB kinase β by protein kinase C isoforms. Mol Cell Biol *19*, 2180–2188.

Lammers, R., Bossenmaier, B., Cool, D. E., Tonks, N. K., Schlessinger, J., Fischer, E. H., and Ullrich, A. (1993). Differential activities of protein tyrosine phosphatases in intact cells. J Biol Chem *268*, 22456–22462.

Langlois, W. J., Sasaoka, T., Saltiel, A. R., and Olefsky, J. M. (1995). Negative feedback regulation and desensitization of insulin- and epidermal growth factor-stimulated p21ras activation. J Biol Chem *270*, 25320–25323.

Lauro, D., Kido, Y., Castle, A. L., Zarnowski, M. J., Hayashi, H., Ebina, Y., and Accili, D. (1998). Impaired glucose tolerance in mice with a targeted impairment of insulin action in muscle and adipose tissue. Nat Genet *20*, 294–298.

Lavan, B. E., Lane, W. S., and Lienhard, G. E. (1997). The 60-kDa phosphotyrosine protein in insulin-treated adipocytes is a new member of the insulin receptor substrate family. J Biol Chem *272*, 11439–11443.

Le Roith, D., and Zick, Y. (2001). Recent advances in our understanding of insulin action and insulin resistance. Diabetes Care *24*, 588–597.

Lee, J., Pilch, P. F., Shoelson, S. E., and Scarlata, S. F. (1997). Conformational changes of the insulin receptor upon insulin binding and activation as monitored by fluorescence spectroscopy. Biochemistry *36*, 2701–2708.

Li, J., DeFea, K., and Roth, R. A. (1999). Modulation of insulin receptor substrate-1 tyrosine phosphorylation by an Akt/phosphatidylinositol 3-kinase pathway. J Biol Chem *274*, 9351–9356.

Liu, Y. F., Paz, K., Herschkovitz, A., Alt, A., Tennenbaum, T., Sampson, S. R., Ohba, M., Kuroki, T., LeRoith, D., and Zick, Y. (2001). Insulin stimulates PKCζ-mediated phosphorylation of insulin receptor substrate-1 (IRS-1). A self-attenuated mechanism to negatively regulate the function of IRS proteins, J Biol Chem *276*, 14459–14465.

Locksley, R. M., Killeen, N., and Lenardo, M. J. (2001). The TNF and TNF receptor superfamilies: integrating mammalian biology. Cell *104*, 487–501.

Lonnqvist, F., Nordfors, L., and Schalling, M. (1999). Leptin and its potential role in human obesity. J Intern Med *245*, 643–652.

Maegawa, H., Hasegawa, M., Sugai, S., Obata, T., Ugi, S., Morino, K., Egawa, K., Fujita, T., Sakamoto, T., Nishio, Y., et al. (1999). Expression of a dominant negative SHP-2 in transgenic mice induces insulin resistance. J Biol Chem *274*, 30236–30243.

Martin, A. G., San-Antonio, B., and Fresno, M. (2001). Regulation of nuclear factor kappa B transactivation. Implication of phosphatidylinositol 3-kinase and protein kinase C ζ in c-Rel activation by tumor necrosis factor α. J Biol Chem *276*, 15840–15849.

Matthaei, S., Stumvoll, M., Kellerer, M., and Haring, H. U. (2000). Pathophysiology and pharmacological treatment of insulin resistance. Endocr Rev *21*, 585–618.

Milarski, K. L., and Saltiel, A. R. (1994). Expression of catalytically inactive Syp phosphatase in 3T3 cells blocks stimulation of mitogen-activated protein kinase by insulin. J Biol Chem *269*, 21239–21243.

Moller, D. E. (2000). Potential role of TNF-α in the pathogenesis of insulin resistance and type 2 diabetes. Trends Endocrinol Metab *11*, 212–217.

Montagnani, M., Esposito, D. L., and Quon, M. J. (2001). Insulin-stimulated activation of eNOS requires IRS-1 to couple signaling from the insulin receptor to PI 3-kinase pathways. Diabetes *50*, A298–A299.

Montagnani, M., and Quon, M. J. (2000). Insulin action in vascular endothelium: potential mechanisms linking insulin resistance with hypertension. Diabetes Obes Metab *2*, 285–292.

Mothe, I., and Van Obberghen, E. (1996). Phosphorylation of insulin receptor substrate-1 on multiple serine residues, 612, 632, 662, and 731, modulates insulin action. J Biol Chem *271*, 11222–11227.

Muller, G., Ayoub, M., Storz, P., Rennecke, J., Fabbro, D., and Pfizenmaier, K. (1995). PKCζ is a molecular switch in signal transduction of TNF-α, bifunctionally regulated by ceramide and arachidonic acid. EMBO J *14*, 1961–1969.

Muoio, D. M., Dohm, G. L., Fiedorek, F. T., Jr., Tapscott, E. B., Coleman, R. A., and Dohn, G. L. (1997). Leptin directly alters lipid partitioning in skeletal muscle. Diabetes *46*, 1360–1363.

Myers, M. G., Jr., and White, M. F. (1996). Insulin signal transduction and the IRS proteins. Annu Rev Pharmacol Toxicol *36*, 615–658.

Nystrom, F. H., and Quon, M. J. (1999). Insulin signalling: metabolic pathways and mechanisms for specificity. Cell Signal *11*, 563–574.

Paz, K., Hemi, R., LeRoith, D., Karasik, A., Elhanany, E., Kanety, H., and Zick, Y. (1997). A molecular basis for insulin resistance. Elevated serine/threonine phosphorylation of IRS-1 and IRS-2 inhibits their binding to the juxtamembrane region of the insulin receptor and impairs their ability to undergo insulin-induced tyrosine phosphorylation. J Biol Chem *272*, 29911–29918.

Paz, K., Liu, Y. F., Shorer, H., Hemi, R., LeRoith, D., Quon, M. J., Kanety, H., Seger, R., and Zick, Y. (1999). Phosphorylation of insulin receptor substrate-1 (IRS-1) by protein kinase B positively regulates IRS-1 function. J Biol Chem *274*, 28816–28822.

Pederson, T. M., Kramer, D. L., and Rondinone, C. M. (2001). Serine/threonine phosphorylation of IRS-1 triggers its degradation: possible regulation by tyrosine phosphorylation. Diabetes *50*, 24–31.

Peraldi, P., and Spiegelman, B. (1998). TNF-α and insulin resistance: summary and future prospects. Mol Cell Biochem *182*, 169–175.

Pessin, J. E., Thurmond, D. C., Elmendorf, J. S., Coker, K. J., and Okada, S. (1999). Molecular basis of insulin-stimulated GLUT4 vesicle trafficking. Location! Location! Location!. J Biol Chem *274*, 2593–2596.

Quon, M. J., Butte, A. J., Zarnowski, M. J., Sesti, G., Cushman, S. W., and Taylor, S. I. (1994a). Insulin receptor substrate 1 mediates the stimulatory effect of insulin on GLUT4 translocation in transfected rat adipose cells. J Biol Chem *269*, 27920–27924.

Quon, M. J., Chen, H., Ing, B. L., Liu, M. L., Zarnowski, M. J., Yonezawa, K., Kasuga, M., Cushman, S. W., and Taylor, S. I. (1995). Roles of 1-phosphatidylinositol 3-kinase and ras in regulating translocation of GLUT4 in transfected rat adipose cells. Mol Cell Biol *15*, 5403–5411.

Quon, M. J., Guerre-Millo, M., Zarnowski, M. J., Butte, A. J., Em, M., Cushman, S. W., and Taylor, S. I. (1994b). Tyrosine kinase-deficient mutant human insulin receptors (Met1153—>Ile) overexpressed in transfected rat adipose cells fail to mediate translocation of epitope-tagged GLUT4. Proc Natl Acad Sci USA *91*, 5587–5591.

Ravichandran, L. V., Chen, H., Li, Y., and Quon, M. J. (2001a). Phosphorylation of ptp1b at ser(50) by akt impairs its ability to dephosphorylate the insulin receptor. Mol Endocrinol *15*, 1768–1780.

Ravichandran, L. V., Esposito, D. L., Chen, J., and Quon, M. J. (2001b). Protein kinase C-ζ phosphorylates insulin receptor substrate-1 and impairs its ability to activate phosphatidylinositol 3-kinase in response to insulin. J Biol Chem *276*, 3543–3549.

Reaven, G. M. (1995). Pathophysiology of insulin resistance in human disease. Physiol Rev *75*, 473–486.

Reaven, G. M., Hollenbeck, C., Jeng, C. Y., Wu, M. S., and Chen, Y. D. (1988). Measurement of plasma glucose, free fatty acid, lactate, and insulin for 24 h in patients with NIDDM. Diabetes *37*, 1020–1024.

Roach, P., Zick, Y., Formisano, P., Accili, D., Taylor, S. I., and Gorden, P. (1994). A novel human insulin receptor gene mutation uniquely inhibits insulin binding without impairing posttranslational processing. Diabetes *43*, 1096–1102.

Roth, R. A., and Beaudoin, J. (1987). Phosphorylation of purified insulin receptor by cAMP kinase. Diabetes *36*, 123–126.

Rui, L., Aguirre, V., Kim, J. K., Shulman, G. I., Lee, A., Corbould, A., Dunaif, A., and White, M. F. (2001). Insulin/IGF-1 and TNF-α stimulate phosphorylation of IRS-1 at inhibitory Ser307 via distinct pathways. J Clin Invest *107*, 181–189.

Saltiel, A. R. (2001). You are what you secrete. Nat Med *7*, 887–888.

Segui, B., Cuvillier, O., Adam-Klages, S., Garcia, V., Malagarie-Cazenave, S., Leveque, S., Caspar-Bauguil, S., Coudert, J., Salvayre, R., Kronke, M., and Levade, T. (2001). Involvement of FAN in TNF-induced apoptosis. J Clin Invest *108*, 143–151.

Shimabukuro, M., Higa, M., Zhou, Y. T., Wang, M. Y., Newgard, C. B., and Unger, R. H. (1998). Lipoapoptosis in beta-cells of obese prediabetic fa/fa rats. Role of serine palmitoyltransferase overexpression. J Biol Chem *273*, 32487–32490.

Shulman, G. I. (2000). Cellular mechanisms of insulin resistance. J Clin Invest *106*, 171–176.

Skolnik, E. Y., and Marcusohn, J. (1996). Inhibition of insulin receptor signaling by TNF: potential role in obesity and non-insulin-dependent diabetes mellitus. Cytokine Growth Factor Rev *7*, 161–173.

Songyang, Z., Shoelson, S. E., Chaudhuri, M., Gish, G., Pawson, T., Haser, W. G., King, F., Roberts, T., Ratnofsky, S., Lechleider, R. J., et al. (1993). SH2 domains recognize specific phosphopeptide sequences. Cell 72, 767–778.

Spiegelman, B. M., and Flier, J. S. (1996). Adipogenesis and obesity: rounding out the big picture. Cell 87, 377–389.

Spiegelman, B. M., and Flier, J. S. (2001). Obesity and the regulation of energy balance. Cell 104, 531–543.

Stadtmauer, L., and Rosen, O. M. (1986). Increasing the cAMP content of IM-9 cells alters the phosphorylation state and protein kinase activity of the insulin receptor. J Biol Chem 261, 3402–3407.

Staubs, P. A., Reichart, D. R., Saltiel, A. R., Milarski, K. L., Maegawa, H., Berhanu, P., Olefsky, J. M., and Seely, B. L. (1994). Localization of the insulin receptor binding sites for the SH2 domain proteins p85, Syp, and GAP. J Biol Chem 269, 27186–27192.

Steppan, C. M., Bailey, S. T., Bhat, S., Brown, E. J., Banerjee, R. R., Wright, C. M., Patel, H. R., Ahima, R. S., and Lazar, M. A. (2001). The hormone resistin links obesity to diabetes. Nature 409, 307–312.

Takayama, S., White, M. F., Lauris, V., and Kahn, C. R. (1984). Phorbol esters modulate insulin receptor phosphorylation and insulin action in cultured hepatoma cells. Proc Natl Acad Sci USA 81, 7797–7801.

Tao, J., Malbon, C. C., and Wang, H. Y. (2001a). Gα(i2) enhances insulin signaling via suppression of protein-tyrosine phosphatase 1B. J Biol Chem 276, 39705–39712.

Tao, J., Malbon, C. C., and Wang, H. Y. (2001b). Insulin stimulates tyrosine phosphorylation and inactivation of protein-tyrosine phosphatase 1B in vivo. J Biol Chem 276, 29520–29525.

Tartaglia, L. A. (1997). The leptin receptor. J Biol Chem 272, 6093–6096.

Taylor, S. I. (1999). Deconstructing type 2 diabetes. Cell 97, 9–12.

Taylor, S. I., and Arioglu, E. (1998). Syndromes associated with insulin resistance and acanthosis nigricans. J Basic Clin Physiol Pharmacol 9, 419–439.

Tirosh, A., Potashnik, R., Bashan, N., and Rudich, A. (1999). Oxidative stress disrupts insulin-induced cellular redistribution of insulin receptor substrate-1 and phosphatidylinositol 3-kinase in 3T3-L1 adipocytes. A putative cellular mechanism for impaired protein kinase B activation and GLUT4 translocation. J Biol Chem 274, 10595–10602.

Tonks, N. K. (1990). Protein phosphatases: key players in the regulation of cell function. Curr Opin Cell Biol 2, 1114–1124.

Tremblay, F., Lavigne, C., Jacques, H., and Marette, A. (2001). Defective insulin-induced GLUT4 translocation in skeletal muscle of high fat-fed rats is associated with alterations in both Akt/protein kinase B and atypical protein kinase C (ζ/λ) activities. Diabetes 50, 1901–1910.

Tuttle, R. L., Gill, N. S., Pugh, W., Lee, J. P., Koeberlein, B., Furth, E. E., Polonsky, K. S., Naji, A., and Birnbaum, M. J. (2001). Regulation of pancreatic beta-cell growth and survival by the serine/threonine protein kinase Akt1/PKBα. Nat Med 7, 1133–1137.

Uysal, K. T., Wiesbrock, S. M., and Hotamisligil, G. S. (1998). Functional analysis of tumor necrosis factor (TNF) receptors in TNF-α-mediated insulin resistance in genetic obesity. Endocrinology 139, 4832–4838.

Van Antwerp, D. J., Martin, S. J., Verma, I. M., and Green, D. R. (1998). Inhibition of TNF-induced apoptosis by NF-κ B. Trends Cell Biol 8, 107–111.

van der Geer, P., Hunter, T., and Lindberg, R. A. (1994). Receptor protein-tyrosine kinases and their signal transduction pathways. Annu Rev Cell Biol 10, 251–337.

Venkatesan, A. M., Dunaif, A., and Corbould, A. (2001). Insulin resistance in polycystic ovary syndrome: progress and paradoxes. Recent Prog Horm Res 56, 295–308.

Walton, K. M., and Dixon, J. E. (1993). Protein tyrosine phosphatases. Annu Rev Biochem 62, 101–120.

Watson, R. T., Shigematsu, S., Chiang, S. H., Mora, S., Kanzaki, M., Macara, I. G., Saltiel, A. R., and Pessin, J. E. (2001). Lipid raft microdomain compartmentalization of TC10 is required for insulin signaling and GLUT4 translocation. J Cell Biol 154, 829–840.

Way, J. M., Gorgun, C. Z., Tong, Q., Uysal, K. T., Brown, K. K., Harrington, W. W., Oliver, W. R., Jr., Willson, T. M., Kliewer, S. A., and Hotamisligil, G. S. (2001). Adipose tissue resistin expression is severely suppressed in obesity and stimulated by peroxisome proliferator-activated receptor gamma agonists. J Biol Chem 276, 25651–25653.

White, M. F. (1998). The IRS-signaling system: a network of docking proteins that mediate insulin and cytokine action. Recent Prog Horm Res 53, 119–138.

Withers, D. J., Gutierrez, J. S., Towery, H., Burks, D. J., Ren, J. M., Previs, S., Zhang, Y., Bernal, D., Pons, S., Shulman, G. I., et al. (1998). Disruption of IRS-2 causes type 2 diabetes in mice. Nature 391, 900–904.

Yamauchi, K., Milarski, K. L., Saltiel, A. R., and Pessin, J. E. (1995). Protein-tyrosine-phosphatase SHPTP2 is a required positive effector for insulin downstream signaling. Proc Natl Acad Sci USA 92, 664–668.

Yamauchi, T., Kamon, J., Waki, H., Terauchi, Y., Kubota, N., Hara, K., Mori, Y., Ide, T., Murakami, K., Tsuboyama-Kasaoka, N., et al. (2001). The fat-derived hormone adiponectin reverses insulin resistance associated with both lipoatrophy and obesity. Nat Med 7, 941–946.

Yin, M. J., Yamamoto, Y., and Gaynor, R. B. (1998). The anti-inflammatory agents aspirin and salicylate inhibit the activity of I(κ)B kinase-β. Nature 396, 77–80.

Yki-Jarvinen, H. (1992). Glucose toxicity. Endocr Rev 13, 415–431.

Yuan, M., Konstantopoulos, N., Lee, J., Hansen, L., Li, Z. W., Karin, M., and Shoelson, S. E. (2001). Reversal of obesity- and diet-induced insulin resistance with salicylates or targeted disruption of Ikkβ. Science 293, 1673–1677.

Zeng, G., Nystrom, F. H., Ravichandran, L. V., Cong, L. N., Kirby, M., Mostowski, H., and Quon, M. J. (2000). Roles for insulin receptor, PI3-kinase, and Akt in insulin-signaling pathways related to production of nitric oxide in human vascular endothelial cells. Circulation 101, 1539–1545.

Zeng, G., and Quon, M. J. (1996). Insulin-stimulated production of nitric oxide is inhibited by wortmannin. Direct measurement in vascular endothelial cells. J Clin Invest 98, 894–898.

Zhou, L., Chen, H., Lin, C. H., Cong, L. N., McGibbon, M. A., Sciacchitano, S., Lesniak, M. A., Quon, M. J., and Taylor, S. I. (1997). Insulin receptor substrate-2 (IRS-2) can mediate the action of insulin to stimulate translocation of GLUT4 to the cell surface in rat adipose cells. J Biol Chem 272, 29829–29833.

Zhou, L., Chen, H., Xu, P., Cong, L. N., Sciacchitano, S., Li, Y., Graham, D., Jacobs, A. R., Taylor, S. I., and Quon, M. J. (1999). Action of insulin receptor substrate-3 (IRS-3) and IRS-4 to stimulate translocation of GLUT4 in rat adipose cells. Mol Endo *13*, 505–514.

Zick, Y. (2001). Insulin resistance: a phosphorylation-based uncoupling of insulin signaling. Trends Cell Biol *11*, 437–441.

Zick, Y., Grunberger, G., Podskalny, J. M., Moncada, V., Taylor, S. I., Gorden, P., and Roth, J. (1983). Insulin stimulates phosphorylation of serine residues in soluble insulin receptors. Biochem Biophys Res Commun *116*, 1129–1135.

CHAPTER 6

DYSFUNCTION OF G PROTEIN-REGULATED PATHWAYS AND ENDOCRINE DISEASES

WILLIAM F. SIMONDS

Metabolic Diseases Branch, National Institute of Diabetes and Digestive and Kidney Diseases, National Institutes of Health, Bethesda, Maryland

INTRODUCTION

The endocrine regulation of human development and metabolic homeostasis requires hormone action on target tissues mediated by a variety of hormone-specific signal transduction pathways. Physiological signaling by different hormones may employ cell surface or nucleocytoplasmic receptors and one or more signal transduction mechanisms including chemical or ionic second messengers, kinase cascades, or translocation and/or activation of transcriptional regulators. The effect of a hormone on the target tissue ultimately results from altered protein activity, synthesis, or release, sometimes also associated with an altered rate of cellular differentiation or division. Medically important endocrine disorders frequently result when hormonal pathways are either overly activated (by excessive levels of hormone or by unregulated activation of the hormonal signal) or inadequately stimulated (by lack of hormone or unresponsive hormonal signaling components). This chapter reviews the clinical manifestations, pathophysiology, and approaches to treatment of endocrine diseases resulting from the dysfunction of signal-transducing G proteins, the G protein-coupled receptors (GPCRs) at the cell surface that regulate their activity, and downstream effectors of G proteins. First, a brief review of the important features of G protein signaling provides a context for the reader to better interpret the dysfunctional signaling states described in the subsequent disease-oriented sections.

Signal Transduction and Human Disease, Edited by Toren Finkel and J. Silvio Gutkind
ISBN 0-471-02011-7   Copyright © 2003 John Wiley & Sons, Inc.

OVERVIEW OF G PROTEIN STRUCTURE AND FUNCTION

Signal-transducing heterotrimeric G proteins are positioned at the inner (cytoplasmic) face of the plasma membrane, where they can interact with membrane-spanning GPCR and effector molecules. The human genome encodes many hundreds of GPCRs and includes those activated by hormones, pheromones, neurotransmitters, mineral ions, and sensory stimuli (e.g., photons of light, odorants, tastants). Analysis of the family of GPCR suggests that they all contain seven hydrophobic transmembrane (TM) domains flanked by an extracellular N-terminal domain and an intracellular C-terminal domain (Wess, 1998). Serpentine, 7-TM, and heptahelical receptors are all synonyms for GPCRs.

G protein heterotrimers consist of $G\alpha$, $G\beta$, and $G\gamma$ subunits at a $1:1:1$ stoichiometry. Lipid modification of the $G\alpha$ subunits by myristoylation and/or palmitoylation and of the $G\gamma$ subunits by isoprenylation helps anchor the G protein heterotrimer to the plasma membrane. The $G\alpha$ subunits bind guanine nucleotides with high affinity and contain an intrinsic GTP hydrolytic activity. The ability of $G\alpha$ subunits to bind guanine nucleotides arises from their homology with other members of a GTP-binding protein superfamily including smaller proteins such as $p21^{ras}$, Rab, Ran, Ral, Rac, Rho, and EF-Tu. $G\beta$ and $G\gamma$ subunits form a very tight, noncovalent heterodimer and function as a single entity (the $G\beta\gamma$ complex) throughout the G protein signaling cycle. The crystal structure of two different G protein heterotrimers has been determined (Wall et al., 1995; Lambright et al., 1996).

G proteins function in the context of two interrelated cycles: a cycle of subunit association and dissociation and a cycle of GTP binding and hydrolysis (Fig. 6.1). Activation of G proteins by GPCRs in response to extracellular stimuli results from GPCR-G protein interaction causing the release of GDP from the guanine nucleotide binding site on the $G\alpha$ subunit. This then allows the binding to $G\alpha$ of GTP from the abundant intracellular pool. With GTP bound, $G\alpha$ undergoes conformational changes in three polypeptide "switch" regions (Hamm and Gilchrist, 1996) resulting in the dissociation of the $G\alpha$ subunit from the $\beta\gamma$ complex. Both the $G\alpha$ subunit (in GTP-bound form) and the free $G\beta\gamma$ complex are then able to regulate the function of downstream effector molecules until GTP on the $G\alpha$ subunit is hydrolyzed to GDP. With GDP in place, $G\alpha$ reverts in conformation to again exhibit high affinity for $G\beta\gamma$ and the heterotrimer is reformed, terminating effector signaling. In some cells regulator of G protein signaling (RGS) proteins are present that can bind to specific $G\alpha$ subunits to accelerate GTP hydrolysis and more rapidly terminate both $G\alpha$ and $G\beta\gamma$ signaling. Some RGS proteins may physically block $G\alpha$-effector interaction, and at least in certain cases RGS protein-$G\alpha$ interaction appears to generate effector signals (see Hepler, 1999; De Vries et al., 2000; Rossand Wilkie, 2000 for reviews).

The diversity of G protein subunits helps to satisfy the biological imperative for specificity of coupling between G proteins and both

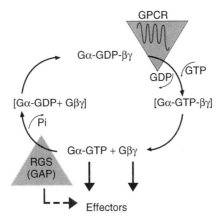

Figure 6.1. The G protein signaling cycle. G proteins operate in the context of two interrelated cycles: a cycle of subunit association and dissociation and a cycle of GTP binding and hydrolysis. Activated receptors (GPCRs) interact with G protein heterotrimers and cause dissociation of GDP from the nucleotide binding site on Gα. This allows the binding of intracellular GTP to the Gα subunit, which then adopts an activated conformation and loses affinity for the Gβγ complex. Both Gα -GTP and Gβγ are then free to activate or inhibit downstream effectors, such as PLC-β, adenylyl cyclase, ion channels, and several MAP kinase pathways. Some effectors are regulated by Gα, some by Gβγ, and some by both. The GTPase activity intrinsic to the Gα subunit hydrolyzes GTP to GDP. In its GDP-bound form, Gα loses its active conformation and regains its high affinity for Gβγ. Isoforms of regulators of G protein signaling (RGS) proteins are present in many cell types and accelerate the G protein turn-off reaction. The RGS proteins bind to the Gα subunit and act as GTPase-activating proteins (GAP). In certain cases, at least, Gα-RGS protein interactions appear to generate effector signals (Hart et al., 1998; Fukuhara et al., 1999) (dashed arrow).

GPCRs and effectors. Genes for 16 Gα subunits are known and give rise via alternative RNA processing to at least 20 mature Gα subunits with differential tissue expression (Table 6.1). Four homology-based subfamilies of Gα subunits can be recognized: the Gs subfamily, whose members stimulate adenylyl cyclase (AC); the Gi subfamily, which includes Gi_{1-3} and Gz, which inhibit AC; the Gq subfamily, whose members activate phospholipase C-β (PLC-β); and the G_{12} subfamily, whose members interact with RGS domain-containing Rho exchange factors (Kozasa et al., 1998; Fukuhara et al., 1999; Table 6.1). Genes encoding 5 Gβ isoforms and 12 different Gγ subunits are known in humans, creating the possibility of a great combinatorial diversity of Gβγ heterodimers. Effectors of Gβγ complexes include ion channels, isoforms of AC, isoforms of PLC-β, and MAP kinase (MAPK) pathways (Marinissen and Gutkind, 2001).

TABLE 6.1. Diversity and Expression of G Protein-α Subunits

Gα Subfamily	Gα Subunit	Expression	Effector Pathway
G_s	G_s	Ubiquitous	Stimulation of adenylyl cyclases
	G_{olf}	Olfactory neuroepithelium	
		Basal ganglia	
		Testis	
		Pancreatic islets	
G_i	G_{t1}	Retinal rods	Stimulation of cyclic nucleotide phosphodiesterase
		Taste cells (bitter)	
	G_{t2}	Retinal cones	
		Pancreatic islets	
	G_{gust}	Taste cells (bitter, sweet)	
		Foregut epithelium	
	G_{i1}	Neural > other tissues	Inhibition of adenylyl cyclases
	G_{i2}	Ubiquitous	
	G_{i3}	Other tissues > neural	
	G_z	Brain, retina, platelets	
		Adrenal glands	
		Pancreatic islets	
	G_o	Heart	Inhibition of adenylyl cyclase I
		Neural	Rap1 GAP (binds as G_o-GDP)
			GRIN1, GRIN2
G_q	G_q	Ubiquitous	Stimulation of phospholipase C-β
	G_{11}	Ubiquitous	
	G_{14}	Spleen	
		Kidney	
		Testis	
		Lung	
	$G_{15/16}$	Spleen	
		Thymus	
		Bone marrow	
		Lung	
G_{12}	G_{12}	Ubiquitous	Binding and activation (G_{13}) of RGS domain-containing Rho exchange factors
	G_{13}	Ubiquitous	

MECHANISMS OF G PROTEIN-REGULATED SIGNALING DYSFUNCTION IN ENDOCRINE DISEASE

Acquired or inherited alterations of G protein subunits, GPCRs, or downstream effector molecules all have the potential to disrupt G protein-regulated signaling. Acquired nongenetic alterations of such signaling elements, through infectious and autoimmune mechanisms, for example, can cause human disease. The pathogenesis of cholera and pertussis (whooping cough), for example, involves bacterial exotoxins that covalently modify Gα subunits and disrupt G protein signaling to the detriment of the host (Farfel et al., 1999). Autoimmunity to a GPCR mediates the endocrine manifestations of Graves disease, an acquired disorder that is the most common cause of thyrotoxicosis (Weetman, 2000). In this disease antibodies to the receptor for thyroid-stimulating hormone (TSH), called thyroid-stimulating immunoglobulins, activate the receptor in the absence of TSH, resulting in the unregulated synthesis and secretion of thyroid hormone by the thyroid gland. The manifestations of Graves disease include hyperthyroid symptoms, in common with other causes of thyrotoxicosis (see below), and Graves disease-specific findings such as ophthalmopathy and dermopathy (Weetman, 2000). Although the importance of the autoimmune process in the pathogenesis of Graves and many other endocrine diseases is indisputable, the remainder of this chapter focuses on mutations in genes encoding GPCRs, G protein subunits, and effectors that result in endocrine dysfunction.

Endocrine disease can result from either acquired or inherited gene mutations that result in loss or gain of gene function (Table 6.2). The effect of mutations that cause loss of gene function can be to disrupt gene promoter or other regulatory sequences, to impair RNA transcript splicing, or to alter the protein coding region. Detrimental alterations in the latter category include truncation mutations and missense/insertion mutations that impair protein-protein or protein-ligand interactions as well as mutations that impair protein folding and/or stability, subcellular localization, and/or posttranslational protein modification. Acquired gene mutations (i.e., somatic or nongermline) in GPCRs, G proteins, or G protein-regulated effectors that cause loss of function (LOF) have not yet been associated with sporadic endocrine disease. This is most likely because of compensation at the cellular level by redundant signaling elements in the affected cells and/or at the level of the organ or organism by the reserve of unaffected endocrine tissue.

In contrast examples of inherited (i.e., germline) LOF mutation in genes involved in G protein signaling giving rise to endocrine disease are numerous (Table 6.2). The disease phenotype and inheritance pattern depend on the particular gene affected. Dominant inheritance of inactivating mutations in the calcium-sensing receptor (CaSR), a GPCR important in parathyroid and kidney function, results from haploinsufficiency and causes familial hypocalciuric hypercalcemia (FHH) in simple heterozygotes (Brown, 2000). The disease neonatal severe primary hyperparathyroidism results from biallelic inactivation of the same CaSR gene and is inherited recessively (as homozygous or compound heterozygous mutations). Inherited LOF of the *GNAS1* gene encoding Gsα can cause the phenotype of Albright hereditary osteodystrophy (AHO) either with or without an associated state of resistance to parathyroid and other hormones (Weinstein and Yu, 1999). Because the *GNAS1* gene is imprinted in a tissue-specific fashion, maternal inheritance of the inactivating Gsα mutation gives rise to pseudohypoparathyroidism (PHP) with AHO and hormone resistance whereas paternal inheritance of the mutant *GNAS1* allele results in

TABLE 6.2. Sporadic and Inherited Gene Mutations Affecting G Protein-Regulated Signaling Pathways in Endocrine Disease

Genetic mechanism/effect on signaling pathway	Endocrine Disease	G Protein, Receptor, or Effector Alteration	Type of Gene Alteration	References
Sporadic gain-of-function	Acromegaly	$G\alpha_s$ activation	Missense mut	Landis et al., 1989
	Pituitary corticotroph adenoma	$G\alpha_s$ activation	Missense mut	Williamson et al., 1995; Riminucci et al., 2002
	McCune–Albright Syndrome	$G\alpha_s$ activation	Missense mut	Weinstein et al., 1991
	Hyperfunctioning thyroid nodules	$G\alpha_s$ activation	Missense mut	Lyons et al., 1990
	Thyroid cancer	$G\alpha_s$ activation	Missense mut	Suarez et al., 1991; Goretzki et al., 1992
	Ovarian sex cord stromal tumor	$G\alpha_{i2}$ activation	Missense mut	Lyons et al., 1990; but see Shen et al., 1996; Ichikawa et al., 1996
	Adrenocortical adenoma	$G\alpha_{i2}$ activation	Missense mut	Lyons et al., 1990; but see Gicquel et al., 1995; Reincke et al., 1993
	Hyperfunctioning thyroid nodules	TSH receptor activation	Missense mut	Parma et al., 1993
	Leydig cell tumor	LH receptor activation	Missense mut	Liu et al., 1999
Inherited loss of function	Pseudohypoparathyroidism type Ia	$G\alpha_s$ inactivation	Inactivating mut	Patten et al., 1990; Weinstein and Yu, 1999
	Pseudohypoparathyroidism type Ib	$G\alpha_s$ inactivation	Imprinting abnormality	Liu et al., 2000
	X-linked nephrogenic diabetes insipidus	V_2 vasopressin receptor inactivation	Inactivating mut (hemizygous)	van den Ouweland et al., 1992; Pan et al., 1992; Rosenthal et al., 1992

	Disease	Defect	Mutation	Reference
	Isolated glucocorticoid deficiency	ACTH receptor inactivation	Inactivating mut (cmpnd heterozygous; homozygous)	Tsigos et al., 1993; Clark et al., 1993
	Isolated growth hormone deficiency	GHRH receptor inactivation	Inactivating mut (homozygous)	Baumann and Maheshwari, 1997
	Familial hypocalciuric hypercalcemia	Calcium-sensing receptor inactivation	Inactivating mut (heterozygous)	Pollak et al., 1993
	Neonatal severe hyperparathyroidism	Calcium-sensing receptor inactivation	Inactivating mut (homozygous)	Pollak et al., 1993
	Familial hypothyroidism	TSH receptor inactivation	Inactivating mut (homozygous)	Abramowicz et al., 1997
	Familial Leydig cell hypoplasia	LH receptor inactivation	Inactivating mut (homozygous)	Kremer et al., 1995
	Familial ovarian dysgenesis	FSH receptor inactivation	Inactivating mut (homozygous)	Gromoll et al., 1996
Inherited gain of function	Testitoxicosis (in setting of pseudohypoparathyroidism type Ia)	$G\alpha_s$ activation/inactivation	Missense mut	Iiri et al., 1994
	Familial male precocious puberty	LH receptor activation	Missense mut	Shenker et al., 1993
	Autosomal dominant hyperthyroidism (nonautoimmune)	TSH receptor activation	Missense mut	Duprez et al., 1994
	Jansen's metaphyseal chondrodysplasia	PTH/PTHrP receptor activation	Missense mut	Schipani et al., 1995
	Autosomal dominant hypocalcemia	Calcium-sensing receptor activation	Missense mut	Pollak et al., 1994
	Carney complex	Protein kinase A R1α regulatory subunit inactivation	Inactivating mut (heterozygous)	Kirschner et al., 2000a; Casey et al., 2000

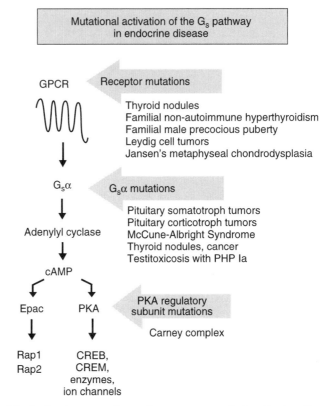

Figure 6.2. Activating mutations of components of the G_s signaling pathway result in a variety of endocrine diseases. On the left is shown a schematic diagram indicating the flow of information from activated G_s-coupled receptors (GPCRs) through adenylyl cyclase and the second messenger cyclic adenosine 3′, 5′-monophosphate (cAMP) to Epac [exchange protein directly activated by cAMP (de Rooij et al., 1998)], a regulator of the low-molecular-weight GTP-binding proteins Rap1 and Rap2, and protein kinase A (PKA) whose kinase targets include the transcription factors cAMP-response element binding protein (CREB) and cAMP-response element modulator (CREM) as well as various enzymes and ion channels. Not shown are other putative effector targets of $G_s\alpha$ including the Src and Hck non-receptor tyrosine kinases (Ma et al., 2000). On the right are indicated the G_s pathway signaling components (arrows), which when mutated can result in the sporadic or inherited endocrine diseases listed. PHP Ia, pseudohypoparathyroidism type Ia. Details and references for the individual endocrine diseases listed can be found in the text and Table 6.2.

AHO without hormone resistance or PHP (reviewed by Weinstein et al., 2001).

Gain of function (GOF) mutation of genes in G protein signaling pathways can give rise to disease in any of several endocrine systems (Table 6.2, Fig. 6.2). In theory, the effect of mutations that cause gain of gene function can be to cause excessive and/ or ectopic gene expression or, if they occur in the protein coding region, to alter protein structure in a way to cause constitutive signaling activation. Sporadic mutation of Gsα that results in loss of intrinsic GTPase activity and causes constitutive Gsα activation can cause acromegaly (see below), autonomous thyroid nodules, or the McCune–Albright syndrome (MAS). MAS is a pediatric disorder of gonadotropin-independent precocious puberty caused gonadal hyperfunction, café-au-lait skin pigmentation, and polyostotic fibrous dysplasia (reviewed in Weinstein et al., 2001). Fibrous dysplasia without precocious puberty can also result from similar somatic activating *GNAS1* mutation (Alman et al., 1996; Bianco et al., 2000). No inherited diseases due to activating G protein α subunit mutation have been found, most likely because of disruption of embryogenesis by such constitutive signaling. Somatic GOF mutation in the GPCR for luteinizing hormone (LH) can cause sporadic Leydig cell tumors (Liu et al., 1999), whereas inherited activating mutations in the LH receptor result in familial male precocious puberty (Shenker et al., 1993).

More detailed descriptions of three endocrine diseases resulting from sporadic or inherited GOF mutations in the Gs signaling pathway follow (Fig. 6.2). Examples of diseases caused by mutation at the level of the GPCR (nonautoimmune autosomal dominant hyperthyroidism), the G protein Gs (acromegaly), and the effector protein (Carney complex) illustrate the range of disease phenotypes that can result from disruption of a single G protein-regulated signaling pathway. The clinical manifestations, pathophysiology, and approaches to treatment of these three very different endocrine diseases are explored.

AUTOSOMAL DOMINANT HYPERTHYROIDISM (NONAUTOIMMUNE)

Clinical Manifestations of Nonautoimmune Autosomal Dominant Hyperthyroidism

Like other forms of hyperthyroidism the signs and symptoms of nonautoimmune autosomal dominant hyperthyroidism reflect the diffuse effects of excessive thyroid hormone action on the metabolic, cardiovascular, nervous, endocrine, and other systems. Metabolic manifestations include weight loss, heat intolerance, weakness, and fatigue. Cardiovascular signs and symptoms include palpitations, tachycardia and/or atrial fibrillation with a difficult to control ventricular response, and angina pectoris. Nervousness, irritability, restlessness, fine resting tremor, hyperreflexia, and muscle cramps are all manifestations of hyperthyroidism. Menstrual irregularities in women with thyrotoxicosis are common. In men decreased libido, impotence, gynecomastia, and lowered sperm count may accompany hyperthyroidism. Other signs and symptoms of hyperthyroidism include increased sweating, moist, warm skin, fine hair, frequent, loose bowel movements, and a characteristic "stare" due to widening of the palpebral fissures.

The diagnosis of nonautoimmune autosomal dominant hyperthyroidism rests on a set of both positive and negative clinical and laboratory findings. The age of onset of hyperthyroid symptoms in patients from families with nonautoimmune autosomal dominant hyperthyroidism varies from infancy to adulthood, even within the same family. In affected patients, findings specific to autoimmune goitrous hyperthyroidism (Graves disease) are absent. These include Graves exophthalmos and pretibial dermopathy. Patients with Graves disease sometimes have associated systemic autoimmune disorders such as pernicious anemia and myasthenia gravis, uncommon findings in nonautoimmune autosomal dominant hyperthyroidism. Hormone testing of patients with primary hyperthyroidism of any etiology reveals elevated free thyroxine and free thyroxine index associated with suppressed TSH. Laboratory findings typical of Graves disease such as stimulating anti-TSH receptor immunoglobulins and elevated antithyroglobulin and antimicrosomal antibodies are usually absent in nonautoimmune autosomal dominant hyperthyroidism. Radioactive iodine uptake and scan can be useful in the differential diagnosis of thyrotoxicosis. Hyperthyroidism due to subacute thyroiditis or ingestion of exogenous thyroid hormone can be distinguished from nonautoimmune autosomal dominant hyperthyroidism by the suppressed radioiodine uptake in the former conditions. In the latter disease, furthermore, the thyroid scan typically shows diffuse homogeneous uptake of tracer throughout the thyroid.

Dysfunction of G Protein-Regulated Pathways In Nonautoimmune Autosomal Dominant Hyperthyroidism

After the demonstration in hyperfunctioning thyroid adenomas of somatic activating point mutations in the gene for the G protein-coupled TSH receptor (Parma et al., 1993), Vassart and coworkers identified germline activating TSH receptor mutations in affected patients from families with nonautoimmune autosomal dominant hyperthyroidism (Duprez et al., 1994). When studied by transfection in vitro, cells expressing the mutant receptors showed constitutive elevation of cAMP levels compared to cells with wild-type receptors at all doses of transfected DNA (Duprez et al., 1994). The mutant TSH receptors from sporadic hyperfunctioning thyroid adenomas showed similar constitutive activation of G protein regulated cAMP pathways (Parma et al., 1993). In human thyrocytes cAMP activates iodine trapping and promotes thyroid hormone secretion and cell growth. The ability of germline point mutations in GPCR to cause constitutive activation of G_s-dependent pathways was previously demonstrated in the case of the luteinizing hormone receptor in familial male precocious puberty by Shenker and coworkers (Shenker et al., 1993).

Missense mutations resulting in nonautoimmune autosomal dominant hyperthyroidism map to different domains of the TSH receptor in different kindreds. Mutations in several kindreds map to the sixth transmembrane spanning segment (TM6), the segment of the GPCR in which the first activating mutations were demonstrated in vitro (Ren et al.,

1993). Such TM6 mutations include F629L (Fuhrer et al., 1997), T632I (Kopp et al., 1997), P639S (Khoo et al., 1999), and N650Y (Tonacchera et al., 1996). In other kindreds with nonautoimmune autosomal dominant hyperthyroidism the mutations map elsewhere in the TSH receptor including TM1 [G431S (Biebermann et al., 2001)], TM2 [M463V (Lee et al., 2002)], TM3 [S505R (Tonacchera et al., 1996), V509A (Duprez et al., 1994)], TM5 [V597F (Alberti et al., 2001)], the third extracellular loop [A623V (Schwab et al., 1997)], and TM7 [N670S (Tonacchera et al., 1996), C672Y (Duprez et al., 1994)]. When tested in vitro these mutant TSH receptors share the ability to constitutively activate G protein-regulated cAMP production. Sporadic cases of congenital nonautoimmune hyperthyroidism have been associated with similar germline activating de novo TSH receptor mutations in the extracellular domain [S281N (Gruters et al., 1998)], the second extracellular loop [I568T (Tonacchera et al., 2000)], and TM5 [V597L (Esapa et al., 1999)].

Treatment of Nonautoimmune Autosomal Dominant Hyperthyroidism

The goal of management of patients with nonautoimmune autosomal dominant hyperthyroidism is a euthyroid state. Short-term and prompt relief of hyperthyroid symptoms can be achieved with the administration of propranolol. This nonselective β-adrenergic receptor blocker relieves the anxiety, tachycardia, palpitations, diaphoresis, and tremor associated with hyperthyroidism (of any etiology) but does not affect thyroid hormone synthesis or secretion. Treatment is usually initiated at 10 mg daily and then titrated upward as needed for symptomatic relief, often to a dosage of 20 mg four times daily. Larger doses are occasionally required. Treatment of nonautoimmune autosomal dominant hyperthyroidism with thiourea drugs such as methimazole and propylthiouracil provides unsatisfactory long-term control but can be beneficial over the short-term, such as rendering the patient euthyroid before surgery.

Long-term resolution of hyperthyroid symptoms in nonautoimmune autosomal dominant hyperthyroidism often requires a more aggressive approach than in cases of sporadic hyperthyroidism because of the strong tendency toward recurrent goiter and hyperthyroidism associated with the germline TSH receptor mutations. Thus near-total thyroidectomy in place of partial thyroidectomy has been recommended as the surgery of choice (Fuhrer et al., 1997; Schwab et al., 1997), and thyroid ablation with [131I] radioiodine has often been required in this setting to eliminate remnant thyroid tissue in the postoperative setting. Radioactive iodine should not be given to pregnant women. Radioactive iodine can frequently be given while symptomatic relief is being provided with propranolol, because the β-adrenergic antagonist does not block iodine uptake or thyroid hormone synthesis in thyrocytes. The dosage of propranolol can then be tapered after radioiodine ablation as the hyperthyroxinemia slowly resolves.

ACROMEGALY

Clinical Manifestations of Acromegaly

Acromegaly is a syndrome of bony and soft tissue overgrowth that results from excessive circulating growth hormone (GH) occurring after puberty (Melmed et al., 1995; Barkan, 1998). If the GH excess occurs before epiphysial closure, tall stature and gigantism result (Eugster and Pescovitz, 1999). The cause of this hormonal excess in the vast majority of cases is a GH-secreting pituitary adenoma, frequently with tumor size >1 cm (macroadenoma). The metabolic effects of the GH are mediated in large part by release of insulin-like growth factor I (IGF-I) from the liver and other tissues. The signs and symptoms of acromegaly include enlargement of the hands, feet, jaw, and skull and abnormal overgrowth of internal organs. Patients may notice increasing glove and ring size caused by hand enlargement and finger widening. Cutaneous changes in acromegaly include hyperhydrosis and oily skin. The GH-induced changes also give the hands a distinctive doughy, moist character that can be detected on handshake. Associated carpal tunnel syndrome is common. Increasing foot width may necessitate changing show size. Growth of the mandible causes protrusion of the jaw (prognathism) and dental malocclusion. Enlargement of the bones of the skull and paranasal sinuses may alter facial appearance over time and result in increased hat size. Pharyngeal overgrowth may result in obstructive sleep apnea, and laryngeal enlargement produces a characteristic deepening of the voice. Left ventricular hypertrophy, cardiomegaly, and frequently associated hypertension contribute to increased incidence of congestive heart failure and other cardiovascular morbidity. Arthritis and arthralgias are common in acromegaly, especially affecting the large joints (hips and shoulders). Impaired carbohydrate tolerance and hypogonadism are also frequently seen in acromegaly. Several centers have noted a significant increase in adenomatous colonic polyps and colorectal cancer in acromegalic patients and recommend surveillance colonoscopy (Delhougne et al., 1995; Jenkins et al., 1997), although this association has been recently challenged by others (Renehan et al., 2000).

Biochemical testing to confirm the diagnosis of acromegaly includes the demonstration of fasting elevations in IGF-I and a failure of GH suppression after an oral glucose challenge (Melmed et al., 1995; Barkan, 1998; Giustina et al., 2000). Associated findings may include fasting hyperglycemia and elevations in serum prolactin or phosphorus. The diagnosis of acromegaly is associated with a pituitary mass lesion demonstrable by MRI in more than 90% of patients (Saeki et al., 1999; Marro et al., 1997).

Dysfunction of G Protein-Regulated Pathways in Acromegaly

Hypothalamic GH-releasing hormone (GHRH) acting through a receptor coupled to Gs and utilizing cAMP as a second messenger normally regulates release of GH from pituitary somatotrophs. Studying the in vitro properties of GH-secreting adenomas removed from patients with acromegaly, Vallar and coworkers identified a subset of tumors (group 2) with increased basal GH secretion and elevated intracellular cAMP levels that did not respond further to GHRH treatment (Vallar et al., 1987). The remaining GH-secreting tumors (group 1) had low levels of basal cAMP and GH secretion but showed significant stimulation by GHRH treatment. The presence of a constitutively activated Gs regulatory protein in the group 2 tumors was inferred from these biochemical studies, and indeed subsequent investigation found activating missense mutations in tumor-derived Gsα cDNA from four of four group 2 tumors (Landis et al., 1989). These mutations encoded $Arg^{201} \rightarrow Cys$ (R201C), $Arg^{201} \rightarrow His$ (R201H), and $Gln^{227} \rightarrow Arg$ (Q227R) changes in Gsα leading to its constitutive activation (Landis et al., 1989). Series of GH-secreting adenomas analyzed in various centers demonstrated mutations in residues 201 or 227 of Gsα in 4% to 53% of tumors (Yang et al., 1996; Yoshimoto et al., 1993; Shi et al., 1998; Kim et al., 2001; Hosoi et al., 1993; Landis et al., 1990; Lyons et al., 1990; Johnson et al., 1999). As noted above, acromegaly associated with Arg^{201} mutations in Gsα can be a component of MAS (Weinstein et al., 2001).

The mechanism of activation of Gsα by mutation of Arg^{201} and Gln^{227} has been inferred from structural and biochemical studies to result from loss of GTPase activity. The structure of $G_s\alpha$, like that of $G_i\alpha$ and $G_t\alpha$, consists of an α-helical domain bound through two linking peptides to a p21 Ras-like domain containing a six-stranded β-sheet (Sunahara et al., 1997). The Ras-like domain contains the high-affinity guanine nucleotide-binding site. Three flexible switch elements (switches I–III) in Gα subunits adopt different conformations in the Gα-GDP vs. Gα-GTP states and provide surfaces for conformation-dependent interaction of Gα with Gβγ complexes, RGS proteins, and effector molecules (Hamm and Gilchrist, 1996). Switch I corresponds to one of the two linker peptides joining the Gα helical domain with the Ras-like domain and contains Arg^{201} in $G_s\alpha$ (Sunahara et al., 1997). This arginine residue is thought to facilitate GTP hydrolysis by stabilizing a transition-state pentavalent phosphate (Coleman et al., 1994; Sondek et al., 1994) and is the same arginine in $G_s\alpha$ covalently modified by cholera toxin resulting in constitutive signal activation (Bourne et al., 1989). Mutation of this residue in vitro also leads to diminished GTPase activity and constitutive activation of $G_s\alpha$ (Freissmuth and Gilman, 1989).

The switch II region in the Ras-like domain of $G_s\alpha$ contains Gln^{227}, which corresponds to Gln^{61} in p21-ras, site of GTPase-inactivating oncogenic point mutations (Sunahara et al., 1997). The role of this glutamine in GTP hydrolysis is not fully understood, but structural studies suggest it may stabilize the orientation of a catalytic water molecule (Coleman et al., 1994). Mutation of this glutamine in $G_s\alpha$ can be shown in vitro to inhibit GTPase activity and constitutively activate AC (Landis et al., 1989).

Treatment of Acromegaly

The goal of treatment in acromegaly is to reduce GH levels to <1 ng/ml 2 hours after a standard oral glucose load (75 g) and to normalize IGF-I levels relative to age- and sex-matched controls (Giustina et al., 2000).

Surgical, radiological, and medical treatment modalities directed at GH-secreting pituitary adenomas are available. Transsphenoidal pituitary microsurgery is the preferred initial treatment for most patients with acromegaly. As preoperative imaging and neurosurgical techniques have improved, the rate of successful pituitary surgery in acromegaly has risen significantly (Ahmed et al., 1999). In a study with a mean follow-up period of 16 years, nearly 20% of acromegalic patients developed recurrence after initially successful surgery, whereas 40% of patients treated with surgery only remained cured of their disease (Biermasz et al., 2000). Poor prognostic factors for surgical cure in acromegaly include macroadenomas >20 mm and preoperative GH levels >50 ng/ml (Shimon et al., 2001). Radiotherapy plays an important adjunctive role in the treatment of acromegaly, especially in cases of large or invasive tumors, although there is a typically a long delay between treatment and GH response. Partial or complete hypopituitarism is a common late sequel to pituitary irradiation for acromegaly (Barrande et al., 2000). Pharmacologic treatment of acromegaly with D_2 dopamine receptor agonists has shown mixed results. The parenteral administration of long-acting somatostatin analogs has been shown to reduce GH and IGF-I levels (Kendall-Taylor et al., 2000; Chanson et al., 2000) and to shrink tumors in previously untreated acromegalic patients (Amato et al., 2002). Preliminary studies with the novel GH antagonist pegvisomant appear promising (Trainer et al., 2000; van der Lely et al., 2001).

CARNEY COMPLEX

Clinical Manifestations of Carney Complex

Carney complex was first described in 1985 as the constellation of "myxomas, spotty skin pigmentation, and endocrine overactivity" (Carney et al., 1985) and is a multiple endocrine neoplasia syndrome with numerous nonendocrine manifestations (Stratakis, 2001). Carney complex is inherited in an autosomal dominant fashion (Carney et al., 1986). Previously designated LAMB (lentigines, atrial myxomas, and blue nevi; Rhodes et al., 1984) and NAME (nevi, atrial myxoma, and ephelides; Vidaillet et al., 1984) syndromes are now widely recognized to be manifestations of Carney complex. Carney complex is a highly penetrant disorder, and the diagnosis can be made at birth (Stratakis et al., 2001).

The endocrine tumors most commonly associated with the Carney complex include primary pigmented nodular adrenocortical disease (PPNAD), pituitary GH adenomas, testicular tumors of the large-cell calcifying Sertoli cell type (LCCSCT), and thyroid nodules or cancer. PPNAD is found in more than 25% of patients with Carney complex (Stratakis et al., 2001) and most frequently manifests as Cushing syndrome of hypercortisolism with truncal obesity, hypertension, edema, easy fatigability, hirsutism, and violaceous abdominal striae. Diagnosis of PPNAD in a patient with Cushing syndrome is best made by the Liddle test, in which the urinary free cortisol response to dexamethasone administration is monitored (Stratakis et al., 1999). The diagnosis of PPNAD is greatly aided by recognition of other manifestations of Carney complex in the patient or first-degree relatives. The clinical manifestations of pituitary GH-secreting adenomas in the setting of Carney complex are similar to those from other causes of acromegaly and were detailed in the section above. LCCSCT in Carney complex is frequently bilateral and most often presents as an asymptomatic testicular mass (Washecka et al., 2002), although these sex cord stromal testicular tumors are occasionally associated with the development of gynecomastia (Tanaka et al., 1999; Stratakis et al., 2001). Testicular ultrasound provides a sensitive noninvasive screening tool for LCCSCT in men at risk (Premkumar et al., 1997). The vast majority of LCCSCT in Carney complex are benign (Kratzer et al., 1997). Thyroid gland abnormalities are common in sporadic and familial Carney complex and range from benign solid and cystic lesions to carcinoma (Stratakis et al., 1997).

The nonendocrine manifestations of Carney complex are varied and range from the clinically silent to the highly morbid. The most common clinical manifestation is spotty skin pigmentation that can include ephelides (freckles), lentigines, and blue and other nevi (moles) (Stratakis et al., 2001). A rare tumor of the peripheral nervous system, the psammomatous melanotic schwannoma (PMS), is present in Carney complex (Carney, 1990); such nerve sheath tumors are malignant about 10% of the time (Watson et al., 2000). Cutaneous myxomas are present in one-third of patients with Carney complex at the time of diagnosis (Stratakis et al., 2001). The most morbid feature of Carney complex is the presence of cardiac myxomas in more than one-half of patients; the cardiac myxomas can be multicentric, present metachronously, and result in embolic stroke or intracardiac obstruction with heart failure (Carney, 1985; Stratakis et al., 2001). Echocardiogram can confirm the diagnosis or detect clinically silent myxomatous lesions in those at risk for Carney complex.

Dysfunction of G Protein-Regulated Pathways in Carney Complex

Linkage analysis of families with Carney complex has demonstrated genetic heterogeneity with the trait in some families mapping to chromosome 2p16 (Stratakis et al., 1996) and in others to 17q22-24 (Casey et al., 1998). Stratakis and coworkers discovered mutations in the gene for the R1α regulatory subunit of protein kinase A (PKA) (*PRKAR1A*) in several families mapping to the 17q locus (Kirschner et al., 2000a). Subsequent analysis of other sporadic and familial cases suggested that mutations in *PRKAR1A* are likely to account for some 40% of all cases of Carney complex (Kirschner et al., 2000b). Another group working independently confirmed the finding of *PRKAR1A* mutations in Carney complex and familial isolated cardiac myxoma patients (Casey et al., 2000). The *PRKAR1A* gene mutations in Carney complex were universally inactivating mutations resulting in a null allele (2342,2341).

PKA is a major downstream effector of the $G_s\alpha$-AC pathway. In the resting state the PKA holoenzyme is an inactive tetramer consisting of two regulatory subunits, which bind cAMP, and two catalytic subunits. The cAMP generated from activation of AC binds to the PKA regulatory subunits, causing them to dissociate from the catalytic subunits, which in their free state are enzymatically active. Four PKA regulatory subunits with tissue-specific patterns of expression and differential ability to interact with the Akap (PKA anchoring) proteins (Michel and Scott, 2002) are known: R1α, R1β, R2α, and R2β (Tasken et al., 1997). Compensatory increases in the other regulatory subunits result from selective loss of specific regulatory subunit isoforms (Burton et al., 1997; Amieux et al., 1997). The overall level of PKA activity in the cell depends on the tissue-specific ratio of regulatory subunit isoforms (Tasken et al., 1997).

Analysis of PKA activity in tumors from *PRKAR1A* mutation-positive Carney complex patients demonstrated similar basal activity but increased cAMP-stimulated activity compared to similar tumors from non-Carney complex patients (Kirschner et al., 2000a). It was hypothesized that in Carney complex patients with null mutations of *PRKAR1A*, as in animal models deficient in specific PKA regulatory subunit isoforms (Burton et al., 1997; Amieux et al., 1997), compensatory increases in the other regulatory subunits may result in dysregulation of PKA in particular tissues with excessive responsiveness to cAMP (Kirschner et al., 2000a). Analysis of tumor tissue from *PRKAR1A* mutation-positive Carney complex patients demonstrated loss of the wild-type allele (Kirschner et al., 2000a). This observation, together with the knowledge that cAMP-driven pathways promote cell growth in endocrine and certain other tissues affected in the Carney complex, suggests that the *PRKAR1A* gene likely functions as a classic tumor suppressor (Stratakis, 2001).

Treatment of Carney Complex

The goal of management of patients with Carney complex is the early detection of clinically important manifestations at a presymptomatic stage through regular clinical and biochemical screening. A detailed approach to such screening was recently detailed by Stratakis and co-workers (Stratakis et al., 2001). As noted above, the most clinically significant manifestation of Carney complex is the development of cardiac myxomas that may be multifocal and recurrent. Screening with annual echocardiogram is recommended. Screening tests for the development of PPNAD and acromegaly should include at a minimum measurement of urinary free cortisol and serum IGF-1 levels, respectively. In male patients at risk for Carney complex testicular ultrasonography for the early detection of LCCSCT and associated testicular pathology is recommended. Gene mutational testing at *PRKAR1A* locus is not routinely recommended for patients with Carney complex (Stratakis et al., 2001).

Treatment options vary for the clinically significant complications of Carney complex. When detected early, cardiac myxomas can be safely excised in both the pediatric and adult age groups (Schaff and Mullany, 2000). The PMS seen in Carney complex, which occur primarily in the posterior nerve roots and alimentary tract (Carney, 1990; Carney and Stratakis, 1998), can be treated surgically or microneurosurgically depending on their location (Watson et al., 2000). The treatment of GH-secreting pituitary adenomas in the setting of Carney complex is the same as in other types of acromegaly and is discussed in a preceding section. For Cushing syndrome due to PPNAD associated with Carney complex, a form of primary adrenal hyperplasia, total adrenalectomy is the treatment of choice (Shenoy et al., 1984; Grant et al., 1986). Successful laparoscopic adrenalectomy for Cushing syndrome in Carney complex has been reported (Shichman et al., 1999). Testicular Sertoli cell tumors associated with Carney complex can be treated by excisional biopsy or managed by surveillance only because of the very low rate of LCCSCT malignancy (Washecka et al., 2002).

CONCLUSION

Most of the endocrine diseases recognized to date to result from disrupted G protein-regulated signaling pathways involve the G_s signaling pathway (Table 6.2). The diverse clinical manifestations of the three endocrine diseases described above resulting from GOF of three different signaling elements in the G_s signaling pathway reflect how cellular expression patterns and protein-protein interactions of the affected signaling elements influence the disease expression. Considering the large numbers of GPCRs, G proteins, and effectors encoded in the human genome it is easy to imagine the pleiotropic disease expressions that might result from gain or loss of function of any particular component of a G protein-regulated signaling pathway.

From the foregoing examples of endocrine diseases resulting from disrupted G protein-regulated signaling pathways, it is evident that a wide gulf exists between our often detailed knowledge of the molecular pathophysiology of a disease and the frequently conventional means of diagnosis and treatment that persist in practice today. This gulf represents a major opportunity for the development of specific and novel diagnostic and therapeutic tools in the years ahead. The knowledge that many dozens of clinically valuable drugs in use today act at GPCRs would seem to justify pursuit of potentially novel therapeutics targeted at G

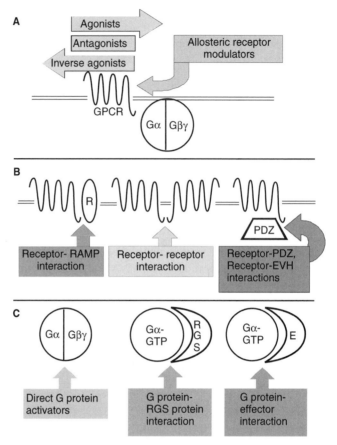

Figure 6.3. Targets in G protein signaling pathways for future drug discovery with implications for the therapy of endocrine and other diseases. **A.** G protein coupled receptors (GPCRs) can be targeted both at the ligand binding and allosteric regulatory sites. Agonists promote coupling of GPCR with heterotrimeric G proteins (Gα-Gβγ), whereas inverse agonists promote receptor-G protein uncoupling. (Neutral) antagonists occupy the ligand binding site without affecting the state of receptor-G protein coupling. Receptor modulators bind allosterically and potentiate or inhibit the effects of ligand binding. **B.** Three non-G protein interactions of GPCR as potential targets of future therapeutics. These include receptor interactions with accessory proteins such as receptor activity modifying protein (RAMP), receptor-receptor interactions mediating homo- or heterodimerization, and interactions between specific sequence motifs in the C-terminal tail of GPCR with postsynaptic density-95/Discs large/zona occludens-1 (PDZ) or enabled/vasodilator-stimulated phosphoprotein (VASP) homology (EVH) domains. Not shown are other non-G protein interactions of GPCR that are potential drug targets including those with G protein receptor kinases (GRKs) and β-arrestins (Pierce and Lefkowitz, 2001). **C.** Potential targets for drug development at the level of the G protein, including direct G protein activators, interactions between Gα and RGS proteins, and interactions between Gα and effector proteins. Not illustrated are other potential drug targets at the level of the G protein, including interactions between Gα and GDP-dissociation

protein-regulated signaling pathways. Strategies for the development of such novel therapeutics are emerging from our expanding understanding of the mechanisms and complexities of G protein signaling, facilitated by an enlarging repertoire of analytical tools and model systems (Fig. 6.3).

The discovery and identification of novel agonists and antagonists acting at the ligand binding site of GPCRs have been aided by advances in computer modeling and structural biology coupled with new methods of high-throughput screening of candidate ligands. Models of GPCR generated by comparison of primary sequence data from many homologous receptors (Baldwin et al., 1997) can now be refined using the three-dimensional coordinates of crystallized rhodopsin determined by X-ray diffraction (Palczewski et al., 2000) to facilitate rational drug design (Muller, 2000). The study of inverse agonists [also termed "negative antagonists" (Costa and Herz, 1989)] acting at GPCR, compounds with negative intrinsic activity for a given biological response, has been spurred by the recognition that such agents frequently possess desirable pharmacologic properties. Identification of inverse agonists has been facilitated by the development of screening methods employing constitutively active GPCRs (Weiner et al., 2001; Chen et al., 2000). High-throughput screening in yeast of ligands for heterologous GPCRs introduced into *Saccharomyces cerevisiae* utilizes components of the endogenous pheromone-signaling pathway and has facilitated drug discovery efforts (Pausch, 1997; Broach and Thorner, 1996). Ligand screening in high-throughput format has also facilitated the characterization of "orphan" GPCRs, that is, apparent GPCRs encoded in the human genome recognized from homologies evident in sequence analysis but for whom the endogenous ligand is initially unknown (Klein, 2000).

The recognition of compounds that act allosterically to potentiate or inhibit agonist-induced responses has opened up another arena of GPCR-targeted drug discovery suitable for high-throughput screening (Ijzerman et al., 2001). Novel "calcimimetic" agents acting allosterically to enhance calcium-stimulated responses mediated by the CaSR are promising agents in the medical treatment of primary and secondary hyperparathyroidism (Nemeth and Fox, 1999). Such calcimimetics have also been used to treat parathyroid cancer (Collins et al., 1998). High-throughput screening identified two classes of allosteric enhancers of metabotropic glutamate type 1 receptors (mGluR1) with potentiating activity on synaptically evoked mGluR1 responses in brain neurons (Knoflach et al., 2001). Other types of allosteric modulators active at GPCR have been identified, with activity at single (Musser et al., 1999;

---

inhibitor proteins containing the G-protein regulatory (GPR) or GoLoco motif (Peterson et al., 2000; Natochin et al., 2001; Kimple et al., 2002) and interactions between Gβγ and its effector proteins. See text for more details and references.

Birdsall et al., 2001) and multiple (Fawzi et al., 2001) receptor types. The potential benefits of GPCR-targeted allosteric agents in the treatment of endocrine diseases resulting from G protein signaling dysfunction are many. These could include potentiation of GPCR signals inhibitory to hormone release, such as the example of CaSR-targeted drugs in the treatment of parathyroid tumors cited above, as well as enhancement of GPCR signals promoting desirable signals impaired by disease or age-mediated LOF.

Apart from the ligand-binding site, potential new targets for drug development are emerging from increasing awareness of novel interactions that can regulate signaling by GPCRs. Homo- or heterodimerization of GPCRs is now known to regulate the expression, targeting, ligand binding and/or signaling of many receptors, including $GABA_B$, opioid, CaSR, mGluR5, and others (Angers et al., 2002). The site of protein-protein interaction comprising the GPCR dimerization interface is a potential target of pharmacologic manipulation. Accessory proteins have been described for some GPCRs that are required as chaperones for receptor expression and can modulate the binding specificity at the ligand-binding site. The first such receptor activity modifying protein (RAMP) was identified as a single transmembrane protein required for cell surface expression of the calcitonin-receptor-like receptor (CRLR) producing a functional receptor for calcitonin gene-related peptide (CGRP) (McLatchie et al., 1998). Several isoforms of RAMPs have been cloned, and coexpression of different RAMP isoforms can change the pharmacologic profile of the CRLR from a functional CGRP receptor to an adrenomedullin receptor (Aldecoa et al., 2000). Protein-protein interaction between the GPCR and RAMP governs this switch in the pharmacologic phenotype (Hilairet et al., 2001), providing another potential target for novel therapeutics. Regulation of non-G protein interactions in the cytoplasm involving the C terminus of GPCRs represents another opportunity for drug discovery. Examples of such interactions include the C tail of certain mGluR isoforms with the adapter protein Homer (Brakeman et al., 1997) that is mediated by a specific polyproline motif in the former and an enabled/VASP homology domain in the latter (Beneken et al., 2000). Homer-mGluR interaction governs receptor clustering and subsynaptic targeting (Tadokoro et al., 1999; Ciruela et al., 2000). Interactions between PDZ domain-containing proteins and the C-terminal tails of β-adrenergic, 5-hydroxytryptamine, and other receptors are critical for a variety of receptor functions including signal transduction, endocytosis, and subcellular localization (Becamel et al., 2002; Hall et al., 1998).

Downstream of GPCRs are a number of targets potentially amenable to pharmacologic intervention. G proteins themselves have been proposed as drug targets in light of the ability of the wasp venom peptide mastoparan and certain other small molecules to directly stimulate or inhibit G protein function (Holler et al., 1999). Whereas RGS proteins can terminate G protein signaling by acting as GAPs for Gα subunits and/or acting as effector antagonists, certain RGS proteins can also propagate or transduce the signal from the activated Gα subunit (De Vries et al., 2000). These properties have led to the consideration of RGS proteins as promising new drug targets (Zhong and Neubig, 2001). The action of protein kinases specific for the agonist-liganded form of GPCR (G protein receptor kinases, or GRKs) leads to the binding of arrestin proteins to many GPCRs, initiating a process of desensitization, receptor internalization, and G-protein-independent signaling (Ferguson, 2001). The interactions of GRKs and β-arrestins (Pierce and Lefkowitz, 2001) with GPCRs are potential targets of future therapeutic drugs aimed at promoting or blocking desensitization and related processes. Finally, the interface between G proteins and their effectors could be a target of future drug development. Peptides derived from PLC-β1, for example, have been used to block interaction between Gαq and its effector (Paulssen et al., 1996). Crystal structures of G protein-effector complexes, such as that between Gαs and AC (Tesmer et al., 1997), may provide a three-dimensional template for the rational design of such drugs.

These examples illustrate the many future avenues for drug development that may expand and improve therapy options for endocrine and other diseases resulting from dysfunction in G protein signaling pathways. Progress in the structural biology of signaling proteins fed by informatics (Hurley et al., 2002) combined with accelerated analysis of genetically tractable lower eukaryotes with defined genomic sequence will increase our awareness of physiologically relevant protein-protein partnering and the molecular details of such interactions. The list of potential targets for highly specific and rational therapeutics will therefore only keep growing for the foreseeable future.

REFERENCES

Abramowicz, M. J., Duprez, L., Parma, J., Vassart, G., and Heinrichs, C. (1997). Familial congenital hypothyroidism due to inactivating mutation of the thyrotropin receptor causing profound hypoplasia of the thyroid gland. J Clin Invest *99*, 3018–3324.

Ahmed, S., Elsheikh, M., Strattonm I. M., Page, R. C., Adams, C. B., and Wass, J. A. (1999). Outcome of transsphenoidal surgery for acromegaly and its relationship to surgical experience. Clin Endocrinol (Oxf) *50*, 561–557.

Alberti, L., Proverbio, M. C., Costagliola, S., Weber, G., Beck-Peccoz, P., Chiumello, G., and Persani, L. (2001). A novel germline mutation in the TSH receptor gene causes non-autoimmune autosomal dominant hyperthyroidism. Eur J Endocrinol *145*, 249–254.

Aldecoa, A., Gujer, R., Fischer, J. A., and Born, W. (2000). Mammalian calcitonin receptor-like receptor/receptor activity modifying protein complexes define calcitonin gene-related peptide and adrenomedullin receptors in *Drosophila* Schneider 2 cells. FEBS Lett *471*, 156–160.

Alman, B. A., Greel, D. A., and Wolfe, H. J. (1996). Activating mutations of Gs protein in monostotic fibrous lesions of bone. J Orthop Res *14*, 311–335.

Amato, G., Mazziotti, G., Rotondi, M., Iorio, S., Doga, M., Sorvillo, F., Manganella, G., Di Salle, F., Giustina, A., and Carella, C. (2002). Long-term effects of lanreotide SR and octreotide LAR on tumour shrinkage and GH hypersecretion in patients with previously untreated acromegaly. Clin Endocrinol (Oxf) 56, 65–71.

Amieux, P. S., Cummings, D. E., Motamed, K., Brandon, E. P., Wailes, L. A., Le, K., Idzerda, R. L., and McKnight, G. S. (1997). Compensatory regulation of RIα protein levels in protein kinase A mutant mice. J Biol Chem 272, 3993–3398.

Angers, S., Salahpour, A., and Bouvier, M. (2002). Dimerization: an emerging concept for G protein-coupled receptor ontogeny and function. Annu Rev Pharmacol Toxicol 42, 409–435.

Baldwin, J. M., Schertler, G. F., and Unger, V. M. (1997). An α-carbon template for the transmembrane helices in the rhodopsin family of G-protein-coupled receptors. J Mol Biol 272, 144–164.

Barkan, A. L. (1998). New options for diagnosing and treating acromegaly. Cleve Clin J Med 65, 343, 347–343, 39.

Barrande, G., Pittino-Lungo, M., Coste, J., Ponvert, D., Bertagna, X., Luton, J. P., and Bertherat, J. (2000). Hormonal and metabolic effects of radiotherapy in acromegaly: long-term results in 128 patients followed in a single center. J Clin Endocrinol Metab 85, 3779–3785.

Baumann, G., and Maheshwari, H. (1997). The Dwarfs of Sindh: severe growth hormone (GH) deficiency caused by a mutation in the GH-releasing hormone receptor gene. Acta Paediatr Suppl 423, 33–38.

Becamel, C., Alonso, G., Galeotti, N., Demey, E., Jouin, P., Ullmer, C., Dumuis, A., Bockaert, J., and Marin, P. (2002). Synaptic multiprotein complexes associated with 5-HT(2C) receptors: a proteomic approach. EMBO J 21, 2332–2342.

Beneken, J., Tu, J. C., Xiao, B., Nuriya, M., Yuan, J. P., Worley, P. F., and Leahy, D. J. (2000). Structure of the Homer EVH1 domain-peptide complex reveals a new twist in polyproline recognition. Neuron 26, 143–154.

Bianco, P., Riminucci, M., Majolagbe, A., Kuznetsov, S. A., Collins, M. T., Mankani, M. H., Corsi, A., Bone, H. G., Wientroub, S., Spiegel, A. M., Fisher, L. W., and Robey, P. G. (2000). Mutations of the GNAS1 gene, stromal cell dysfunction, and osteomalacic changes in non-McCune-Albright fibrous dysplasia of bone. J Bone Miner Res 15, 120–128.

Biebermann, H., Schoneberg, T., Hess, C., Germak, J., Gudermann, T., and Gruters, A. (2001). The first activating TSH receptor mutation in transmembrane domain 1 identified in a family with nonautoimmune hyperthyroidism. J Clin Endocrinol Metab 86, 4429–4433.

Biermasz, N. R., van Dulken, H., and Roelfsema, F. (2000). Ten-year follow-up results of transsphenoidal microsurgery in acromegaly. J Clin Endocrinol Metab 85, 4596–4602.

Birdsall, N. J., Lazareno, S., Popham, A., and Saldanha, J. (2001). Multiple allosteric sites on muscarinic receptors. Life Sci 68, 2517–2224.

Bourne, H. R., Landis, C. A., and Masters, S. B. (1989). Hydrolysis of GTP by the α-chain of Gs and other GTP binding proteins. Proteins 6, 222–230.

Brakeman, P. R., Lanahan, A. A., O'Brien, R., Roche, K., Barnes, C. A., Huganir, R. L., and Worley, P. F. (1997). Homer: a protein that selectively binds metabotropic glutamate receptors. Nature 386, 284–228.

Broach, J. R., and Thorner, J. (1996). High-throughput screening for drug discovery. Nature *384*, 14–16.

Brown, E. M. (2000). Familial hypocalciuric hypercalcemia and other disorders with resistance to extracellular calcium. Endocrinol Metabol Clin North Am *29*, 503–522.

Burton, K. A., Johnson, B. D., Hausken, Z. E., Westenbroek, R. E., Idzerda, R. L., Scheuer, T., Scott, J. D., Catterall, W. A., and McKnight, G. S. (1997). Type II regulatory subunits are not required for the anchoring-dependent modulation of Ca^{2+} channel activity by cAMP-dependent protein kinase. Proc Natl Acad Sci USA *94*, 11067–11172.

Carney, J. A. (1985). Differences between nonfamilial and familial cardiac myxoma. Am J Surg Pathol *9*, 53–55.

Carney, J. A. (1990). Psammomatous melanotic schwannoma. A distinctive, heritable tumor with special associations, including cardiac myxoma and the Cushing syndrome. Am J Surg Pathol *14*, 206–222.

Carney, J. A., Gordon, H., Carpenter, P. C., Shenoy, B. V., and Go, V. L. (1985). The complex of myxomas, spotty pigmentation, and endocrine overactivity. Medicine (Baltimore) *64*, 270–283.

Carney, J. A., Hruska, L. S., Beauchamp, G. D., and Gordon, H. (1986). Dominant inheritance of the complex of myxomas, spotty pigmentation, and endocrine overactivity. Mayo Clin Proc *61*, 165–172.

Carney, J. A., and Stratakis, C. A. (1998). Epithelioid blue nevus and psammomatous melanotic schwannoma: the unusual pigmented skin tumors of the Carney complex. Semin Diagn Pathol *15*, 216–224.

Casey, M., Mah, C., Merliss, A. D., Kirschner, L. S., Taymans, S. E., Denio, A. E., Korf, B., Irvine, A. D., Hughes, A., Carney, J. A., Stratakis, C. A., and Basson, C. T. (1998). Identification of a novel genetic locus for familial cardiac myxomas and Carney complex. Circulation *98*, 2560–2566.

Casey, M., Vaughan, C. J., He, J., Hatcher, C. J., Winter, J. M., Weremowicz, S., Montgomery, K., Kucherlapati, R., Morton, C. C., and Basson, C. T. (2000). Mutations in the protein kinase A R1α regulatory subunit cause familial cardiac myxomas and Carney complex. J Clin Invest *106*, R31–R38.

Chanson, P., Boerlin, V., Ajzenberg, C., Bachelot, Y., Benito, P., Bringer, J., Caron, P., Charbonnel, B., Cortet, C., Delemer, B., Escobar-Jimenez, F., Foubert, L., Gaztambide, S., Jockenhoevel, F., Kuhn, J. M., Leclere, J., Lorcy, Y., Perlemuter, L., Prestele, H., Roger, P., Rohmer, V., Santen, R., Sassolas, G., Scherbaum, W. A., Schopohl, J., Torres, E., Varela, C., Villamil, F., and Webb, S. M. (2000). Comparison of octreotide acetate LAR and lanreotide SR in patients with acromegaly. Clin Endocrinol (Oxf) *53*, 577–586.

Chen, G., Way, J., Armour, S., Watson, C., Queen, K., Jayawickreme, C. K., Chen, W. J., and Kenakin, T. (2000). Use of constitutive G protein-coupled receptor activity for drug discovery. Mol Pharmacol *57*, 125–134.

Ciruela, F., Soloviev, M. M., Chan, W. Y., and McIlhinney, R. A. (2000). Homer-1c/Vesl-1L modulates the cell surface targeting of metabotropic glutamate receptor type 1α: evidence for an anchoring function. Mol Cell Neurosci *15*, 36–50.

Clark, A. J., McLoughlin, L., and Grossman, A. (1993). Familial glucocorticoid deficiency associated with point mutation in the adrenocorticotropin receptor. Lancet *341*, 461–442.

Coleman, D. E., Berghuis, A. M., Lee, E., Linder, M. E., Gilman, A. G., and Sprang, S. R. (1994). Structures of active conformations of $G_{i\alpha1}$ and the mechanism of GTP hydrolysis. Science *265*, 1405–1412.

Collins, M. T., Skarulis, M. C., Bilezikian, J. P., Silverberg, S. J., Spiegel, A. M., and Marx, S. J. (1998). Treatment of hypercalcemia secondary to parathyroid carcinoma with a novel calcimimetic agent. J Clin Endocrinol Metab *83*, 1083–1088.

Costa, T., and Herz, A. (1989). Antagonists with negative intrinsic activity at delta opioid receptors coupled to GTP-binding proteins. Proc Natl Acad Sci USA *86*, 7321–735.

de Rooij, J., Zwartkruis, F. J., Verheijen, M. H., Cool, R. H., Nijman, S. M., Wittinghofer, A., and Bos, J. L. (1998). Epac is a Rap1 guanine-nucleotide-exchange factor directly activated by cyclic AMP. Nature *396*, 474–447.

De Vries, L., Zheng, B., Fischer, T., Elenko, E., and Farquhar, M. G. (2000). The regulator of G protein signaling family. Annu Rev Pharmacol Toxicol *40*, 235–271.

Delhougne, B., Deneux, C., Abs, R., Chanson, P., Fierens, H., Laurent-Puig, P., Duysburgh, I., Stevenaert, A., Tabarin, A., and Delwaide, J. (1995). The prevalence of colonic polyps in acromegaly: a colonoscopic and pathological study in 103 patients. J Clin Endocrinol Metab *80*, 3223–3326.

Duprez, L., Parma, J., Van Sande, J., Allgeier, A., Leclere, J., Schvartz, C., Delisle, M. J., Decoulx, M., Orgiazzi, J., Dumont, J., and Vassart, G. (1994). Germline mutations in the thyrotropin receptor gene cause non-autoimmune autosomal dominant hyperthyroidism. Nature Genet *7*, 396–401.

Esapa, C. T., Duprez, L., Ludgate, M., Mustafa, M. S., Kendall-Taylor, P., Vassart, G., and Harris, P. E. (1999). A novel thyrotropin receptor mutation in an infant with severe thyrotoxicosis. Thyroid *9*, 1005–1110.

Eugster, E. A., and Pescovitz, O. H. (1999). Gigantism. J Clin Endocrinol Metab *84*, 4379–4484.

Farfel, Z., Bourne, H. R., and Iiri, T. (1999). The expanding spectrum of G protein diseases. N Engl J Med *340*, 1012–1020.

Fawzi, A. B., Macdonald, D., Benbow, L. L., Smith-Torhan, A., Zhang, H., Weig, B. C., Ho, G., Tulshian, D., Linder, M. E., and Graziano, M. P. (2001). SCH-202676: An allosteric modulator of both agonist and antagonist binding to G protein-coupled receptors. Mol Pharmacol *59*, 30–37.

Ferguson, S. S. (2001). Evolving concepts in G protein-coupled receptor endocytosis: the role in receptor desensitization and signaling. Pharmacol Rev *53*, 1–24.

Freissmuth, M., and Gilman, A. G. (1989). Mutations of GS α designed to alter the reactivity of the protein with bacterial toxins. Substitutions at ARG187 result in loss of GTPase activity. J Biol Chem *264*, 21907–21914.

Fuhrer, D., Wonerow, P., Willgerodt, H., and Paschke, R. (1997). Identification of a new thyrotropin receptor germline mutation (Leu629Phe) in a family with neonatal onset of autosomal dominant nonautoimmune hyperthyroidism. J Clin Endocrinol Metab *82*, 4234–428.

Fukuhara, S., Murga, C., Zohar, M., Igishi, T., and Gutkind, J. S. (1999). A novel PDZ domain containing guanine nucleotide exchange factor links heterotrimeric G proteins to Rho. J Biol Chem *274*, 5868–5879.

Gicquel, C., Dib, A., Bertagna, X., Amselem, S., and Le Bouc, Y. (1995). Onco-genic mutations of α-Gi2 protein are not determinant for human adrenocor-tical tumourigenesis. Eur J Endocrinol *133*, 166–172.

Giustina, A., Barkan, A., Casanueva, F. F., Cavagnini, F., Frohman, L., Ho, K., Veldhuis, J., Wass, J., Von Werder, K., and Melmed, S. (2000). Criteria for cure of acromegaly: a consensus statement. J Clin Endocrinol Metab *85*, 526–559.

Goretzki, P. E., Lyons, J., Stacy-Phipps, S., Rosenau, W., Demeure, M., Clark, O. H., McCormick, F., Roher, H. D., and Bourne, H. R. (1992). Mutational acti-vation of RAS and GSP oncogenes in differentiated thyroid cancer and their biological implications. World J Surg *16*, 576–581.

Grant, C. S., Carney, J. A., Carpenter, P. C., and Van Heerden, J. A. (1986). Primary pigmented nodular adrenocortical disease: diagnosis and management. Surgery *100*, 1178–1184.

Gromoll, J., Simoni, M., Nordhoff, V., Behre, H. M., De Geyter, C., and Nieschlag, E. (1996). Functional and clinical consequences of mutations in the FSH receptor. Mol Cell Endocrinol *125*, 177–182.

Gruters, A., Schoneberg, T., Biebermann, H., Krude, H., Krohn, H. P., Dralle, H., and Gudermann, T. (1998). Severe congenital hyperthyroidism caused by a germ-line neo mutation in the extracellular portion of the thyrotropin recep-tor. J Clin Endocrinol Metab *83*, 1431–1146.

Hall, R. A., Premont, R. T., Chow, C. W., Blitzer, J. T., Pitcher, J. A., Claing, A., Stoffel, R. H., Barak, L. S., Shenolikar, S., Weinman, E. J., Grinstein, S., and Lefkowitz, R, J. (1998). The β_2-adrenergic receptor interacts with the Na^+/H^+-exchanger regulatory factor to control Na^+/H^+ exchange. Nature *392*, 626–630.

Hamm, H. E., and Gilchrist, A. (1996). Heterotrimeric G proteins. Curr Opin Cell Biol *8*, 189–196.

Hart, M. J., Jiang, X. J., Kozasa, T., Roscoe, W., Singer, W. D., Gilman, A. G., Sternweis, P. C., and Bollag, G. (1998). Direct stimulation of the guanine nucleotide exchange activity of p115 RhoGEF by $G\alpha_{13}$. Science *280*, 2112–2114.

Hepler, J. R. (1999). Emerging roles for RGS proteins in cell signalling. Trends Pharmacol Sci *20*, 376–382.

Hilairet, S., Foord, S. M., Marshall, F. H., and Bouvier, M. (2001). Protein-protein interaction and not glycosylation determines the binding selectivity of heterodimers between the calcitonin receptor-like receptor and the receptor activity-modifying proteins. J Biol Chem *276*, 29575–29581.

Holler, C., Freissmuth, M., and Nanoff, C. (1999). G proteins as drug targets. Cell Mol Life Sci *55*, 257–270.

Hosoi, E., Yokogoshi, Y., Hosoi, E., Horie, H., Sano, T., Yamada, S., and Saito, S. (1993). Analysis of the Gs α gene in growth hormone-secreting pituitary adenomas by the polymerase chain reaction-direct sequencing method using paraffin-embedded tissues. Acta Endocrinol (Copenh) *129*, 301–336.

Hurley, J. H., Anderson, D. E., Beach, B., Canagarajah, B., Ho, Y. S., Jones, E., Miller, G., Misra, S., Pearson, M., Saidi, L., Suer, S., Trievel, R., and Tsujishita, Y. (2002). Structural genomics and signaling domains. Trends Biochem Sci *27*, 48–53.

Ichikawa, Y., Yoshida, S., Suzuki, H., Nishida, M., Tsunoda, H., Kubo, T., Miwa, M., and Uchida, K. (1996). Mutation analysis of gonadotropin receptor and G protein genes in various types of human ovarian tumors. Jpn J Clin Oncol *26*, 298–302.

Iiri, T., Herzmark, P., Nakamoto, J. M., Van Dop, C., and Bourne, H. R. (1994). Rapid GDP release from $G_{s\alpha}$ in patients with gain and loss of endocrine function. Nature *371*, 164–168.

Ijzerman, A., Kourounakis, A., and van der Klein, P. (2001). Allosteric modulation of G protein-coupled receptors. Farmaco *56*, 67–70.

Jenkins, P. J., Fairclough, P. D., Richards, T., Lowe, D. G., Monson, J., Grossman, A., Wass, J. A., and Besser, M. (1997). Acromegaly, colonic polyps and carcinoma. Clin Endocrinol (Oxf) *47*, 17–22.

Johnson, M. C., Codner, E., Eggers, M., Mosso, L., Rodriguez, J. A., and Cassorla, F. (1999). Gps mutations in Chilean patients harboring growth hormone-secreting pituitary tumors. J Pediatr Endocrinol Metab *12*, 381–337.

Kendall-Taylor, P., Miller, M., Gebbie, J., Turner, S., and al-Maskari, M. (2000). Long-acting octreotide LAR compared with lanreotide SR in the treatment of acromegaly. Pituitary *3*, 61–65.

Khoo, D. H., Parma, J., Rajasoorya, C., Ho, S. C., and Vassart, G. (1999). A germline mutation of the thyrotropin receptor gene associated with thyrotoxicosis and mitral valve prolapse in a Chinese family. J Clin Endocrinol Metab *84*, 1459–1162.

Kim, H. J., Kim, M. S., Park, Y. J., Kim, S. W., Park, D. J., Park, K. S., Kim, S. Y., Cho, B. Y., Lee, H. K., Jung, H. W., Han, D. H., Lee, H. S., and Chi, J. G. (2001). Prevalence of Gs α mutations in Korean patients with pituitary adenomas. J Endocrinol *168*, 221–226.

Kimple, R. J., Kimple, M. E., Betts, L., Sondek, J., and Siderovski, D. P. (2002). Structural determinants for GoLoco-induced inhibition of nucleotide release by Gα subunits. Nature *416*, 878–881.

Kirschner, L. S., Carney, J. A., Pack, S. D., Taymans, S. E., Giatzakis, C., Cho, Y. S., Cho-Chung, Y. S., and Stratakis, C. A. (2000a). Mutations of the gene encoding the protein kinase A type I-α regulatory subunit in patients with the Carney complex. Nature Genet *26*, 89–92.

Kirschner, L. S., Sandrini, F., Monbo, J., Lin, J. P., Carney, J. A., and Stratakis, C. A. (2000b). Genetic heterogeneity and spectrum of mutations of the PRKAR1A gene in patients with the Carney complex. Hum Mol Genet *9*, 3037–3046.

Klein, I. (2000). Validation of genomics-derived drug targets using yeast. Drug Discov Today *5*, 37–38.

Knoflach, F., Mutel, V., Jolidon, S., Kew, J. N., Malherbe, P., Vieira, E., Wichmann, J., and Kemp, J. A. (2001). Positive allosteric modulators of metabotropic glutamate 1 receptor: characterization, mechanism of action, and binding site. Proc Natl Acad Sci USA *98*, 13402–13407.

Kopp, P., Jameson, J. L., and Roe, T. F. (1997). Congenital nonautoimmune hyperthyroidism in a nonidentical twin caused by a sporadic germline mutation in the thyrotropin receptor gene. Thyroid *7*, 765–770.

Kozasa, T., Jiang, X. J., Hart, M. J., Sternweis, P. M., Singer, W. D., Gilman, A. G., Bollag, G., and Sternweis, P. C. (1998). p115 RhoGEF, a GTPase activating protein for $G\alpha_{12}$ and $G\alpha_{13}$. Science *280*, 2109–2111.

Kratzer, S. S., Ulbright, T. M., Talerman, A., Srigley, J. R., Roth, L. M., Wahle, G. R., Moussa, M., Stephens, J. K., Millos, A., and Young, R. H. (1997). Large cell calcifying Sertoli cell tumor of the testis: contrasting features of six malignant and six benign tumors and a review of the literature. Am J Surg Pathol *21*, 1271–1280.

Kremer, H., Kraaij, R., Toledo, S. P., Post, M., Fridman, J. B., Hayashida, C. Y., van Reen, M., Milgrom, E., Ropers, H. H., and Mariman, E. (1995). Male pseudo-hermaphroditism due to a homozygous missense mutation of the luteinizing hormone receptor gene. Nat Genet 9, 160–114.

Lambright, D. G., Sondek, J., Bohm, A., Skiba, N. P., Hamm, H. E., and Sigler, P. B. (1996). The 2.0 Å crystal structure of a heterotrimeric G protein. Nature 379, 311–319.

Landis, C. A., Harsh, G., Lyons, J., Davis, R. L., McCormick, F., and Bourne, H. R. (1990). Clinical characteristics of acromegalic patients whose pituitary tumors contain mutant G_s protein. J Clin Endocrinol Metab 71, 1416–1420.

Landis, C. A., Masters, S. B., Spada, A., Pace, A. M., Bourne, H. R., and Vallar, L. (1989). GTPase inhibiting mutations activate the α chain of Gs and stimulate adenylyl cyclase in human pituitary tumours. Nature 340, 692–666.

Lee, Y. S., Poh, L., and Loke, K. Y. (2002). An activating mutation of the thyrotropin receptor gene in hereditary non-autoimmune hyperthyroidism. J Pediatr Endocrinol Metab 15, 211–225.

Liu, G., Duranteau, L., Carel, J. C., Monroe, J., Doyle, D. A., and Shenker, A. (1999). Leydig-cell tumors caused by an activating mutation of the gene encoding the luteinizing hormone receptor. N Engl J Med 341, 1731–1176.

Liu, J., Litman, D., Rosenberg, M. J., Yu, S., Biesecker, L. G., and Weinstein, L. S. (2000). A GNAS1 imprinting defect in pseudohypoparathyroidism type IB. J Clin Invest 106, 1167–1174.

Lyons, J., Landis, C. A., Harsh, G., Vallar, L., Grunewald, K., Feichtinger, H., Duh, Q. Y., Clark, O. H., Kawasaki, E., and Bourne, H. R. (1990). Two G protein oncogenes in human endocrine tumors. Science 249, 655–669.

Ma, Y. C., Huang, J. Y., All, S., Lowry, W., and Huang, X. Y. (2000). Src tyrosine kinase is a novel direct effector of G proteins. Cell 102, 635–646.

Marinissen, M. J., and Gutkind, J. S. (2001). G-protein-coupled receptors and signaling networks: emerging paradigms. Trends Pharmacol Sci 22, 368–376.

Marro, B., Zouaoui, A., Sahel, M., Crozat, N., Gerber, S., Sourour, N., Sag, K., and Marsault, C. (1997). MRI of pituitary adenomas in acromegaly. Neuroradiology 39, 394–339.

McLatchie, L. M., Fraser, N. J., Main, M. J., Wise, A., Brown, J., Thompson, N., Solari, R., Lee, M. G., and Foord, S. M. (1998). RAMPs regulate the transport and ligand specificity of the calcitonin-receptor-like receptor. Nature 393, 333–339.

Melmed, S., Ho, K., Klibanski, A., Reichlin, S., and Thorner, M. (1995). Clinical review 75: Recent advances in pathogenesis, diagnosis, and management of acromegaly. J Clin Endocrinol Metab 80, 3395–3402.

Michel, J. J., and Scott, J. D. (2002). Akap mediated signal transduction. Annu Rev Pharmacol Toxicol 42, 235–257.

Muller, G. (2000). Towards 3D structures of G protein-coupled receptors: a multidisciplinary approach. Curr Med Chem 7, 861–888.

Musser, B., Mudumbi, R. V., Liu, J., Olson, R. D., and Vestal, R. E. (1999). Adenosine A1 receptor-dependent and -independent effects of the allosteric enhancer PD 81,723. J Pharmacol Exp Ther 288, 446–454.

Natochin, M., Gasimov, K. G., and Artemyev, N. O. (2001). Inhibition of GDP/GTP exchange on G α subunits by proteins containing G-protein regulatory motifs. Biochemistry 40, 5322–5538.

Nemeth, E. F., and Fox, J. (1999). Calcimimetic compounds: A direct approach to controlling plasma levels of parathyroid hormone in hyperparathyroidism. Trends Endocrinol Metab *10*, 66–71.

Palczewski, K., Kumasaka, T., Hori, T., Behnke, C. A., Motoshima, H., Fox, B. A., Le Trong, I., Teller, D. C., Okada, T., Stenkamp, R. E., Yamamoto, M., and Miyano, M. (2000). Crystal structure of rhodopsin: A G protein-coupled receptor. Science *289*, 739–745

Pan, Y., Metzenberg, A., Das, S., Jing, B., and Gitschier, J. (1992). Mutations in the V2 vasopressin receptor gene are associated with X-linked nephrogenic diabetes insipidus. Nat Genet *2*, 103–116.

Parma, J., Duprez, L., Van Sande, J., Cochaux, P., Gervy, C., Mockel, J., Dumont, J., and Vassart, G. (1993). Somatic mutations in the thyrotropin receptor gene cause hyperfunctioning thyroid adenomas. Nature *365*, 649–651.

Patten, J. L., Johns, D. R., Valle, D., Eil, C., Gruppuso, P. A., Steele, G., Smallwood, P. M., and Levine, M. A. (1990). Mutation in the gene encoding the stimulatory G protein of adenylate cyclase in Albright's hereditary osteodystrophy. N Engl J Med *322*, 1412–1449.

Paulssen, R. H., Woodson, J., Liu, Z., and Ross, E. M. (1996). Carboxyl-terminal fragments of phospholipase C-β1 with intrinsic Gq GTPase-activating protein (GAP) activity. J Biol Chem *271*, 26622–26669.

Pausch, M. H. (1997). G-protein-coupled receptors in *Saccharomyces cerevisiae*: high-throughput screening assays for drug discovery. Trends Biotechnol *15*, 487–494.

Peterson, Y. K., Bernard, M. L., Ma, H., Hazard, S., Graber, S. G., and Lanier, S. M. (2000). Stabilization of the GDP-bound conformation of Giα by a peptide derived from the G-protein regulatory motif of AGS3. J Biol Chem *275*, 33193–33196.

Pierce, K. L., and Lefkowitz, R. J. (2001). Classical and new roles of β-arrestins in the regulation of G-protein-coupled receptors. Nat Rev Neurosci *2*, 727–733.

Pollak, M. R., Brown, E. M., Chou, Y.-H. W., Hebert, S. C., Marx, S. J., Steinmann, B., Levi, T., Seidman, C. E., and Seidman, J. G. (1993). Mutations in the human Ca^{2+}-sensing receptor gene cause familial hypocalciuric hypercalcemia and neonatal severe hyperparathyroidism. Cell *75*, 1297–1303.

Pollak, M. R., Brown, E. M., Estep, H. L., McLaine, P. N., Kifor, O., Park, J., Hebert, S. C., Seidman, C. E., and Seidman, J. G. (1994). Autosomal dominant hypocalcaemia caused by a Ca^{2+}-sensing receptor gene mutation. Nat Genet *8*, 303–337.

Premkumar, A., Stratakis, C. A., Shawker, T. H., Papanicolaou, D. A., and Chrousos, G. P. (1997). Testicular ultrasound in Carney complex: report of three cases. J Clin Ultrasound *25*, 211–224.

Reincke, M., Karl, M., Travis, W., and Chrousos, G. P. (1993). No evidence for oncogenic mutations in guanine nucleotide-binding proteins of human adrenocortical neoplasms. J Clin Endocrinol Metab *77*, 1419–1422.

Ren, Q., Kurose, H., Lefkowitz, R. J., and Cotecchia, S. (1993). Constitutively active mutants of the α_2-adrenergic receptor. J Biol Chem *268*, 16483–16487.

Renehan, A. G., Bhaskar, P., Painter, J. E., O'Dwyer, S. T., Haboubi, N., Varma, J., Ball, S. G., and Shalet, S. M. (2000). The prevalence and characteristics of colorectal neoplasia in acromegaly. J Clin Endocrinol Metab *85*, 3417–3424.

Rhodes, A. R., Silverman, R. A., Harrist, T. J., and Perez-Atayde, A. R. (1984). Mucocutaneous lentigines, cardiomucocutaneous myxomas, and multiple blue nevi: the "LAMB" syndrome. J Am Acad Dermatol *10*, 72–82.

Riminucci, M., Collins, M. T., Lala, R., Corsi, A., Matarazzo, P., Gehron Robey, P., and Bianco, P. (2002). An R201H activating mutation of the GNAS1 (Gsα) gene in a corticotroph pituitary adenoma. Mol Pathol *55*, 58–60.

Rosenthal, W., Seibold, A., Antaramian, A., Lonergan, M., Arthus, M. F., Hendy, G. N., Birnbaumer, M., and Bichet, D. G. (1992). Molecular identification of the gene responsible for congenital nephrogenic diabetes insipidus. Nature *359*, 233–225.

Ross, E. M., and Wilkie, T. M. (2000). GTPase-activating proteins for heterotrimeric G proteins: Regulators of G protein signaling (RGS) and RGS-like proteins. Annu Rev Biochem *69*, 795–827.

Saeki, N., Iuchi, T., Isono, S., Eda, M., and Yamaura, A. (1999). MRI of growth hormone-secreting pituitary adenomas: factors determining pretreatment hormone levels. Neuroradiology *41*, 765–771.

Schaff, H. V., and Mullany, C. J. (2000). Surgery for cardiac myxomas. Semin Thorac Cardiovasc Surg *12*, 77–88.

Schipani, E., Kruse, K., and Juppner, H. (1995). A constitutively active mutant PTH-PTHrP receptor in Jansen-type metaphyseal chondrodysplasia. Science *268*, 98–100.

Schwab, K. O., Gerlich, M., Broecker, M., Sohlemann, P., Derwahl, M., and Lohse, M. J. (1997). Constitutively active germline mutation of the thyrotropin receptor gene as a cause of congenital hyperthyroidism. J Pediatr *131*, 899–904.

Shen, Y., Mamers, P., Jobling, T., Burger, H. G., and Fuller, P. J. (1996). Absence of the previously reported G protein oncogene (gip2) in ovarian granulosa cell tumors. J Clin Endocrinol Metab *81*, 4159–4461.

Shenker, A., Laue, L., Kosugi, S., Merendino, J. J., Jr., Minegishi, T., and Cutler, G. B., Jr. (1993). A constitutively activating mutation of the luteinizing hormone receptor in familial male precocious puberty. Nature *365*, 652–664.

Shenoy, B. V., Carpenter, P. C., and Carney, J. A. (1984). Bilateral primary pigmented nodular adrenocortical disease. Rare cause of the Cushing syndrome. Am J Surg Pathol *8*, 335–344.

Shi, Y., Tang, D., Deng, J., and Su, C. (1998). Detection of gsp oncogene in growth hormone-secreting pituitary adenomas and the study of clinical characteristics of acromegalic patients with gsp-positive pituitary tumors. Chin Med J (Engl) *111*, 891–884.

Shichman, S. J., Herndon, C. D., Sosa, R. E., Whalen, G. F., MacGillivray, D. C., Malchoff, C. D., and Vaughan, E. D. (1999). Lateral transperitoneal laparoscopic adrenalectomy. World J Urol *17*, 48–53.

Shimon, I., Cohen, Z. R., Ram, Z., and Hadani, M. (2001). Transsphenoidal surgery for acromegaly: endocrinological follow-up of 98 patients. Neurosurgery *48*, 1239–1243.

Sondek, J., Lambright, D. G., Noel, J. P., Hamm, H. E., and Sigler, P. B. (1994). GTPase mechanism of G proteins from the 1.7-Å crystal structure of transducin α-GDP-AIF-4. Nature *372*, 276–229.

Stratakis, C. A. (2001). Clinical genetics of multiple endocrine neoplasias, Carney complex and related syndromes. J Endocrinol Invest *24*, 370–383.

Stratakis, C. A., Carney, J. A., Lin, J. P., Papanicolaou, D. A., Karl, M., Kastner, D. L., Pras, E., and Chrousos, G. P. (1996). Carney complex, a familial multiple neoplasia and lentiginosis syndrome. Analysis of 11 kindreds and linkage to the short arm of chromosome 2. J Clin Invest 97, 699–705.

Stratakis, C. A., Courcoutsakis, N. A., Abati, A., Filie, A., Doppman, J. L., Carney, J. A., and Shawker, T. (1997). Thyroid gland abnormalities in patients with the syndrome of spotty skin pigmentation, myxomas, endocrine overactivity, and schwannomas (Carney complex). J Clin Endocrinol Metab 82, 2037–2043.

Stratakis, C. A., Kirschner, L. S., and Carney, J. A. (2001). Clinical and molecular features of the Carney complex: diagnostic criteria and recommendations for patient evaluation. J Clin Endocrinol Metab 86, 4041–4046.

Stratakis, C. A., Sarlis, N., Kirschner, L. S., Carney, J. A., Doppman, J. L., Nieman, L. K., Chrousos, G. P., and Papanicolaou, D. A. (1999). Paradoxical response to dexamethasone in the diagnosis of primary pigmented nodular adrenocortical disease. Ann Intern Med 131, 585–591.

Suarez, H. G., du Villard, J. A., Caillou, B., Schlumberger, M., Parmentier, C., and Monier, R. (1991). gsp mutations in human thyroid tumours. Oncogene 6, 677–669.

Sunahara, R. K., Tesmer, J. J. G., Gilman, A. G., and Sprang, S. R. (1997). Crystal structure of the adenylyl cyclase activator $G_{s\alpha}$. Science 278, 1943–1947.

Tadokoro, S., Tachibana, T., Imanaka, T., Nishida, W., and Sobue, K. (1999). Involvement of unique leucine-zipper motif of PSD-Zip45 (Homer 1c/vesl-1L) in group 1 metabotropic glutamate receptor clustering. Proc Natl Acad Sci USA 96, 13801–13806.

Tanaka, Y., Sano, K., Ijiri, R., Tachibana, K., Kato, K., and Terashima, K. (1999). A case of large cell calcifying Sertoli cell tumor in a child with a history of nasal myxoid tumor in infancy. Pathol Int 49, 471–446.

Tasken, K., Skalhegg, B. S., Tasken, K. A., Solberg, R., Knutsen, H. K., Levy, F. O., Sandberg, M., Orstavik, S., Larsen, T., Johansen, A. K., Vang, T., Schrader, H. P., Reinton, N. T., Torgersen, K. M., Hansson, V., and Jahnsen, T. (1997). Structure, function, and regulation of human cAMP-dependent protein kinases. Adv Second Messenger Phosphoprotein Res 31, 191–204.

Tesmer, J. J. G., Sunahara, R. K., Gilman, A. G., and Sprang, S. R. (1997). Crystal structure of the catalytic domains of adenylyl cyclase in a complex with $G_{s\alpha}$•GTPγS. Science 278, 1907–1916.

Tonacchera, M., Agretti, P., Rosellini, V., Ceccarini, G., Perri, A., Zampolli, M., Longhi, R., Larizza, D., Pinchera, A., Vitti, P., and Chiovato, L. (2000). Sporadic nonautoimmune congenital hyperthyroidism due to a strong activating mutation of the thyrotropin receptor gene. Thyroid 10, 859–863.

Tonacchera, M., Van Sande, J., Cetani, F., Swillens, S., Schvartz, C., Winiszewski, P., Portmann, L., Dumont, J. E., Vassart, G., and Parma, J. (1996). Functional characteristics of three new germline mutations of the thyrotropin receptor gene causing autosomal dominant toxic thyroid hyperplasia. J Clin Endocrinol Metab 81, 547–554.

Trainer, P. J., Drake, W. M., Katznelson, L., Freda, P. U., Herman-Bonert, V., van der Lely, A. J., Dimaraki, E. V., Stewart, P. M., Friend, K. E., Vance, M. L., Besser, G. M., Scarlett, J. A., Thorner, M. O., Parkinson, C., Klibanski, A., Powell, J. S., Barkan, A. L., Sheppard, M. C., Malsonado, M., Rose, D. R., Clemmons, D. R., Johannsson, G., Bengtsson, B. A., Stavrou, S., Kleinberg, D. L., Cook, D. M., Phillips, L. S., Bidlingmaier, M., Strasburger, C. J., Hackett, S.,

Zib, K., Bennett, W. F., and Davis, R. J. (2000). Treatment of acromegaly with the growth hormone-receptor antagonist pegvisomant. N Engl J Med *342*, 1171–1177.

Tsigos, C., Arai, K., Hung, W., and Chrousos, G. P. (1993). Hereditary isolated glucocorticoid deficiency is associated with abnormalities of the adrenocorticotropin receptor gene. J Clin Invest *92*, 2458–2461.

Vallar, L., Spada, A., and Giannattasio, G. (1987). Altered Gs and adenylate cyclase activity in human GH-secreting pituitary adenomas. Nature *330*, 566–558.

van den Ouweland, A. M., Dreesen, J. C., Verdijk, M., Knoers, N. V., Monnens, L. A., Rocchi, M., and van Oost, B. A. (1992). Mutations in the vasopressin type 2 receptor gene (AVPR2) associated with nephrogenic diabetes insipidus. Nat Genet *2*, 99–102.

van der Lely, A. J., Hutson, R. K., Trainer, P. J., Besser, G. M., Barkan, A. L., Katznelson, L., Klibanski, A., Herman-Bonert, V., Melmed, S., Vance, M. L., Freda, P. U., Stewart, P. M., Friend, K. E., Clemmons, D. R., Johannsson, G., Stavrou, S., Cook, D. M., Phillips, L. S., Strasburger, C. J., Hackett, S., Zib, K. A., Davis, R. J., Scarlett, J. A., and Thorner, M. O. (2001). Long-term treatment of acromegaly with pegvisomant, a growth hormone receptor antagonist. Lancet *358*, 1754–1779.

Vidaillet, H. J., Jr., Seward, J. B., Fyke, E., and Tajik, A. J. (1984). NAME syndrome (nevi, atrial myxoma, myxoid neurofibroma, ephelides): a new and unrecognized subset of patients with cardiac myxoma. Minn Med *67*, 695–666.

Wall, M. A., Coleman, D. E., Lee, E., Iñiguez-Lluhi, J. A., Posner, B. A., Gilman, A. G., and Sprang, S. R. (1995). The structure of the G protein heterotrimer $G_{i\alpha 1}\beta_1\gamma_2$. Cell *83*, 1047–1058.

Washecka, R., Dresner, M. I., and Honda, S. A. (2002). Testicular tumors in Carney's complex. J Urol *167*, 1299–1302.

Watson, J. C., Stratakis, C. A., Bryant-Greenwood, P. K., Koch, C. A., Kirschner, L. S., Nguyen, T., Carney, J. A., and Oldfield, E. H. (2000). Neurosurgical implications of Carney complex. J Neurosurg *92*, 413–448.

Weetman, A. P. (2000). Graves' disease. N Engl J Med *343*, 1236–1248.

Weiner, D. M., Burstein, E. S., Nash, N., Croston, G. E., Currier, E. A., Vanover, K. E., Harvey, S. C., Donohue, E., Hansen, H. C., Andersson, C. M., Spalding, T. A., Gibson, D. F., Krebs-Thomson, K., Powell, S. B., Geyer, M. A., Hacksell, U., and Brann, M. R. (2001). 5-Hydroxytryptamine 2A receptor inverse agonists as antipsychotics. J Pharmacol Exp Ther *299*, 268–276.

Weinstein, L. S., Shenker, A., Gejman, P. V., Merino, M. J., Friedman, E., and Spiegel, A. M. (1991). Activating mutations of the stimulatory G protein in the McCune-Albright syndrome. N Engl J Med *325*, 1688–1695.

Weinstein, L. S., and Yu, S. (1999). The role of genomic imprinting of Gα in the pathogenesis of Albright hereditary osteodystrophy. Trends Endocrinol Metab *10*, 81–85.

Weinstein, L. S., Yu, S. H., Warner, D. R., and Liu, J. (2001). Endocrine manifestations of stimulatory G protein α-subunit mutations and the role of genomic imprinting. Endocr Rev *22*, 675–705.

Wess, J. (1998). Molecular basis of receptor/G-protein-coupling selectivity. Pharmacol Ther *80*, 231–264.

Williamson, E. A., Ince, P. G., Harrison, D., Kendall-Taylor, P., and Harris, P. E. (1995). G-protein mutations in human pituitary adrenocorticotrophic hormone-secreting adenomas. Eur J Clin Invest *25*, 128–131.

Yang, I., Park, S., Ryu, M., Woo, J., Kim, S., Kim, J., Kim, Y., and Choi, Y. (1996). Characteristics of gsp-positive growth hormone-secreting pituitary tumors in Korean acromegalic patients. Eur J Endocrinol *134*, 720–776.

Yoshimoto, K., Iwahana, H., Fukuda, A., Sano, T., and Itakura, M. (1993). Rare mutations of the Gs α subunit gene in human endocrine tumors. Mutation detection by polymerase chain reaction-primer-introduced restriction analysis. Cancer *72*, 1386–1393.

Zhong, H., and Neubig, R. R. (2001). Regulator of G protein signaling proteins: novel multifunctional drug targets. J Pharmacol Exp Ther *297*, 837–845.

CHAPTER 7

BACTERIAL REGULATION OF THE CYTOSKELETON

JEREMY W. PECK, DORA C. STYLIANOU, and
PETER D. BURBELO
Lombardi Cancer Center, Georgetown University Medical Center,
Washington, D.C.

INTRODUCTION

The ability of bacterial pathogens to colonize, multiply, and avoid destruction in mammals is of major scientific and health concern. Accumulating evidence indicates that successful infections by many bacterial pathogens such as those of *Listeria monocytogenes, Shigella flenerxi, Salmonella typhimurium*, and *Clostridium difficile* require dramatic changes in the actin cytoskeleton of host cells during several phases of the infection and disease processes. These host actin changes require proteins encoded in the bacterial genome, often on plasmids or in discrete regions of chromosomal DNA called "pathogenicity" islands. The mechanisms by which these bacterial proteins affect the host actin cytoskeleton are diverse; some bacterial proteins are anchored on the surface of the invading bacteria, some are injected into host cells before pathogen internalization takes place, and still others are secreted by the bacteria and taken up by the host via endocytosis. These bacterial proteins can hijack the mammalian actin cytoskeleton in a number of different ways including subverting the normal function of actin-binding proteins, directly modifying actin, or interfering with the normal regulators of the actin cytoskeleton. In this chapter we review recent studies that increase our understanding of how bacterial pathogens alter the actin cytoskeleton and how these alterations facilitate bacterial entry and spread. We also delineate the rationale behind potential new areas for drug discovery aimed at blocking these pathogen-induced actin cytoskeletal changes.

Signal Transduction and Human Disease, Edited by Toren Finkel and
J. Silvio Gutkind
ISBN 0-471-02011-7   Copyright © 2003 John Wiley & Sons, Inc.

THE ACTIN CYTOSKELETON: A TARGET FOR
BACTERIAL PATHOGENS

The cytoskeleton is a network of subcellular structures involved in a variety of biological processes including cell shape changes, cell migration, phagocytosis, endocytosis, and cytokinesis. The cytoskeleton itself is a dynamic structure mainly because of the assembly and disassembly of its three major components: actin, microtubules, and intermediate filaments. Actin, the major component of the cytoskeleton, constitutes 5% of the total cellular protein and exists in two forms, G-actin and F-actin. G-actin or monomeric actin is a globular protein that can be assembled into long filaments called F-actin (Fig. 7.1). Structures containing F-actin are rigid and support the cell membrane, thus giving cells their characteristic shape. F-actin fibers can be assembled in different ways to create a variety of cellular structures including filopodia, thin actin filaments that protrude from the cell; lamellipodia, bands of cortical F-actin; and stress fibers, cables of F-actin that traverse the cell. When no longer needed by the cell, these structures can be recycled back to pools of G-actin and used to form new F-actin structures. A wide range of regulatory proteins controls the dynamic turnover of actin. Some of the best-understood regulators of the actin cytoskeleton sequester monomers of actin, whereas others create nucleation sites or regulate the polymerization of actin monomers at the ends of growing actin filaments.

Although there are many different actin binding and regulatory proteins, pathogenic bacteria are able to hijack the actin cytoskeleton by targeting only a small subset of these regulators. One critical regulator of actin nucleation is the Arp2/3 complex (Fig. 7.1). This highly conserved complex of proteins consists of seven subunits (Machesky et al., 1994; Welch et al., 1997; Winter et al., 1997) that nucleate the formation of new

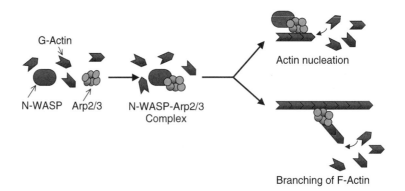

Figure 7.1. *G-actin can polymerize into long filaments of F-actin or can form branches on preexisting filaments.* Binding to N-WASP induces the actin nucleation activity of the Arp2/3 complex. After its activation, the Arp2/3 complex nucleates actin at the growing end of F-actin filaments or at the sides of preexisting actin filaments to form branches. F-actin and branched actin filaments are the building blocks for distinct structures of the actin cytoskeleton.

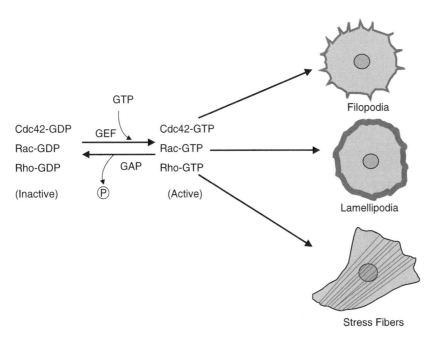

Figure 7.2. *Small Rho GTPases switch between an inactive, GDP-bound state and an active, GTP-bound state and induce distinct actin structures within the cell.* Inactive Cdc42, Rho, and Rac become activated by GEF proteins that facilitate the exchange of the bound GDP with a GTP. Small Rho GTPases are converted to their inactive state by GAPs, which increase their intrinsic GTPase activity to facilitate the hydrolysis of the bound GTP to GDP. Activation of different small Rho GTPases results in the formation of distinct actin structures.

actin filaments from the sides of preexisting filaments (Blanchoin et al., 2000). An important role for the Arp2/3 complex in the formation of filopodia and lamellipodia and in phagocytosis has been well established (for review, see Welch, 1999). The Arp2/3 complex plays a potent role in actin polymerization, and its activity is strictly controlled by the binding of additional proteins, including members of the WASP family of proteins (for review, see Mullins and Machesky, 2000).

Another major class of actin regulators targeted by bacterial pathogens is the Rho family of small GTPases (Fig. 7.2). Rho GTPases are best known for their ability to integrate extracellular signals by controlling the activity of many downstream kinases and by regulating various actin-binding proteins. Rho GTPases have been described as molecular switches because of their ability to cycle between an active state, which occurs when they are bound to GTP, and an inactive state, which occurs when they are bound to GDP. The activation state of these molecular switches is controlled both by the intrinsic hydrolytic ability of these GTPases, converting GTP to GDP, and by a number of other molecules that modulate the on or off state by several distinct mechanisms (see below). The most extensively characterized members of the Rho GTPase family are Cdc42, Rac, and Rho. Although structurally related (for review, see Hall, 1998), these GTPases have distinct biolog-ical effects (Fig. 7.2). Cdc42 promotes filopodia formation (Kozma et al., 1995; Nobes and Hall, 1994), Rac promotes lamellipodia formation (Ridley et al, 1992), and Rho induces the formation of stress fibers (Ridley and Hall, 1992). In addition to these well-established roles, Rho, Rac, and Cdc42 are also involved in regulating the actin polymerization required for phagocytosis (for review, see Chimini and Chavrier, 2000) as well as the formation, maintenance, and regulation of tight junctions between cells (for review, see Braga, 2000).

Rho GTPases affect the actin cytoskeleton by activating and/or alter-ing the cellular localization of a variety of effector proteins. Two of these effector proteins, N-WASP and WAVE, can directly bind and activate the Arp2/3 complex downstream of Cdc42 and Rac, respectively, resulting in the nucleation of new actin filaments (Machesky and Insall, 1999; Rohagti et al., 1999). N-WASP can directly bind and activate the Arp2/3 complex only after it binds Cdc42 and is activated. This is because Cdc42 binding to N-WASP induces a conformational change in N-WASP that reveals a domain that can bind and activate the Arp2/3 complex (Rohagti et al., 1999). An N-WASP-related protein, WAVE, is essential for the induction of Rac-mediated membrane ruffling and also recruits and acti-vates the Arp2/3 complex (Miki et al., 1998). WAVE, however, is not acti-vated by directly binding to Rac; rather, it binds IRS-p53, which interacts with Rac (Miki et al., 2000). Thus, after the conversion of Rac to its GTP-bound state, it binds to IRS-p53, which then binds to WAVE, activating the Arp2/3 binding activity of WAVE and leading to actin polymeriza-tion and the formation of membrane ruffles. Other actin changes down-stream of Rho also require unique effector proteins including ROCK and Diaphanous. ROCK is a serine/threonine effector kinase that is activated after Rho binding and stimulates phosphorylation of the myosin light chain, leading to increased contractility (Kimura et al., 1996). Another Rho effector protein, Diaphanous, also acts in concert to regulate stress fiber formation by recruiting the actin-monomer binding protein profilin (Watanabe et al., 1997; Watanabe et al., 1999).

The idea that Rho GTPases can integrate different kinds of extracel-lular signals to produce a "coherent" response of the actin cytoskeleton emerged from studies showing that the activation state of these GTPases is determined by a large number of regulatory proteins (Van Aelst and D'Souza-Schorey, 1997). In particular, two large families of proteins reg-ulate the on or off state of the Rho GTPases (Fig. 7.2). One of these fam-ilies, the GTPase activating proteins (GAPs), can increase the rate of conversion of the active, GTP-bound state to the inactive, GDP-bound state by enhancing the intrinsic GTPase activity of Rho proteins (Diek-mann et al., 1991). The second family, the guanine nucleotide exchange factors (GEFs), can convert Rho GTPases to their active form by cat-alyzing the dissociation of GDP and thereby facilitating the binding of GTP (Hart et al., 1992). Because the Rho family of proteins is active only in the GTP-bound form, its ability to interact with downstream effector proteins is controlled by GAPs and GEFs. Moreover, it is generally thought that the ability of Rho GTPases to mediate actin cytoskeletal reorganization is ultimately regulated by mechanisms controlling the spatial distributions and effective concentrations of GAPs and GEFs, as well as other factors.

LISTERIA MONOCYTOGENES: A SINGLE BACTERIAL PROTEIN ACTA SUBVERTS TWO HOST SIGNALING PATHWAYS

Listeria monocytogenes is a gram-positive, rod-shaped bacterium that can be isolated from soil, vegetation, and many animal reservoirs. *Listeria monocytogenes* infection (otherwise known as listeriosis) causes invasive syndromes such as meningitis, sepsis, chorioamnionitis, and still birth because of the ability of this bacterium to cross intestinal, blood, brain, and placental barriers. Individuals with deficient T cell-mediated immunity are particularly susceptible to infection with *Listeria monocytogenes* and have a high mortality rate. Each year in the United States there are about 500 deaths due to listeriosis (Mead et al., 1999). Outbreaks of listeriosis are usually due to contaminated foods such as coleslaw, pasteurized milk, soft cheeses, pork, and a variety of prepared foods. The treatment of choice is intravenous administration of ampicillin or penicillin, often in combination with an aminoglycoside for synergy.

Much work has been devoted to understanding how *Listeria monocytogenes* functions as a virulent intracellular pathogen. In cultured mammalian cells, *Listeria monocytogenes* was shown to invade cells by actively rearranging the actin cytoskeleton at the site of attachment and to use a second kind of actin-based mechanism to spread from cell to cell (for review, see Cossart and Bierne, 2001). Here we review only those actin cytoskeletal changes involved in the spread of *Listeria* from cell to cell that occur after it has been engulfed into an intracellular vacuole and has escaped from this vacuole into the cytoplasm. Pioneering studies by Tilney and Portnoy (1989) demonstrated that intracellular *Listeria* is able to move about the cytoplasm by generating actin structures composed of cross-linked actin, commonly referred to as actin tails or actin comets. These actin tails enable *Listeria* to "rocket" to the plasma membrane, where they induce the formation of long, thin membranous protrusions that penetrate into neighboring cells. These protrusions are then pinched off from the host cell, which results in the release of the bacteria from the enclosed double membrane into the cytoplasm of the newly infected cell. After a short delay, the bacteria replicate, generate new actin tails, and infect additional cells. Because the rates of tail formation and bacterial speed are strictly correlated, it has been suggested that the force of propulsion is provided by actin polymerization itself (Theriot et al., 1992) and that this system may be useful for understanding how actin polymerization can generate force.

Interestingly, only a single bacterial protein is required to mediate actin rocketing by *Listeria*. The gene encoding this protein was isolated by mutagenizing a pathogenic *Listeria* strain and screening for mutants with reduced pathogenicity (Domann et al., 1992; Kocks et al., 1992). Immunofluorescence studies revealed that this protein, ActA, is normally displayed on the bacterial surface and is present in greater amounts on one end of the bacterium (Kocks et al., 1993). Transfection experiments in kidney epithelial and HeLa cells demonstrate that ActA is alone sufficient to induce polymerization of G-actin to F-actin (Pistor et al., 1994; Friederich et al., 1995). Several studies demonstrated ActA to be a virulence factor essential for pathogenicity. For example, a pathogenic *Listeria* strain carrying a transposon inserted into the ActA gene showed no difference in growth in bacterial media but was avirulent in mice (Kocks et al., 1992). Its LD_{50} in mice was more than 4 logs higher than the pathogenic parental strain. These results were confirmed by another laboratory using an in-frame ActA deletion mutant, where the LD_{50} in this deletion strain increased by three orders of magnitude (Brundage et al., 1993). Furthermore, in tissue culture, actin tail formation was undetectable and the ability of these mutant bacteria to spread from cell to cell was greatly impaired.

ActA does not itself polymerize actin but rather recruits to the bacterial surface at least two factors that normally promote actin assembly. By using a reconstitution assay with purified cellular components, it was found that neither *Listeria* bacteria nor purified ActA protein could alone induce actin polymerization. However, *Listeria* or ActA can induce actin polymerization by interacting with the host Arp2/3 complex (Welch et al., 1998) (Fig. 7.3). Normally, Arp2/3 will nucleate actin polymerization only when bound to activated N-WASP or certain other host molecules. Thus ActA subverts the normal signaling pathways by both

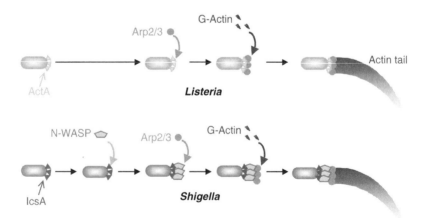

Figure 7.3. *Listeria and Shigella move by utilizing the actin cytoskeleton of the host cell. Listeria* and *Shigella* express the surface proteins ActA and IscA, respectively, preferentially at one end of the bacterium. ActA binds to and activates the Arp2/3 complex, whereas IscA binds to N-WASP, which then activates the Arp2/3 complex. Subsequent actin polymerization by the Arp2/3 complex at one end of the bacterium leads to the formation of an actin tail that propels the bacterium through the cytosol of the host cell.

binding and activating the final component of the actin nucleation pathway, the Arp2/3 complex. The charged N terminus of ActA, which is essential for actin polymerization, is able to bind to and activate the Arp2/3 complex both in vitro and in vivo (Skoble et al, 2000; Pistor et al., 2000; Zalevsky et al., 2001; May et al., 1999). The N terminus of ActA may have additional functions because it also binds to G-actin (Skoble et al., 2000; Zalevsky et al., 2001).

The central region of ActA is involved in actin assembly at the bacterial surface. This region contains proline-rich repeats and binds VASP (Chakraborty et al., 1995; Pistor et al., 1995). VASP is a cytoskeletal protein that is involved in actin bundling and actin filament elongation (for review, see Reinhard et al., 2001). Whereas the interaction between ActA and Arp2/3 is essential for actin tail-dependent intracellular motility, several studies suggest that interaction between ActA and VASP significantly increases *Listeria* movement within host cells (Smith et al., 1996; Niebuhr et al., 1997). It is thought that binding of VASP links the actin tail to the bacterium surface, allowing directional polymerization (Laurent et al., 1999). *Listeria* carrying deletions within the proline-rich VASP binding site of ActA move at rates 30% of wild type and show a log decrease in pathogenicity in mice (Smith et al., 1996). Together these results suggest that *Listeria* is able to move efficiently through the host's cytoplasm by the surface display of a single protein that subverts two major signaling pathways that normally influence actin dynamics.

SHIGELLA TARGETS N-WASP, AN ACTIVATOR OF THE ARP2/3 COMPLEX

Shigella is a rod-shaped gram-negative bacterium that can invade and colonize the mucosa of the human colon. Invasive infections by *Shigella* can quickly turn into a disease (shigellosis) whose classic symptoms are fevers, intestinal cramps, and dysentery (severe diarrhea with blood and mucus). Globally, it is estimated that annually there are at least 140 million cases of shigellosis, mostly in children of ages 1 to 5 of developing countries, resulting in approximately 600,000 deaths (Kotloff et al., 1999). *Shigella* is a natural pathogen for humans and a few other primates and is typically transmitted in crowded, unsanitary conditions via the fecal-oral route. All four known *Shigella* species, *S. dysenteriae*, *S. flexneri*, *S. boydii*, and *S. sonnei*, can infect and cause illness in humans, even when as few as 10 to 100 microorganisms are orally administered to adult volunteers (Sansonetti, 2001). Shigellosis is the principal bacterial cause of dysentery and should be considered whenever a person has bloody diarrhea. The first line of treatment for mild to moderate shigellosis is oral rehydration. The role of antibiotic therapy is variable and depends on the severity of the disease. Typically, antibiotics are helpful in reducing the severity and duration of illness in cases of bloody diarrhea. Resistance to many antibiotics such as chloramphenicol and tetracyclines is universal, and many cases now involve resistance to ampicillin.

The current drugs of choice in drug-resistant infections are nalidixic acid and 4-fluoroquinolones.

Although much is known about the pathogenesis of shigellosis, most of our understanding of the molecular mechanisms involved is inferred from in vitro studies with cell culture and cell-free extracts. This is especially true for our understanding of the steps and signals involved in the initial bacterial entry into host cells and their translocation through host cells into neighboring cells. The pathogenesis of shigellosis involves the invasion of three populations of cells in the intestinal barrier: M cells, epithelial cells, and resident macrophages. A simplified overview of how this pathogenesis develops would start with *Shigella* invasion of the apical surface of M cells (Sansonetti, 2001). Translocation across M cells allows invasion of adjacent epithelial cells through their basolateral surfaces and invasion of the resident macrophages in the mucosa-associated lymphoid follicles. Invasion of macrophages quickly results in apoptosis, and invasion of epithelial cells contributes to inflammation that facilitates further invasion and destruction of the epithelium. Several steps in this pathogenesis require bacterial proteins that act to remodel the actin cytoskeleton of host cells. Changes in the actin cytoskeleton are required for *Shigella* invasion of M cells, epithelial cells, and resident macrophages. Movement in and translocation across M cells and epithelial cells requires additional kinds of actin cytoskeletal alterations. Movement of polymorphic neutrophils to and through the infected sites as part of the inflammatory response as well as phagocytosis of the infecting bacteria both require changes in the host cell actin cytoskeleton. Thus alterations in the host actin cytoskeleton are involved in almost every stage of the disease. This section focuses on how *Shigella* controls its intracellular movement by subverting actin polymerization, whose study gives us additional insights in the normal mechanism for controlling actin cytoskeletal dynamics.

After ingestion, *Shigella* gains access to the intestinal barrier of the colon, in part because of its ability to survive the low pH in the gut. The spread of *Shigella* from cell to cell resembles the intracellular movement of *Listeria* in two ways. *Shigella* assembles and is propelled by actin tails that are virtually indistinguishable from those induced by *Listeria*. Furthermore, a single gene from either bacterium is sufficient to reconstruct both actin tail formation and actin rocketing. The *Shigella* gene, VirG, was first identified by Makino et al. (1986), who localized it to a 4-kb region on a 230-kb virulence plasmid that was previously shown to be required for cell-to-cell spread and the development of clinical disease. A subsequent study identified IcsA/VirG, as a 120-kDa outer bacterial membrane protein that organizes and regulates actin rocketing (Bernardini et al., 1989). This study also showed that small molecule inhibitors of actin polymerization completely blocked *Shigella* cell-to-cell transmission.

Subsequent analysis revealed that IscA was preferentially localized to a single pole of the bacterium, suggesting that, like *Listeria*, the polarized polymerization on the surface of *Shigella* may generate the force required for bacterial motility (Goldberg et al., 1993). Transfer of the IcsA gene to bacteria such as nonpathogenic *Escherichia coli* that normally do not induce actin polymerization allowed these bacteria to polymerize actin and propel themselves in some host cells (Goldberg and Theriot, 1995; Kocks et al., 1995). The region of IcsA required for actin polymerization is an α-helical region located between amino acids 53 and 758 of this 1102-amino acid protein (Goldberg et al., 1993; Suzuki et al., 1995). Early attempts to identify how IscA directs actin tail formation revealed that a number of different cytoskeletal proteins were associated with the actin tails of *Shigella*, but none of these was capable of producing actin polymerization. It was not until 1998 that Suzuki et al. demonstrated that N-WASP, a Cdc42 GTPase effector protein, was the host protein subverted by *Shigella* for actin tail formation (Fig. 7.3). Immunofluorescence studies found that N-WASP localized with IcsA to one pole of the bacterium and that overexpressing a dominant negative N-WASP mutant protein greatly reduced actin tail formation (Suzuki et al., 1998). Additional experiments showed that the α-helical region of IcsA directly interacted with N-WASP (Suzuki et al., 1998). Finally, immunodepletion of N-WASP from cell-free *Xenopus* egg extracts with an anti-N-WASP antibody blocked actin tail formation by IcsA.

How might IcsA binding to N-WASP induce actin polymerization in infected cells? Because N-WASP normally induces actin polymerization by binding to and activating the Arp2/3 complex after the binding of Cdc42 and other molecules, one might expect that the interaction of IcsA with N-WASP would also enable N-WASP to stimulate actin polymerization via the Arp2/3 complex. This expectation was confirmed (Egile et al., 1999) (Fig. 7.3). Closer examination of the N-WASP-IcsA complex revealed the presence of additional host cell signaling molecules that include Grb2 and WIP (Moreau et al., 2000). These results might suggest that Cdc42, the small GTPase that normally binds to and activates N-WASP, is not involved in *Shigella* movement in vivo. In support of this model, Mounier et al. did not observe any role for Cdc42 in *Shigella* intracellular movement. In these experiments, cells expressing dominant negative Rho GTPases or inhibitors of the GTPases did not alter actin tail formation induced by *Shigella* in host cells (Mounier et al., 1999).

As in many biological pathways, *Shigella* motility is likely to be more complicated and involve many more factors than outlined here. To identify such factors, reconstitution assays have been used (Goldberg and Theriot, 1995; Kocks et al., 1995; Suzuki et al., 1998; Egile et al., 1999; Loisel et al., 1999). Although these studies show that additional host proteins, such as actin depolymerizing factors and capping proteins, are required for efficient movement in vitro, it is not yet clear whether all of these factors are also required in vivo (Loisel et al., 1999). These reconstitution assays also suggest that the molecular mechanisms controlling actin rocketing in *Shigella* and *Listeria*, although similar in many details, are not identical. For example, *Shigella* but not *Listeria* requires the presence of VASP for efficient actin rocketing (Egile et al., 1999). This short review of intracellular *Listeria* and *Shigella* motility highlights the importance of the Arp2/3 complex in controlling actin dynamics and also illustrates how the complexity of the actin nucleation and polymerization machinery offers a variety of opportunistic targets for subversion by microorganisms.

SALMONELLA INVASION: BACTERIAL EFFECTOR PROTEINS REGULATE CDC42/RAC SIGNALING AND DIRECTLY MODIFY ACTIN CYTOSKELETON DYNAMICS

Salmonella are very diverse gram-negative bacteria containing over 2300 distinguishable species (Edwards, 1999). Some of these species are pathogenic, causing typhoid fever, focal systemic infection, and septicemia, but most commonly, diarrhea. Infection by *Salmonella* is usually initiated by oral ingestion, requires an ability of the bacteria to evade acid death in the stomach, and occurs in the distal ileum and colon where the bacteria cross the mucosal barrier. Although many different *Salmonella* species infect a wide variety of hosts, they all appear to use similar mechanisms for invading the host's intestines (for review, see Galan, 1999).

Before intracellular movement, bacterial pathogens such as *Listeria*, *Shigella*, and *Salmonella* must initially gain entry into host cells, which they accomplish by subverting actin polymerization pathways by another mechanism. *Salmonella* gains entry into host cells by inducing a phagocytosis-like process, called macropinocytosis, in cells that do not normally phagocytose (for review, see Galan, 1999). Phagocytosis, the uptake of particulate matter greater than 0.5 μm, involves the localized formation of microspikes, which coalesce into membrane ruffles, engulfing the particulate matter and then pinching off to form an intracellular vacuole. Microbes either escape from this intracellular vacuole or are delivered by it to death and destruction. In epithelial cells, *Salmonella* can escape from such vacuoles into the cytoplasm to infect host cells (for review, see Galan, 1999).

Progress in identifying and analyzing the mechanism used by *Salmonella* to gain entry into host cells has been rapid because of the generation of mutated virulence genes in pathogenic strains. Studies with *Salmonella* mutants revealed that invasion requires a number of proteins encoded in a pathogenicity island mapped to centisome 63 on the *Salmonella* chromosome (for review, see Galan, 1999). Analysis of this locus revealed that it encodes structural proteins for a type III contact-induced secretion apparatus and several additional proteins, some of which are secreted and translocated into the host membrane or cytoplasm after *Salmonella* attaches to the host cell. The mechanisms and signals regulating this contact-dependent secretion are largely unknown. However, the functions for several of the translocated proteins, including SopE/SopE2, SopB, SipA, SipC, and SptP, in inducing actin cytoskeleton changes that facilitate bacterial invasion have been determined (Fig. 7.4).

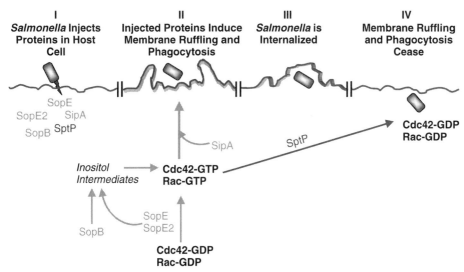

I
***Salmonella* Injects Proteins in Host Cell**

II
Injected Proteins Induce Membrane Ruffling and Phagocytosis

III
***Salmonella* is Internalized**

IV
Membrane Ruffling and Phagocytosis Cease

Figure 7.4. *Salmonella injects the effectors SopE, SopE2, SopB, SipA, and SptP into the host cell to regulate uptake of the bacterium.* After injection of these *Salmonella* proteins by a type III secretion system, SopE and SopE2 act as GEFs to activate Cdc42 and Rac in addition to producing inositol intermediates. SopB, however, indirectly activates Cdc42 by producing lipid intermediates necessary for Cdc42 activation. The activated Cdc42 and Rac induce actin rearrangements that lead to phagocytosis of the bacterium while SipA acts to stabilize the newly formed F-actin structures. After uptake of *Salmonella* is complete, SptP utilizes its GAP activity to inactivate Cdc42 and Rac and thus terminate the process of phagocytosis.

Clues to how *Salmonella* enters host cells were derived from studies showing that inhibitors of actin polymerization blocked *Salmonella* uptake into cells (Kihlstrom and Nilsson, 1977; Bukholm, 1984). Later it was determined that *Salmonella* attachment to the outside membrane of host cells induces a dramatic localized plasma membrane ruffling, resulting in engulfment and internalization of the bacterium (Francis et al., 1993). Roles for Rho GTPases in participating in *Salmonella*-induced actin changes were first demonstrated by experiments showing that over-expression of dominant negative or dominant active mutants of Cdc42 in cultured cells blocked and increased phagocytosis of *Salmonella*, respectively (Chen et al., 1996). Subsequent studies showed that SopE was sufficient for initiating membrane ruffling of the host cells (Fig. 7.4) (Hardt et al., 1998). SopE appears to be sufficient for this activity, because microinjection of recombinant SopE into cultured cells induces membrane ruffling within 5 minutes. Biochemical analysis revealed that SopE directly binds to and activates two related host cell targets, Rac and Cdc42 (Hardt et al., 1998). Because overexpression of either Rac or Cdc42 induces these effects, the localized activation of these signaling molecules by translocated SopE can account for initiating the membrane ruffling that results in the entry of *Salmonella*. SopE activates Rac and

Cdc42 via its action as a GEF (Hardt et al., 1998). Because SopE catalyzes the exchange of GDP for GTP in inactive Rac and Cdc42, it increases the intracellular level of the active, GTP-bound form of Rac and Cdc42. Although SopE functions as a GEF, it has little amino acid sequence similarity to any known cellular GEF. Thus it may be possible to design SopE inhibitors that do not significantly interfere with host GEFs. However, the therapeutic value of such inhibitors may be limited for several reasons. Although present in many pathogenic *Salmonella* strains, SopE is absent from some (Prager et al., 2000), perhaps because it is encoded in a temperate bacteriophage and not in the centrosome 63 pathogenic island. In addition, although SopE is sufficient for inducing membrane ruffles it is not necessary for *Salmonella* invasion. That is, a pathogenic strain of *Salmonella* engineered to contain a mutant SopE gene can still invade cultured epithelial cells (Zhou et al., 2001). Furthermore, the cellular effector proteins involved in membrane ruffling downstream of SopE-activated Cdc42 and Rac have not yet been identified. However, the phenotype of cells infected with *Salmonella* SopE mutants suggests the possibility that metabolites of membrane-bound phospholipids and soluble inositol phosphates are also likely involved in *Salmonella*-induced membrane ruffling (Fig. 7.4). Specifically, the phenotype of such cells suggests that SopE activates both phospholipase C (PLC), possibly as a direct result of activating Cdc42, and inositol phosphate phosphatases, by an unknown mechanism (Zhou et al., 2001).

SopE is not essential for *Salmonella* invasion of cultured epithelial cells because two other *Salmonella* effector proteins, SopE2 and SopB, can also initiate the host cell actin cytoskeletal rearrangements and membrane ruffling that lead to bacterial entry (Zhou et al., 2001). Roles for SopE2 and SopB in *Salmonella* internalization have been established only as a result of constructing numerous multiple mutants of pathogenic strains, most often *Salmonella typhimurium*, and then examining their invasive behavior in model cell culture systems. SopE2 is a recently described SopE-homologous protein (Bakshi et al., 2000, Stender et al., 2000), encoded in all 280 phylogenetically well-characterized or pathogenic *Salmonella* strains examined (Prager et al., 2000). Although SopE2 is needed for entry of *Salmonella dublin* into host cells, it does so by acting only as a GEF for Cdc42 (Fig. 7.4) (Stender et al., 2000). The preferred explanation for why SopE2 promotes bacterial invasion and is normally not observed is that it may not be expressed at a sufficiently high level to overcome the activity of another *Salmonella* pathogenic effector protein, SptP, whose function is the opposite of the presumed function of SopE2. SptP functions to inactivate Cdc42 and Rac and is described in more detail below.

SopB is a third *Salmonella* pathogenic effector protein, also translocated into host cells by the type III secretion system, which was originally characterized as a protein possessing inositol phosphatase activity and required for the virulence of *Salmonella dublin* (Norris et al., 1998). SopB was not considered a likely candidate for inducing the membrane ruffling leading to bacterial entry, because it was already known to be involved in later stages of the infection process and because its effect on membrane ruffling and bacterial internalization was masked by the activity of SopE (Zhou et al., 2001). In these studies, invasion assays showed that the internalization efficiency of SopB single-mutant bacteria is reduced by 50%. However, a SopE mutant also exhibits a 50% reduction in internalization, whereas the double mutant lacking both SopB and SopE showed a 95% drop in invasion efficiency (Zhou et al., 2001). Similarly, host cells transfected with plasmids encoding wild-type SopB, but not with plasmids encoding the phosphatase-defective SopB, induced actin cytoskeleton changes resembling the membrane ruffling induced during *Salmonella* infection (Zhou et al., 2001). The SopB-dependent membrane changes appear to require Cdc42 activity because they did not occur when the host cells were transfected with a plasmid expressing a dominant negative Cdc42. The authors of these elegant genetic studies offer an explanation for why SopB-dependent bacterial internalization appears to require both Cdc42 activation and a SopB-generated inositol phosphate. They suggest that SopB-generated inositol phosphate acts upstream of Cdc42 (Fig. 7.4) (Zhou et al., 2001). Thus *Salmonella* can trigger its internalization by three different effector proteins, which may act exclusively on upstream components of signaling cascades that control actin cytoskeleton rearrangements.

The *Salmonella* effector proteins SopE, SopE2, and SopB influence actin cytoskeletal dynamics indirectly, by activating upstream regulators of Cdc42 and Rac. In contrast, the *Salmonella* pathogenic effectors SipA and SipC can alter the actin cytoskeleton during bacterial entry by interacting directly with actin. The altered phenotypes of SipA and SipC mutants on the invasiveness of *Salmonella* strains and the known in vitro capabilities of purified SipA and SipC proteins lead to a more detailed picture of how *Salmonella* might orchestrate its own rapid entry into host cells. The combined activities of SipA and SipC can theoretically allow new actin polymerization in vivo under conditions in which actin polymerization in epithelial cells would not normally occur. Thus SipA and SipC promote membrane ruffling in the normally nonphagocytotic epithelial cells (Fig. 7.4). In vitro, SipA lowers the critical concentration of G-actin needed for the spontaneous polymerization of G-actin into F-actin by a factor of 10 (Zhou et al., 1999a). In vitro, SipC has the ability to nucleate actin fiber formation, allowing actin filament polymerization to begin, without the initial lag phase that reflects the kinetic barrier of actin nucleation (Hayward and Koronakis, 1999). The in vitro combination of SipA and SipC allows SipC to nucleate actin polymerization at significantly lower concentrations of G-actin than those required with SipC alone (McGhie et al., 2001). If these reactions also occur in vivo, localized translocation of SipA and SipC into host epithelial cells at the sites of *Salmonella* attachment could lead to the rapid net accumulation of actin filament, without the prior recruitment of monomeric G-actin.

Both SipA and SipC also have in vitro effects on the bundling of actin fibrils. There has been considerable speculation about the possible in vivo consequences of the individual or collective in vitro effects of SipA and SipC, including that they could generate host cell-independent stable networks of F-actin bundles (McGhie et al., 2001), that they could both drive and support membrane ruffle formation (Galan and Zhou, 2000), and, more specifically, that they might favor the outward extension of host cell protrusions involved in *Salmonella* invasion (Bourdet-Sicard and Van Nhieu, 1999). SipA can bind F-actin in a 1-to-1 ratio, resulting in three identifiable consequences: It stabilizes F-actin by blocking spontaneous depolymerization; it has a morphologic effect on F-actin fibers, making them appear straighter in electron micrographs, and it enhances the bundling ability of a host cell actin-binding protein, T-plastin (Zhou et al., 1999b). SipC can bundle actin filaments in the absence of any additional host cell proteins (Hayward and Koronakis, 1999). SipC-dependent activities may occur near or at membrane surfaces because SipC can associate with lipid bilayers (Hayward and Koronakis, 1999). Most or all of these in vitro activities may also occur in vivo because semipermeabilized cultured Swiss 3T3 cells treated with purified recombinant SipA and SipC proteins develop regions of actin condensation as well as filopodia-like and lamellipodia-like structures (McGhie et al., 2001).

The initiation of *Salmonella* invasion of epithelial cells has been described as a two-step process, in which one step involves Cdc42 and Rac activation by SopE, SopE2, and/or SopB and the other step involves the effects of SipA and SipC (Fig. 7.4). However, there is disagreement about which step occurs first. One model proposes that Cdc42/Rac activation is the first step because T-plastin/fimbrin, the host cell protein whose actin-bundling ability is enhanced by SipA, is recruited to membrane ruffles in a *Salmonella* SipA mutant and, more importantly, this recruitment is Cdc42-dependent (Zhou et al., 1999b). The other model emphasizes the ability of SipA and SipC to nucleate F-actin polymerization at concentrations of G-actin below the critical concentration (McGhie et al., 2001). One way to resolve this uncertainty would be to determine either the origin or the local concentration of G-actin that becomes polymerized into the earliest-appearing actin rearrangements that are induced after host cell attachment of wild-type pathogenic strains.

In contrast to SopE/SopE2, SopB, SipA, and SipC, which are all involved in inducing actin polymerization needed for *Salmonella* phagocytosis, SptP actually functions to terminate the actin changes induced by the invading bacterium. The function of SptP explains the results first observed by Takeuchi (1967) that the actin cytoskeleton of host cells recovers shortly after *Salmonella* infection. Biochemical analysis revealed that SptP contains GAP activity, efficiently converting the active, GTP-bound forms of Rac and Cdc42 to their inactive, GDP-bound forms (Fu and Galan, 1999). This property of SptP can explain why the actin rearrangements associated with membrane ruffling and phagocytosis stop after *Salmonella* entry. Although SptP does not share extensive amino acid sequence homology with any known cellular GAPs, there is a small region of homology between the N-terminal region of SptP and many eukaryotic GAP proteins, which contain a conserved arginine that is essential for the catalytic activity of GAP proteins (Fu and Galan, 1999; Stebbins and Galan, 2000). Site-directed mutagenesis of this arginine completely inactivates the GAP activity of SptP (Fu and Galan, 1999). SptP might provide a selective evolutionary advantage for pathogenic *Salmonella*. This is because such a protein might allow host cells to recover from the unnatural production of ruffles and phagocytosis and avoid cell death by allowing host cell gene expression to occur that is needed for efficient *Salmonella* growth (Galan and Zhou, 2000).

As described in this section, it is clear that actin and Rho GTPase signaling pathways are major targets for *Salmonella* virulence. Furthermore, a great deal of redundancy of the *Salmonella* effector proteins is required for entry and some of these proteins have multiple targets and effects. For example, both SopE and SopE2 act as direct activators of Cdc42 and Rac and also independently alter inositol intermediates that can also activate these GTPases. Intriguingly, SipA has effects on actin cytoskeleton dynamics and is also associated with the clinical manifestations of *Salmonella* infections. A recent study indicates that highly purified SipA can induce a proinflammatory response in epithelial cells that involves the transepithelial migration of neutrophils into the intestinal lumen (Lee et al., 2000). Is this effect of SipA directly or indirectly related to its effects on actin cytoskeleton dynamics or is this caused by a yet undiscovered function of this protein? Future studies on SipA and other *Salmonella* proteins will likely yield additional insights into the complexity of *Salmonella* virulence and host entry.

CLOSTRIDIAL TOXINS MODIFY ACTIN AND THE RHO GTPASES

Clostridia are gram-positive, spore-forming bacteria. This bacterial genus consists of over 100 species, of which a small portion is pathogenic in humans and animals (Hatheway, 1990). Clostridial infections include botulism, localized wound contamination, and diarrhea. For instance, *Clostridium perfringens* and *Clostridium novyi* cause wound lesions that can rapidly lead to gas gangrene, whereas *Clostridium botulinum* and *Clostridium tetani* produce neurotoxins that cause rare paralytic or spastic diseases. One of the most clinically important species, involved in 25% of all cases of antibiotic-associated diarrhea, is *Clostridium difficile* (Bartlett, 1996). Although antibiotics are usually the first line of treatment for bacterial infections, in the case of *Clostridium difficile*-mediated diarrhea, such treatment must be discontinued, because antibiotics disrupt the normal flora of the colon and thus facilitate *Clostridium difficile* colonization. However, certain antibiotics such as metronidazole or oral vancomycin are used in moderate or severe cases of *Clostridium difficile*-induced diarrhea.

Unlike the other pathogens discussed so far, *Clostridia* do not enter or move within the cytoplasm of host cells but rather secrete toxins and enzymes that act as virulence factors to promote colonization and pathogenesis (Hatheway, 1990). Here we review only those clostridial toxins that act as poisons to alter the normal dynamics of actin in host cells. Mechanistically these toxins fall into two types: toxins that alter actin directly and toxins that modify Rho GTPases. In both cases, the toxins secreted by these bacteria are internalized by receptor-mediated endocytosis and are later released from acidic endosomes into the cytoplasm. Within the cytoplasm of host cells, these toxins inactivate either actin polymerization or Rho GTPases.

Although it was known that bacterial toxins such as cholera toxin could mediate pathogenicity by catalyzing the ADP-ribosylation of host proteins, two laboratories made the exciting discovery that G-actin is ADP-ribosylated by the *Clostridium botulinum* C2 toxin (Oshihi and Tsuyama, 1986; Aktories et al., 1986). Thus the toxicity of this toxin in mammalian cells is easily explained by the inhibition of actin polymerization within cells caused by the transfer of an ADP ribose to G-actin. Subsequent studies revealed that toxins derived from other *Clostridium* species including *Clostridium perfringens* (Vandekerckhove et al., 1987), *Clostridium spiroforme* (Popoff and Boquet, 1988), and *Clostridium difficile* (Popoff et al., 1988) also ADP-ribosylated G-actin (Table 7.1). Analysis of these actin-modifying toxins revealed that they are structurally similar and are composed of two subunits. Furthermore, the

TABLE 7.1. Clostridial Toxins that Target Actin or Rho GTPases

Toxin	Organism(s)	Function
Clostridial C2 toxins	*Clostridium spiroforme* *Clostridium difficile* *Clostridium botulinum*	Block actin polymerization, nucleation, and ATPase activity associated with G-actin by ADP-ribosylation of Arg-177.
Clostridial iota toxins	*Clostridium perfringens*	Block actin polymerization, nucleation, and ATPase activity associated with G-actin and completely depolymerize the microfilament system by ADP-ribosylation of both G and F-actin
TCdA (toxin A) TCdB (toxin B) lethal toxin α-toxin	*Clostridium difficile* *Clostridium difficile* *Clostridium sordellii* *Clostridium novyi*	Inactivate Rho GTPases by glucosylation of Thr-37 in Rho and Thr-35 in Rac and Cdc42.
C3	*Clostridium botulinum*	Inactivates the Rho isoforms A, B, and C by ADP-ribosylation of Asn-41.

region on actin that was ribosylated was identified as the arginine residue at position 177 (Vandekerckhove et al., 1987; Vandekerckhove et al., 1988). It also appears that the *Clostridium perfringens iota* toxin has broader substrate specificity. This toxin ADP-ribosylates G-actin as well as F-actin and consequently shows more potency in disrupting the actin cytoskeleton (Vandekerckhove et al., 1987). The identification of actin as a target of various clostridial toxins is not surprising. The targeting of the host cell actin cytoskeleton allows *Clostridia* to disrupt the cell-cell contacts of epithelial cells and possibly inactivate secondary targets such as host immune cells, thus allowing for more efficient colonization and spread of these bacteria.

Another means of achieving disruption of the actin cytoskeleton by bacterial pathogens is to inhibit the activity of Rho GTPases that normally regulate actin polymerization. This mechanism of action was confirmed by the discovery that the C3 toxin from *Clostridium botulinum* ADP-ribosylates a G protein of 21 kDa (Rubin et al., 1988). Additional studies showed that ADP-ribosylation by the C3 toxin is specific for Rho isoforms (RhoA, RhoB, and RhoC, but not other related GTPases such as Rac or Cdc42 (Sekine et al., 1989; Braun et al., 1989; Aktories et al., 1989) (Table 7.1). Characterization of the C3 toxin revealed that, unlike the binary toxins that ADP-ribosylate actin, this toxin is a 23-kDa single-chain toxin. Biochemical studies also identified the site of ADP-ribosylation in Rho proteins as asparagine residue 41 (Sekine et al., 1989). The consequence of Rho ADP-ribosylation does not appear to alter the interaction of Rho with its effector proteins. but rather it may inhibit its conversion of to the GTP-bound state by GEFs (Sehr et al., 1998).

Another long-recognized pathogenic *Clostridium* species is *Clostridium difficile*, which is a frequent cause of antibiotic-induced diarrhea. The pathogenicity associated with *Clostridium difficile* infection is entirely due to the production of two toxins, which can produce marked intestinal inflammation, designated toxin A and toxin B. Initial characterization of these toxins revealed that they were capable of causing the disruption of stress fibers and cell adhesion, leading to cell rounding (Thelestam and Bronnegard, 1980; Wedel et al., 1983; Pothoulakis et al., 1986). Purification of these toxins revealed that they were single polypeptide chains of approximately 300 kDa in size and were 60% homologous to each other (for review, see Aktories and Just, 1995). The mechanism involved in the ability of these toxins to disrupt the actin cytoskeleton was solved when Aktories's group showed that these toxins inactivated various members of the Rho GTPase family (Just et al., 1995a; Just et al., 1995b). Unlike the ADP-ribosylation activity associated with other toxins, toxin A and toxin B use an entirely different mechanism. This inactivation of Rho proteins involves monoglucosylation, in which a simple glucose molecule derived from UDP-glucose is attached to threonine residue 37 in Rho and to threonine residue 35 in Rac and Cdc42 (Just et al., 1995a; Just et al., 1995b) (Table 7.1). In contrast to ADP-ribosylation of Rho isoforms by C3 toxin, there is evidence that monoglucosylation blocks the interaction of Rho with downstream effector proteins resulting in the disruption of the normal regulation of the actin cytoskeletal (Sehr et al., 1998).

Structural analysis of these large toxins revealed that their N terminus contains the glucosyltransferase activity (Hofmann et al., 1997). Studies with *Clostridium sordelli* and *Clostridium novyi* showed that these species also secrete toxins that inactivate Rho proteins (Table 7.1). Interestingly, these toxins show some differences from those of *Clostridium difficile*. Although structurally quite similar to the toxins from *Clostridium difficile*, the hemorrhagic toxin from *Clostridium sordelli* is more lethal, perhaps because in addition to its ability to modify Rho GTPases it can also modify the Ras subfamily of GTPase proteins (Genth et al., 1996; Popoff et al., 1996; Just et al., 1996). The αtoxin from *Clostridium novyi* covalently attaches an *N*-acetylglucosamine to threonine 37 of Rho and to threonine 35 of Rac and Cdc42, thereby inactivating them (Selzer et al., 1996).

Rho GTPases are major targets of the *Clostridium difficile* toxins A and B and contribute to disruption or loosening of the normal intestinal epithelial tight junctions between cells. resulting in increased permeability (Hecht et al., 1988; Moore et al., 1990; Riegler et al., 1995). However, the activities of these toxins are clearly more complicated. In particular, cytotoxicity due to toxin B causes damage to and exfoliation of superficial epithelial cells (Riegler et al., 1995). Studies show that toxin B first damages the mitochondria and then induces alterations in the actin cytoskeleton (He et al., 2000). In contrast, toxin A appears to have a greater effect on electrophysiological alterations of colonic tissue leading to enterotoxicity (Castagliuolo et al., 1994; Pothoulakis et al., 1994; Castagliuolo et al., 1998). The enterotoxicity of toxin A is due to its ability to induce a necroinflammatory response in the many cell types of the intestine (for review see Pothoulakis and LaMont, 2001). Although the general mechanism of toxin A-induced toxicity involves recruitment of immune regulatory cells and release of cytokines and neuropeptides, the exact details of this response are not clearly understood. In particular, it is not known whether Rho inactivation by these clostridial toxins is involved in the inflammatory response or whether this activity is independent of Rho activity. Despite the gaps in our understanding of these molecular mechanisms, the studies outlined here have significant implications for our understanding of the pathogenesis associated with *Clostridium difficile* infections.

CONCLUSIONS AND FUTURE AREAS FOR DRUG DEVELOPMENT

This review has focused on the molecular aspects of how many bacterial pathogens alter the actin cytoskeleton of host cells in order to promote infection and spread. Although these bacteria utilize specific proteins to alter the host actin cytoskeleton, the mechanisms involved are quite diverse. Some bacterial proteins, such as those of *Listeria* and

Shigella, are anchored on the surface of the invading bacteria, whereas others, such as the bacterial effector proteins of *Salmonella*, are injected into host cells to facilitate entry. Furthermore, other pathogenic bacteria, for instance, *Clostridia*, secrete toxins that directly inactivate actin or indirectly alter the actin cytoskeleton by inactivating Rho GTPases. Additional studies not covered in this review suggest that other pathogenic bacteria and even viruses, including *Neisseria, Yersinia*, enteropathogenic *Escherichia coli* (EPEC), and the vaccinia virus, also subvert the actin cytoskeleton to promote infection and spread. Future studies will likely discover new pathways and additional intricacies of how these and other pathogens target the actin cytoskeleton of host cells.

The exploitation of the host actin cytoskeleton by so many different pathogens suggests the possibility that drugs may be designed to inhibit the activity of the bacterial proteins involved in these processes. However, to date this has not been the case, possibly because many of the mechanisms used by bacteria to alter the host actin cytoskeleton have only been discovered in recent years. Other significant obstacles for this strategy revolve around the diversity and innate complexity of these pathways. Because distinct proteins and different mechanisms are used by bacterial pathogens, it is unlikely that a single drug will have cross-species specificity. This obstacle contrasts with the success of broad-spectrum antibiotics that inhibit cell wall or protein synthesis in a wide range of pathogenic bacteria. A second problem involves the role of multiple bacterial proteins that are redundant in function. Thus the loss of one protein may not be sufficient to inhibit infection, because of compensation by another protein. *Salmonella*, for example, utilize multiple proteins with similar and sometimes overlapping function to promote actin polymerization and entry into host cells, so that loss of any one protein is usually not sufficient to prevent bacterial entry. Specific immunotherapy, rather than drug therapy, shows more promise in blocking bacterial entry and spread. Indeed, many of these proteins are involved in virulence and have already been deleted in strains of bacteria to generate attenuated vaccines. Finally, the increasing resistance of bacterial pathogens to present-day antibiotics and the lack of other obvious bacterial targets will likely propel research into the direction of developing novel therapeutic agents that block the entry and spread of pathogenic bacteria.

REFERENCES

Aktories, K., Barmann, M., Ohishi, I., Tsuyama, S., Jakobs, K. H., and Haberma, E. (1986). *Botulinum* C2 toxin ADP-ribosylates actin. Nature *322*, 390–392.

Aktories, K., Braun, U., Rosener, S., Just, I., and Hall, A. (1989). The rho gene product expressed in *E. coli* is a substrate of *botulinum* ADP-ribosyltransferase C3. Biochem Biophys Res Commun *158*, 209–213.

Aktories, K., and Just, I. (1995). Monoglucosylation of low-molecular mass GTP-binding Rho proteins by clostridial cytotoxin. Trends Cell Biol *5*, 441–443.

Bakshi, C. S., Singh, V. P., Wood, M. W., Jones, P. W., Wallis, T. S., and Galyov, E. E. (2000). Identification of SopE2, a *Salmonella* secreted protein which is highly homologous to SopE and involved in bacterial invasion of epithelial cells. J Bacteriol *182*, 2341–2344.

Bartlett, J. G. (1996). Management of *Clostridium difficile* infection and other antibiotic-associated diarrhoeas. Eur J Gastroenterol Hepatol *11*, 1054–61.

Bernardini, M. L., Mounier, J., d'Hauteville, H., Coquis-Rondon, M., and Sansonetti, P. J. (1989). Identification of icsA, a plasmid locus of *Shigella flexneri* that governs bacterial intra- and intercellular spread through interaction with F-actin. Proc Natl Acad Sci USA *86*, 3867–3871.

Blanchoin, L., Pollard, T. D., and Mullins, R. D. (2000). Interactions of ADF/cofilin, Arp2/3 complex, capping protein and profilin in remodeling of branched actin filament networks. Curr Biol *10*, 1273–1282.

Bourdet-Sicard, R., and Tran Van Nhieu, G. (1999). Actin reorganization by SipA and *Salmonella* invasion of epithelial cells. Trends Microbiol *7*, 309–310.

Braga, V. (2000). Epithelial cell shape: cadherins and small GTPases. Exp Cell Res *261*, 83–90.

Braun, U., Habermann, B., Just, I., Aktories, K., and Vandekerckhove, J. (1989). Purification of the 22 kDa protein substrate of *botulinum* ADP-ribosyltransferase C3 from porcine brain cytosol and its characterization as a GTP-binding protein highly homologous to the rho gene product. FEBS Lett *243*, 70–76.

Brundage, R. A., Smith, G. A., Camilli, A., Theriot, J. A., and Portnoy, D. A. (1993). Expression and phosphorylation of the *Listeria monocytogenes* ActA protein in mammalian cells. Proc Natl Acad Sci USA *90*, 11890–11894.

Bukholm, G. (1984). Effect of cytochalasin B and dihydrocytochalasin B on invasiveness of entero-invasive bacteria in Hep-2 cell cultures. Acta Pathol Microbiol Immunol Scand *92*, 145–149.

Castagliuolo, I., Keates, A. C., Wang, C. C., Pasha, A., Valenick, L., Kelly, C. P., Nikulasson, S. T., LaMont, J. T., and Pothoulakis C. (1998). *Clostridium difficile* toxin A stimulates macrophage inflammatory protein-2 production in rat intestinal epithelial cells. J Immunol *160*, 6039–6045.

Castagliuolo, I., LaMont, J. T., Letourneau, R., Kelly, C., O'Keane, J. C., Jaffer, A., Theoharides, T. C., and Pothoulakis, C. (1994). Neuronal involvement in the intestinal effects of *Clostridium difficile* toxin A and *Vibrio cholerae* enterotoxin in rat ileum. Gastroenterology *107*, 657–665.

Chakraborty. T., Ebel, F., Domann, E., Niebuhr, K., Gerstel, B., Pistor, S., Temm-Grove, C. J., Jockusch, B. M., Reinhard, M., Walter, U., and Wehland, J. (1995). A focal adhesion factor directly linking intracellularly motile *Listeria monocytogenes* and *Listeria ivanovii* to the actin-based cytoskeleton of mammalian cells. EMBO J *14*, 1314–1321.

Chen, L. M., Hobbie, S., and Galan, J. E. (1996). Requirement of CDC42 for *Salmonella*-induced cytoskeletal and nuclear responses. Science *274*, 2115–2118.

Chimini, G., and Chavrier, P. (2000). Function of Rho family proteins in actin dynamics during phagocytosis and engulfment. Nature *2*, 191–196.

Cossart, P., and Bierne, H. (2001). The use of host cell machinery in the pathogenesis of *Listeria monocytogenes*. Curr Opin Immunol *13*, 96–103.

Diekmann, D., Brill, S., Garrett, M. D., Totty, N., Hsuan, J., Monfries, C., Hall, C., Lim, L., and Hall, A (1991). Bcr encodes a GTPase-activating protein for p21rac. Nature *351*, 400–402.

Domann, E., Wehland, J., Rohde, M., Pistor, S., Hartl, M., Goebel, W., Leimeister-Wachter, M., Wuenscher, M., and Chakraborty, T. (1992). A novel bacterial virulence gene in *Listeria monocytogenes* required for host cell microfilament interaction with homology to the proline-rich region of vinculin. EMBO J *11*, 1981–1990.

Edwards, B. H. (1999). *Salmonella* and *Shigella* species. Clin Lab Med *19*, 469–487.

Egile, C., Loisel, T. P., Laurent, V., Li, R., Pantaloni, D., Sansonetti, P. J., and Carlier, M. F. (1999). Activation of the CDC42 effector N-WASP by the *Shigella flexneri* IcsA protein promotes actin nucleation by Arp2/3 complex and bacterial actin-based motility. J Cell Biol *146*, 1319–1332.

Francis, C. L., Ryan, T. A., Jones, B. D., Smith, S. J., and Falkow, S. (1993). Ruffles induced by *Salmonella* and other stimuli direct macropinocytosis of bacteria. Nature *364*, 639–642.

Friederich, E., Gouin, E., Hellio, R., Kocks, C., Cossart, P., and Louvard, D. (1995). Targeting of *Listeria monocytogenes* ActA protein to the plasma membrane as a tool to dissect both actin-based cell morphogenesis and ActA function. EMBO J *14*, 2731–2744.

Fu, Y., and Galan, J. E. (1999). A *Salmonella* protein antagonizes Rac-1 and Cdc42 to mediate host-cell recovery after bacterial invasion. Nature *401*, 293–297.

Galan, J., and Zhou, D. (2000). Striking a balance: modulation of the actin cytoskeleton by *Salmonella*. Proc Natl Acad Sci USA *97*, 8754–8761.

Galan. J. E. (1999). Interaction of *Salmonella* with host cells through the centisome 63 type III secretion system. Curr Opin Microbiol *2*, 46–50.

Genth, H., Hofmann, F., Selzer, J., Rex, G., Aktories, K., and Just, I. (1996). Difference in protein substrate specificity between hemorrhagic toxin and lethal toxin from *Clostridium sordellii*. Biochem Biophys Res Commun *229*, 370–374.

Goldberg, M. B., Barzu, O., Parsot, C., and Sansonetti, P. J. (1993). Unipolar localization and ATPase activity of IcsA, a *Shigella flexneri* protein involved in intracellular movement. J Bacteriol *175*, 2189–2196.

Goldberg, M. B., and Theriot, J. A. (1995). *Shigella flexneri* surface protein IcsA is sufficient to direct actin-based motility. Proc Natl Acad Sci USA *92*, 6572–6576.

Hall. A/ (1998). Rho GTPases and the actin cytoskeleton. Science *279*, 509–514.

Hardt, W. D., Chen, L. M., Schuebel, K. E., Bustelo, X. R., and Galan, J. E. (1998). *S. typhimurium* encodes an activator of Rho GTPases that induces membrane ruffling and nuclear responses in host cells. Cell *93*, 815–826.

Hart, M. J., Maru, Y., Leonard, D., Witte, O. N., Evans, T., and Cerione, R. A. (1992). A GDP dissociation inhibitor that serves as a GTPase inhibitor for the Ras-like protein CDC42Hs. Science *258*, 812–815.

Hatheway, C. L. (1990). Toxigenic *Clostridia*. Clin Microbiol Rev *3*, 66–98.

Hayward, R. D., and Koronakis, V. (1999). Direct nucleation and bundling of actin by the SipC protein of invasive *Salmonella*. EMBO J *18*, 4926–4934

He, D., Hagen, S. J., Pothoulakis, C., Chen, M., Medina, N. D., Warny, M., and LaMont, J. T. (2000). *Clostridium difficile* toxin A causes early damage to mitochondria in cultured cells. Gastroenterology *119*, 139–150.

Hecht, G., Pothoulakis, C., LaMont, J. T., and Madara, J. L. (1988). *Clostridium difficile* toxin A perturbs cytoskeletal structure and tight junction permeability of cultured human intestinal epithelial monolayers. J Clin Invest *82*, 1516–1524.

Hofmann, F., Busch, C., Propens, U., Just, I., and Aktories, K. (1997). Localization of the glucosyltransferase activity of *Clostridium difficile* toxin B to the N-terminal part of the holotoxin. J Biol Chem *272*, 11074–11078.

Just, I., Selzer, J., Wilm, M., von Eichel-Streiber, C., Mann, M., and Aktories, K. (1995a). Glucosylation of Rho proteins by *Clostridium difficile* toxin B. Nature *375*, 500–503.

Just, I., Wilm, M., Selzer, J., Rex, G., von Eichel-Streiber, C., Mann, M., and Aktories, K. (1995b). The enterotoxin from *Clostridium difficile* (ToxA) monoglucosylates the Rho proteins. J Biol Chem *270*, 13932–13936.

Just, I., Selzer, J., Hofmann, F., Green, G. A., and Aktories, K. (1996). Inactivation of Ras by *Clostridium sordellii* lethal toxin-catalyzed glucosylation. J Biol Chem *271*, 10149–10153.

Kihlstrom, E., and Nilsson, L. (1977). Endocytosis of *Salmonella typhimurium* 395 MS and MR10 by HeLa cells. Acta Pathol Microbiol Scand *85*, 322–328.

Kimura, K., Ito, M., Amano, M., Chihara, K., Fukata, Y., Nakafuku, M., Yamamori, B., Feng, J., Nakano, T., Okawa, K., Iwamatsu, A., and Kaibuchi, K. (1996). Regulation of myosin phosphatase by Rho and Rho-associated kinase (Rho-kinase). Science *273*, 245–248.

Kocks, C., Gouin, E., Tabouret, M., Berche, P., Ohayon, H., and Cossart, P. (1992). *L. monocytogenes*-induced actin assembly requires the actA gene product, a surface protein. Cell *68*, 521–531.

Kocks, C., Hellio, R., Gounon, P., Ohayon, H., and Cossart, P. (1993). Polarized distribution of *Listeria monocytogenes* surface protein ActA at the site of directional actin assembly. J Cell Sci *105*, 699–710.

Kocks, C., Marchand, J. B., Gouin, E., d'Hauteville, H., Sansonetti, P. J., Carlier, M. F., and Cossart, P. (1995). The unrelated surface proteins ActA of *Listeria monocytogenes* and IcsA of *Shigella flexner*i are sufficient to confer actin-based motility on *Listeria innocua* and *Escherichia coli* respectively. Mol Microbiol *18*, 413–423.

Kotloff, K. L., Winickoff, J. P., Ivanoff, B., Clemens, J. D., Swerdlow, D. L., Sansonetti, P. J., Adak, G. K., and Levine, M. M. (1999). Global burden of *Shigella* infections: implications for vaccine development and implementation of control strategies. Bull World Health Organ *77*, 651–666.

Kozma, R., Ahmed, S., Best, A., and Lim, L. (1995). The Ras-related protein Cdc42Hs and bradykinin promote formation of peripheral actin microspikes and filopodia in Swiss 3T3 fibroblasts. Mol Cell Biol *4*, 1942–1952.

Laurent, V., Loisel, T. P., Harbeck, B., Wehman, A., Grobe, L., Jockusch, B. M., Wehland, J., Gertler, F. B., and Carlier, M. F. (1999). Role of proteins of the Ena/VASP family in actin-based motility of *Listeria monocytogenes*. J Cell Biol *144*, 1245–1258.

Lee, C. A., Silva, M., Siber, A. M., Kelly, A. J., Galyov, E., and McCormick, B. A. (2000). A secreted *Salmonella* protein induces a proinflammatory response in epithelial cells, which promotes neutrophil migration. Proc Natl Acad Sci USA *97*, 12283–12288.

Loisel, T. P., Boujemaa, R., Pantaloni, D., and Carlier, M. F. (1999). Reconstitution of actin-based motility of *Listeria* and *Shigella* using pure proteins. Nature *401*, 613–616.

Machesky. L. M., Atkinson, S. J., Ampe, C., Vandekerckhove, J., and Pollard, T. D. (1994). Purification of a cortical complex containing two unconventional actins from *Acanthamoeba* by affinity chromatography on profilin-agarose. J Cell Biol *127*, 107–115.

Machesky, L. M. and Insall, R. H. (1999). Signaling to actin dynamics. J Cell Biol *146*, 267–272.

Makino, S., Sasakawa, C., Kamata, K., Kurata, T., and Yoshikawa, M. (1986). A genetic determinant required for continuous reinfection of adjacent cells on large plasmid in *S. flexneri* 2a. Cell *46*, 551–555.

May R.C., Hall M.E., Higgs H.N., Pollard T.D., Chakraborty T., Wehland J., Machesky L. M., and Sechi A. S. (1999). The Arp2/3 complex is essential for the actin-based motility of *Listeria monocytogenes*. Curr Biol *9*, 759–762.

McGhie, E. J., Hayward, R.D., and Koronakis, V. (2001). Cooperation between actin-binding proteins of invasive *Salmonella*: SipA potentiates SipC nucleation and bundling of actin. EMBO J *20*, 2131–2139

Mead, P. S., Slutsker, L., Griffin, P. M., and Tauxe, R. V. (1999). Food-related illness and death in the United States. Reply to Dr. Hedberg. Emerg Infect Dis *6*, 841–842.

Miki, H., Sasaki, T., Takai, Y., and Takenawa, T. (1998). Induction of filopodium formation by a WASP-related actin depolymerizing protein N-WASP. Nature *391*, 93–96.

Miki, H., Yamaguchi, H,, Suetsugu, S., and Takenawa, T. (2000). IRSp53 is an essential intermediate between Rac and WAVE in the regulation of membrane ruffling. Nature *408*, 732–735.

Moore, R., Pothoulakis, C., LaMont, J. T., Carlson, S., and Madara, J. L. (1990). *C. difficile* toxin A increases intestinal permeability and induces Cl⁻ secretion. Am J Physiol *259*, 165–172.

Moreau, V., Frischknecht, F., Reckmann, I., Vincentelli, R., Rabut, G., Stewart, D., and Way, M. (2000). A complex of N-WASP and WIP integrates signalling cascades that lead to actin polymerization. Nat Cell Biol *2*, 441–448.

Mounier, J., Laurent, V., Hall, A., Fort, P., Carlier, M. F., Sansonetti, P. J., and Egile, C. (1999). Rho family GTPases control entry of *Shigella flexneri* into epithelial cells but not intracellular motility. J Cell Sci *112*, 2069–2080.

Mullins, R. D., and Machesky, L. M. (2000). Actin assembly mediated by Arp2/3 complex and WASP family proteins. Meth Enzymol *325*, 214–237.

Niebuhr, K., Ebel, F., Frank, R., Reinhard, M., Domann, E., Carl, U. D., Walter, U., Gertler, F. B., Wehland, J., and Chakraborty, T. (1997). A novel proline-rich motif present in ActA of *Listeria monocytogenes* and cytoskeletal proteins is the ligand for the EVH1 domain, a protein module present in the Ena/VASP family. EMBO J *16*, 5433–5444.

Nobes, C. D., and Hall, A. (1994). Rho, Rac and Cdc42 GTPases regulate the assembly of multi-molecular focal complexes associated with actin stress fibers, lamellipodia and filopodia. Cell *81*, 53–62.

Norris, F. A., Wilson, M. P., Wallis, T. S., Galyov, E. E., and Majerus, P. W. (1998). SopB, a protein required for virulence of *Salmonella dublin*, is an inositol phosphate phosphatase. Proc Natl Acad Sci USA *95*, 14057–14059.

Ohishi, I., and Tsuyama, S. (1986). ADP-ribosylation of nonmuscle actin with component I of C2 toxin. Biochem Biophys Res Commun *136*, 802–806.

Pistor, S., Chakraborty, T., Niebuhr, K., Domann, E., and Wehland, J. (1994). The ActA protein of *Listeria monocytogenes* acts as a nucleator inducing reorganization of the actin cytoskeleton. EMBO J *13*, 758–763.

Pistor, S., Chakraborty, T., Walter, U., and Wehland J. (1995). The bacterial actin nucleator protein ActA of *Listeria monocytogenes* contains multiple binding sites for host microfilament proteins. Curr Biol *5*, 517–525.

Pistor, S., Grobe, L., Sechi, A. S., Domann, E., Gerstel, B., Machesky, L. M., Chakraborty, T., and Wehland, J. (2000). Mutations of arginine residues within the 146-KKRRK-150 motif of the ActA protein of *Listeria monocytogenes* abolish intracellular motility by interfering with the recruitment of the Arp2/3 complex. J Cell Sci *113*, 3277–3287.

Popoff, M. R., and Boquet, P. (1988). *Clostridium spiroforme* toxin is a binary toxin which ADP-ribosylates cellular actin. Biochem Biophys Res Commun *152*, 1361–1368.

Popoff, M. R., Chaves-Olarte, E., Lemichez, E., von Eichel-Streiber, C., Thelestam, M., Chardin, P., Cussac, D., Antonny, B., Chavrier, P., Flatau, G., Giry, M., de Gunzburg, J., and Boquet, P. (1996). Ras, Rap, and Rac small GTP-binding proteins are targets for *Clostridium sordellii* lethal toxin glucosylation. J Biol Chem *271*, 10217–10224.

Popoff, M. R., Rubin, E. J., Gill, D. M., and Boquet, P. (1988). Actin-specific ADP-ribosyltransferase produced by a *Clostridium difficile* strain. Infect Immun *56*, 2299–2306. Pothoulakis, C., Barone, L.M., Ely, R., Faris, B., Clark, M.E., Franzblau, C., and LaMont, J.T. (1986). Purification and properties of *Clostridium difficile* cytotoxin B. J Biol Chem *261*, 1316–1321.

Pothoulakis, C., Castagliuolo, I., LaMont, J. T., Jaffer, A., O'Keane, J. C., Snider, R. M., and Leeman, S. E. (1994). CP-96,345, a substance P antagonist, inhibits rat intestinal responses to *Clostridium difficile* toxin A but not cholera toxin. Proc Natl Acad Sci USA *91*, 947–951.

Pothoulakis, C., and Lamont, J. T. (2001). Microbes and microbial toxins: paradigms for microbial-mucosal interactions. II. The integrated response of the intestine to *Clostridium difficile* toxins. Am J Physiol *280*, G178–G183.

Prager, R., Mirold, S., Tietze, E., Strutz, U., Knuppel, B., Rabsch, W., Hardt, W. D., and Tschape, H. (2000). Prevalence and polymorphism of genes encoding translocated effector proteins among clinical isolates of *Salmonella enterica*. Int J Med Microbiol *290*, 605–617. Reinhard, M., Jarchau, T., and Walter, U. (2001). Actin-based motility: stop and go with Ena/VASP proteins. Trends Biochem Sci *26*, 243–249.

Ridley, A. J., and Hall, A. (1992). The small GTP-binding protein Rho regulates the assembly of focal adhesions and actin stress fibers in response to growth factors. Cell *70*, 389–399.

Ridley, A. J., Paterson, H. F., Johnston, C. L., Diekmann, D., and Hall, A. (1992). The small GTP-binding protein rac regulates growth factor-induced membrane ruffling. Cell *70*, 401–410.

Riegler, M., Sedivy, R., Pothoulakis, C., Hamilton, G., Zacherl, J., Bischof, G., Cosentini, E., Feil, W., Schiessel, R., LaMont, J. T., and Wenzl, E. (1995). *Clostridium difficile* toxin B is more potent than toxin A in damaging human colonic epithelium in vitro. J Clin Invest *95*, 2004–2011.

Rohatgi, R., Ma, L., Miki, H., Lopez, M., Kirchhausen, T., Takenawa, T., and Kirschner, M. W. (1999). The interaction between N-WASP and the

Arp2/3 complex links Cdc42-dependent signals to actin assembly. Cell *97*, 221–231.

Rubin, E. J., Gill, D. M., Boquet, P., and Popoff, M R. (1988). Functional modification of a 21-kilodalton G protein when ADP-ribosylated by exoenzyme C3 of *Clostridium botulinum*. Mol Cell Biol *8*, 418–426.

Sansonetti, P. J. (2001). Microbes and microbial toxins: paradigms for microbial-mucosal interactions. III. Shigellosis: from symptoms to molecular pathogenesis. Am J Physiol *280*, G319–G323.

Sehr, P., Joseph, G., Genth, H., Just, I., Pick, E., and Aktories, K. (1998). Glucosylation and ADP ribosylation of rho proteins: effects on nucleotide binding, GTPase activity, and effector coupling. Biochemistry *37*, 5296–5304.

Sekine, A., Fujiwara, M., and Narumiya, S. (1989). Asparagine residue in the rho gene product is the modification site for *botulinum* ADP-ribosyltransferase. J Biol Chem *264*, 8602–8605.

Selzer, J., Hoffman, F., Rex, G., Wilm, M., Mann, M., Just, I., and Aktories, K. (1996). *Clostridium novyi* α-toxin-catalyzed incorporation of GlcNac into Rho subfamily proteins. J Biol Chem *271*, 25173–25177.

Skoble, J., Portnoy, D. A., and Welch, M. D. (2000). Three regions within ActA promote Arp2/3 complex-mediated actin nucleation and *Listeria monocytogenes* motility. J Cell Biol *150*, 527–538.

Smith, G. A., Theriot, J. A., and Portnoy, D. A. (1996). The tandem repeat domain in the *Listeria monocytogenes* ActA protein controls the rate of actin-based motility, the percentage of moving bacteria, and the localization of vasodilator-stimulated phosphoprotein and profilin. J Cell Biol *135*, 647–660.

Stebbins, C. E., and Galan, J. E. (2000). Modulation of host signaling by a bacterial mimic: structure of the *Salmonella* effector SptP bound to Rac1. Mol Cell *6*, 1449–1460.

Stender, S., Friebel, A., Linder, S., Rohde, M., Mirold, S., and Hardt, W. D. (2000). Identification of SopE2 from *Salmonella typhimurium*, a conserved guanine nucleotide exchange factor for Cdc42 of the host cell. Mol Microbiol *36*, 1206–1221.

Suzuki, T., Lett, M. C., and Sasakawa, C. (1995). Extracellular transport of VirG protein in *Shigella*. J Biol Chem *270*, 30874–30880.

Suzuki, T., Miki, H., Takenawa, T., and Sasakawa, C. (1998). Neural Wiskott-Aldrich syndrome protein is implicated in the actin-based motility of *Shigella flexneri*. EMBO J *17*, 2767–2776.

Takeuchi, A. (1967). Electron microscope studies of experimental *Salmonella* infection. I. Penetration into the intestinal epithelium by *Salmonella typhimurium*. Am J Pathol *50*, 109–136.

Thelestam, M., and Bronnegard. M. (1980). Interaction of cytopathogenic toxin from *Clostridium difficile* with cells in tissue culture. Scand J Infect Dis Suppl 16–29.

Theriot, J. A., Mitchison, T. J., Tilney, L. G., and Portnoy, D. A. (1992). The rate of actin-based motility of intracellular *Listeria monocytogenes* equals the rate of actin polymerization. Nature *357*:257–260.

Tilney, L. G., and Portnoy, D. A. (1989). Actin filaments and the growth, movement, and spread of the intracellular bacterial parasite, *Listeria monocytogenes*. J Cell Biol *109*, 1597–1608.

Van Aelst, L., and D'Souza-Schorey, C. (1997). Rho GTPases and signaling networks. Genes Dev *11*, 2295–2322.

Vandekerckhove, J., Schering, B., Barmann, M., and Aktories, K. (1987). *Clostridium perfringens* iota toxin ADP-ribosylates skeletal muscle actin in Arg-177. FEBS Lett *225*, 48–52.

Vandekerckhove, J., Schering, B., Barmann, M., and Aktories, K. (1988). *Botulinum* C2 toxin ADP-ribosylates cytoplasmic β/γ-actin on arginine 177. J Biol Chem *263*, 696–700.

Watanabe, N., Kato, T., Fujita, A., Ishizaki, T., and Narumiya, S. (1999). Cooperation between mDia1 and ROCK in Rho-induced actin reorganization. Nat Cell Biol *1*, 136–143.

Watanabe, N., Madaule, P., Reid, T., Ishizaki, T., Watanabe, G., Kakizuka, A., Saito, Y., Nakao, K., Jockusch, B. M., and Narumiya, S. (1997). p140mDia, a mammalian homolog of *Drosophila* diaphanous, is a target protein for Rho small GTPase and is a ligand for profilin. EMBO J *16*, 3044–3056.

Wedel, N., Toselli, P., Pothoulakis, C., Faris, B., Oliver, P., Franzblau, C., and LaMont, T. (1983). Ultrastructural effects of *Clostridium difficile* toxin B on smooth muscle cells and fibroblasts. Exp Cell Res *148*, 413–422.

Welch, M. D. (1999). The world according to Arp: regulation of actin nucleation by the Arp2/3 complex. Trends Cell Biol *9*, 423–427.

Welch, M. D., Iwamatsu, A., and Mitchinson, T. J. (1997). Actin polymerization is induced by the Arp2/3 complex at the surface of *Listeria monocytogenes*. Nature *385*, 265–269.

Welch, M. D., Rosenblatt, J., Skoble, J., Portnoy, D. A., and Mitchison, T. J. (1998). Interaction of human Arp2/3 complex and the *Listeria monocytogenes* ActA protein in actin filament nucleation. Science *281*, 105–108.

Winter, D., Podtelejnikov, A. V., Mann, M., and Li, R. (1997). The complex containing actin-related proteins Arp2 and Arp3 is required for the motility and integrity of yeast actin patches. Curr Biol *7*, 519–529.

Zalevsky, J., Grigorova, I., and Mullins, R. D. (2001). Activation of the Arp2/3 complex by the *Listeria* ActA protein. ActA binds two actin monomers and three subunits of the Arp2/3 complex. J Biol Chem *276*:3468–3475.

Zhou, D., Chen, L. M., Hernandez, L., Shears, S. B., and Galan, J. E. (2001). A *Salmonella* inositol polyphosphatase acts in conjunction with other bacterial effectors to promote host cell actin cytoskeleton rearrangements and bacterial internalization. Mol Microbiol *39*, 248–259.

Zhou, D., Mooseker, M. S., and Galan, J. E. (1999a). Role of the *S. typhimurium* actin-binding protein SipA in bacterial internalization. Science *283*, 2092–2095.

Zhou, D., Mooseker, M. S., and Galan, J. E. (1999b). An invasion-associated *Salmonella* protein modulates the actin bundling activity of plastin. Proc Natl Acad Sci USA *96*, 10176–10181.

CHAPTER 8

BACTERIAL TOXINS AND DIARRHEA

WALTER A. PATTON, JOEL MOSS, and MARTHA VAUGHAN
Department of Chemistry, Lebanon Valley College, Annville,
Pennsylvania; Pulmonary-Critical Care Medicine Branch, National
Heart, Lung, and Blood Institute, National Institutes of Health,
Bethesda, Maryland

INTRODUCTION

Bacterial enteric pathogens and the diarrheal diseases they cause have long been the scourge of people in areas in which food and water supplies are jeopardized by poor hygiene and little or no basic sanitation. Even in developed countries, however, natural disasters (e.g., floods) or episodes of inadequate food sanitation permit exposure to pathogenic species of *Vibrio, Escherichia, Clostridium, Campylobacter, Shigella,* or *Salmonella.* Some species of enteric bacteria have devastating effects because of a combination of virulence mechanisms each has evolved to colonize the host and disrupt intestinal function via malabsorption or diarrhea. The mechanisms used by these enteric pathogens may include one or more of the following: microbial attachment, effacement of intestinal epithelium, invasion of epithelial cells, and production of toxins, secreted enterotoxins, and/or cytotoxins (Guerrant et al., 1999).

Toxins can be defined as substances that have an undesirable pathophysiological effect on a cell or organism. Unlike simple chemical toxins, toxins produced by enteropathogenic bacteria are complex biomolecules, of seemingly intelligent design, that exploit existing host organismal anatomy, cellular processes, and the finest details of molecular structure to produce varied clinical manifestations from inflammation to death. Toxins secreted by *Vibrio cholerae* and specific strains of enterotoxigenic *Escherichia coli* are well known to be responsible for the clinical manifestations of cholera and travelers' diarrhea, respectively. These toxins, cholera toxin (CT) and *E. coli* heat-labile enterotoxin-1 (LT-1), are the focus of this review.

Signal Transduction and Human Disease, Edited by Toren Finkel and
J. Silvio Gutkind
ISBN 0-471-02011-7   Copyright © 2003 John Wiley & Sons, Inc.

THE VIRULENCE OF PATHOGENIC VIBRIO

Pathogenic organisms of the genus *Vibrio* can be classified in two groups: those responsible for gastrointestinal infections and those responsible for extraintestinal infections. *Vibrio* species that cause gastrointestinal diseases include *V. mimicus*, *V. parahaemolyticus*, *V. fluvialis*, *V. furnissii*, *V. hollisae*, *V. metschnikovii*, as well as perhaps the most studied of all diarrheal pathogens, *V. cholerae* O1, *V. cholerae* O139, and *V. cholerae* non-O1 (reviewed by Shinoda, 1999). Virulence factors characteristic of these pathogenic strains can include enterotoxins, cytotoxins, hemolysins, hemagglutinins, proteases, neuraminadases, siderophores, pili, and flagellae. *Vibrio* are gram-negative bacilli, whose *cholerae* species are categorized phenotypically as classic or El Tor. The biotypes are classified serologically by the presence or absence of the lipopolysaccharide O antigen. Heterogeneity of the O antigen results in subcategorization into Inaba and Ogawa serotypes. The leading cause of the estimated 5 to 7 million cases of cholera thought to occur in the world each year is *V. cholerae* O1, El Tor biotype, Ogawa serotype (Ryan and Calderwood, 2001).

Cholera results from ingestion of *V. cholerae* present in contaminated food or water. The severity of disease is related to many factors: the size of the inoculum, the biotype ingested, the gastric acidity of the individual receiving the inoculum, and, in individuals from endemic areas, the level of immunity. With an inoculum as small as 10^4 organisms, an individual with gastric hypoacidity can excrete up to 10^{13} *V. cholerae* organisms in the rice water stool (Ryan and Calderwood, 2000). The large number of organisms excreted per infected individual is certainly one reason that an outbreak of cholera can rapidly become an epidemic and, subsequently, a pandemic.

In 1884, Koch's recognition of *V. cholerae* as the microbial pathogen was the first insight into the nature of cholera. Much later observations that diarrhea could be caused by cell-free filtrates lent credence to the notion of a noncellular toxin as the causative agent (De, 1959; Dutta et al., 1959), Since those observations and the purification of a toxin from *V. cholerae* (Finkelstein and LoSpalluto, 1969), it has become clear that the major cause of the diarrhea of cholera is the proteinaceous cholera toxin (CT). CT was shown early to increase intracellular cAMP concentrations and is now known to inhibit NaCl absorption and stimulate Cl⁻ secretion as part of initiation of a massive efflux of fluid and electrolyte from the polarized epithelial cells that line the human intestine (reviewed by Lahiri, 2000).

CT is not the only factor that makes *V. cholerae* such a virulent pathogen. Two of the known virulence factors are a colonization factor termed toxin-coregulated pilus (TCP) and CTX, a large genetic element containing the CT gene that can be transferred to nontoxic *Vibrio* via the lysogenic filamentous bacteriophage CTXφ. The Ace and Zot gene products, whose genes are also a part of the CTX element and were initially thought to be accessory toxins, now appear (on the basis of sequence similarity to other known proteins) to be protein components required for the assembly of filamentous phage (Faruque et al., 1998). Other proteins such as hemolysin or hemagglutinin protease may also play a role in the virulence of *V. cholerae* (Shinoda, 1999).

Structural and Biochemical Features of Cholera Toxin

Similar to other cellular toxins, CT is a protein with an enzymatic activity that ultimately proves detrimental to the cell or organism that encounters it (Fig. 8.1). CT is a multimeric protein belonging to the AB_5 family of toxins (reviewed by Merrit and Hol, 1995). A close relative of CT is *E. coli* heat-labile enterotoxin-1 (LT-1). CT and LT-1 are similar in many ways: ~80% sequence identity, strikingly similar three-dimensional crystal structure, and analogous effects on cellular machinery. Both toxins cause diarrhea, although the effects of LT are less severe. These highly conserved characteristics led to the postulate that CT and LT-1 evolved from a common ancestor (Yamamoto et al., 1984; Lee et al., 1991). In the AB_5 toxins, the enzymatic A subunit is noncovalently associated with a pentamer of B subunits that bind and localize the holotoxin complex to membrane receptor sites/molecules before, and requisite for, toxin internalization and action on cells. Structures, primary or three-dimensional, of other AB_5 toxins (e.g., pertussis, tetanus, and botulinum toxins) have little similarity to those of CT or LT-1.

The A subunits of CT and LT-1 are composed of 240 amino acids, and each B subunit contains 103 amino acids. Sequences of both the A- and B subunits of CT and LT-1 are >82% identical (reviewed recently by de Hann and Hirst, 2000). In the toxin A subunit, a catalytic A_1 chain (22 kDa) is linked by a single disulfide bond to an A_2 chain (5 kDa); the 11.6-KDa B subunits assemble into a ringlike homopentamer that binds to a cell surface receptor. The B pentamer of CT binds almost exclusively to ganglioside GM1, whereas the B pentamer of LT-1 can bind to GM-1, other GM1-like gangliosides, and cell surface glycoproteins with poly-lactosylaminoglycan moieties (Lencer et al., 1999).

Crystal structures revealed that the intact AB_5 holotoxin structure of CT or LT is formed by insertion of the A_2 chain into the central pore of the ringlike B pentamer (Sixma et al., 1993; Zhang et al., 1995b) (Fig. 8.2); however, minor differences are noted throughout the structures. The portion of the LT A_2 subunit that is anchored in the B pentamer is seen as an elongated chain in the LT crystal structure (Sixma et al., 1993), whereas the same portion of CT is an extended helix (Zhang et al., 1995b). The reader is referred to the primary articles and several substantive review articles for further details of the toxin structures (Merritt et al., 1994a, 1994b, 1994c, 1995, 1997, 1998; Verlinde et al., 1994; Bastiaens et al., 1996; Van den Akker et al., 1997).

Gill (1975) first reported the requirement for NAD in CT-induced activation of adenylyl cyclase as a mechanism for accumulation of intracellular cAMP, which led to the recognition that CT is an enzyme displaying arginine-specific ADP- ribosyltransferase activity (Moss and

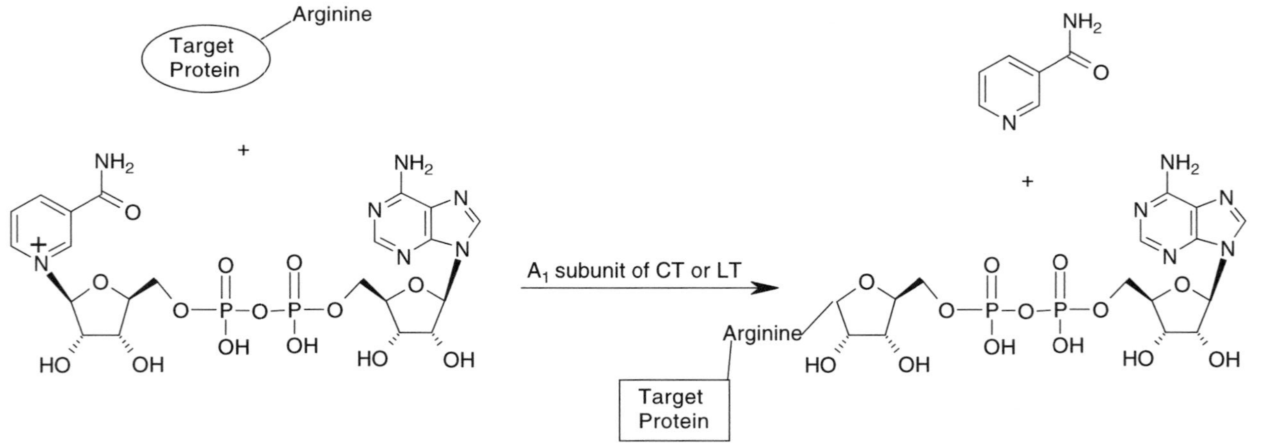

Figure 8.1. *Toxin-catalyzed ADP-ribosylation.* Cholera toxin and *E. coli* heat-labile enterotoxins ADP-ribosylate target proteins at arginine residues. These toxins catalyze the formation of α-anomeric product from β-NAD+ (Oppenheimer, 1978); nicotinamide is a side product of this reaction. If an arginine or simple guanidino compound (e.g., agmatine) is unavailable, the toxins utilize water as a nucleophile, resulting in toxin-catalyzed NADase activity. In vitro, ADP-ribosylation factors (ARFs) stimulate toxin ADP-ribosyltransferase activity.

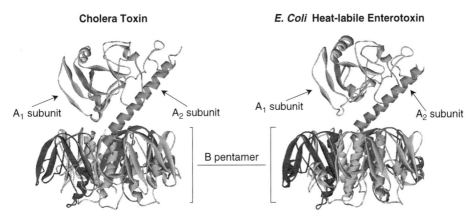

Figure 8.2. *The AB$_5$ Structure of CT and LT Holotoxins.* The near identity of the three-dimensional structures of cholera toxin and *E. coli* heat-labile entero-toxin supports the hypothesis that the toxins arose from a common ancestor. Ganglioside-binding loops at the bottom of the B pentamer anchor the toxoid to a lipid membrane. The carboxyl-terminal end of each A$_2$ chain penetrates the central pore of the B pentamer and thereby noncovalently anchors the enzymatic portion of each toxin complex (A$_1$ chain) to the B pentamer. The crystal struc-ture coordinates for cholera toxin (Protein Data Bank ID# 1XTC) and *E. coli* heat-labile enterotoxin (Protein Data Bank ID# 1LTS) were originally reported by Zhang et al. (1995b) and Sixma et al. (1993) respectively, and were visualized for this figure using Viewer Pro (Accelrys; San Diego, CA).

Vaughan, 1977) (Fig. 8.1). Also shown first for CT was the generation of active toxin by proteolytic cleavage of the CT A chain at Arg 192 (Moss et al., 1976; Mekalanos et al., 1979) followed by a reduction of disulfide bond linking the A$_1$ and A$_2$ chains between Cys residues 187 and 199 (Tomasi et al., 1979; Mekalanos et al., 1979).

Toxin A$_1$ chain is an arginine-specific ADP-ribosyltransferase that catalyzes the transfer of ADP-ribose from NAD to arginine, or other simple guanidine compounds, or, in the absence of an acceptor guanidino moiety, to water (reviewed by Patton et al., 2000) (Fig. 8.1). On the basis of sequence alignment and functional characterization, CT and LT, like ADP-ribosylating toxins from *Corynebacterium diph-theriae* (diptheria toxin or DT) (Carroll and Collier, 1984), *Bordetella pertussis* (pertussis toxin or PT) (Barbieri et al., 1989), and *Pseudomonas* (exotoxin A or ETA) (Carroll and Collier, 1987), contain a critical glu-tamate at or near the NAD$^+$-binding and catalytic sites. In functional studies, LT E112K, a mutant in which lysine replaces glutamate 112 (Tsuji et al., 1990; Tsuji et al., 1991) exhibited greatly reduced ADP-ribosyltransferase activity, although the toxin retained its ability to inter-act in vitro with an activating ADP-ribosylation factor (Moss et al., 1993). On the basis of computer modeling and crystallographic data, the posi-tion of Glu112 in CTA1 and LTA1 is consistent with its role as a critical active site residue (Sixma et al., 1993; Domenighini et al., 1994; Zhang et al., 1995b).

STUDIES OF CHOLERA TOXIN ACTION REVEAL IMPORTANT CELLULAR MOLECULES AND PROCESSES

In the broad field of toxicology, questions regarding mechanisms of toxin actions have often opened new avenues of investigation that have led to a wealth of new information on details of cellular processes; this is certainly true of CT and LT. Not only have several new classes of molecules been discovered as a direct result of studies of CT action (i.e., ADP-ribosylation factors and mammalian ADP-ribosyltransferases), but in addition, we now understand how to treat effectively the diseases associated with *Vibrio* and enterotoxigenic *E. coli* infection, as well as the utility of the toxins as vaccine adjuvants (Snider, 1995).

ADP-Ribosylation Factors (ARFs) and Molecules Relevant to ARF Function in Cells

ADP-Ribosylation Factors. After the recognition of CT as an ADP-ribosyltransferase, it was found that cholera toxin-catalyzed ADP-ribosylation of $G_{\alpha s}$, the endogenous substrate for CT, was stimulated by tissue factors and GTP (reviewed in Patton et al., 2000). One factor, later named ADP-ribosylation factor or ARF, was identified as a 21.5-kDa protein doublet on SDS-PAGE (Kahn and Gilman, 1984) that supported the GTP-dependent cholera toxin-catalyzed ADP-ribosylation of $G_{\alpha s}$ (Kahn and Gilman, 1986). ARF could bind GTP (now known to be critical for all ARF activities), GDP, and nonhydrolyzable analogs of both but was incapable of hydrolyzing GTP or binding adenine nucleotides (Kahn and Gilman, 1986). In what later would prove to be a clue to other ARF properties and function, a membrane-bound (Tsai et al., 1987) and two soluble (Tsai et al., 1988) ARFs were identified in bovine brain.

In vitro, ARFs were shown to be allosteric activators of CTA1 (Tsai et al., 1987; Tsai et al., 1988; Noda et al., 1990) by lowering the K_m for both NAD and guanidino-containing ADP-ribose acceptor (Noda et al., 1990). Enhancement of this effect, as well as increase in V_{max} and the binding of GTP by ARF, could be achieved by the addition of lipid and/or detergent (Noda et al., 1990; Bobak et al., 1990; Murayama et al., 1993). Another clue to the normal cellular role of ARF was revealed when it was shown that GTP binding by ARF facilitates its tight binding to lipid membranes (Kahn et al., 1991; Regazzi et al., 1991; Walker et al., 1992). In a later model, a loose association of ARF with a membrane, for example, in the form of ARF-GDP, was converted to a tighter association on GTP binding by ARF (Franco et al., 1995).

Six mammalian ARFs are now known. These are categorized in three classes (ARFs 1–3—Class I; ARF 4 and 5—Class II; ARF6—Class III), based on amino acid and DNA sequences as well as gene structure (Moss and Vaughan, 1999). ARFs are low-molecular-weight (LMW) GTP-binding proteins of ~180 amino acids that contain consensus sequences [common to all LMW and heterotrimeric GTP-binding proteins (Moorman et al., 1999)] for the binding of guanine nucleotides and a glycine in position two of the amino acid sequence that is posttranslationally modified by myristoylation. A role for ARF as a regulator of vesicular trafficking events was first suggested when it was shown that yeast ARF1 was localized in the Golgi and invertase secretion was defective in *arf1⁻ Saccharomyces cerevisiae* (Stearns et al., 1990). Subsequently, a role in ER to Golgi transport (Balch et al., 1992) and association with vesicular coat proteins of non-clathrin-coated, Golgi-derived vesicles (Serafini et al., 1991) was shown to be based on the interaction of ARF with β-COP (Donaldson et al., 1992a; Palmer et al., 1993), via a process that was sensitive to the fungal metabolite brefeldin A (BFA) (Donaldson et al., 1992b). BFA causes disintegration of the Golgi apparatus by blocking the activation of ARF by GEPs (Donaldson et al., 1992b; Helms and Rothman, 1992, see below). ARF was shown also to mediate binding of clathrin adaptor proteins AP-1 (Stamnes and Rothman, 1993; Traub et al., 1993; Dittié et al., 1996; Zhu et al., 1998) and AP-3 (Ooi et al., 1998) to membranes. It now seems clear that ARF1 functions in retrograde vesicular transport between ER and Golgi, as well as in clathrin coat recruitment (reviewed by Wieland and Harter, 1999; Chavrier and Goud, 1999; Wakeham et al., 2000). It is clear that ARF6 functions at the plasma membrane in recycling of endocytic vesicles and regulates receptor-mediated endocytosis, whereas ARFs 4 and 5 are thought to function in the Golgi (reviewed by Takai et al., 2001).

Crystal structures of various ARF molecules, crystallized with different forms of guanine nucleotide (Amor et al., 1994; Greasley et al., 1995; Goldberg, 1998; Ménétrey et al., 2000; Amor et al., 2001), revealed that ARF contains an amino-terminal amphipathic α-helix and a positively charged surface patch that is seemingly important for the interaction with membrane lipids. A possible explanation for the inability of ARF to hydrolyze GTP was also gleaned from the finding that Gln71, previously shown to be important for GTP hydrolysis (Tanigawa et al., 1993; Teal et al., 1994; Zhang et al., 1994; Kahn et al., 1995), is situated too far from the putative gamma-phosphate binding site to allow hydrolysis without a GTPase-activating protein or GAP. Thus far, three functional domains of ARF have been identified: the amino terminus, switch I, and switch II. The amino terminus of ARF appears to function as a GTP- and lipid-dependent structure (Randazzo et al., 1995) that influences the conformation of the so-called switch I and II (segments 38–50 and 70–83, respectively, in ARF1) regions. The switch regions, originally defined in the low molecular weight GTP-binding protein Ras (Takai et al., 2001), are involved in the recognition by regulatory molecules and in the interaction of activated ARF-GTP with effector molecules. It is now clear that CT can interact with ARF in cells and that discrete regions of activated ARF are required for that interaction (see below). For further review of ARF structure as related to function, the reader is directed to additional reviews by Jones et al. (1999a) and Béraud-Dufour and Balch (2001).

ARF Regulators: Guanine Nucleotide Exchange Proteins and GTPase-activating Proteins. *In vivo*, ARFs require the action of guanine nucleotide-exchange proteins (GEPs) to accelerate the release of GDP from inactive ARF to permit GTP binding and activation. All known

ARF GEPs contain sequences homologous to that in yeast phospho-protein Sec7, which is essential for secretion (Franzusoff and Schekman, 1989), and can be divided into two general classes according to their size and BFA sensitivity. The high-molecular weight GEPs include BIG1 and BIG2 (Morinaga et al., 1996; Morinaga et al., 1997), Gea1 and Gea2 (Peyroche et al., 1996), GBF1 (Claude et al., 1999), GNOM (Steinman et al., 1999), and the yeast Sec7 protein (Sata et al., 1998). The GEP activity of all these proteins, except GBF1, is BFA sensitive, and all appear to function in the region of the Golgi. The smaller ARF GEPs (~50 kDa) are BFA-insensitive proteins that contain amino-terminal coiled-coil, central Sec7, and carboxyl-terminal pleckstrin homology (PH) domains. The last can interact with negatively charged phospholipids, for example, phosphatidylinositol 3,4,5-trisphosphate, that thereby modulate or influence GEP activity (Kolanus et al., 1996; Paris et al., 1997). Members of this family include cytohesin-1 (Meacci et al., 1997), cytohesin-2 or ARNO (Chardin et al., 1996). cytohesin-3 or GRP1 (Klarlund et al., 1997), cytohesin-4 (Ogasawara et al., 2000), and the similar EFA6 (Franco et al., 1999). ARF GEPs appear to accelerate GDP release via a glutamate "finger" that displaces Mg^{2+} and the β-phosphate destabilizing the ARF-GDP interaction (Béraud-Dufour et al., 1998; Betz et al., 1998; Goldberg, 1998). GDP displacement allows GTP binding and resulting conformational changes in the amino terminus of ARF, as well as in switch I and switch II.

Inactivation of ARFs requires hydrolysis of bound GTP. Because ARFs have low (undetectable) intrinsic GTPase activity, however, this depends on a GTPase-activating protein (GAP). Among the known GAPs are 1) GAPs 1 and 2 that act on ARFs 1–5 (Randazzo, 1997) and appear to function in the Golgi (Aoe et al., 1997); 2) GIT proteins (Premont et al., 1998; Vitale et al., 2000) that may regulate events at the plasma membrane by acting on ARF6; 3) ASAP proteins (Randazzo et al., 2000) that bind to and are phosphorylated by Src; 4) Glo3p (Dogic et al., 1999), which is involved in ER retrieval; and 5) GCS1 (Poon et al., 1996; Blader et al., 1999), a yeast protein that is involved in actin cytoskeletal organization.

Membrane lipids can modify GAP activity, at least in part by concentrating it at membranes in proximity to ARF-GTP. As first shown for GAP1 (Cukierman et al., 1995), a conserved zinc finger motif is critical for activation of ARF GTPase. In addition, a conserved arginine in an arginine finger structure may contribute by stabilizing the transition state during GTP hydrolysis (Scheffzek et al., 1997). For further review of ARF GEPs, ARF GAPs and their function, the reader is directed to additional articles and reviews (Roth, 1999; Donaldson and Jackson, 2000; Jackson and Casanova, 2000; Renault et al., 2002).

ARF Effectors. After the recognition of in vitro ARF interaction with CT, the subsequent search for physiological functions of ARF in cells resulted in the identification of several proteins that may be regulated by interaction with ARF. These include isoforms of phospholipase

D (PLD), arfaptin, phosphatidylinositol 4-phosphate 5-kinase, GGA proteins, and arfophilin. ARF activation of specific isoforms of PLD (Brown et al., 1993; Cockcroft et al., 1994; Massenburg et al., 1994; Hammond et al., 1997; Colley et al., 1997), which could influence the formation of membrane vesicles by altering membrane lipid composition, has been reviewed by Exton (1999), Jones et al. (1999b), and Waite (1999). Arfaptins (Kanoh et al., 1997) appear to interact with the amino terminus of ARF, can influence ARF activity, and are reported to mediate cross-talk between Rac and ARF signaling pathways (Tarricone et al., 2001). Phosphatidylinositol 4(P) 5-kinase (Honda et al., 1999; Godi et al., 1999; Jones et al., 2000), when activated by ARF, produces phosphatidylinositol 4,5-bisphosphate (PIP_2), which is a regulator of molecules containing PH domains, for example, cytohesins, PLD, and GGA proteins (Boman et al., 2000; Dell'Angelica et al., 2000; Hirst et al., 2000; Dell'Angelica and Payne, 2001). These proteins appear to serve as effector adaptor molecules for clathrin coat assembly, as well as regulating sorting events in the trans-Golgi network (TGN). Arfophilins (Shin et al., 1999; Shin et al., 2001) are putative effectors for both class II and III ARFs.

MECHANISM OF CHOLERA PATHOGENESIS

$G_{\alpha s}$, the α subunit of the heterotrimeric GTP-binding protein that activates adenylyl cyclase, is the major protein modified by CT- and LT-catalyzed ADP-rbosylation in cells. In the basal state, $G_{\alpha s}$-GDP is complexed with $G_{\beta \gamma}$ at the plasma membrane. Agonist binding to its receptor causes the release of bound GDP, followed by formation of $G_{\alpha s}$-GTP, which dissociates from $G_{\beta \gamma}$ and activates adenylyl cyclase. ADP-ribosylation of $G_{\alpha s}$ inhibits its endogenous GTPase activity, resulting in the persistence of active $G_{\alpha s}$-GTP and activated adenylyl cyclase with elevated levels of cAMP that are associated with CT and LT action on cells (reviewed by Kaper et al., 1995). Although there are believed to be some differences between events in polarized and nonpolarized cells, they appear to be very similar overall.

Intact holotoxin (CTA or LTA bound to the CTB or LTB pentamer, respectively) is required for toxin action on cells. The B pentamer was shown to localize the holotoxin to ganglioside GM1-rich caveola-like plasma membrane domains (Orlandi and Fishman, 1998; Wolf et al., 1998) via a GM1-binding site in each of the five B chains (Fishman 1982; Orlandi and Fishman, 1993; Merritt et al., 1994b). This results in endocytosis of the holotoxin via apical endosomes and BFA-sensitive, COPI-mediated retrograde transport from the Golgi to the ER, presumably due to the presence of a Golgi-retrieval sequence (KDEL in CT and RDEL in LT; Majoul et al., 1998) in the toxin A chain (Lencer et al., 1999). Because β-COP in COPI-coated vesicles is known to interact with ARF, it is plausible that CT or LT encounters ARF during the retrograde transport. Recent studies show that the cellular effects of CT (Morinaga et al., 2001) and LT (Zhu and Kahn, 2001) are dependent on class I ARF-dependent membrane trafficking processes in the Golgi that precede cAMP production. Direct interaction of ARF and CT or LT in cells, however, remains to be demonstrated.

Unlike the pH-dependent unfolding and membrane translocation of diphtheria toxin (Falnes and Sandvig, 2000) the detailed mechanism by which CTA (or a portion thereof) enters the cytosol from the ER is not defined. Current models include the possible use of the Sec61p protein translocation channel (Hazes and Read, 1997) or the spontaneous partitioning of naturally hydrophobic toxins into a membrane (as discussed in Lencer et al., 1999).

Information regarding the segment of ARF that interacts with CT or LT is limited. Experiments with an ARF1-ARL1 (ARF-like protein 1) chimera, demonstrated that amino acids 73–181 of ARF are sufficient for the functional interaction of ARF and CTA (Zhang et al., 1995a). In agreement is the more recent report that LTA$_1$ interacts with a region of LTA$_2$ containing a sequence of five amino acids, four of which are identical in a five-amino acid segment of switch II in ARF3 (Zhu et al., 2001). As was shown in that study with a two-hybrid system, LTA$_1$ interacts with ARF through the same amino acids that it uses to interact with LTA$_2$. Other work with a two-hybrid system showed the interaction of ARF6 with somewhat different residues in CT and led the authors to speculate that a conformational change is necessary for ARF to bind to the somewhat buried residues (Jobling and Holmes, 2000); some of the CT amino acids identified in this study are the same conserved residues (V97 and Y104) that were identified in LT by functional assay analysis (Stevens et al., 1999).

The reason for the greater severity of diarrhea in CT than in LT intoxication has long been of interest. Recent studies mapped structural differences in CT and LT to a 10-amino acid sequence in the A2 fragment that resides in the central pore of the B pentamer (Rodighiero et al., 1999). It was shown, with CT and LT chimeric toxins, that the CT sequence results in greater holotoxin stability, which is thought to be important during toxin uptake and transport. In the effects of CT and LT in vivo, the role of the enteric nervous system and the contribution of other cells or substances (e.g., 5-hydroxytryptamine; Turvill et al., 1998) are of interest (Farthing, 2000).

CLOSING REMARKS

Toxin-catalyzed ADP-ribosylation of $G_{\alpha s}$ and the subsequent activation of adenylyl cyclase causes increased cAMP levels in cells. These events make CT useful as a molecular tool for probing cAMP-dependent processes in cells. As discussed above, a great deal is now understood about the cellular processes involved in CT and LT intoxication. In these times when sophisticated molecular approaches to treatment bring promise to a variety of diseases, antimicrobial therapy (e.g., ciprofloxacin), new vaccines, and, perhaps the simplest approach, oral rehydration therapy are all used to effectively combat cholera and the devastation it brings (Butler, 2001).

REFERENCES

Amor, J. C., Harrison, D., Kahn, R. A., and Ringe, D. (1994). Structure of the human ADP-ribosylation factor 1 complexed with GDP. Nature *372*, 704–708.

Amor, J. C., Horton, J. R., Zhu, X., Wang, Y., Sullards, C., Ringe, D., Cheng, X., and Kahn, R. A. (2001). Structures of yeast ARF1 and ARL1. J Biol Chem *276*, 42477–42484.

Aoe, T., Cukierman, E., Lee, A., Cassel, D., Peters, P. J., and Hsu, V. W. (1997). The KDEL receptor, ERD2, regulates intracellular traffic by recruiting a GTPase-activating protein for ARF1. EMBO J *16*, 7305–7316.

Balch, W. E., Kahn, R. A., and Schwaninger, R. (1992). ADP-ribosylation factor is required for vesicular trafficking between the endoplasmic reticulum and the *cis*-Golgi compartment. J Biol Chem *267*, 13053–13061.

Barbieri, J. T., Mende-Mueller, L. M., Rappuoli, R., and Collier, R. J. (1989). Photolabeling of Glu-129 of the S-1 subunit of pertussis toxin with NAD. Infect Immun *57*, 3549–3554.

Bastiaens, P. I., Majoul, I. V., Verveer, P. J., Soling, H. D., and Jovin, T. M. (1996). Imaging the intracellular trafficking and state of the AB5 quaternary structure of cholera toxin. EMBO J *15*, 4246–4253.

Béraud-Dufour, S., and Balch, W. E. (2001). Structural and functional organization of ADP-ribosylation factor (ARF) proteins. Methods Enzymol *329*, 245–247.

Béraud-Dufour, S., Robineau, S., Chardin, P., Paris, S., Chabre, M., Cherfils, I., and Antonny, B. (1998). A glutamic finger in the guanine nucleotide exchange factor ARNO displaces Mg^{2+} and the β-phosphate to destabilize GDP on ARF1. EMBO J *17*, 3651–3659.

Betz, S. F., Schnuchel, A., Wang, H., Olejniczak, E. T., Meadows, R. P., Lipsky, B. P., Harris, E. A. S., Staunton, D. E., and Fesik, S. W. (1998). Solution structure of the cytohesin-1 (B2-1): Sec7 domain and its interaction with the GTPase ADP-ribosylation factor 1. Proc Natl Acad Sci USA *95*, 7909–7914.

Blader, I. J., Cope, M. J., Jackson, T. R., Profit, A. A., Greenwood, A. F., Drubin, D. G., Prestwich, G. D., and Thiebert, A. B. (1999). GCSI, an ARF guanosine triphosphate-activating protein in *Saccharomyces cerevisiae* is required for normal actin organization in vivo and stimulates actin polymerization in vitro. Mol Biol Cell *10*, 581–596.

Bobak, D. A., Bliziotes, M. M., Noda, M., Tsai, S.-C., Adamik, R., and Moss, J. (1990). Mechanisms of activation of cholera toxin by ADP-rbosylation factor (ARF): Both low- and high-affinity interactions of ARF with guanine nucleotides promote toxin activation. Biochemistry *29*, 855–861.

Boman, A. I., Zhang, C.-J., Zhu, X., and Kahn, R. A. (2000). A family of ARF effectors that can alter membrane transport through the trans-Golgi. Mol Biol Cell *11*, 1241–1255.

Brown, H. A., Gutowski, S., Moomaw, C. P., Slaughter, C., and Sternweis, P. C. (1993). ADP-ribosylation factor, a small GTP-dependent regulatory protein, stimulates phospholipase D activity. Cell *75*, 1137–1144.

Butler, T. (2001). New developments in the understanding of cholera. Curr Gastroenterol Rep *4*, 315–321.

Carroll, S. F., and Collier, R. J. (1984). NAD binding site of diphtheria toxin: Identification of a residue within the nicotinamide subsite by photochemical modification with NAD. Proc Natl Acad Sci USA *81*, 3307–3311.

Carroll, S. F., and Collier, R. J. (1987). Active site of *Pseudomonas aeruginosa* enterotoxin A. J Biol Chem *262*, 8707–8711.

Chardin, P., Paris, S., Antonny, B., Robineau, S., Béraud-Dufour, S., Jackson, C. L., and Chabre, M. (1996). A human exchange factor for ARF contains Sec7- and pleckstrin homology domains. Nature *384*, 481–484.

Chavrier, P., and Goud, B. (1999). The role of ARF and Rab GTPase in membrane transport. Curr Opin Cell Biol *11*, 446–475.

Claude, A., Zhao, B. P., Kuziemsky, C. E., Dahan, S., Berger, S. J., Yan, J. P., Arnold, A. D., Sullivan, E. M., and Melançon, P. (1999). GBF1. A novel golgi-associated BFA-resistant guanine nucleotide exchange factor that displays specificity for ADP-ribosylation factor 5. J Cell Biol *146*, 71–84.

Cockcroft, S., Thomas, G. M. H., Fensome, A., Geny, B., Cunningham, E., Gout, I., Hiles, I., Totty, N. F., Truong, O., and Hsuan, J. J. (1994). Phospholipase D: A downstream effector of ARF in granulocytes. Science *263*, 523–526.

Colley, W. C., Sung, T. C., Roll, R., Jenco, J., Hammond, S. M., Altshuller, Y., Bar-Sagi, D., Morris, A. J., and Frohman, M. A. (1997). Phospholipase D2, a distinct phospholipase D isoform with novel regulatory properties that provokes cytoskeletal reorganization. Curr Biol *7*, 191–201.

Cukierman, E., Huber, I., Rotman, M., and Cassel, D. (1995). The ARF1 GTPase-activating protein: Zinc finger motif and Golgi localization. Science *270*, 1999–2002.

de Hann, L., and Hirst, T. R. (2000). Cholera toxin and related enterotoxins: a cell biological and immunological perspective. J Nat Toxins *9*, 281–297.

De, S. N. (1959). Enterotoxity of bacteria-free culture filtrates of *Vibrio cholerae*. Nature *183*, 1533–1534.

Dell'Angelica, E. C., and Payne, G. S. (2001). Intracellular cycling of lysosomal enzyme receptors: cytoplasmic tails' tales. Cell *106*, 395–398.

Dell'Angelica, E. C., Puertollano, R., Mullins, C., Aguilar, R. C., Vargas, J. D., Hartnell, L. M., and Bonifacino, J. S. (2000). GGAs: A family of ADP-ribosylation factor-binding proteins related to adaptors and associated with the Golgi complex. J Cell Biol *149*, 81–94.

Dittié, A. S., Hajibagheri, N., and Tooze, S. A. (1996). The AP-1 adaptor complex binds to immature secretory granules from PC-12 cells, and is regulated by ADP-ribosylation factor. J Cell Biol *132*, 523–536.

Dogic, D., de Chassey, B., Pick, E., Cassel, D., Lefkir, Y., Hennecke, S., Cosson, P., and Letourneur, F. (1999). The ADP-ribosylation factor GTPase-activating protein Glo3p is involved in ER retrieval. Eur J Cell Biol *78*, 305–310.

Domenighini, M., Magagnoli, C., Pizza, M., and Rappuoli, R. (1994). Common features of the NAD-binding and catalytic site of ADP-ribosylating toxins. Mol Microbiol *14*, 41–50.

Donaldson, J. G., and Jackson, C. L. (2000). Regulators and effectors of the ARF GTPases. Curr Opin Cell Biol *12*, 475–482.

Donaldson, J. G., Cassel, D., Kahn, R. A., and Klausner, R. D. (1992a). ADP-ribosylation factor, a small GTP-binding protein, is required for binding of the coatomer protein β-COP to Golgi membranes. Proc Natl Acad Sci USA *89*, 6408–6412.

Donaldson, J. G., Finazzi, D., and Klausner, R. D. (1992b). Brefeldin A inhibits Golgi-membrane catalyzed exchange of guanine nucleotide onto ARF protein. Nature *360*, 350–352.

Dutta, N. K., Panse, M. W., Kulkarni, D. R. (1959). Role of cholera toxin in experimental cholera. J Bacteriol *78*, 594–595.

Exton, J. H. (1999). Regulation of phospholipase D. Biochim Biophys Acta *1439*, 121–133.

Falnes, P. O., and Sandvig, K. (2000). Penetration of protein toxins into cells. Curr Opin Cell Biol *12*, 407–413.

Farthing, M. I. G. (2000). Enterotoxins and the enteric nervous system—a fatal attraction. Int J Med Microbiol *290*, 491–496.

Faruque, S. M., Albert, J. M., and Mekalanos, J. J. (1998). Epidemiology, genetics and ecology of toxigenic *Vibrio cholerae*. Microbiol Mol Biol Rev *62*, 1301–1314.

Finkelstein, R. A., and LoSpalluto, J. J. (1969). Pathogenesis of experimental cholera: Preparation and isolation of choleragen and choleragenoid. J Exp Med *130*, 185–202.

Fishman, P. H. (1982). Role of membrane gangliosides in the binding and action of bacterial toxins. J Membr Biol *69*, 85–97.

Franco, M., Chardin, P., Chabre, M., and Paris, S. (1995). Myristoylation of ADP-ribosylation factor 1 facilitates nucleotide exchange at physiological Mg^{2+} levels. J Biol Chem *270*, 1337–1341.

Franco, M., Peters, P. J., Boretto, J., van Donselaar, E., Neri, A., D'Souza-Schorey, C., and Chavrier, P. (1999). EFA6, a Sec7 domain-containing exchange factor for ARF6, coordinates membrane recycling and cytoskeletal organization. EMBO J *18*, 1480–1491.

Franzuosoff, A., and Schekman, R. (1989). Functional compartments of the yeast Golgi apparatus are defined by the Sec7 mutation. EMBO J *8*, 2695–2702.

Gill, D. M. (1975). Involvement of nicotinamide adenine dinucleotide in the action of cholera toxin *in vitro*. Proc Natl Acad Sci *72*, 2064–2068.

Godi, A., Pertile, P., Meyers, R., Marra, P., Di Tullio, G., Iurisci, C., Luini, A., Corda, D., and De Matteis, M. A. (1999). ARF mediates recruitment of PtdIns-4-OH kinase-β and stimulates synthesis of PtdIns $(4,5)P_2$ on the Golgi complex. Nat Cell Biol *1*, 280–287.

Goldberg, J. (1998). Structural basis for activation of ARF GTPase: mechanisms of guanine nucleotide-exchange and GTP-myristoyl switching. Cell *95*, 237–248.

Greasley, S. E., Jhoti, H., Teahan, C., Solari, R., Fensome, A., Thomas, G. M. H., Cockcroft, S., and Bax, B. (1995). The structure of rat ADP-ribosylation factor-1 (ARF-1): complexed to GDP determined from two different crystal forms. Nat Struct Biol *2*, 797–806.

Guerrant, R. L., Steiner, T. S., Lima, A. A. M., and Bobak, D. A. (1999). How intestinal bacteria cause disease. J Infect Dis *179*, S331–S337.

Hammond, S. M., Jenco, J. M., Nakashima, S., Cadwallader, K., Gu, Q., Nozawa, Y., Prestwich, G. D., Frohman, M. A., and Morris, A. J. (1997). Characterization of two alternately spliced forms of phospholidase D1. Activation of the purified enzymes by phosphatidylinositol 4,5-bisphosphate, ADP-ribosylation factor, and Rho family monomeric GTP-binding proteins and protein kinase C-α. J Biol Chem *272*, 3860–3868.

Hazes, B., and Read, R. J. (1997). Accumulating evidence suggests that several AB-toxins subvert the endoplasmic reticulum-associated protein degradation pathway to enter target cells. Biochemistry 36, 11051–11054.

Helms, J. B., and Rothman, J. E. (1992). Inhibition by brefeldin A of a Golgi membrane enzyme that catalyzes exchange of guanine nucleotide bound to ARF. Nature 360, 352–354.

Hirst, J., Liu, W. W. Y., Bright, N. A., Totty, N., Seaman, M. N. J., and Robinson, M. S. (2000). A family of proteins with γ-adaptin and VHS domains that facilitate trafficking between the trans-Golgi network and the vacuole/lysosome. J Cell Biol 149, 67–80.

Honda, A., Nogami, M., Yokozeki, T., Nakamura, H., Watanabe, H., Kawamoto, K., Nakayama, K., Morris, A. J., and Frohman, M. A. (1999). Phosphatidylinositol 4-phosphate 5-kinase α is a downstream effector of the small G protein ARF6 in membrane ruffle formation. Cell 99, 521–532.

Jackson, C. L., and Cassanova, J. E. (2000). Turning on ARF: the Sec7 family of guanine nucleotide-exchange factors. Trends Cell Biol 10, 60–67.

Jobling, M. G., and Holmes, R. K. (2000). Identification of motifs in cholera toxin A1 polypeptide that are required for its interaction with human ADP-ribosylation factor 6 in a bacterial two-hybrid system. Proc Natl Acad Sci USA 97, 14662–14667.

Jones, D., Bax, B., and Cockcroft, S. (1999a). ADP-ribosylation factor GTPases in signal transduction and membrane traffic: independent functions? Biochem Soc Trans 27, 642–647.

Jones, D., Morgan, C., and Cockcroft, S. (1999b). Phospholipase D and membrane traffic: Potential roles in regulated exocytosis, membrane delivery and vesicle budding. Biochim Biophys Acta 1439, 229–244.

Jones, D. H., Morris, J. B., Morgan, C. P., Kondo, H., Irvine, R. F., and Cockcroft, S. (2000). Type 1 phosphatidylinositol 4-phosphate 5-kinase directly interacts with ADP-ribosylation factor 1 and is responsible for phosphatidylinositol 4,5 bisphosphate synthesis at the Golgi compartment. J Biol Chem 275, 13962–13966.

Kahn, R. A., Clark, J., Rulka, C., Stearns, T., Zhang, C.-J., Randazzo, P. A., Terui, T., and Cavenagh, M. (1995). Mutational analysis of Saccharomyces cerevisiae ARF1. J Biol Chem 270, 143–150.

Kahn, R. A., and Gilman, A. G. (1984). Purification of a protein cofactor required for ADP-ribosylation of the stimulatory regulatory component of adenylate cyclase by cholera toxin. J Biol Chem 259, 6228–6234.

Kahn, R. A., and Gilman, A. G. (1986). The protein cofactor necessary for ADP-ribosylation of Gs by cholera toxin is itself a GTP-binding protein. J Biol Chem 261, 7906–7911.

Kahn, R. A., Kern, F., Clark, J., Gelmann, E. P., and Rulka, C. (1991). Human ADP-ribosylation factors: A functionally conserved family of GTP-binding proteins. J Biol Chem 266, 2606–2614.

Kanoh, H., Williger, B.-T., and Exton, J. H. (1997). Arfaptin 1, a putative cytosolic target protein of ADP-ribosylation factor, is recruited to Golgi membranes. J Biol Chem 272, 5421–5429.

Kaper, J. B., Morris, J. G., and Levine, M. M. (1995). Cholera. Clin Microbiol Rev 8, 48–86.

Klarlund, J. K., Guilherme, A., Holik, J. J., Virbasius, J. V., Chawla, A., and Czech, M. P. (1997). Signaling by phosphoinositide-3,4,5-trisphosphate through proteins containing pleckstrin and Sec7 homology domains. Science *275*, 1927–1930.

Koch, R. (1984). An address on cholera and its bacillus. Br Med J *30*, 403–407; 453–459.

Kolanus, W., Nagel, W., Schiller, B., Zeitlman, L., Goda, S., Stockinger, H., and Seed, B. (1996). α L β 2 integrin/LFA-1 binding to ICAM-1 induced by cytohesin-1, a cytoplasmic regulatory molecule. Cell *86*, 233–242.

Lahiri, S. S. (2000). Bacterial toxins—An overview. J Nat Toxins *4*, 381–408.

Lee, C. M., Chang, P. P., Tsai, S. C., Adamik, R., Price, S. R., Kunz, B. C., Moss, J., Twiddy, E. M., Holmes, R. K. (1991). Activation of *Excherichia coli* heat-labile enterotoxins by native and recombinant adenosine diphosphate-ribosylation factors, 20-kD guanine nucleotide-binding proteins. J Clin Invest *87*, 1780–1786.

Lencer, W. I., Hirst, T. R., and Holmes, R. K. (1999). Membrane traffic and the cellular uptake of cholera toxin. Biochim Biophys Acta *1450*, 177–190.

Majoul, I., Sohn, K., Wieland, F. T., Pepperkok, R., Pizza, M., Hillemann, J., Soling, H. D. (1998). KDEL receptor (Erd2p)-mediated retrograde transport of the cholera toxin A subunit from the Golgi involves COPI, p23, and the COOH terminus of Erd2p. J Cell Biol *143*, 601–612.

Massenburg, D., Han, J.-S., Liyanage, M., Patton, W. A., Rhee, S. G., Moss, J., and Vaughan, M. (1994). Activation of rat brain phospholipase D by ADP-ribosylation factors 1, 5, and 6: Separation of ADP-ribosylation factor-dependent and oleate-dependent enzymes. Proc Natl Acad Sci USA *91*, 11718–11722.

Meacci, E., Tsai, S. C., Adamik, R., Moss, J., and Vaughan, M. (1997). Cytohesin-1, a cytosolic guanine nucleotide-exchange protein for ADP-ribosylation factor. Proc Natl Acad Sci USA *94*, 1745–1748.

Mekalanos, J. J., Collier, R. J., and Romig, W. R. (1979). Enzymatic activity of cholera toxin: II. Relationships to proteolytic processing, disulfude bond reduction, and submit composition. J Biol Chem *254*, 5855–5861.

Ménétrey, J., Macia, E., Pasqualato, S., Franco, M., and Cherfils, J. (2000). Structure of ARF6-GDP suggests a basis for guanine nucleotide exchange factors specificity. Nat Struct Biol *7*, 466–469.

Merritt, E. A., and Hol, W. G. (1995). AB$_5$ toxins. Curr Opin Struct Biol *5*, 165–171.

Merritt, E. A., Kuhn, P., Sarfaty, S., Erbe, J. L., Holmes, R. K., and Hol, W. G. (1998). The 1.25 Å resolution refinement of the cholera toxin B-pentamer: evidence of peptide backbone strain at the receptor-binding site. J Mol Biol *282*, 1043–1059.

Merritt, E. A., Pronk, S. E., Sixma, T. K., Kalk, K. H., van Zanten, B. A. M., and Hol, W. G. J. (1994a). Structure of partially-activated *E. coli* heat-labile enterotoxin (LT): at 2.6 Å resolution. FEBS Lett *337*, 88–92.

Merritt, E. A., Sarfaty, S., Jobling, M. G., Chang, T., Holmes, R. K., Hirst, T. R., and Hol, W. G. (1997). Structural studies of receptor binding by cholera toxin mutants. Protein Sci *6*, 1516–1528.

Merritt, E. A., Sarfaty, S., Pizza, M., Rappuoli, R., and Hol, W. G. J. (1995). Mutation of a buried residue causes loss of activity but no conformational change in the heat-labile enterotoxin of *Escherichia coli*. Struct Biol *2*, 269–272.

Merritt, E. A., Sarfaty, S., Van Den Akker, F., L'Hoir, C., Martial, J. A., and Hol, W. G. J. (1994b). Crystal structure of cholera toxin B-pentamer bound to receptor G$_{M1}$ pentasaccharide. Protein Sci *3*, 166–175.

Merritt, E. A., Sixma, T. K., Kalik, K. H., van Zanten, B. A. M., and Hol, W. G. J. (1994c). Galactose-binding site in *Escherichia coli* heat-labile enterotoxin (LT) and cholera toxin (CT). Mol Microbiol *13*, 745–753.

Moorman, J., Patton, W. A., Moss, J., and Bobak, D. A. (1999). Structure and function of Ras superfamily small GTPases. *In*: J. Moss (ed) *LAM and Other Diseases Characterized by Smooth Muscle Proliferation*. New York, Marcel Dekker, pp. 441–478.

Morinaga, N., Kaihou, Y., Vitale, N., Moss, J., and Noda, M. (2001). Involvement of ADP-ribosylation factor 1 in cholera toxin-induced morphological changes in Chinese hamster ovary cells. J Biol Chem *276*, 22838–22843.

Morinaga, N., Moss, J., and Vaughan, M. (1997). Cloning and expression of a cDNA encoding a brovine brain brefeldin A-sensitive guanine nucleotide-exchange protein for ADP-ribosylation factor. Proc Natl Acad Sci USA *94*, 12926–12931.

Morinaga, N., Tsai, S.-C., Moss, J., and Vaughan, M. (1996). Isolation of a brefeldin A-inhibited guanine nucleotide-exchange protein for ADP-ribosylation factor (ARF) 1 and ARF3 that contains a Sec7-like domain. Proc Natl Acad Sci USA *93*, 12856–12860.

Moss, J., Manganiello, V. C., and Vaughan, M. (1976). Hydrolysis of nicotinamide adenine dinucleotide by choleragen and it's a protomer: Possible role in the activation of adenylate cyclase. Proc Natl Acad Sci USA *73*, 4424–4427.

Moss, J., Stanley, S. J., Vaughan, M., and Tsuji, T. (1993). Interaction of ADP-ribosylation factor with *Escherichia coli* enterotoxin that contains an inactivating lysine-112 substitution. J Biol Chem *268*, 6383–6387.

Moss, J., and Vaughan, M. (1977). Mechanism of action of choleragen: Evidence for ADP-ribosyltransferase activity with arginine as an acceptor. J Biol Chem *252*, 2455–2457.

Moss, J., and Vaughan, M. (1998). Molecules in the ARF orbit. J Biol Chem *273*, 21431–21434.

Moss, J., and Vaughan, M. (1999). Activation of toxin ADP-ribosyltransferases by eukaryotic ADP-ribosylation factors. Mol Cell Biochem *193*, 153–157.

Murayama, T., Tsai, S.-C., Adamik, R., Moss, J., and Vaughan, M. (1993). Effects of temperature on ADP-ribosylation factor stimulation of cholera toxin activity. Biochemistry *32*, 561–566.

Noda, M., Tsai, S.-C., Adamik, R., Moss, J., and Vaughan, M. (1990). Mechanism of cholera toxin activation by a guanine nucleotide-dependent 19-kDa protein. Biochim Biophys Acta *1034*, 195–199.

Ogasawa, M., Kim, S.-C., Adamik, R., Togawa, A., Ferrans, V. J., Takeda, K., Kirby, M., Moss, J., and Vaughan, M. (2000). Similarities in function and gene structure of cytohesin-4 and cytohesin-1, guanine nucleotide-exchange proteins for ADP-ribosylation factors. J Biol Chem *275*, 3221–3230.

Ooi, C. E., Dell'Angelica, E. C., and Bonifacino, J. S. (1998). ADP-ribosylation factor 1 (ARF1) regulates recruitment of the AP-3 adaptor complex to membranes. J Cell Biol *142*, 391–402.

Oppenheimer, N. J. (1978). Structural determination and sterospecificity of the choleragen-catalyzed reaction of NAD$^+$ with guanidines. J Biol Chem *253*, 4907–4910.

Orlandi, P. A., and Fishman, P. H. (1993). Orientation of cholera toxin bound to target cells. J Biol Chem *268*, 17038–17044.

Orlandi, R. A., and Fishman, P. H. (1998). Filipin-dependent inhibition of cholera toxin: evidence for toxin internalization and activation through caveolae-like domains. J Cell Biol *141*, 905–915.

Palmer, D. J., Helms, J. B., Beckers, C. J. M., Orci, L., and Rothman, J. E. (1993). Binding of coatomer to Golgi membranes requires ADP-ribosylation factor. J Biol Chem *268*, 12083–12089.

Paris, S., Béraud-Dufour, S., Robineau, S., Bigay, J., Antonny, B., Chabre, M., and Chardin, P. (1997). Role of protein-phospholipid interactions in the activation of ARF1 by the guanine nucleotide exchange factor ARNO. J Biol Chem *272*, 22221–22226.

Patton, W. A., Vitale, N., Moss, J., and Vaughan, M. (2000). Mechanism of cholera toxin action: ADP-ribosylation factors as stimulators of cholera toxin-catalyzed ADP-ribosylation and effectors in intracellular vesicular trafficking events. *In*: K. Aktories and I. Just (eds): *Handbook of Experimental Pharmacology*, Vol 145, Bacterial Protein Toxins, Berlin, Springer-Verlag, pp. 133–165.

Peyroche, A., Paris, S., and Jackson, C. L. (1996). Nucleotide exchange on Arf mediated by yeast Gea1 protein. Nature *384*, 479–481.

Poon, P. P., Wang, X., Rotman, M., Huber, I., Cukierman, E., Cassel, D., Singer, R. A., and Johnston, G. C. (1996). *Saccharomyces cerevisiae* Gcs1 is an ADP-ribosylation factor GTPase-activating protein. Proc Natl Acad Sci USA *93*, 10074–10077.

Premont, R. T., Claing, A., Vitale, N., Freeman, J. L. R., Pitcher, J. A., Patton, W. A., Moss, J., Vaughan, M., and Lefkowitz, R. J. (1998). β-2-adrenergic receptor regulation by GIT1, a G protein-coupled receptor kinase-associated ADP-ribosylation factor GTPase-activating protein. Proc Natl Acad Sci USA *95*, 14082–14087.

Randazzo, P. A. (1997). Resolution of two ADP-ribosylation factor 1 GTPase-activating proteins from rat liver. Biochem J *324*, 413–419.

Randazzo, P. A., Andrade, J., Miura, K., Brown, M. T., Long, Y. O., Stauffer, S., Roller, P., and Cooper, J. A. (2000). The ARF GTPase-activating protein ASAP1 regulates the actin cytoskeleton. Proc Natl Acad Sci USA *97*, 4011–4016.

Randazzo, P. A., Terui, T., Sturch, S., Fales, H. M., Ferrige, A. G., and Kahn, R. A. (1995). The myristoylated amino terminus of ADP-ribosylation factor 1 is a phospholipid- and GTP-sensitive switch. J Biol Chem *270*, 14809–14815.

Regazzi, R., Ullrich, S., Kahn, R. A., and Wolheim, C. B. (1991). Redistribution of ADP-ribosylation factor during stimulation of permeabilized cells with GTP analogues. Biochem J *275*, 639–644.

Renault, L., Christova, P., Guibert, B., Pasqualato, S., and Cherfils, J. (2002). Mechanism of domain closure of Sec7 domains and role in BFA sensitivity. Biochemistry *41*, 3605–3612.

Rodighiero, C., Aman, A. T., Kenny, M. J., Moss, J., Lencer, W. I., and Hirst, T. R. (1999). Structural basis for the differential toxicity of cholera toxin and *Escherichia coli* heat-labile enterotoxin. J Biol Chem *274*, 3962–3969.

Roth, M. G. (1999). Snapshots of ARF1: Implications for mechanisms of activation and inactivation. Cell *97*, 149–152.

Ryan, E. T., and Calderwood, S. B. (2000). Cholera vaccines. Clin Infect Immun *31*, 561–565.

Ryan, E. T., and Calderwood, S. B. (2001). Cholera vaccines. J Travel Med *8*, 82–91.

Sata, M., Donaldson, J. G., Moss, J., and Vaughan, M. (1998). Brefeldin A-inhibited guanine nucleotide-exchange activity of Sec7 domain from yeast Sec7 with yeast and mammalian ADP-ribosylation factors. Proc Natl Acad Sci USA *95*, 4204–4208.

Scheffzek, K., Ahmadian, M. R., Kabsch, W., Wiesmuller, L., Lautwein, A., Schmitz, F., and Wittinghofer, A. (1997). The Ras-Ras GAP complex: structural basis for GTPase activation and its loss in oncogenic Ras mutants. Science *277*, 333–338.

Serafini, T., Orci, L., Amherdt, M., Brunner, M., Kahn, R. A., and Rothman, J. E. (1991). ADP-ribosylation factor is a subunit of the coat of Golgi-derived COP-coated vesicles: A novel role for a GTP-binding protein. Cell *67*, 239–253.

Shin, O. H., Couvillon, A. D., and Exton, J. H. (2001). Arfophilin is a common target of both class II and class III ADP-ribosylation factors. Biochemistry *40*, 10846–10852.

Shin, O. H., Ross, A. H., Mihai, I., and Exton, J. H. (1999). Identification of Arfophilin, a target protein for GTP-bound Class II ADP-ribosylation factors. J Biol Chem *274*, 36609–36615.

Shinoda, S. (1999). Protein toxins produced by pathogenic *Vibrios*. J Nat Tox *8*, 259–269.

Sixma, T. K., Kalk, K. H., Van Zanten, B. A. M., Dauter, Z., Kingma, J., Witholt, B., and Hol, W. G. J. (1993). Refined structure of *Escherichia coli* heat-labile enterotoxin, a close relative of cholera toxin. J Mol Biol *230*, 890–918.

Snider, D. P. (1995). The mucosal adjuvant activities of ADP-ribosylating bacterial enterotoxins. Crit Rev Immunol *15*, 317–348.

Stamnes, M. A., and Rothman, J. E. (1993). The binding of AP-1 clathrin adaptor particles to Golgi membranes requires ADP-ribosylation factor, a small GTP-binding protein. Cell *73*, 999–1005.

Stearns, T., Willingham, M. C., Botstein, D., and Kahn, R. A. (1990). ADP-ribosylation factor is functionally and physically associated with the Golgi complex. Proc Natl Acad Sci USA *87*, 1238–1242.

Steinmann, T., Geldner, N., Grebe, M., Mangold, S., Jackson, C. L., Paris, S., Gälweiler, L., Palme, K., and Jürgens, G. (1999). Coordinated polar localization of auxin efflux carrier PIN1 by GNOM ARF GEF. Science *286*, 316–318.

Stevens, L. A., Moss, J., Vaughan, M., Pizza, M., and Rappuoli, R. (1999). Effects of site-directed mutagenesis of *Escherichia coli* heat-labile enterotoxin on ADP-ribosyltransferase activity and interaction with ADP-ribosylation factors. Infect Immun *67*, 259–265.

Takai, Y., Sasaki, T., and Matozaki, T. (2001). Small GTP-binding proteins. Physiol Rev *81*, 153–208.

Tanigawa, G., Orci, L., Amherdt, M., Ravazzola, M., Helms, J., and Rothman, J. (1993). Hydrolysis of bound GTP by ARF protein triggers uncoating of Golgi-derived COP-coated vesicles. J Cell Biol *123*, 1365–1371.

Tarricone, C., Xiao, B., Justin, N., Walker, P. A., Rittinger, K., Gamblin, S. J., Smerdon, S. J. (2001). The structural basis of arfaptin-mediated cross-talk between Rac and Arf signalling pathways. Nature *411*, 215–219.

Teal, S., Hsu, V., Peters, P., Klausner, R., and Donaldson, J. (1994). An activating mutation in ARF1 stabilizes coatomer binding to Golgi membranes. J Biol Chem *269*, 3135–3138.

Tomasi, M., Battistini, A., Araco, A., Roda, G., and D'Agnolo, G. (1979). The role of the reactive disulfide bond in the interaction of cholera-toxin function regions. Eur J Biochem *93*, 621–627.

Traub, L. M., Ostrom, J. A., and Kornfeld, S. (1993). Biochemical dissection of AP-1 recruitment onto Golgi membranes. J Cell Biol *123*, 561–573.

Tsai, S.-C., Noda, M., Adamik, R., Chang, P. P., Chen, H.-C., Moss, J., and Vaughan, M. (1988). Stimulation of choleragen enzymatic activities by GTP and two soluble proteins purified from bovine brain. J Biol Chem *263*, 1768–1772.

Tsai, S.-C., Noda, M., Adamik, R., Moss, J., and Vaughan, M. (1987). Enhancement of choleragen ADP-ribosyltransferase activities by guanyl nucleotides and a 19-kDa membrane protein. Proc Natl Acad Sci USA *84*, 5139–5142.

Tsuji, T., Inoue, T., Miyama, A., Noda, M. (1991). Glutamic acid-112 of the A subunit of heat-labile enterotoxin from enterotoxigenic *Escherichia coli* is important for ADP-ribosyltransferase activity. FEBS Letters *291*, 319–321.

Tsuji, T., Inoue, T., Miyama, A., Okamoto, K., Honda, T., and Miwatani, T. (1990). A single amino acid substitution in the A subunit of *Escherichia coli* enterotoxin results in a loss of its toxic activity. J Biol Chem *265*, 22520–22525.

Turvill, J. L., Mourad, F. H., and Farthing, M. J. (1998). Crucial role for 5-HT in cholera toxin but not *Escherichia coli* heat-labile enterotoxin-intestinal secretion in rats. Gastroenterology *115*, 883–890.

Van den Akker, F., Feil, I. K., Roach, C., Platas, A. A., Merritt, E. A., and Hol, W. G. (1997). Crystal structure of heat-labile enterotoxin from *Escherichia coli* with increased thermostability introduced by an engineered disulfide bond in the A subunit. Protein Sci *6*, 2644–2649.

Verlinde, C. I. M. J., Merritt, E. A., Van den Akker, F., Kim, H., Feil, I., Delboni, L. F., Mande, S., Sarfaty, S., Petra, P. H., and Hol, W. G. J. (1994). Protein crystallography and infectious diseases. Protein Sci *3*, 1670–1686.

Vitale, N., Patton, W. A., Moss, J., Vaughan, M., Lefkowitz, P. J., and Premont, R. T. (2000). GIT proteins, a novel family of phosphatidylinositol 3,4,5-triphosphate-stimulated GTPase-activating proteins for ARF6. J Biol Chem *275*, 13901–13906.

Waite, M. (1999). The PLD superfamily: Insights into catalyst. Biochim Biophys Acta *1439*, 187–197.

Wakeham, D. E., Ybe, J. A., Brodsky, F. M., and Hwang, P. K. (2000). Molecular structures of proteins involved in vesicle coat formation. Traffic *1*, 393–398.

Walker, M. W., Bobak, D. A., Tsai, S.-C., Moss, J., and Vaughan, M. (1992). GTP but not GDP analogues promote association of ADP-ribosylation factors, 20-kDa protein activators of cholera toxin, with phospholipids and PC-12 cell membranes. J Biol Chem *267*, 3230–3235.

Wieland, F., and Harter, C. (1999). Mechanisms of vesicle formation: Insights from the COP system. Curr Opin Cell Biol *11*, 440–446.

Wolf, A. A., Jobling, M. G., Wimer-Mackin, S., Ferguson-Maltzman, M., Madara, J. L., Holmes, R. K., and Lencer, W. I. (1998). Ganglioside structure dictates signal transduction by cholera toxin and association with caveolae-like membrane domains in polarized epithelia. J Cell Biol *141*, 917–927.

Yamamoto, T., Nakazawa, T., Miyata, T., Kaji, A., and Yokota, T. (1984). Evolution and structure of two ADP-ribosylation enterotoxins, *Escherichia coli* heat-labile enterotoxin and cholera toxin. FEBS Lett *169*, 241–246.

Zhang, C.-J., Rosenwald, A. G., Willingham, M. C., Skuntz, S., Clark, J., and Kahn, R. (1994). Expression of a dominant allele of human ARF1 inhibits membrane traffic in vivo. J Cell Biol *124*, 289–300.

Zhang, G.-F., Patton, W. A., Lee, F.-J. S., Liyanage, M., Han, J.-S., Rhee, S. G., Moss, J., and Vaughan, M. (1995a). Different ARF domains are required for the activation of cholera toxin and phospholipase D. J Biol Chem *270*, 21–24.

Zhang, R.-G., Scott, D. L., Westbrook, M. L., Nance, S., Spangler, B. D., Spangler, B. D., Shipley, G. G., and Westbrook, E. M. (1995b). The three-dimensional crystal structure of cholera toxin. J Mol Biol *9*, 1323–1337.

Zhu, Y., Traub, L. M., and Kornfeld, S. (1998). ADP-ribosylation factor 1 transiently activates high-affinity adaptor protein complex AP-1 binding sites on Golgi membranes. Mol Biol Cell *9*, 1323–1337.

Zhu, X., and Kahn, R. A. (2001). The *Escherichia coli* heat-labile toxin binds to Golgi membranes and alters Golgi and cell morphologies using ADP-ribosylation factor-dependent processes. J Biol Chem *276*, 25014–25021.

Zhu, X., Kim, E., Boman, A. L., Hodel, A., Cieplak, W., and Kahn, R. A. (2001). ARF binds the C-terminal region of the *Escherichia coli* heat-labile toxin (LTA$_1$) and competes for the binding of LTA$_2$. Biochemistry *40*, 4560–4568.

CHAPTER 9

MOLECULAR BASIS OF SEVERE COMBINED IMMUNODEFICIENCY: LESSONS FROM CYTOKINE SIGNALING PATHWAYS

ROBERTA VISCONTI, FABIO CANDOTTI, and
JOHN J. O'SHEA
Istituto di Endocrinologia ed Oncologia Sperimentale "G. Salvatore"
del Consiglio Nazionale delle Ricerche, Napoli, Italy; Genetics and
Molecular Biology Branch, National Human Genome Research
Institute, National Institutes of Health, Bethesda, Maryland; Molecular
Immunology and Inflammation Branch, National Institute of Arthritis,
Musculoskeletal and Skin Diseases, National Institutes of Health,
Bethesda, Maryland

INTRODUCTION

The immune system is a potent, tightly regulated, and highly specific defense apparatus higher organisms have developed to protect themselves from infectious organisms. Nonetheless, numerous examples indicate that mammalian development can be entirely normal in the absence of the immune system; indeed, before birth individuals with even very severe immunodeficiencies are developmentally unaffected. However, after birth the failure of one or more of its components can have severe, sometimes fatal, consequences. Thus it is not surprising that more than 95 different primary immunodeficiencies have been identified, encompassing defects in lymphocytes, phagocytes, and complement proteins. Perhaps more impressively, though, within the past few years more than 70 separate genes have been identified whose mutation causes immunodeficiency. These discoveries have been made by using both candidate gene and positional cloning approaches. In some cases, the generation of gene-targeted mice preceded the identification of human mutations, whereas in other cases, the reverse was true.

Signal Transduction and Human Disease, Edited by Toren Finkel and
J. Silvio Gutkind
ISBN 0-471-02011-7   Copyright © 2003 John Wiley & Sons, Inc.

Classically, primary immunodeficiencies are grouped on the basis of the transmission modalities and/or of the main abnormality (Fischer et al., 1997; Rosen et al., 1995; WHO Scientific Group, 1995). We can therefore recognize syndromes that affect specific or adaptive immunity, due to defects in lymphocyte maturation or differentiation, and syndromes that affect innate immunity, which include deficits of various host defense mechanisms and comprise abnormalities of complement proteins or natural killer (NK) and phagocytes. The defects in specific immunity can be further subclassified on the basis of the mainly involved cell type (Fischer, 2001; Buckley, 2000). A clear distinction, however, is not always possible because of the interactions among immune components. Moreover, the clinical features and the immune defects often overlap among groups so that a single syndrome can be caused by mutations in several genes and defects in a single gene manifest themselves with several quite different clinical phenotypes. Nonetheless, recent outstanding advances in molecular biology have facilitated classification of immunodeficiencies on the basis of the type of genetically altered molecules involved and of their function in the immune cells (Table 9.1).

Importantly, the immunodeficiency syndromes can paradigmatically highlight how the understanding of the molecular defects causing these diseases has helped the comprehension of some of immune system key functions. In fact, the systematic study of naturally occurring mutations in patients with immunodeficiency syndromes has helped in elucidating the basic mechanisms of antigen receptors and cytokine signal transduction. Simultaneously, the improved knowledge of cytokine and antigen receptor action has resulted in a deeper comprehension of the immunodeficiencies' pathophysiological mechanisms, opening a new perspective in the diagnosis, genetic counseling, and therapy for these diseases. Most importantly, though, therapy for these disorders, ranging from replacement therapy to bone marrow transplantation and gene therapy, has also moved at a rapid pace (Buckley and Schiff, 1991; Porta and Friedrich, 1998; Buckley et al., 1993; Buckley et al., 1999; Hacein-Bey-Abina et al., 2002; Cavazzana-Calvo et al., 2001; Fischer et al., 2000; Cournoyer and Caskey, 1993). This field, therefore, provides outstanding examples of the power of molecular medicine, with tremendous opportunities for integration between basic and clinical science. We focus on this aspect, providing, in particular, examples in which mutations affecting signaling by γ_c binding cytokines result in immunodeficiency. Several excellent reviews provide a comprehensive discussion of the genetic and pathophysiological bases and of the clinical heterogeneity of primary immunodeficiencies (Fischer, 2001; Buckley, 2000). We therefore highlight the insights and the recent advances, including the implications for gene therapy, with particular emphasis on mutations in molecules involved in γ_c cytokine signaling pathways. However, the lessons learned from these disorders are applicable to most primary immunodeficiencies.

TABLE 9.1. Mechanisms Underlying Primary Immunodeficiencies

Disease or Syndrome	Mutant Gene(s)	Defect or Phenotype
Deficiencies of antigen receptors and signaling		
Defects of genes of the CD3 complex	CD3 γ or ε chain	T cell deficiency
Autosomal recessive agammaglobulinemia	μ Chain; surrogate light chain; Igα; BLNK	Absence of B cells
Selective Ig deficiency	Ig heavy chains	Absence of Ig isotypes
Ig with only λ chains	κ Chain	κ Chain deficiency
X-linked recessive agammaglobulinemia	Btk	Absence of B cells
Autosomal recessive SCID	p56^{lck}; CD45	T$^+$ B$^+$ NK$^+$ SCID
MHC class I antigen deficiency	TAP1; TAP2	Impaired MHC class I expression and CD8$^+$ T cell development
MHC class II antigen deficiency	RFXAP; CIITA; RFX5; RFXANK	Impaired MHC class I expression and CD4$^+$ T cell development
CD8 lymphopenia	Zap-70	CD8$^+$ T cell deficiency
Deficiencies of cytokine receptors and signaling		
X-linked SCID	Common cytokine receptor γ chain	T$^-$ B$^+$ NK$^-$ SCID
Autosomal recessive SCID	IL-7 receptor α chain	T$^-$ B$^+$ NK$^+$ SCID
CD25 deficiency	IL-2 receptor α chain	Lymphoproliferative T cell deficiency with autoimmunity
Autosomal recessive SCID	Jak3	T$^-$ B$^+$ NK$^-$ SCID
Atypical mycobacterial infections	IFN-γR1; IFN-γR2; IL-12p40; IL-12Rβ1; Stat1	Atypical mycobacterial, Salmonella viral infection environmental mycobacteria
Hyper IgM syndromes		
HIGM1	CD40L	Very low IgG and IgA, normal to increased IgM. Absent germinal centers
HIGM2	AID	Very low IgG and IgA, normal to increased IgM. Enlarged germinal centers. Lymphoid hyperplasia
HIGM3	CD40	Phenocopy of HIGM1
HIGM-ED	NEMO	Hypogammaglobulinemia, ectodermic dysplasia. T and NK cell defects
Autoimmune lymphoproliferative syndrome (ALPS)	Fas Fas-ligand Caspase 10	Apoptosis defect. Expansion of CD4-/CD8- TcR+ T cells, CD3+ T cells and B cells
Deficiencies in accessory and adhesion molecules		
X-linked lymphoproliferative disease	SH2D1A	Lymphoproliferative disease after EBV infection
Wiskott–Aldrich syndrome	WASP	Immunodeficiency with thrombocytopenia and eczema
Leukocyte adhesion deficiency I	β2 integrin	Hyperleukocytosis-recurrent cutaneous and mucosal infections
Leukocyte adhesion deficiency II	GDP-fucose transporter	Infections, Short Stature, Mental Retardation
Deficiencies in DNA rearrangement and repair		
Omenn syndrome; defective V(D)J recombination	Rag1; Rag2	T$^-$ B$^-$ NK$^+$ SCID
SCID with radiosensitivity	Artemis	T$^-$ B$^-$ SCID
Ataxia telangiectasia	ATM	Combined immunodeficiency with cerebellar ataxia and oculocutaneous teleangiectasias
Nijmegen breakage syndrome	Nibrin DNA ligase IV	Immunodeficiency SCID
Metabolic defects		
Autosomal recessive SCID	Adenosine deaminase	T$^-$ B$^-$ NK$^-$ SCID

γ_c AND THE MOLECULAR BASIS OF SEVERE COMBINED IMMUNODEFICIENCY

Lymphocyte development is a multistep process by which mature B and T cells are generated from undifferentiated progenitors. This process is tightly regulated by a panoply of cytokines so that defects at various stages result in immunodeficiency. Disorders affecting relatively undifferentiated progenitor cells lead to defects in both T and B cell functions. This can result in inadequate humoral and cell-mediated immune responses and, therefore, in the most severe forms of primary immunodeficiencies, termed severe combined immunodeficiency (SCID). One cause of SCID associated with virtual lack of circulating mature T and B lymphocytes (T⁻ B⁻ SCID) is due to mutations of either RAG1 or RAG2, the lymphoid-restricted recombinase activating genes 1 and 2, which mediate the rearrangement of the immunoglobulin and T cell antigen receptor genes (Mombaerts et al., 1992; Shinkai et al., 1992; Schwarz et al., 1996). Moreover, a newly identified player in antigen receptor rearrangements, ARTEMIS, has been recently linked to a subgroup of T⁻ B⁻ SCID patients with increased radiosensitivity (Moshous et al., 2001). Additionally, SCID may also be due to metabolic defects such as adenosine deaminase deficiency, resulting in declining numbers of both T and B cells. Other defects are in antigen presentation or signaling via the T cell receptor. For instance, Zap-70 mutations also cause SCID, but this is characterized by the absence of CD8⁺ T cells.

Approximately 50% of SCID patients lack NK and peripheral, mature T cells, but present with a normal or increased proportion of B cells; therefore, they are also described as suffering from T⁻ B⁺ NK⁻ SCID. However, B cells are not functioning properly and no specific antibodies are produced on antigenic challenge. This could be explained in part by the lack of the T cell helper action; however, an intrinsic B cell defect may also exist, because B cell function typically does not improve after successful bone marrow transplantation with T cell engraftment. Typically, the disease has an early and severe onset, presenting with frequent episodes of pneumonia, diarrhea, otitis, and cutaneous infections, as well as opportunistic infections. Unless promptly diagnosed and treated, the disease is often fatal within the first years of life.

It has been recognized for some time that most T⁻ B⁺ NK⁻ SCID patients are males, suggesting an X-linked inheritance of the disease. In 1993 Noguchi et al. first demonstrated that mutation of the interleukin-2 receptor (IL-2R)γ chain was responsible for this disease (Noguchi et al., 1993b), now termed SCIDX1. This was later confirmed by others (Puck et al., 1993b). The gene maps to Xq13, the locus associated with SCIDX1 (Puck et al., 1993a). Mutations identified in the IL-2Rγ in different SCIDX1 patients span the entire gene and include nonsense, missense, and frameshift mutations as well as deletions of entire exons (Leonard, 2000), but with a few notable exceptions phenotype/genotype correlations have not been convincingly demonstrated. Many mutations lead to a nonfunctional IL-2Rγ protein, unable to properly transduce IL-

2-mediated signals. However, it was immediately clear that defects in IL-2 signaling did not explain the clinical manifestations of patients with SCIDX1. That is, IL-2 knockout mice exhibited defects in T cell homeostasis and autoimmunity but also indicated that IL-2 is dispensable for normal thymic development (Schorle et al., 1991). Moreover, patients with primary immunodeficiency caused by defective IL-2 production and IL-2R mutations were identified and they also had normal T cell development (DiSanto et al., 1990; Weinberg and Parkman, 1990; Sharfe et al., 1997b). These observations suggested that defective IL-2 signaling was not responsible for the complex phenotype of the SCIDX1 patients, and it was hypothesized that IL-2Rγ was a common receptor subunit for other cytokines, at least one of which was necessary for early T cell development. Indeed, it was soon demonstrated that the IL-2Rγ is also a component of the IL-4, IL-7, IL-9, IL-15, and IL-21 receptors (Kondo et al., 1993; Russell et al., 1993; Noguchi et al., 1993a; Kondo et al., 1994; Kimura et al., 1995; Giri et al., 1994; Asao et al., 2001), and, therefore, it was renamed "common cytokine receptor γ chain" (γ_c cytokines) (Leonard et al., 1994); the cytokines binding this receptor are also termed "γ_c cytokines. " γ_c belongs to the type I cytokine receptor superfamily and contains a WSXWS motif and four highly conserved cysteine residues in the extracellular domain. It is constitutively expressed on B, T, NK, and myeloid cells, as well as on erythroblasts (Sugamura et al., 1996). IL-2, IL-4, IL-7, IL-9, IL-15, and IL-21 bind γ_c in association with at least one other chain that specifically binds the various ligands (Gadina et al., 2001). Remarkably, IL-15R shares with IL-2R not only the γ_c chain but also the IL-2Rβ chain (Giri et al., 1994), explaining why the two cytokines have many overlapping biological activities.

Thus, as we will discuss in detail, a profound deficiency in signaling by γ_c cytokines, molecules with redundant but also specific functions, clearly explains the various defects observed in SCIDX1 patients (Sugamura et al., 1996).

IL-2 is produced solely by activated T lymphocytes and its synthesis is rapidly and potently induced after antigen presentation to resting T cells. IL-2 is the major T cell growth factor (Morgan et al., 1976) but has also other important functions, including enhancement of immunoglobulin synthesis in B cells (Mingari et al., 1984) and augmentation of cytolytic activity of NK and lymphokine-activated killer (LAK) cells (Siegel et al., 1987). Paradoxically, IL-2 has also been shown to be indispensable for constraining lymphoid growth by promoting activation-induced cell death and maintaining peripheral tolerance. As indicated above, IL-2 knockout mice T cells develop normally but have disturbed peripheral homeostasis, leading to lymphoadenopathy and splenomegaly due to uncontrolled proliferation of CD4[+] T cells and to severe autoimmune phenomena, as hemolytic anemia and ulcerative colitis (Schorle et al., 1991). Remarkably, the same autoimmune manifestations have also been observed in humans and mice with mutations of the IL-2Rα and IL-2Rβ subunits (Sharfe et al., 1997a; Willerford et al., 1995; Suzuki et al., 1995), arguing that the autoimmune diseases observed in SCIDX1 patients are the consequence of the lack of IL-2-mediated signals (Nelson, 2002).

IL-4 is produced by activated $CD4^+$ T cells, mast cells, and basophils (Paul, 1997). IL-4 was initially identified as the major growth factor for B cells (Yokota et al., 1986), in which it is also essential for immunoglobulin class switch, enhancing the production and secretion of IgG_1 and IgE (Lee et al., 1986). Moreover, IL-4 induces expression of class II major histocompatibility complex (MHC) molecules on B cells. Finally, it has been demonstrated that IL-4 has also the ability of promoting T and mast cell proliferation (Paul, 1997). However, IL-4 knockout mice also have peripheral mature T cells in the normal ranges (Kuhn et al., 1991). Thus, as clearly demonstrated by the knockout mice phenotypes (Schorle et al., 1991; Kuhn et al., 1991) and by the evidence that IL-2-deficient patients also have T cells in the periphery (DiSanto et al., 1990; Weinberg and Parkman, 1990), neither IL-2 nor IL-4 is necessary for T cell development.

These findings suggest that the complete lack of T cells in SCIDX1 patients is due to another cytokine. First identified as a pre-B cell growth factor (Namen et al., 1988a; Namen et al., 1988b), IL-7 is now recognized to be required for thymocyte growth, survival, and differentiation (Murray et al., 1989; Watson et al., 1989). Indeed, the phenotype of IL-7- and IL-7Rα-deficient mice (von Freeden-Jeffry et al., 1995; Peschon et al., 1994) substantiated that IL-7 is necessary for T cell development and the production of early T cell progenitors in the thymus. $IL-7^{-/-}$ and $IL-7R\alpha^{-/-}$ mice exhibit a 20-fold reduction in thymocyte number, with a block at the double negative stage, resulting in a profound deficiency in the peripheral T cell pool. Moreover, $T^- B^+ NK^+$ SCID patients harboring mutations in the gene encoding the IL-7Rα chain have recently been described (Puel et al., 1998), demonstrating the essential functions of IL-7 in humans. Of note, lately, it has been shown that IL-7 plays an essential role in regulating not only thymocyte but also mature T cell functions, as it is required for mediating the homeostasis of naive and memory $CD8^+$ T cells (Schluns et al., 2000).

Because IL-4 and IL-7 are both important for B cell functions, defective IL-4 and IL-7 signaling might be a contribution to the defective B cell function observed in SCIDX1 patients. It is interesting to note that SCIDX1 B cells are able to produce IgE in response to IL-4 stimulation because of γ_c-independent IL-4 signaling (He and Malek, 1995; Obiri et al., 1995; Oakes et al., 1996; Matthews et al., 1997).

Because both IL-7Rα-deficient mice and humans have NK cells (Peschon et al., 1994; Puel et al., 1998), the NK cell deficiency observed in SCIDX1 patients has been attributed to defective IL-15-induced signaling. IL-15 is a key cytokine in promoting NK cell differentiation and activation (Fehniger and Caligiuri, 2001; Carson et al., 1994; Cavazzana-Calvo et al., 1996; Leclercq et al., 1996; Puzanov et al., 1996), and its nonredundant role in inducing NK cell differentiation was evident in IL-15Rα knockout mice, which completely lack NK cells (Lodolce et al., 1998).

Recently, the physiological functions of another γ_c cytokine have been explored in vivo. IL-21R knockout mice have been generated, demonstrating that IL-21 limits NK cell responses while promoting antigen-specific T cell activation (Kasaian et al., 2002). These data suggest that IL-21 might function in vivo as a cytokine regulating the transition from innate to adaptive immunity.[note added in proof]

In conclusion, the complex phenotype of SCIDX1 patients can be fully explained by defective signaling by different cytokines sharing γ_c in their membrane receptors. It is, however, worth mentioning that a subgroup of patients has been described with γ_c mutations leading to a less severe degree of deficiency in cellular and humoral immunity than that seen in SCIDX1 (Schmalstieg et al., 1995). These patients are described as suffering X-linked combined immunodeficiency disease (XCID). They clearly show that the clinical phenotype associated with primary immunodeficiencies can be amazingly unpredictable. Thus more than one gene can cause similar immunodeficiency, and a single gene can have rather different clinical presentations.

Today about 100 different γ_c mutations have been observed in SCIDX1 and XCID patients, making these diseases a remarkable model of disorders affecting signaling. We now focus on one of the more relevant signaling pathways induced by cytokine receptors, highlighting the insights derived from the study of SCID patient-derived mutations.

SCID DUE TO JAK3 MUTATIONS

It had been noted that a proportion of the $T^- B^+ NK^-$ SCID patients were females, suggesting that a defect in a gene inherited in an autosomal manner, and thus distinct from γ_c, was responsible for this disease. The γ_c-containing cytokine receptors lack intrinsic kinase activity but are noncovalently associated with cytoplasmic tyrosine kinases belonging to the Janus family, Jak1 and Jak3 (O'Shea et al., 2002; Aringer et al., 1999; Ihle et al., 1998). Whereas Jak1 is broadly expressed and is used by many different cytokine receptors (Leonard and O'Shea, 1998), Jak3 is predominantly present in hematopoietic cells (Kawamura et al., 1994) and uniquely binds γ_c (Miyazaki et al., 1994). Thus mutations in Jak3 were sought and finally identified in SCID patients (Macchi et al., 1995; Russell et al., 1995). Clinically, Jak3-deficient patients have a phenotype identical to that of SCIDX1 patients, supporting the contention that γ_c transduces all its physiologically relevant signals through Jak3 and, on the other hand, the main function of Jak3 is to mediate γ_c-induced signals. In patients with autosomal recessive SCID more than 30 different mutations spanning the entire Jak3 molecule have been identified (Notarangelo et al., 2000a; Notarangelo et al., 2000b; Candotti et al., 1997), the systematic analysis of the effect of these mutations on Jak3 function having improved our understanding of how this tyrosine kinase works.

Many mutations in the Jak3 gene greatly affect expression of the protein. However, in a number of patients residual and sometimes normal expression levels of the protein have been described. In almost all cases, however, Jak3 protein is not functional, that is, IL-2-induced Jak3 tyrosine phosphorylation and activation cannot be detected (Notarangelo et al., 2000a; Notarangelo et al., 2000b; Candotti et al., 1997). Three-dimensional structural data for any of the Jaks are still lacking, but, on the basis of sequence comparison and similarity among the family members, seven regions of homology have been identified and named Janus homology (JH) domains 1–7 in a C-terminal to N-terminal direction. The Janus kinases were named after the "two-faced" Roman god because of a unique structural feature: They have a kinase and a pseudokinase domain (Wilks et al., 1991). The JH1 domain comprises the tyrosine kinase domain. It is homologous to other tyrosine kinase domains and highly conserved among the Jak family members; in fact, it allowed the identification of the Jaks by PCR-based strategy or low-stringency hybridization (Wilks, 1989; Krolewski et al., 1990; Harpur et al., 1992; Kawamura et al., 1994; Rane and Reddy, 1994). The JH2 domain is highly homologous to the JH1 domain but, lacking many key residues required for phospho-transferase activity, is not catalytically functional and has therefore been termed the pseudokinase domain. For instance, it lacks an aspartic acid that serves as the proton acceptor typically conserved in the catalytic loop of both tyrosine and serine kinases; it is also missing the third glycine in the Gly-X-Gly-X-X-Gly motif in subdomain I. Finally, the pseudokinase domain lacks the conserved Phe in the Asp-Phe-Glu motif that binds ATP. No other protein tyrosine kinases contain such a domain; it is, however, highly conserved in *Drosophila*, teleost, and mammalian Jaks, suggesting that it serves important, though poorly understood functions. However, some relevant insights on the role of the Jak JH2 domain have been provided by the analysis of naturally occurring mutations in these regions. That is, a number of mutations in the Jak3 pseudokinase domain have been identified in SCID patients (Candotti et al., 1997; Chen et al., 2000). The functional effects of these mutations have been evaluated with both Epstein–Barr virus-transformed B cell lines from patients and heterologous systems. This mutant Jak3 protein was normally expressed, but its kinase activity was completely dysregulated, being undetectable in vitro. Moreover, Jak3 mutants were unresponsive to IL-2 stimulation, even though they could normally bind γ_c. However, surprisingly, the mutated Jak3 appeared to be constitutively and hyper-tyrosine phosphorylated compared with wild-type Jak3. Thus it was hypothesized and, finally, demonstrated that the physiological role of the JH2 pseudokinase domain is to regulate kinase activity and therefore substrate phosphorylation by directly interacting with the JH1 kinase domain. Further confirming this, it has been shown that, whereas the wild-type Jak3 pseudokinase domain modestly inhibited the Jak3-mediated signaling pathway, the mutated pseudokinase domain from the two patients studied had an increased capacity of inhibiting kinase activity, which likely contributes to disease pathogenesis. The potential role of the Jaks pseudokinase domain as a regulator of kinase activity is further supported by the finding that a mutation in the JH2 domain can hyperactivate the *Drosophila* Jak pathway, causing leukemia in flies (Harrison et al., 1995; Luo et al., 1997). Remarkably, the corresponding mutation introduced in murine Jak2 also resulted in increased autophosphorylation of Jak2 in transfected cells (Luo et al., 1997). Thus the in vitro biological studies on patient-derived mutated Jak3 proteins allowed the identification of the molecular defect underlying the disease pathogenesis. At the same time, the in vitro study of Jak3 naturally occurring mutations has greatly improved our understanding of Jak regulation; in this case it has provided the first evidence that the JH1 and JH2 domains interact and that the likely function of the unique and still superficially understood JH2 domain is to tightly regulate catalytic activity. Only the solution of the three-dimensional structure of the Jaks will definitely establish whether the pseudokinase and kinase domains are closely positioned in the three-dimensional structure of the Jaks.

Finally, recent data have demonstrated an additional mechanism for autosomal SCID pathogenesis. A patient with a single amino acid substitution in the JH7 domain of Jak3 has been identified (Cacalano et al., 1999). This mutation prevents kinase-receptor interaction. A more accurate molecular analysis (Chen et al., 1997) has resulted in the mapping and delimitation of the region of interaction between Jak3 and γ_c. The N-terminal portion of Jak3 is required for receptor binding, the JH7–JH5 domain, a band four-point-one, ezrin, radixin, moesin (FERM) homology domain (Girault et al., 1999), interacting with the γ_c Box1 domain. Very interestingly, this has been confirmed by the identification of a patient-derived mutation in the membrane proximal region of the intracellular domain of γ_c. It has been shown that this mutation caused SCIDX1 by disrupting the association of γ_c with Jak3 and therefore γ_c-mediated signaling (Russell et al., 1994; Schmalstieg et al., 1995; Chen et al., 1997). However, recently analysis of SCID patient-derived mutations in the Jak3 FERM region has demonstrated a novel role for this domain (Zhou et al., 2001). Patient-derived FERM mutations, in fact, not only impair γ_c/Jak3 association but also abrogate Jak3 in vitro catalytic activity. A more accurate molecular analysis utilizing constructs harboring the SCID-associated FERM mutations has shown that the Jak3 FERM and kinase domains associate and reciprocally influence each other's function and structure. Thus, in SCID patients with FERM mutations, two mechanisms contribute to the disease pathogenesis: impaired γ_c/Jak3 association and inactivation of catalytic activity.

Of note, although Jak3 is inducibly expressed in monocytes (Musso et al., 1995), no defects in myeloid function have been reported so far in Jak3-deficient patients (Villa et al., 1996).

Jak3-deficient mice have also been generated, and they are also immunodeficient (Nosaka et al., 1995; Park et al., 1995; Thomis et al., 1995). Their phenotype is completely concordant to that of γ_c-deficient mice (Cao et al., 1995; DiSanto et al., 1995), further demonstrating that γ_c-induced responses depend on Jak3-mediated signals. However, the phenotype of both γ_c- and Jak3-deficient mice is distinct from γ_c- and Jak3-deficient patients, suggesting different functions of the γ_c cytokines in the two species. Mice lacking γ_c or Jak3 show a profound block in B cell development at the pre-B stage, suggesting a role for IL-7 or for a still-unknown γ_c-binding cytokine as an essential growth factor for B cell development in mice but not humans. In contrast, although the thymus of these knockout mice is small, T cell maturation progresses relatively normally, suggesting that an IL-7-independent pathway leading to T cell development exists in mice. Indeed, it was recently shown that IL-3, a γ_c- and Jak3-independent cytokine, can promote T cell development in Jak3$^{-/-}$ mice (Brown et al., 1999). However, the γ_c- and Jak3-deficient mouse T cells express activation markers, further confirming the key role of IL-2 in controlling T cell homeostasis.

Remarkably, a family was very recently reported whose Jak3 mutations are associated with the persistence of circulating activated T cells (Frucht et al., 2001). The Jak3 mutations, resulting in unexpected and variable phenotypes among the affected family members, were associated with minimal amounts of functional Jak3 expression, suggesting that residual Jak3 activity accounted for the maturation of thymocytes but was insufficient to sustain IL-2-mediated homeostasis of peripheral T cells.

OVERVIEW OF CYTOKINE SIGNALING

As demonstrated by humans and mice with mutations in γ_c cytokines, their receptors, and Jak3, these molecules have essential functions in host defense. Because of their importance, much work has focused on how they regulate cell function. Among the different signaling pathways initiated by cytokines and mediated by the Jaks, the pathway leading to the activation of the Stat (signal transducer and activator of transcription) family of transcription factors is particularly exciting in that it is a rapid mechanism by which signals can be transduced from the membrane directly to the nucleus. A vast amount of information now available on the Jak/Stat pathway demonstrates its importance in controlling key cellular processes as different as development, differentiation, proliferation, and transformation (O'Shea et al., 2002; Gadina et al., 2001; Leonard and O'Shea, 1998).

First discovered as essential for interferon (IFN) signal transduction (Velazquez et al., 1992; Silvennoinen et al., 1993; Watling et al., 1993), the Jak family is now known to be composed in mammals of four members: Jak1, Jak2, Jak3, and Tyk2 (Leonard and O'Shea, 1998). Jaks are constitutively bound to cytokine receptors; although a ligand-inducible augmentation has been demonstrated in several receptor systems, the mechanism underlying this phenomenon has not been clarified yet (Leonard and O'Shea, 1998). On ligand binding, the cytokine receptors form homo- or heterodimers that, in turn, result in Jak juxtaposition and transphosphorylation (O'Shea, 1997). The transphosphorylation of Jaks has at least two major consequences. First, it allows activation of catalytic activity by phosphorylation of activation loop tyrosine residues. Second, this modification also creates the basis for a mechanism of classic negative feedback of signaling. The phosphorylated tyrosine residues in Jaks are bound by the SH2 domain of a recently identified family of proteins, alternatively named Jab, SOCS, SSI, and CIS. The binding of these proteins to Jaks results in inhibition of Jak kinase activity and of Jak-mediated signaling (Endo et al., 1997; Naka et al., 1997; Starr et al., 1997).

Activated Jaks phosphorylate tyrosine residues on cytokine receptors, allowing the recruitment of molecules with SH2 or phosphotyrosine binding (PTB) domains. Among the different molecules that bind phosphorylated receptors, the Stats are particularly important in explaining cytokine actions. In mammals, seven Stat family members have been identified: Stat1, Stat2, Stat3, Stat4, Stat5a, Stat5b, and Stat6 (Fu et al., 1992; Akira et al., 1994; Zhong et al., 1994; Yamamoto et al., 1994; Wakao et al., 1994; Liu et al., 1995; Hou et al., 1994; Quelle et al., 1995).

Stats are a family of latent, cytosolic, SH2-containing transcription factors (Ihle, 2001; Leonard, 2001; Bromberg and Darnell, 2000; Hoey and Grusby, 1999), highly conserved through evolution, as Stat homologs have been identified in *Drosophila* (Sweitzer et al., 1995; Yan et al., 1996), in *Dictyostelium*, and in *Caenorhabditis elegans*.

The mechanism leading to Stat activation on cytokine stimulation has been elucidated in detail elsewhere. Stats bind via SH2 domain phosphotyrosine residues on activated cytokine receptors (Greenlund et al., 1994; Heim et al., 1995; Stahl et al., 1995). Once bound to the receptors, the Stats are themselves phosphorylated on a conserved tyrosine residue by the Jaks (Schindler et al., 1992; Shuai et al., 1993; Shual et al., 1993; Greenlund et al., 1994). After tyrosine phosphorylation, the Stats can homo- or heterodimerize by virtue of the interaction between the SH2 domain of one Stat and the phosphotyrosine of another. Thus the Stats SH2 domain plays two important roles: receptor docking and Stat dimerization. Stat dimers then translocate to the nucleus, a process that is regulated by both nuclear import and export (McBride et al., 2000; Melen et al., 2001; McBride et al., 2002); nuclear import is mediated by importin α and Ran-GTP (McBride et al., 2002; Sekimoto et al., 1996). Once in the nucleus, the Stats directly bind DNA, therefore regulating gene transcription.

Although the activation of the Stats is deeply understood, we have much less insight on how they are turned off. Several different mechanisms have been hypothesized, but only a few have been convincingly demonstrated. It has been shown that Stat1 is ubiquitinated after IFN-γ-dependent activation (Kim and Maniatis, 1996), suggesting that this may be a general mechanism by which cytokine signaling is terminated. Moreover, Stat transcriptional activity may be inhibited by a family of recently identified proteins, called the protein inhibitor of activated Stat (PIAS) (Chung et al., 1997; Liu et al., 1998).

Unlike the Jaks, the Stats have been crystallized (Becker et al., 1998; Chen et al., 1998; Vinkemeier et al., 1998), providing us with much information on their structural features. DNA binding is provided by a region, defined as the DNA binding domain, located in the central portion of most of the Stat proteins; Stat2, indeed, also requires Stat1 and p48 for a stable interaction with the DNA (Bluyssen and Levy, 1997). Moreover, two Stat dimers, interacting through a N-terminal dimerization domain, can recognize with higher affinity two imperfect binding sites present in tandem on many Stat-responsive promoters (Vinkemeier et al., 1996; Xu et al., 1996; Vinkemeier et al., 1998; John et al., 1999).

In the C-terminal portion of most of the Stats (missed in Stat2 and Stat6), where the transcriptional activation domain has been mapped, a conserved site of serine phosphorylation has been identified. It has been shown that this site is phosphorylated in vivo by the MAPK family of serine/threonine kinases, the functional effects of this additional modification having not been completely clarified yet (Decker and Kovarik, 2000).

Stat crystal structure has also revealed the presence of coiled-coil domains mediating Stat interaction with the coactivator protein p300/CBP (Bhattacharya et al., 1996; Zhang et al., 1996; Horvai et al., 1997; Pfitzner et al., 1998) as well as with other transcription factors (Zhu et al., 1999).

As with the Jaks, a single Stat member can be activated by different cytokines. However, not all the cytokine receptors can activate all the Stats. Thus a certain degree of specificity of activation does exist and is ensured by the fact that the SH2 domains of the various Stats differ sufficiently so that they recognize with distinct affinity the phosphorylated residues on the different receptor subunits. The cytokines whose receptors share γ_c activate Stat3, Stat5a, and Stat5b (Lin and Leonard, 2000), with the notable exception of IL-4, which leads to Stat6 activation (Hou et al., 1994). The nonredundant functions of particular Stats in mediating signaling by the various cytokines have been recently highlighted by the phenotype of knockout mice for each of these proteins. Still, it remains to be definitively clarified whether Stat3 plays any essential role in γ_c-activating cytokines signaling because Stat3 gene targeting leads to early embryonic lethality (Takeda et al., 1997). However, Stat3 conditional knockouts in T cells have been generated (Takeda et al., 1998), showing that absence of Stat3 has no effect on lymphoid development and only a small effect on lymphocyte proliferation. The generation of Stat5a and Stat5b knockout mice has demonstrated that these two transcription factors, despite being remarkably homologous, have very diverse functions, Stat5a being mandatory for mammary gland development and lactogenesis (Liu et al., 1997) and Stat5b for sexually dimorphic growth and growth hormone-dependent regulation of liver gene expression (Udy et al., 1997). Stat5a/Stat5b double-knockout mice have also been generated (Teglund et al., 1998), and they have been very informative as to the role played by these two transcription factors in γ_c-activating cytokine signal transduction. T and B cells develop normally in the double-null mice, suggesting that Stat5 proteins have redundant roles in mediating IL-7 signaling. The Stat5a/b$^{-/-}$ T cells, are, however, unable to proliferate in response to IL-2 stimulation (Moriggl et al., 1999) and have an activated phenotype as seen in IL-2-, IL-2R-deficient humans and mice (Sharfe et al., 1997a; Schorle et al., 1991; Willerford et al., 1995; Suzuki et al., 1995), demonstrating the essential role played by the Stat5 proteins in IL-2 signaling.

Nonetheless, although Stat3, Stat5a, and Stat5b are key signaling molecules downstream of γ_c and Jak3, no human diseases have yet been associated with mutations or loss of these molecules. Recently, though, a missense mutation of Stat1 was identified in a patient suffering from atypical mycobacterial infections (Dupuis et al., 2001). Interestingly, however, the Stat1 mutation was not associated with higher susceptibility to viral infections, in contrast to what was observed in mice lacking Stat1 (Durbin et al., 1996; Meraz et al., 1996).[note added in proof] This is so far the only example of mutation in any of the Stat family members causing an immune disease in humans. However, it might be anticipated that mutations in Stat2 and Stat4 could cause immunodeficiencies as suggested by the knockout mice phenotype. Stat2$^{-/-}$ mice, in fact, have defective responses to Type I IFNs and are highly susceptible to viral infections (Park et al., 2000). Stat4$^{-/-}$ mice develop completely normally but have defective cell-mediated immune responses and T helper (Th)1 cell differentiation (Kaplan et al., 1996; Thierfelder et al., 1996), their phenotype being consistent with the abnormalities seen in IL-12- or IL-12R-deficient mice (Magram et al., 1996; Wu et al., 1997; Wu et al., 2000). Moreover, the same defects, causing severe mycobacterial and *Salmonella* infections, have been described in patients harboring mutations in IL-12 and IL-12R (Altare et al., 1998b; Altare et al., 1998a; de Jong et al., 1998). Thus it may be predicted that Stat4 defects in humans may lead to the same pattern of immune defects.

NEW THERAPEUTIC APPROACHES TO SCID

X-linked and autosomal recessive SCID are characterized by severe infections leading to death early in infancy unless successfully treated. Although passive immunization with γ-globulin has been enormously valuable in the past, and still is for patients with deficits of immunoglobulins (Buckley and Schiff, 1991), bone marrow transplantation is now the treatment of choice for most severe immunodeficiencies, the overall survival rate being between 60% and 78% (Porta and Friedrich, 1998; Buckley et al., 1993; Buckley et al., 1999). Pluripotent hematopoietic stem cells are collected from a donor and transplanted in the patient. If the treatment is successful, donor stem cells repopulate the recipient's bone marrow with their differentiating progeny, definitely correcting the immunodeficiency.

However, the transplantation may be associated with severe complications such as host versus graft disease; moreover, the immune reconstitution may not be complete in the transplanted patients. Furthermore, the lack of suitable donors is also problematic. Thus many groups are evaluating alternative forms of therapy. Several primary immunodeficiencies are attractive candidates for gene therapy (Hacein-Bey-Abina et al., 2002; Cavazzana-Calvo et al., 2001; Fischer et al., 2000; Cournoyer and Caskey, 1993). γ_c gene transfer appears to confer a selective growth advantage to transduced lymphoid precursors. Moreover, the transfer of the defective gene into the lymphoid progenitors, able to self-renew and to repopulate the bone marrow of affected children, should in theory lead to a definitive cure of the disease. Until very recently, however, the vectors commonly available for clinical gene therapy failed to efficiently integrate the transgene into the noncycling hematopoietic precursor cells. Recently, though, Cavazzana-Calvo et al. reported the greatly encouraging results of a gene therapy clinical trial for SCIDX1 patients based on an ex vivo approach (Cavazzana-Calvo et al., 2001; Cavazzana-Calvo et al., 2000). In this study CD34$^+$ cells were isolated from the bone marrow of five infant SCIDX1 patients and pre-activated with a cytokine combination to make them cycling and thus permissive for transgene integration. The CD34$^+$ cells were then transduced in vitro with a γ_c-containing retroviral vector and infused in the patients. The transduced T cells typically became detectable 60–90 days after treatment and normalized in number 6 months later. Importantly, T cells appeared to be perfectly functioning and able to normally respond to antigen stimulation. Moreover, in the first two children treated, B cell immunity developed and their NK cells were able to physiologically kill target cells. Thus, in conclusion, this treatment led to normal T and NK cell development from precursor CD34$^+$ γ_c-transduced cells and to complete restoration of T, NK, and B cell immune responses. As a consequence, all the clinical manifestations of SCID disappeared. Today, more than 2 years after the treatment, the responding children are living without any therapy and not suffering any severe infection (Hacein-Bey-Abina et al., 2002). Thus gene transfer has been successfully applied to correction of one form of SCID (Cavazzana-Calvo et al., 2001; Cavazzana-Calvo et al., 2000; Hacein-Bey-Abina et al., 2002), providing proof of the principle that congenital defects of lymphocytes can be cured by replacing the defective gene.[note added in proof] Clearly, other aspects will need to be investigated further in the future, the most important of which is the stability of the phenotype. A longer follow-up period will be necessary to verify if the target of transgene transduction, in these patients was the hematopoietic stem cells with self-renewal capacity, pluripotent hematopoietic progenitors with no or little self-renewal capacity and committed lymphoid progenitors with no self-renewal capacity. The evidence that a small fraction of transduced monocytes and granulocytes were still detectable 6 months after treatment gives hope that immature pluripotent cells have in fact been transduced and that the benefits of the therapy will be long lasting.

In summary, ex vivo gene therapy has been able in some patients to fully correct SCIDX1 defects, suggesting that this therapeutic approach may be of enormous importance in the future. Another obvious candidate for a gene therapy clinical trial is Jak3 deficiency. Reassuringly, it has been shown that Jak3 deficiency in murine models can, indeed, be corrected by ex vivo gene therapy (Bunting et al., 1998). Moreover, Candotti et al. demonstrated that retroviral transduction of a Jak3 cDNA into B cell lines derived from Jak3-deficient SCID patients is sufficient to restore IL-2-mediated cell proliferation (Candotti et al., 1996). Thus Jak3 biological functions can be corrected both in vitro and in an animal model by a gene therapy approach, suggesting that a similar strategy could be applied as therapy for Jak3-deficient SCID patients. It remains, however, to be fully investigated whether uncontrolled Jak3 tyrosine kinase overexpression could lead to any dangerous and deleterious consequence, although, no adverse effects have been noted so far in treated mice.

CONCLUSIONS

Thus mutations in IL-2 (DiSanto et al., 1990; Weinberg and Parkman, 1990), IL-2Rα (Sharfe et al., 1997a), IL-7Rα (Puel et al., 1998), γ$_c$ (Noguchi et al., 1993b; Puck et al., 1993b), and Jak3 (Macchi et al., 1995; Russell et al., 1995) all underlie primary immunodeficiencies in humans. However, mutations in other cytokines, cytokine receptors, and Stat molecules are also associated with disease in humans (Candotti et al., 2002). Patients with mutations of IFN-γ receptor subunit (IFNγR1 or -R2), Stat1, IL-12, or IL-12R present with atypical mycobacterial and *Salmonella* infections (Dupuis et al., 2000; Dupuis et al., 2001; Altare et al., 1998b; Altare et al., 1998a; de Jong et al., 1998). Moreover, a point mutation in the extracellular domain of another type I cytokine receptor, the granulocyte colony-stimulating factor (G-CSF) receptor, was recently identified in a case of severe congenital neutropenia (Ward et al., 1999). Thus investigation of cytokine signaling pathways is clearly a fruitful area of investigation that sheds light on disease in humans.

Over the past few years, there has been enormous progress. Consequently, the understanding we have today of the key steps in cytokine signaling can be used to identify new candidate genes responsible for other ill-defined immunodeficiencies.

In conclusion, in the last few years we have witnessed enormous progress in the understanding of how a cytokine signal is delivered from the cell membrane to the nucleus. This has led to the identification of the molecular basis of many primary immunodeficiencies. Analysis of patient-derived mutations has given us important clues to the structure and function of these genes; as such, primary immunodeficiency diseases are a superb examples of molecular medicine. However, many challenges still remain, because there are diseases associated with unidentified molecular defects, but mutations in cytokines, cytokine receptors, and their signaling pathway components will remain important candidates. Most importantly, these advances are anticipated to improve diagnosis and treatment of individuals affected by these disorders.

Notes added in proof: IL-21 has recently been demonstrated to be critical for immunologlubulin product; this has important implications for SCID. Also, viral infections have recently been found to be associated with Star1 deficiency in humans. Finally, leukemia has unfortunately been recently reported to be associated with gene therapy for X-SCID; thus the enthusiasm for this therapy needs to be tempered by this complication.

REFERENCES

Akira, S., Nishio, Y., Inoue, M., Wang, X. J., Wei, S., Matsusaka, T., Yoshida, K., Sudo, T., Naruto, M., and Kishimoto, T. (1994). Molecular cloning of APRF, a novel IFN-stimulated gene factor 3 p91-related transcription factor involved in the gp130-mediated signaling pathway. Cell 77, 63–71.

Altare, F., Durandy, A., Lammas, D., Emile, J. F., Lamhamedi, S., Le Deist, F., Drysdale, P., Jouanguy, E., Doffinger, R., Bernaudin, F., Jeppsson, O., Gollob, J. A., Meinl, E., Segal, A. W., Fischer, A., Kumararatne, D., and Casanova, J. L. (1998a). Impairment of mycobacterial immunity in human interleukin-12 receptor deficiency. Science 280, 1432–1435.

Altare, F., Lammas, D., Revy, P., Jouanguy, E., Doffinger, R., Lamhamedi, S., Drysdale, P., Scheel-Toellner, D., Girdlestone, J., Darbyshire, P., Wadhwa, M., Dockrell, H., Salmon, M., Fischer, A., Durandy, A., Casanova, J. L., and Kumararatne, D. S. (1998b). Inherited interleukin 12 deficiency in a child with bacille Calmette-Guerin and Salmonella enteritidis disseminated infection. J Clin Invest 102, 2035–2040.

Aringer, M., Cheng, A., Nelson, J. W., Chen, M., Sudarshan, C., Zhou, Y. J., and O'Shea, J. J. (1999). Janus kinases and their role in growth and disease. Life Sci 64, 2173–2186.

Asao, H., Okuyama, C., Kumaki, S., Ishii, N., Tsuchiya, S., Foster, D., and Sugamura, K. (2001). Cutting edge: the common γ-chain is an indispensable subunit of the IL-21 receptor complex. J Immunol 167, 1–5.

Becker, S., Groner, B., and Muller, C. W. (1998). Three-dimensional structure of the Stat3β homodimer bound to DNA. Nature 394, 145–151.

Bhattacharya, S., Eckner, R., Grossman, S., Oldread, E., Arany, Z., D'Andrea, A., and Livingston, D. M. (1996). Cooperation of Stat2 and p300/CBP in signaling induced by interferon-α. Nature 383, 344–347.

Bluyssen, H. A., and Levy, D. E. (1997). Stat2 is a transcriptional activator that requires sequence-specific contacts provided by stat1 and p48 for stable interaction with DNA. J Biol Chem 272, 4600–4605.

Bromberg, J., and Darnell, J. E. J. (2000). The role of STATs in transcriptional control and their impact on cellular function. Oncogene 19, 2468–2473.

Brown, M. P., Nosaka, T., Tripp, R. A., Brooks, J., van Deursen, J. M., Brenner, M. K., Doherty, P. C., and Ihle, J. N. (1999). Reconstitution of early lymphoid proliferation and immune function in Jak3-deficient mice by interleukin-3. Blood 94, 1906–1914.

Buckley, R, H. (2000). Primary immunodeficiency diseases due to defects in lymphocytes. N Engl J Med 343, 1313–1324.

Buckley, R. H., and Schiff, R. I. (1991). The use of intravenous immune globulin in immunodeficiency diseases. N Engl J Med 325, 110–117.

Buckley, R. H., Schiff, S. E., Schiff, R. I., Roberts, J. L., Markert, M. L., Peters, W., Williams, L. W., and Ward, F. E. (1993): Haploidentical bone marrow stem cell transplantation in human severe combined immunodeficiency. Semin Hematol 30, 92–101; discussion 102–104.

Buckley, R. H., Schiff, S. E., Schiff, R. I., Markert, L., Williams, L. W., Roberts, J. L., Myers, L. A., and Ward, F. E. (1999). Hematopoietic stem-cell transplantation for the treatment of severe combined immunodeficiency. N Engl J Med 340, 508–516.

Bunting, K. D., Sangster, M. Y., Ihle, J. N., and Sorrentino, B. P. (1998). Restoration of lymphocyte function in Janus kinase 3-deficient mice by retroviral-mediated gene transfer [see comments]. Nat Med *4*, 58–64.

Cacalano, N. A., Migone, T. S., Bazan, F., Hanson, E., Chen, M., Candotti, F., O'Shea, J. J., and Johnston, J. A. (1999). Autosomal SCID caused by a point mutation in the N-terminus of Jak3: mapping of the Jak3-receptor interaction domain. EMBO J *18*, 1549–1558.

Candotti, F., Notarangelo, L. D., Visconti, R., and O'Shea J. J. (2002). Molecular aspects of primary immunodeficiencies: lessons from cytokine and other signaling pathways. J Clin Invest *109*, 1261–1269.

Candotti, F., Oakes, S. A., Johnston, J. A., Notarangelo, L. D., O'Shea, J. J., and Blaese, R. M. (1996). In vitro correction of JAK3-deficient severe combined immunodeficiency by retroviral-mediated gene transduction. J Exp Med *183*, 2687–2692.

Candotti, F., Oakes, S. A., Johnston, J. A., Giliani, S., Schumacher, R. F, Mella, P., Fiorini, M., Ugazio, A. G., Badolato, R., Notarangelo, L. D., Bozzi, F., Macchi, P., Strina, D., Vezzoni, P., Blaese, R. M., O'Shea, J. J., and Villa, A. (1997). Structural and functional basis for JAK3-deficient severe combined immunodeficiency. Blood *90*, 3996–4003.

Cao, X., Shores, E. W., Hu-Li, J., Anver, M. R., Kelsall, B. L., Russell, S. M., Drago, J., Noguchi, M., Grinberg, A., Bloom, E. T. et al. (1995). Defective lymphoid development in mice lacking expression of the common cytokine receptor γ chain. Immunity *2*, 223–238.

Carson, W. E., Giri, J. G., Lindemann, M. J., Linett, M. L., Ahdieh, M., Paxton, R., Anderson, D., Eisenmann, J., Grabstein, K., and Caligiuri, M. A. (1994). Interleukin (IL) 15 is a novel cytokine that activates human natural killer cells via components of the IL-2 receptor. J Exp Med *180*, 1395–1403.

Cavazzana-Calvo, M., Hacein-Bey, S., de Saint Basile, G., De Coene, C., Selz, F., Le Deist, F., and Fischer, A. (1996). Role of interleukin-2 (IL-2), IL-7, and IL-15 in natural killer cell differentiation from cord blood hematopoietic progenitor cells and from γc transduced severe combined immunodeficiency X1 bone marrow cells. Blood *88*, 3901–3909.

Cavazzana-Calvo, M., Hacein-Bey, S., de Saint Basile, G., Gross, F., Yvon, E., Nusbaum, P., Selz, F., Hue, C., Certain, S., Casanova, J. L., Bousso, P., Deist, F. L., and Fischer, A. (2000). Gene therapy of human severe combined immunodeficiency (SCID)-X1 disease. Science *288*, 669–672.

Cavazzana-Calvo, M., Hacein-Bey, S., Yates, F., de Villartay, J. P., Le Deist, F., and Fischer, A. (2001). Gene therapy of severe combined immunodeficiencies. J Gene Med *3*, 201–206.

Chen, M., Cheng, A., Chen, Y. Q., Hymel, A., Hanson, E. P., Kimmel, L., Minami, Y., Taniguchi, T., Changelian, P. S., and O'Shea, J. J. (1997). The amino terminus of JAK3 is necessary and sufficient for binding to the common γ chain and confers the ability to transmit interleukin 2-mediated signals. Proc Natl Acad Sci USA *94*, 6910–6915.

Chen, M., Cheng, A., Candotti, F., Zhou, Y. J., Hymel, A., Fasth, A., Notarangelo, L. D., and O'Shea, J. J. (2000). Complex effects of naturally occurring mutations in the JAK3 pseudokinase domain: evidence for interactions between the kinase and pseudokinase domains. Mol Cell Biol *20*, 947–956.

Chen, X., Vinkemeier, U., Zhao, Y., Jeruzalemi, D., Darnell, J. E. J., and Kuriyan, J. (1998). Crystal structure of a tyrosine phosphorylated STAT-1 dimer bound to DNA. Cell *93*, 827–839.

Chung, C. D., Liao, J., Liu, B., Rao, X., Jay, P., Berta, P., and Shuai, K. (1997). Specific inhibition of Stat3 signal transduction by PIAS3. Science *278*, 1803–1805.

Cournoyer, D., and Caskey, C. T. (1993). Gene therapy of the immune system. Annu Rev Immunol *11*, 297–329.

Decker, T., and Kovarik, P. (2000). Serine phosphorylation of STATs. Oncogene *19*, 2628–2637.

de Jong, R., Altare, F., Haagen, I. A., Elferink, D. G., Boer, T., van Breda Vriesman, P. J., Kabel, P. J., Draaisma, J. M., van Dissel, J. T., Kroon, F. P., Casanova, J. L., and Ottenhoff, T. H. (1998). Severe mycobacterial and *Salmonella* infections in interleukin-12 receptor-deficient patients. Science *280*, 1435–1438.

DiSanto, J. P., Keever, C. A., Small, T. N., Nicols, G. L., O'Reilly, R. J., and Flomenberg, N. (1990). Absence of interleukin 2 production in a severe combined immunodeficiency disease syndrome with T cells. J Exp Med *171*, 1697–1704.

DiSanto, J. P., Muller, W., Guy-Grand, D., Fischer, A., and Rajewsky, K. (1995). Lymphoid development in mice with a targeted deletion of the interleukin 2 receptor γ chain. Proc Natl Acad Sci USA *92*, 377–381.

Dupuis, S., Dargemont, C., Fieschi, C., Thomassin, N., Rosenzweig, S., Harris, J., Holland, S. M., Schreiber, R. D., and Casanova J. L. (2001). Impairment of mycobacterial but not viral immunity by a germline human STAT1 mutation. Science *293*, 300–303.

Dupuis, S., Doffinger, R., Picard, C., Fieschi, C., Altare, F., Jouanguy, E., Abel, L., and Casanova, J. L. (2000). Human interferon-γ-mediated immunity is a genetically controlled continuous trait that determines the outcome of mycobacterial invasion. Immunol Rev *178*, 129–137.

Dupuis, S., Jouanguy, E., Al-Hajjar, S., Fieschi, C., Al-Mohsen, I. Z., Al-Jumaah, S., Yang, K., Chapgier, A., Eidenschenk, C., Eid, P., Ghonaium, A. A., Tufenkeji, H., Frayha, H., Al-Gazlan, S., Al-Rayes, H., Schreiber, R. D., Gresser, I., Casanova, J. L. Impaired response to interferon-alpha/beta and lethal viral disease in human STAT1 deficiency. Nat Genet 2003 Feb 18 [epub ahead of print]

Durbin, J. E., Hackenmiller, R., Simon, M. C., and Levy, D. E. (1996). Targeted disruption of the mouse Stat1 gene results in compromised innate immunity to viral disease. Cell *84*, 443–450.

Endo, T. A., Masuhara, M., Yokouchi, M., Suzuki, R., Sakamoto, H., Mitsui, K., Matsumoto, A., Tanimura, S., Ohtsubo, M., Misawa, H., Miyazaki, T., Leonor, N., Taniguchi, T., Fujita, T., Kanakura, Y., Komiya, S., and Yoshimura, A. (1997). A new protein containing an SH2 domain that inhibits JAK kinases. Nature *387*, 921–924.

Fehniger, T. A., and Caligiuri, M. A. (2001). Interleukin 15: biology and relevance to human disease. Blood *97*, 14–32.

Fischer, A. (2001). Primary immunodeficiency diseases: an experimental model for molecular medicine. Lancet *357*, 1863–1869.

Fischer, A., Cavazzana-Calvo, M., De Saint Basile, G., De Vollartay, J. P., DiSanto, J. P., Hivroz, C., Rieux-Laucat, F., and Le Deist, F. (1997). Naturally occurring primary deficiencies of the immune system. Annu Rev Immunol *15*, 93–124.

Fischer, A., Hacein-Bey, S., Le Deist, F., Soudais, C., Di Santo, J. P., de Saint Basile, G., and Cavazzana-Calvo, M. (2000). Gene therapy of severe combined immunodeficiencies. Immunol Rev *178*, 13–20.

Frucht, D. M., Gadina, M., Jagadeesh, G. J., Aksentijevich, I., Takada, K., Bleesing, J. J., Nelson, J., Muul, L. M., Perham, G., Morgan, G., Gerritsen, E. J., Schumacher, R. F., Mella, P., Veys, P. A., Fleisher, T. A., Kaminski, E. R., Notarangelo, L. D., O'Shea, J. J., and Candotti, F. (2001). Unexpected and variable phenotypes in a family with JAK3 deficiency. Genes Immun 2, 422–432.

Fu, X. Y., Schindler, C., Improta, T., Aebersold, R., and Darnell, J. E. J. (1992). The proteins of ISGF-3, the interferon α-induced transcriptional activator, define a gene family involved in signal transduction. Proc Natl Acad Sci USA 89, 7840–7843.

Gadina, M., Hilton, D., Johnston, J. A., Morinobu, A., Lighvani, A., Zhou, Y., Visconti, R., and O'Shea, J. J. (2001). Signaling by Type I and II cytokine receptors: ten years after. Curr Opin Immunol 13, 363–373.

Girault, J. A., Labesse, G., Mornon, J. P., and Callebaut, I. (1999). The N-termini of FAK and JAKs contain divergent band 4.1 domains. Trends Biochem Sci 24, 54–57.

Giri, J. G., Ahdieh, M., Eisenman, J., Shanebeck, K., Grabstein, K., Kumaki, S., Namen, A., Park, L. S., Cosman, D., and Anderson, D. (1994). Utilization of the β and γ chains of the IL-2 receptor by the novel cytokine IL-15. EMBO J 13, 2822–2830.

Greenlund, A. C., Farrar, M. A., Viviano, B. L., and Schreiber, R. D. (1994). Ligand-induced IFN γ receptor tyrosine phosphorylation couples the receptor to its signal transduction system (p91). EMBO J 13, 1591–1600.

Hacein-Bey-Abina, S., Le Deist, F., Carlier, F., Bouneaud, C., Hue, C., De Villartay, J. P., Thrasher, A. J., Wulffraat, N., Sorensen, R., Dupuis-Girod, S., Fischer, A., Davies, E. G., Kuis, W., Leiva, L., and Cavazzana-Calvo, M. (2002). Sustained correction of X-linked severe combined immunodeficiency by ex vivo gene therapy. N Engl J Med 346, 1185–1193.

Hacein-Bey-Abina, S., von Kalle C., Schmidt M., Le Deist F., Wulffraat, N., McIntyre, E., Radford, I., Villeval, J. L., Fraser, C. C., Cavazzana-Calvo, M., Fischer, A. A serious adverse event after successful gene therapy for X-linked severe combined immunodeficiency. N Engl J Med 2003 Jan 16;348(3), 255–256.

Harpur, A. G., Andres, A. C., Ziemiecki, A., Aston, R. R., and Wilks, A. F. (1992). JAK2, a third member of the JAK family of protein tyrosine kinases. Oncogene 7, 1347–1353.

Harrison, D. A., Binari, R., Nahreini, T. S., Gilman, M., and Perrimon, N. (1995). Activation of a *Drosophila* Janus kinase (JAK) causes hematopoietic neoplasia and developmental defects. EMBO J 14, 2857–2865.

He, Y. W., and Malek, T. R. (1995). The IL-2 receptor γc chain does not function as a subunit shared by the IL-4 and IL-13 receptors. Implication for the structure of the IL-4 receptor. J Immunol 155, 9–12.

Heim, M. H., Kerr, I. M., Stark, G. R., and Darnell, J. E. J. (1995). Contribution of STAT SH2 groups to specific interferon signaling by the Jak-STAT pathway. Science 267, 1347–1349.

Hoey, T., and Grusby, M. J. (1999). STATs as mediators of cytokine-induced responses. Adv Immunol 71, 145–162.

Horvai, A. E., Xu, L., Korzus, E., Brard, G., Kalafus, D., Mullen, T. M., Rose, D. W., Rosenfeld, M. G., and Glass, C. K. (1997). Nuclear integration of JAK/STAT and Ras/AP-1 signaling by CBP and p300. Proc Natl Acad Sci USA 94, 1074–1079.

Hou, J., Schindler, U., Henzel, W. J., Ho, T. C., Brasseur, M., and McKnight, S. L. (1994). An interleukin-4-induced transcription factor: IL-4 Stat. Science 265, 1701–1706.

Ihle, J. N. (2001). The Stat family in cytokine signaling. Curr Opin Cell Biol *13*, 211–217.

Ihle, J. N., Thierfelder, W., Teglund, S., Stravapodis, D., Wang, D., Feng, J., and Parganas, E. (1998). Signaling by the cytokine receptor superfamily. Ann NY Acad Sci *865*, 1–9.

John, S., Vinkemeier, U., Soldaini, E., Darnell, J. E. J., and Leonard, W. J. (1999). The significance of tetramerization in promoter recruitment by Stat5. Mol Cell Biol 19, 1910–1918.

Kaplan, M. H., Sun, Y. L., Hoey, T., and Grusby, M. J. (1996). Impaired IL-12 responses and enhanced development of Th2 cells in Stat4-deficient mice. Nature *382*, 174–177.

Kasaian, M. T., Whitters, M. J., Carter, L. L., Lowe, L. D., Jussif, J. M., Deng, B., Johnson, K. A., Witek, J. S., Senices, M., Konz, R. F., Wurster, A. L., Donald-son, D. D., Collins, M., Young, D. A., and Grusby, M. J. (2002). IL-21 limits NK cell responses and promotes antigen-specific T cell activation: a mediator of the transition from innate to adaptive immunity. Immunity *16*, 559–569.

Kawamura, M., McVicar, D. W., Johnston, J. A., Blake, T. B., Chen, Y. Q., Lal, B. K., Lloyd, A. R., Kelvin, D. J., Staples, J. E., Ortaldo, J. R. et al. (1994). Mole-cular cloning of L-JAK, a Janus family protein-tyrosine kinase expressed in natural killer cells and activated leukocytes. Proc Natl Acad Sci USA *91*, 6374–6378.

Kim, T. K., and Maniatis, T. (1996). Regulation of interferon-γ-activated STAT1 by the ubiquitin-proteasome pathway. Science *273*, 1717–1719.

Kimura, Y., Takeshita, T., Kondo, M., Ishii, N., Nakamura, M., Van Snick, J., and Sugamura, K. (1995). Sharing of the IL-2 receptor γ chain with the functional IL-9 receptor complex. Int Immunol *7*, 115–120.

Kondo, M., Takeshita, T., Ishii, N., Nakamura, M., Watanabe, S., Arai, K., and Sugamura, K. (1993). Sharing of the interleukin-2 (IL-2) receptor γ chain between receptors for IL-2 and IL-4 [see comments]. Science *262*, 1874–1877.

Kondo, M., Takeshita, T., Higuchi, M., Nakamura, M., Sudo, T., Nishikawa, S., and Sugamura, K. (1994). Functional participation of the IL-2 receptor γ chain in IL-7 receptor complexes. Science *263*, 1453–1454.

Krolewski, J. J., Lee, R., Eddy, R., Shows, T. B., and Dalla-Favera, R. (1990). Iden-tification and chromosomal mapping of new human tyrosine kinase genes. Oncogene *5*, 277–282.

Kuhn, R., Rajewsky, K., and Muller, W. (1991). Generation and analysis of interleukin-4 deficient mice. Science *254*, 707–710.

Leclercq, G., Debacker, V., de Smedt, M., and Plum, J. (1996). Differential effects of interleukin-15 and interleukin-2 on differentiation of bipotential T/natural killer progenitor cells. J Exp Med *184*, 325–336.

Lee, F., Yokota, T., Otsuka, T., Meyerson, P., Villaret, D., Coffman, R., Mosmann, T., Rennick, D., Roehm, N., Smith, C. et al. (1986). Isolation and characteri-zation of a mouse interleukin cDNA clone that expresses B-cell stimulatory factor 1 activities and T-cell- and mast-cell-stimulating activities. Proc Natl Acad Sci USA *83*, 2061–2065.

Leonard, W. J. (2000). X-linked severe combined immunodeficiency: from molec-ular cause to gene therapy within seven years. Mol Med Today *6*, 403–407.

Leonard, W. J. (2001). Role of Jak kinases and STATs in cytokine signal trans-duction. Int J Hematol *73*, 271–277.

Leonard, W. J., Noguchi, M., Russell, S. M., and McBride, O. W. (1994). The molecular basis of X-linked severe combined immunodeficiency: the role of the interleukin-2 receptor γ chain as a common γ chain, γc. Immunol Rev *138*, 61–86.

Leonard, W. J., and O'Shea, J. J. (1998). Jaks and STATs: biological implications. Annu Rev Immunol *16*, 293–322.

Lin, J. X., and Leonard, W. J. (2000). The role of Stat5a and Stat5b in signaling by IL-2 family cytokines. Oncogene *19*, 2566–2576.

Liu, B., Liao, J., Rao, X., Kushner, S. A., Chung, C. D., Chang, D. D., and Shuai, K. (1998). Inhibition of Stat1-mediated gene activation by PIAS1. Proc Natl Acad Sci USA *95*, 10626–10631.

Liu, X., Robinson, G. W., Gouilleux, F., Groner, B., and Hennighausen, L. (1995). Cloning and expression of Stat5 and an additional homologue (Stat5b) involved in prolactin signal transduction in mouse mammary tissue. Proc Natl Acad Sci USA *92*, 8831–8835.

Liu, X., Robinson, G. W., Wagner, K. U., Garrett, L., Wynshaw-Boris, A., and Hennighausen, L. (1997). Stat5a is mandatory for adult mammary gland development and lactogenesis. Genes Dev *11*, 179–186.

Lodolce, J. P., Boone, D. L., Chai, S., Swain, R. E., Dassopoulos, T., Trettin, S., and Ma, A. (1998). IL-15 receptor maintains lymphoid homeostasis by supporting lymphocyte homing and proliferation. Immunity *9*, 669–676.

Luo, H., Rose, P., Barber, D., Hanratty, W. P., Lee, S., Roberts, T. M., D'Andrea, A. D., and Dearolf, C. R. (1997). Mutation in the Jak kinase JH2 domain hyperactivates *Drosophila* and mammalian Jak-Stat pathways. Mol Cell Biol *17*, 1562–1571.

Macchi, P., Villa, A., Gillani, S., Sacco, M. G., Frattini, A., Porta, F., Ugazio, A. G., Johnston, J. A., Candotti, F., O'Shea, J. J. et al. (1995). Mutations of Jak-3 gene in patients with autosomal severe combined immune deficiency (SCID). Nature *377*, 65–68.

Magram, J., Connaughton, S. E., Warrier, R. R., Carvajal, D. M., Wu, C. Y., Ferrante, J., Stewart, C., Sarmiento, U., Faherty, D. A., and Gately, M. K. (1996). IL-12-deficient mice are defective in IFN γ production and type 1 cytokine responses. Immunity *4*, 471–481.

Matthews, D. J., Hibbert, L., Friedrich, K., Minty, A., and Callard, R. E. (1997). X-SCID B cell responses to interleukin-4 and interleukin-13 are mediated by a receptor complex that includes the interleukin-4 receptor α chain (p140) but not the γc chain. Eur J Immunol *27*, 116–121.

McBride, K. M., McDonald, C., and Reich N. C. (2000). Nuclear export signal located within the DNA-binding domain of the STAT1 transcription factor. EMBO J *19*, 6196–6206.

McBride, K. M., Banninger, G., McDonald, C., and Reich, N. C. (2002). Regulated nuclear import of the STAT1 transcription factor by direct binding of importin-α. EMBO J *21*, 1754–1763.

Melen, K., Kinnunen, L., and Julkunen, I. (2001). Arginine/lysine-rich structural element is involved in interferon-induced nuclear import of STATs. J Biol Chem *276*, 16447–16455.

Meraz, M. A., White, J. M., Sheehan, K. C., Bach, E. A., Rodig, S. J., Dighe, A. S., Kaplan, D. H., Riley, J. K., Greenlund, A. C., Campbell, D., Carver-Moore, K., DuBois, R. N., Clark, R., Aguet, M., and Schreiber, R. D. (1996). Targeted disruption of the Stat1 gene in mice reveals unexpected physiologic specificity in the JAK-STAT signaling pathway. Cell *84*, 431–442.

Mingari, M. C., Gerosa, F., Carra, G., Accolla, R. S., Moretta, A., Zubler, R. H., Waldmann, T. A., and Moretta, L. (1984). Human interleukin-2 promotes proliferation of activated B cells via surface receptors similar to those of activated T cells. Nature *312*, 641–643.

Miyazaki, T., Kawahara, A., Fujii, H., Nakagawa, Y., Minami, Y., Liu, Z. J., Oishi, I., Silvennoinen, O., Witthuhn, B. A., Ihle, J. N. et al. (1994). Functional activation of Jak1 and Jak3 by selective association with IL-2 receptor subunits. Science *266*, 1045–1047.

Mombaerts, P., Iacomini, J., Johnson, R. S., Herrup, K, Tonegawa, S., and Papaioannou, V. E. (1992): RAG-1-deficient mice have no mature B and T lymphocytes. Cell *68*, 869–877.

Morgan, D. A., Ruscetti, F. W., and Gallo, R. (1976). Selective in vitro growth of T lymphocytes from normal human bone marrows. Science *193*, 1007–1008.

Moriggl, R., Topham, D. J., Teglund, S., Sexl, V., McKay, C., Wang, D., Hoffmeyer, A., van Deursen, J., Sangster, M. Y., Bunting, K. D., Grosveld, G. C., and Ihle, J. N. (1999). Stat5 is required for IL-2-induced cell cycle progression of peripheral T cells. Immunity *10*, 249–259.

Moshous, D., Callebaut, I., de Chasseval, R., Corneo, B., Cavazzana-Calvo, M., Le Deist, F., Tezcan, I., Sanal, O., Bertrand, Y., Philippe, N., Fischer, A., and de Villartay, J. P. (2001). Artemis, a novel DNA double-strand break repair/V(D)J recombination protein, is mutated in human severe combined immune deficiency. Cell *105*,177–186.

Murray, R., Suda, T., Wrighton, N., Lee, F., and Zlotnik, A. (1989). IL-7 is a growth and maintenance factor for mature and immature thymocyte subsets. Int Immunol *1*, 526–531.

Musso, T., Johnston, J. A., Linnekin, D., Varesio, L., Rowe, T. K., O'Shea, J. J., and McVicar, D. W. (1995). Regulation of JAK3 expression in human monocytes: phosphorylation in response to interleukins 2, 4, and 7. J Exp Med *181*, 1425–1431.

Naka, T., Narazaki, M., Hirata, M., Matsumoto, T., Minamoto, S., Aono, A., Nishimoto, N., Kajita, T., Taga, T., Yoshizaki, K., Akira, S., and Kishimoto, T. (1997). Structure and function of a new STAT-induced STAT inhibitor. Nature *387*, 924–929.

Namen, A. E., Lupton, S., Hjerrild, K., Wignall, J., Mochizuki, D. Y., Schmierer, A., Mosley, B., March, C. J., Urdal, D., and Gillis, S. (1988a). Stimulation of B-cell progenitors by cloned murine interleukin-7. Nature *333*, 571–573.

Namen, A. E., Schmierer, A. E., March, C. J., Overell, R. W., Park, L. S., Urdal, D. L., and Mochizuki, D. Y. (1988b). B cell precursor growth-promoting activity. Purification and characterization of a growth factor active on lymphocyte precursors. J Exp Med *167*, 988–1002.

Nelson, B. H. (2002). Interleukin-2 signaling and the maintenance of self-tolerance. Curr Dir Autoimmun *5*, 92–112.

Noguchi, M., Nakamura, Y., Russell, S. M., Ziegler, S. F., Tsang, M., Cao, X., and Leonard, W. J. (1993a). Interleukin-2 receptor γ chain: a functional component of the interleukin-7 receptor [see comments]. Science *262*, 1877–1880.

Noguchi, M., Yi, H., Rosenblatt, H. M., Filipovich, A. H., Adelstein, S., Modi, W. S., McBride, O. W., and Leonard, W. J. (1993b). Interleukin-2 receptor γ chain mutation results in X-linked severe combined immunodeficiency in humans. Cell *73*, 147–157.

Nosaka, T., van Deursen, J. M., Tripp, R. A., Thierfelder, W. E., Witthuhn, B. A., McMickle, A. P., Doherty, P. C., Grosveld, G. C., and Ihle, J. N. (1995). Defective lymphoid development in mice lacking Jak3. Science *270*, 800–802.

Notarangelo, L. D., Giliani, S., Mazza, C., Mella, P., Savoldi, G., Rodriguez-Perez, C., Mazzolari, E., Fiorini, M., Duse, M., Plebani, A., Ugazio, A. G., Vihinen, M., Candotti, F., and Schumacher, R. F. (2000a). Of genes and phenotypes: the immunological and molecular spectrum of combined immune deficiency. Defects of the γ_c-JAK3 signaling pathway as a model. Immunol Rev *178*, 39–48.

Notarangelo, L. D., Giliani, S., Mella, P., Schumacher, R. F., Mazza, C., Savoldi, G., Rodriguez-Perez, C., Badolato, R., Mazzolari, E., Porta, F., Candotti, F., and Ugazio, A. G. (2000b). Combined immunodeficiencies due to defects in signal transduction: defects of the γ_c-JAK3 signaling pathway as a model. Immunobiology *202*, 106–119.

Oakes, S. A., Candotti, F., Johnston, J. A., Chen, Y. Q., Ryan, J. J., Taylor, N., Liu, X., Hennighausen, L., Notarangelo, L. D., Paul, W. E., Blaese, R. M., and O'Shea, J. J. (1996). Signaling via IL-2 and IL-4 in JAK3-deficient severe combined immunodeficiency lymphocytes: JAK3-dependent and independent pathways. Immunity *5*, 605–615.

Obiri, N. I., Debinski, W., Leonard, W. J., and Puri, R. K. (1995). Receptor for interleukin 13. Interaction with interleukin 4 by a mechanism that does not involve the common γ chain shared by receptors for interleukins 2, 4, 7, 9, and 15. J Biol Chem *270*, 8797–8804.

O'Shea, J. J. (1997). Jaks, STATs, Cytokine signal transduction, and immunoregulation: Are we there yet? Immunity *7*, 1–11.

O'Shea, J. J., Gadina, M., and Schreiber, R. D. (2002). Cytokine signaling in 2002: new surprises in the Jak/Stat pathway. Cell *109* Suppl, S121–S131.

Ozaki, K., Spolski, R., Feng, C. G., Qi, C. F., Cheng, J., Sher, A., Morse, H. C., 3rd, Liu, C., Schwartzberg, P. L., Leonard, W. J. A critical role for IL-21 in regulating immunoglobulin production. Science 2002 Nov 22;*298*(5598), 1630–1634.

Park, C., Li, S., Cha, E., and Schindler, C. (2000). Immune response in Stat2 knockout mice. Immunity *13*, 795–804.

Park, S. Y., Saijo, K., Takahashi, T., Osawa, M., Arase, H., Hirayama, N., Miyake, K., Nakauchi, H., Shirasawa, T., and Saito, T. (1995). Developmental defects of lymphoid cells in Jak3 kinase-deficient mice. Immunity *3*, 771–782.

Paul, W. E. (1997). Interleukin 4: signaling mechanisms and control of T cell differentiation. Ciba Found Symp *204*, 208–216; discussion 216–219.

Peschon, J. J., Morrissey, P. J., Grabstein, K. H., Ramsdell, F. J., Maraskovsky, E., Gliniak, B. C., Park, L. S., Ziegler, S. F., Williams, D. E., Ware, C. B. et al. (1994). Early lymphocyte expansion is severely impaired in interleukin 7 receptor-deficient mice. J Exp Med *180*, 1955–1960.

Pfitzner, E., Jahne, R., Wissler, M., Stoecklin, E., and Groner, B. (1998). p300/CREB-binding protein enhances the prolactin-mediated transcriptional induction through direct interaction with the transactivation domain of Stat5, but does not participate in the Stat5-mediated suppression of the glucocorticoid response. Mol Endocrinol *12*, 1582–1593.

Porta, F., and Friedrich, W. (1998). Bone marrow transplantation in congenital immunodeficiency diseases. Bone Marrow Transplant *21*, Suppl 2: S21–23.

Puck, J. M., Conley, M. E., and Bailey, L. C. (1993a): Refinement of linkage of human severe combined immunodeficiency (SCIDX1) to polymorphic markers in Xq13. Am J Hum Genet *53*, 176–184.

Puck, J. M., Deschenes, S. M., Porter, J. C., Dutra, A. S., Brown, C. J., Willard, H. F., and Henthorn, P. S. (1993b). The interleukin-2 receptor γ chain maps to Xq13.1 and is mutated in X-linked severe combined immunodeficiency, SCIDX1. Hum Mol Genet 2, 1099–1104.

Puel, A., Ziegler, S. F., Buckley, R. H., and Leonard, W. J. (1998). Defective IL7R expression in T⁻B⁺NK⁺ severe combined immunodeficiency. Nat Genet 20, 394–397.

Puzanov, I. J., Bennett, M., and Kumar, V. (1996). IL-15 can substitute for the marrow microenvironment in the differentiation of natural killer cells. J Immunol 157, 4282–4285.

Quelle, F. W., Shimoda, K., Thierfelder, W., Fischer, C., Kim, A., Ruben, S. M., Cleveland, J. L., Pierce, J. H., Keegan, A. D., Nelms, K. et al. (1995). Cloning of murine Stat6 and human Stat6, Stat proteins that are tyrosine phosphorylated in responses to IL-4 and IL-3 but are not required for mitogenesis. Mol Cell Biol 15, 3336–3343.

Rane, S. G., and Reddy, E. P. (1994). JAK3: a novel JAK kinase associated with terminal differentiation of hematopoietic cells. Oncogene 9, 2415–2423.

Rosen, F. S., Cooper, M. D., and Wedgwood, R. J. (1995). The primary immunodeficiencies. N Engl J Med 333, 431–440.

Russell, S. M., Johnston, J. A., Noguchi, M., Kawamura, M., Bacon, C. M., Friedmann, M., Berg, M., McVicar, D. W., Witthuhn, B. A., Silvennoinen, O. et al. (1994). Interaction of IL-2R β and γc chains with Jak1 and Jak3: implications for XSCID and XCID. Science 266, 1042–1045.

Russell, S. M., Keegan, A. D., Harada, N., Nakamura, Y., Noguchi, M., Leland, P., Friedmann, M. C., Miyajima, A., Puri, R. K., Paul, W. E. et al. (1993). Interleukin-2 receptor γ chain: a functional component of the interleukin-4 receptor [see comments]. Science 262, 1880–1883.

Russell, S. M., Tayebi, N., Nakajima, H., Riedy, M. C., Roberts, J. L., Aman, M. J., Migone, T. S., Noguchi, M., Markert, M. L., Buckley, R. H. et al. (1995). Mutation of Jak3 in a patient with SCID: essential role of Jak3 in lymphoid development. Science 270, 797–800.

Schindler, C., Shuai, K., Prezioso, V. R., and Darnell, J. E. J. (1992). Interferon-dependent tyrosine phosphorylation of a latent cytoplasmic transcription factor [see comments]. Science 257, 809–813.

Schluns, K. S., Kieper, W. C., Jameson S. C., and Lefrancois, L. (2000). Interleukin-7 mediates the homeostasis of naive and memory CD8 T cells in vivo. Nat Immunol 1, 426–432.

Schmalstieg, F. C., Leonard, W. J., Noguchi, M., Berg, M., Rudloff, H. E., Denney, R. M., Dave, S. K., Brooks, E. G., and Goldman, A. S. (1995). Missense mutation in exon 7 of the common γ chain gene causes a moderate form of X-linked combined immunodeficiency. J Clin Invest 95, 1169–1173.

Schorle, H., Holtschke, T., Hunig, T., Schimpl, A., and Horak, I. (1991). Development and function of T cells in mice rendered interleukin-2 deficient by gene targeting. Nature 352, 621–624.

Schwarz, K., Gauss, G. H., Ludwig, L., Pannicke, U., Li, Z., Lindner, D., Friedrich, W., Seger, R. A., Hansen-Hagge, T. E., Desiderio, S., Lieber, M. R., and Bartram, C. R. (1996). RAG mutations in human B cell-negative SCID. Science 274, 97–99.

Sekimoto, T., Nakajima, K., Tachibana, T., Hirano, T., and Yoneda, Y. (1996). Interferon-γ-dependent nuclear import of Stat1 is mediated by the GTPase activity of Ran/TC4. J Biol Chem 271, 31017–31020.

Sharfe, N., Dadi, H. K., Shahar, M., and Roifman, C. M. (1997a). Human immune disorder arising from mutation of the α chain of the interleukin-2 receptor. Proc Natl Acad Sci USA *94*, 3168–3171.

Sharfe, N., Shahar, M., and Roifman, C. M., (1997b). An interleukin-2 receptor γ chain mutation with normal thymus morphology. J Clin Invest *100*, 3036–3043.

Shinkai, Y., Rathbun, G., Lam, K. P., Oltz, E. M., Stewart, V., Mendelsohn, M., Charron, J., Datta, M., Young, F., Stall, A. M. et al. (1992). RAG-2-deficient mice lack mature lymphocytes owing to inability to initiate V(D)J rearrangement. Cell *68*, 855–867.

Shuai, K., Stark, G. R., Kerr, I. M., and Darnell, J. E. J. (1993). A single phosphotyrosine residue of Stat91 required for gene activation by interferon-γ [see comments]. Science *261*, 1744–1746.

Shual, K., Ziemiecki, A., Wilks, A. F., Harpur, A. G., Sadowski, H. B., Gilman, M. Z., and Darnell, J. E. J. (1993). Polypeptide signalling to the nucleus through tyrosine phosphorylation of Jak and Stat proteins. Nature *366*, 580–583.

Siegel, J. P., Sharon, M., Smith, P. L., and Leonard, W. J. (1987): The IL-2 receptor β chain (p70): role in mediating signals for LAK, NK, and proliferative activities. Science *238*, 75–78.

Silvennoinen, O., Ihle, J. N., Schlessinger, J., and Levy, D. E. (1993). Interferon-induced nuclear signalling by Jak protein tyrosine kinases. Nature *366*, 583–585.

Stahl, N., Farruggella, T. J., Boulton, T. G., Zhong, Z., Darnell, J. E. J., and Yancopoulos, G. D. (1995). Choice of STATs and other substrates specified by modular tyrosine-based motifs in cytokine receptors. Science *267*, 1349–1353.

Starr, R., Willson, T. A., Viney, E. M., Murray, L. J., Rayner, J. R., Jenkins, B. J., Gonda, T. J., Alexander, W. S., Metcalf, D., Nicola, N. A., and Hilton, D. J. (1997). A family of cytokine-inducible inhibitors of signalling. Nature *387*, 917–921.

Sugamura, K., Asao, H., Kondo, M., Tanaka, N., Ishii, N., Ohbo, K., Nakamura, M., and Takeshita, T. (1996). The interleukin-2 receptor γ chain: its role in the multiple cytokine receptor complexes and T cell development in XSCID. Annu Rev Immunol *14*, 179–205.

Suzuki, H., Kundig, T. M., Furlonger, C., Wakeham, A., Timms, E., Matsuyama, T., Schmits, R., Simard, J. J., Ohashi, P. S., Griesser, H. et al. (1995). Deregulated T cell activation and autoimmunity in mice lacking interleukin-2 receptor β. Science *268*, 1472–1476.

Sweitzer, S. M., Calvo, S., Kraus, M. H., Finbloom, D. S., and Larner, A. C. (1995). Characterization of a Stat-like DNA binding activity in *Drosophila melanogaster*. J Biol Chem *270*, 16510–16513.

Takeda, K., Kaisho, T., Yoshida, N., Takeda, J., Kishimoto, T., and Akira, S. (1998). Stat3 activation is responsible for IL-6-dependent T cell proliferation through preventing apoptosis: generation and characterization of T cell- specific Stat3-deficient mice. J Immunol *161*, 4652–4660.

Takeda, K., Noguchi, K., Shi, W., Tanaka, T., Matsumoto, M., Yoshida, N., Kishimoto, T., and Akira, S. (1997). Targeted disruption of the mouse Stat3 gene leads to early embryonic lethality. Proc Natl Acad Sci USA *94*, 3801–3804.

Teglund, S., McKay, C., Schuetz, E., van Deursen, J. M., Stravopodis, D., Topham, D. J., Wang, D., Brown, M., Bodner, S., Grosveld, G., and Ihle, J. N. (1998). STAT5a and STAT5b proteins have essential and nonessential, or redundant, roles in cytokine responses. Cell *93*, 841-850.

Thierfelder, W. E., van Deursen, J. M., Yamamoto, K., Tripp, R. A., Sarawar, S. R., Carson, R. T., Sangster, M. Y., Vignali, D. A., Doherty, P. C., Grosveld, G. C., and Ihle, J. N. (1996). Requirement for Stat4 in interleukin-12-mediated responses of natural killer and T cells. Nature *382*, 171–174.

Thomis, D. C., Gurniak, C. B., Tivol, E., Sharpe, A. H., and Berg, L. J. (1995). Defects in B lymphocyte maturation and T lymphocyte activation in mice lacking Jak3. Science *270*, 794–797.

Udy, G. B., Towers, R. P., Snell, R. G., Wilkins, R. J., Park, S.-H., Ram, P. A., Waxman, D. J., and Davey, H. W. (1997). Requirement of STAT5b for sexual dimorphism of body growth rates and liver gene expression. Proc Natl Acad Sci USA *94*, 7239–7244.

Velazquez, L., Fellous, M., Stark, G. R., and Pellegrini, S. (1992). A protein tyrosine kinase in the interferon α/β-signaling pathway. Cell *70*, 313–322.

Villa, A., Sironi, M., Macchi, P., Matteucci, C., Notarangelo, L. D., Vezzoni, P., and Mantovani, A. (1996). Monocyte function in a severe combined immunodeficient patient with a donor splice site mutation in the Jak3 gene. Blood *88*, 817–823.

Vinkemeier, U., Cohen, S. L., Moarefi, I., Chait, B. T., Kuriyan, J., and Darnell, J. E. J. (1996). DNA binding of in vitro activated Stat1 α, Stat1 β and truncated Stat1: interaction between NH_2-terminal domains stabilizes binding of two dimers to tandem DNA sites. EMBO J *15*, 5616–5626.

Vinkemeier, U., Moarefi, I., Darnell, J. E. J., and Kuriyan, J. (1998): Structure of the amino-terminal protein interaction domain of STAT-4. Science *279*, 1048–1052.

von Freeden-Jeffry, U., Vieira, P., Lucian, L. A., McNeil, T., Burdach, S. E., and Murray, R. (1995). Lymphopenia in interleukin (IL)-7 gene-deleted mice identifies IL-7 as a nonredundant cytokine. J Exp Med *181*, 1519–1526.

Wakao, H., Gouilleux, F., and Groner, B. (1994). Mammary gland factor (MGF) is a novel member of the cytokine regulated transcription factor gene family and confers the prolactin response [published erratum appears in (1995) EMBO J *14*, 854–855]. EMBO J *13*, 2182–2191.

Ward, A. C., van Aesch, Y. M., Gits, J., Schelen, A. M., de Koning, J. P., van Leeuwen, D., Freedman, M. H., and Touw, I. P. (1999). Novel point mutation in the extracellular domain of the granulocyte colony-stimulating factor (G-CSF) receptor in a case of severe congenital neutropenia hyporesponsive to G-CSF treatment. J Exp Med *190*, 497–508.

Watling, D., Guschin, D., Muller, M., Silvennoinen, O., Witthuhn, B. A., Quelle, F. W., Rogers, N. C., Schindler, C., Stark, G. R., Ihle, J. N. et al. (1993). Complementation by the protein tyrosine kinase JAK2 of a mutant cell line defective in the interferon-γ signal transduction pathway [see comments]. Nature *366*, 166–170.

Watson, J. D., Morrissey, P. J., Namen, A. E., Conlon, P. J., and Widmer, M. B. (1989). Effect of IL-7 on the growth of fetal thymocytes in culture. J Immunol *143*, 1215–1222.

Weinberg, K., and Parkman, R. (1990). Severe combined immunodeficiency due to a specific defect in the production of interleukin-2. N Engl J Med *322*, 1718–1723.

WHO Scientific Group (1995). Primary immunodeficiency diseases. Clin Exp Immunol *99* (Suppl. 1). 1–24.

Wilks, A. F. (1989). Two putative protein-tyrosine kinases identified by application of the polymerase chain reaction. Proc Natl Acad Sci USA *86*, 1603–1607.

Wilks, A. F., Harpur, A. G., Kurban, R. R., Ralph, S. J., Zurcher, G., and Ziemiecki, A. (1991). Two novel protein-tyrosine kinases, each with a second phosphotransferase-related catalytic domain, define a new class of protein kinase. Mol Cell Biol 11, 2057–2065.

Willerford, D. M., Chen, J., Ferry, J. A., Davidson, L., Ma, A., and Alt, F. W. (1995). Interleukin-2 receptor α chain regulates the size and content of the peripheral lymphoid compartment. Immunity *3*, 521–530.

Wu, C., Ferrante, J., Gately, M. K., and Magram, J. (1997). Characterization of IL-12 receptor β1 chain (IL-12Rβ1)-deficient mice: IL-12Rβ1 is an essential component of the functional mouse IL-12 receptor. J Immunol *159*, 1658–1665.

Wu, C., Wang, X., Gadina, M., O'Shea, J. J., Presky, D. H., and Magram J., (2000). IL-12 receptor β2 (IL-12Rβ2)-deficient mice are defective in IL-12-mediated signaling despite the presence of high affinity IL-12 binding sites. J Immunol *165*, 6221–6228.

Xu, X., Sun, Y. L., and Hoey, T. (1996). Cooperative DNA binding and sequence-selective recognition conferred by the STAT amino-terminal domain [see comments]. Science *273*, 794–797.

Yamamoto, K., Quelle, F. W., Thierfelder, W. E., Kreider, B. L., Gilbert, D. J., Jenkins, N. A., Copeland, N. G., Silvennoinen, O., and Ihle, J. N. (1994). Stat4, a novel γ interferon activation site-binding protein expressed in early myeloid differentiation. Mol Cell Biol *14*, 4342–4349.

Yan, R., Small, S., Desplan, C., Dearolf, C. R., and Darnell, J. E. J. (1996). Identification of a Stat gene that functions in *Drosophila* development. Cell *84*, 421–430.

Yokota, T., Otsuka, T., Mosmann, T., Banchereau, J., DeFrance, T., Blanchard, D., De Vries, J. E., Lee, F., and Arai, K. (1986). Isolation and characterization of a human interleukin cDNA clone, homologous to mouse B-cell stimulatory factor 1, that expresses B-cell- and T-cell-stimulating activities. Proc Natl Acad Sci USA *83*, 5894–5898.

Zhang, J. J., Vinkemeier, U., Gu, W., Chakravarti, D., Horvath, C. M., and Darnell, J. E. J. (1996). Two contact regions between Stat1 and CBP/p300 in interferon γ signaling. Proc Natl Acad Sci USA *93*, 15092–15096.

Zhong, Z., Wen, Z., and Darnell, J. E. J. (1994): Stat3 and Stat4: members of the family of signal transducers and activators of transcription. Proc Natl Acad Sci USA *91*, 4806–4810.

Zhou, Y. J., Chen, M., Cusack, N. A., Kimmel, L. H., Magnuson, K. S., Boyd, J. G., Lin, W., Roberts, J. L., Lengi, A., Buckley, R. H., Geahlen, R. L., Candotti, F., Gadina, M., Changelian, P. S., and O'Shea, J. J. (2001). Unexpected effects of FERM domain mutations on catalytic activity of Jak3: structural implication for Janus kinases. Mol Cell *8*, 959–969.

Zhu, M., John, S., Berg, M., and Leonard, W. J. (1999): Functional association of Nmi with Stat5 and Stat1 in IL-2- and IFNγ-mediated signaling. Cell *96*, 121–130.

CHAPTER 10

MAST CELL-RELATED DISEASES: GENETICS, SIGNALING PATHWAYS, AND NOVEL THERAPIES

MICHAEL A. BEAVEN and THOMAS R. HUNDLEY

Laboratory of Molecular Immunology, National Heart, Lung, and Blood Institute, National Institutes of Health, Bethesda, Maryland

INTRODUCTION

Mast Cells and Allergy

Mast cells and blood basophils are responsible for a variety of inflammatory allergic disorders such as allergic rhinitis, dermatitis, asthma, and food allergies as well as catastrophic anaphylactic reactions to insect stings and some drugs. Allergic diseases affect a surprisingly large proportion of the population. In the United States, the CDC reports that about 20%, 7%, and 2% of the population suffer from allergic rhinitis, asthma, and anapylactic hypersensitivities, respectively (CDC, 1998). Similar numbers are reported for other developed countries (Holgate, 1999), where these diseases are also prevalent (Stewart et al., 2001). A common refrain in the literature is that atopic disease primarily afflicts the young and that its incidence has increased dramatically over the past three decades, especially in young children. Nevertheless, adults are affected. Estimates of occupational asthma in the adult working population range from 5% to as high as 15% in several countries including the United States (Meyer et al., 2001; Mapp, 2001; Toren et al., 2000). Although advances have been made in our understanding of the pathogenesis of these diseases, and as a consequence effective medications are available, there is need for improvement because treatment is antisymptomatic rather than curative.

Signal Transduction and Human Disease, Edited by Toren Finkel and J. Silvio Gutkind
ISBN 0-471-02011-7   Copyright © 2003 John Wiley & Sons, Inc.

Mast cells and basophils are uniquely capable of responding to immunoglobulin E (IgE)-directed antigens because of the presence of several hundred thousand Fc receptors with high affinity for IgE (FcεRI) on the surface of these cells. On binding of antigen to the receptor-bound IgE, these cells release a wide array of inflammatory mediators through several mechanisms. These include rapid secretion of granules that contain preformed inflammatory mediators, the rapid generation of lipid mediators (eicosanoids) from arachidonic acid, and the activation of genes responsible for the production of various chemokines and cytokines (see Table 10.1). Receptors for complement-derived products (C3a and C5a), IgG (via Fcγ receptors), adenosine, and other ligands modulate the responsiveness of mast cells to immunological stimuli (reviewed in Colten, 1994) (Table 10.2). Some subpopulations of mast cells are directly stimulated by bacterial products (Leal-Berumen et al., 1994; Malaviya et al., 1999), polybasic compounds, or drugs through the direct activation of heterotrimeric G proteins (Mousli et al., 1990).

The organ tissues targeted by allergens are particularly enriched in mast cells. These tissues include skin, mucosa of airways and gastrointestinal tract, adventitia of blood vessels, and myocardium. Atopic disease may target one organ, such as skin (atopic eczema and the urticarias) and airways (rhinitis and asthma), or multiple organs (food allergies and anaphylaxis). The constellation of clinical symptoms will depend on the tissue targeted by allergen, the subtype of mast cell present in the tissue (see **Mast Cell Proliferation, Differentiation, and Heterogeneity** below), and the types of cells recruited after mast cell activation.

Role of Mast Cells in Innate and Acquired Immunity

The existence of the mast cells, and their presence in almost all tissues of the body, appears paradoxical in that the disorders noted above have no apparent benefit to the host. However, several studies show that mice deficient in mast cells rapidly succumb to fulminant bacterial infections in lung and body cavities (Echtenacher et al., 1996; Malaviya et al., 1996) or are unable to rid themselves of parasite infestation (Lantz et al., 1998). These studies also show that release of inflammatory cytokines, particularly tumor necrosis factor α (TNF-α), from mast cells and the associated recruitment of neutrophils are an important component of the protective action of mast cells against infestation (Wedemeyer et al., 2000). The activation of mast cells may result from bacterial activation of the complement system (Lantz et al., 1998) or by direct action of bacterial products on "pattern recognition receptors" that recognize structural patterns in molecules common to microorganisms. Toll-like receptors (TLRs) have been identified as such receptors (Supajatura et al., 2001; McCurdy et al., 2001). Therefore, mast cells might provide a first-line defense mechanism against acute bacterial and parasitic infections, usually referred to as innate immunity (Wedemeyer et al., 2000; Malaviya and Abraham, 2001; Mekori and Metcalfe, 2000). If so, mast cell-related allergies might be regarded as a hypersensitivity of the innate

TABLE 10.1. Inflammatory Mediators Released from Mast Cells

Mediator	Effects
Preformed: In granules	
Histamine[a]	Induces smooth muscle contraction, mucus secretion, vascular permeability, cytokine production in T cells
Mast cell proteases[a]	Cleave basement membrane and activate other proteinases
Proteoglycans[a]	Increased endothelial cell permeability and protease activity
Synthesized de novo	
Lipid mediators	
PGD2[a]	Bronchoconstriction, increased vascular permeability, vasodilation
LTC4	Bronchoconstriction, vasodilation
LTB4	Chemotaxis of neutrophils and eosinophils, adhesion of leukocytes to endothelium
PAF	Vasodilation, increased vascular permeability, neutrophil adhesion
Cytokines and Chemokines	
IL-1	Activation of lymphocytes, enhances mast cell growth
IL-2	Induces proliferation of T cells
IL-3[a]	Promotes proliferation of mast cells
IL-4[a]	Stimulates IgE production, promotes proliferation of mast cells, fibroblasts, and T_H2 lymphocytes
IL-5[a]	Promotes proliferation of eosinophils
IL-6[a]	Promotes proliferation of B cells, potentiates IL-4-induced IgE production
IL-8[a]	Neutrophil chemotaxis
IL-10[a]	Promotes proliferation of mast cells
IL-13[a]	Stimulates IgE production and expression of adhesion factors in endothelial cells
L-16	Chemoattractant for CD4+ T lymphocytes
TNF-α[a]	Activation of macrophages, inhibits apoptosis of neutrophils and eosinophils, expression of adhesion molecules in endothelial cells
TGF-β	Mast cell chemoattractant, inhibits apoptosis of T cells
NGF	Inhibits apoptosis and stimulates synthesis of mediators in mast cells, proliferation of T and B cells
PDGF	Stimulates airway smooth muscle proliferation
GM-CSF	Inhibits apoptosis of eosinophils
VPF/VEGF	Increases vascular permeability
FGF	Costimulatory for cytokine production in lymphocytes, angiogenesis
C-C chemokines	Chemoattractant: monocytes, lymphocytes, basophils, eosinophils

[a] Indicates that mast cells are the exclusive or major source of these products in asthmatic lungs.
For original citations see Gordon (1997), O'Sullivan (1999), and Page et al. (2001).

immune system toward otherwise benign environmental proteins. The reasons for this hypersensitivity are unclear, but an underlying factor is the predisposition of afflicted individuals to produce IgE against environmental antigens, a condition referred to as atopy.

TABLE 10.2. Receptors on Human Mast Cells

Receptor	Ligand	Function	References
Activation Fc receptors			
FcεRI[a]	IgE/Ag[b]	Mediator release	Novak et al., 2001
FcγRI (induced by IFN-γ)	IgG$_1$/Ag	Mediator release	Woolhiser et al., 2001
FcγRIII (induced by cytokines)	IgG$_3$/Ag	Mediator release	Katz and Lobell, 1995
Other activating receptors			
C5aR[c] and C3aR[d]	C3a, C5a	Recruit and activate MC	Fureder et al., 1995
			Hartmann et al., 1997
Adenosine receptors, A$_{2B}$/A$_3$	Adenosine	Synergizes Ag activation of MC	Tilley et al., 2000
Toll-like receptors	Bacterial products	Stimulates cytokine but not histamine/eicosanoid release	Supajatura et al., 2001 McCurdy et al., 2001
Inhibitory receptors (ITIM bearing)			
FcγRIIβ[c]	IgGAg complexes	Coaggregation with FcεRI inhibits mediator release	Ott and Cambier, 2000
Gp49B1	?	"	"
PIR-B	?	"	"
SIRPα1	IAP/CD47	"	"
MAFA	Oligosaccharides	Inhibits FcεRI-induced mediator release	"
Cytokine receptors			
KIT	SCF	Commitment to MC lineage and expansion in conjunction with other cytokines	Ashman, 1999 Taylor and Metcalfe, 2000 Austen and Boyce, 2001
IL3R[c]	IL-3	Prevalent on MC progenitors MC proliferation	Valent et al., 2001b Ochi et al., 1999
IL4R[a]	IL-4	Comitogenic/upregulates FcεRI expression and function	Ochi et al., 2000 Ochi et al., 1999
IL5Rα	IL-5	Comitogenic/augments MC cytokine production	Ochi et al., 2000 Ochi et al., 1999
IL6R[c]	IL-6	Comitogenic	Saito et al., 1996 Ochi et al., 1999
IL9R[c]	IL-9	Increased expression of MC Proteases and FcεRIα	Eklund et al., 1993 Louahed et al., 1995
IL10R[c]	IL-10	Increased expression of MC proteases	Xia et al., 1996
GM-CSFR[c]	GM-CSF	Mitogenic	Ochi et al., 1999
Chemotactic factor receptors[f]			
CCR3	RANTES	Chemotaxis	Bochner and Schleimer, 2001
	MCP-3 and -4/ eotaxin		Austen and Boyce, 2001 Ochi et al., 1999
CCR1/2/5 (Induced by Ag and SCF)	RANTES/ MIP-1α/MCP-1	Chemotaxis	Oliveira and Lukacs, 2001
CXCR2/4	IL-8/GROα	Chemotaxis? (Expressed in immature MC)	Ochi et al., 1999

TABLE 10.2. (*Continued*)

Receptor	Ligand	Function	References
Adhesion molecules[c,f]			
Integrins of the β1 (VLA) family αvβ3	Fibronectin/ Laminin/ VCAM-1 Vitronectin	Adhesion of MC and enhanced mediator release	Bochner and Schleimer, 2001 Houtman et al., 2001 Bochner and Schleimer, 2001
ICAM-1, -2, and -3	LFA-1/Mac-1	Enhanced mediator release	"
Pgp1 (CD44)	Hyaluronic acid	MC recruitment?	"
Siglec-8	Sialic acid	MC recruitment?	"

[a] Genes for these receptors/ligands have disease-linked polymorphisms (see Table 10.3).

[b] Abbreviations: Ag, antigen; MC, mast cell; R, receptor.

[c] Variably expressed according to mast cell phenotype and cell maturation (Valent et al., 2001b).

[d] Indirect evidence: Chemotactic response to C3a as well as C5a inhibited by pertussis toxin (Hartmann et al., 1997).

[e] Assumed from the effects of cytokines on mast cell growth, maturation, and expression of MC proteins (reviewed in Austen and Boyce, 2001).

[f] Basophils contain additional chemotactic receptors and adhesion molecules (Bochner and Schleimer, 2001) but lack KIT (Valent et al., 2001b).

The atopic predisposition probably has genetic components, as discussed below, but it is certainly influenced by environmental factors as well. In addition to increased air pollution, the so-called hygiene hypothesis has been advanced to explain the increasing prevalence of mast cell-related allergies, especially in children, in the last few decades (Holgate, 1999). This hypothesis holds that reduced exposure to infectious diseases, which typically provoke T helper (T_H)1-type response (i.e., IgG mediated), through use of vaccines, antibiotics, and other measures has shifted the bias of the immune system toward a T_H2-type response (i.e., IgE mediated) (Lappin and Cambell, 2000) to environmental antigens in much of the population. Nonetheless, human mast cells can be activated by IgG-dependent mechanisms via FcγRI and may be involved in IgG-dependent as well as IgE-dependent inflammatory diseases (Woolhiser et al., 2001).

Mast Cell Proliferation, Differentiation, and Heterogeneity

Tissue mast cells originate from circulating CD34$^+$ hematopoietic progenitor cells. Within tissues the progenitor mast cells proliferate and differentiate under the regulation of growth factors and cytokines including stem cell factor (SCF), which is the ligand for the tyrosine kinase receptor KIT (Taylor and Metcalfe, 2000; Ashman, 1999; Wedemeyer and Galli, 2000). KIT, a member of the type III family of tyrosine kinase growth receptors, is encoded by the gene *c-kit* and is expressed in various cell types, among them hematopoietic stem cells, gametocytes, and melanocytes (Ashman, 1999). Other factors augment mast cell proliferation or promote maturation. These include interleukin (IL)-3, IL-4, IL-5, IL-6, IL-9, IL-10, and nerve growth factor (Austen and Boyce, 2001; Mekori and Metcalfe, 2000; Wedemeyer and Galli, 2000) (Table 10.2). For example, mast cells can be derived from bone marrow cells or human $CD34^+$ blood cells by culture in SCF and IL-3. Maturation of the blood-derived human mast cells can be further promoted by addition of IL-6 at later stages of culture (Chaves-Dias et al., 2001). SCF also enhances the functions of mature mast cells such as their adhesion to extracellular matrix proteins (Lorentz et al., 2002) and responses to antigen (Wedemeyer and Galli, 2000). The central role of SCF and complementary roles of other cytokines in determining mast cell lineage and phenotype are reviewed in detail by Austen and Boyce (Austen and Boyce, 2001).

Mast cells assume heterogeneous characteristics in different tissues. The differences lie in the nature of granule constituents, morphology, and sensitivity to stimulants and growth factors. Human mast cells have been categorized as M_T and M_{TC} subtypes based on the type of mast cell-specific proteases present within the mast cell granules. M_T cells exclusively express tryptase and are located in intestinal mucosa and lung alveoli. M_{TC} cells express tryptase, chymase, and other proteases and are found predominantly in skin and intestinal submucosa (Schwartz and Kepley, 1994; Baumgartner and Beaven, 1996). Rodent mast cells have been broadly categorized as mucosal and connective tissue (or serosal) mast cells on the basis of anatomic site, granule constituents, and reactivity to chemical stimulants (Ennerback, 1986). Connective tissue mast cells, as for example in skin, typically express heparin in granules, whereas mucosal mast cells in lung express chondroitin sulfate. Both cell types can be readily distinguished by histochemical stains (Ennerback, 1986). Also, connective tissue mast cells, but not mucosal mast cells, rapidly degranulate in response to polybasic compounds, typified by mastoparan and compound 48/80, as well as therapeutic/-diagnostic agents that are known to induce anaphylaxis-like (called anaphylactoid) reactions (Forsythe and Ennis, 2000). The form that the mast cell assumes within a particular tissue is thought to be determined by the cytokine milieu within the tissue under normal or pathological conditions.

Mastocytosis

Another clinical aberration of the mast cell system is mastocytosis, in which there is an abnormal increase in mast cells in tissues. Although the rubric cutaneous or systemic is often used to describe subtypes of this disease, mastocytosis is notably heterogeneous in its manifestations and probably etiology (Metcalfe and Akin, 2001; Valent et al., 2001a). The pathogenic factors underlying this group of disorders are not entirely clear, but a number of genetic mutations that lead to defects in signaling processes have been linked to this disease. A substantial number of adult patients with mastocytosis carry activating mutations in KIT, and some, with more limited disease, carry an activating mutation in the gene for the IL-4 receptor α-chain.

Scope of Chapter

In the following sections we describe the clinical manifestations of mast cell-related diseases, the possible genetic links to these disorders, the effects on signaling pathways, and how such information may lead to new therapeutic approaches. We focus on the mast cell because this cell contributes to both the acute and late phases of the allergic inflammatory reaction to IgE-directed antigens. Although basophils contribute to the acute IgE-mediated allergic reaction, particularly anaphylaxis, these cells probably have a diminished role compared with mast cells in the late phase of the allergic reaction. Otherwise, there is considerable overlap in the biology of the mast cell and basophil (Wedemeyer and Galli, 2000). Also, the therapeutic strategies discussed here for allergic diseases apply to both cell types.

MANIFESTATIONS OF MAST CELL-RELATED DISORDERS

Allergic Disorders and Role of Mast Cell Mediators

Asthma. Allergic asthma is most prevalent in children, less prevalent in adults, and rare in the elderly. Asthma is a reactive inflammatory response of the lung airways to allergens and chemical irritants. The diagnostic features are reversible bronchial hyperreactivity to such stimulants and airway inflammation. In the acute phase of the disease, bronchoconstriction is accompanied by increased airway mucus secretion, hyperemia, and leakage of plasma proteins. Later, there is a more pronounced inflammatory phase, called the late-phase response, caused by recruitment of eosinophils, macrophages, and lymphocytes, which are detectable by bronchoalveolar lavage (Wedemeyer and Galli, 2000; Page et al., 2001). Persistent allergic inflammation can lead to chronic tissue changes (chronic phase) with epithelial thickening, deposition of collagen beneath the epithelium, hypertrophy of airway smooth muscles, and proliferation of blood vessels. In some patients this progresses into irreversible decline in lung function even with aggressive therapy (Pare et al., 1997). Death is usually associated with the late and chronic phases of asthma when airways are already restricted by inflammation and obstructed by mucus.

In addition to human studies, mouse models of atopic asthma have been widely utilized to study the pathogenesis of asthma (Leong and Huston, 2001). Although the mouse models do not reproduce the chronic phase of inflammation found in humans, they have provided insight into the immunopathology underlying the early- and late-phase response to antigen. As noted above, various types of cells participate in the asthmatic response (Busse and Lemanske, 2001). Of these, the mast cell has a preeminent role in orchestrating the initial and late-phase responses (Wedemeyer and Galli, 2001; Page et al., 2001). Its role in the chronic inflammatory state is probably less crucial but still important. Of paramount importance in the initial response is the secretion of histamine and proteases from granules and the rapid de novo synthesis of eicosanoids such as the prostaglandins (PGs), primarily PGD_2, and the cysteinyl leukotrienes (LT) LTC_4, LTD_4, and LTE_4 (Page et al., 2001) (Table 10.1). Histamine induces bronchial smooth muscle contraction and increased vascular permeability via histamine H_1 receptors and mucus secretion via H_2 receptors. The cysteinyl leukotrienes are also potent bronchoconstrictors and induce increased vascular permeability (Page et al., 2001; Bingham and Austen, 2000).

The late-phase response is initiated, at least in part, by the release of chemokines and cytokines from stimulated mast cells over the course of several hours. Release of these factors is believed by some workers to be essential for the recruitment and function of other effector cells, particularly eosinophils and T cells. The latter cells (Wedemeyer and Galli, 2000), and even bronchial smooth muscle cells (Page et al., 2001), are additional sources of cytokines and inflammatory mediators and are likely to be important additional effector cells during the late-phase reaction. The above scenario, although still debated (Page et al., 2001; Leong and Huston, 2001), suggests that the mast cell-leukocyte cascade is crucial for the amplification and perpetuation of the IgE-mediated inflammation in lung airways (Wedemeyer and Galli, 2000).

The contributions of mast cell mediators, specifically in the late-phase and chronic inflammatory stages of asthma, are less clear because of the participation of eosinophils and T cells in these processes. Mast cell-derived eicosanoids could conceivably play a role in late-stage events (Bingham and Austen, 2000). The initial production of eicosanoids in stimulated mast cells is followed several hours later by sustained production of PGD_2 due to induction of the prostaglandin synthase cyclooxygenase-2 (Reddy et al., 1999 and citations therein). Expression of cycloxygenase-2 continues for at least 24 h, and the mast cell is probably a major source of PGD_2 in the late-phase reaction. Nevertheless, interest is focused on the 5-lipoxygenase products of arachidonic acid, namely the cysteinyl leukotrienes (Drazen et al., 1999a), which are produced by both mast cells and eosinophils (Seymour et al., 2001). The ratio of PGD_2 to cysteinyl leukotrienes produced varies significantly (from 1:1 to 12:1) among different populations of mast cells (Austen and Boyce, 2001). Those in lung release 10-fold more LTC_4 than those in skin (Benyon et al., 1989). This may account for the fact that 5-lipoxygenase inhibitors are effective antiasthmatic drugs in some patients (Barnes, 1999). Mast cell tryptase (and potentially other mast cell-specific proteases) increases airways smooth muscle responsiveness to constrictors and stimulates fibroblast proliferation. Tryptase inhibitors reduce late-phase inflammation in allergen-induced asthma in animal models (Clark et al., 1995; Oh et al., 2002) and less so in humans (Krishna et al., 2001), although more potent inhibitors are in development (Barnes, 1999). As discussed below, these studies and past clinical experience with other mediator-antagonists suggest that cytokines, leukotrienes, and mast cell tryptase are important mediators in the inflammatory phase of asthma whereas histamine and the prostaglandins are not (Barnes, 1999).

Atopic Rhinitis. Allergic rhinitis (hay fever) is the most common of the atopic diseases. Although less catastrophic than severe asthma or anaphylaxis, the sleeplessness and fatigue accompanying hay fever can disrupt the patient's performance and well being. The primary symptoms include sneezing, itching, rhinorrhea, and nasal congestion. The disease may progress to or be associated with allergic asthma. Allergic rhinitis has been categorized as seasonal or perennial (Skoner, 2001). Seasonal allergic rhinitis is easily diagnosed because its rapid onset and cessation correlate with the seasonal exposure to pollen. A positive pollen skin test would confirm the diagnosis. The diagnosis may be obscured if the patient also becomes sensitized to nonseasonal allergens such as cigarette smoke and perennial allergic rhinitis ensues. Symptoms may then be difficult to distinguish from chronic sinusitis and vasomotor (idiopathic) rhinitis unless accompanied by elevated total serum IgE and positive allergen skin tests. As in asthma, a rapid "early-phase" response to allergen, characterized by itching, sneezing, and watery rhinorrhea, is followed some hours later by a "late-phase" response of nasal congestion associated with nasal tissue inflammation. The immunopathology resembles that in asthma but with some differences. The efficacy of antihistamine drugs, alone or in conjunction with leukotriene antagonists (the effects are additive), in the treatment of allergic rhinitis suggests that mast cell histamine and leukotrienes are important inflammatory mediators throughout the disease process (Lipworth, 2001). There is also good evidence that nasal mast cells upregulate local nasal IgE production through release of IL-4/IL-13 and may be more significant source of these T_H2-type cytokines than T cells in nasal tissue (Pawankar et al., 2000). High IgE levels, in turn, promote mast cell survival and upregulate expression of FcεRI (Asai et al., 2001), thereby amplifying responses with continued exposure to allergen.

Atopic Eczema. Atopic eczema is most common in young children, of whom 15–20% may be affected. The patients often have a personal or family history of asthma or hay fever. The disease is usually of early onset (before 3 years of age) and manifests itself as itchy skin rashes, particularly on the extremities in young children and flexural surfaces in adults (Thestrup-Pederson, 2000). The disease may result in craniofacial changes (the atopic face) with suborbital and anterior neck folds and hypopigmented patches. Symptoms include dry skin, excoriations, and other diverse symptoms due to physical damage, infection (bacterial or viral), or contact with chemical irritants. The skin is sensitive to irritants and may remain so even when the main symptoms of atopic eczema have disappeared. Occupational contact dermatitis is often observed in patients with a previous history of atopic dermatitis. Otherwise, the disease spontaneously disappears in the majority of children, although it persists through the teenage years into adulthood in some patients (Turvey, 2001).

By definition, atopic eczema has an atopic background that can be detected by elevated serum IgE levels, presence of allergen-specific IgE, or positive skin tests to allergens in many patients. The inflamed skin contains activated T lymphocytes and eosinophils as well as increased numbers of mast cells and dendritic cells (Thestrup-Pederson, 2000) associated with a T_H2 response (Anderson, 2001). There is accumulating evidence of a link to food and air-borne allergens (Sicherer and Sampson, 1999; Schafer et al., 1999; Turvey, 2001). Up to 40% of children with moderate to severe atopic eczema have allergies to ingested food allergens (Sicherer and Sampson, 1999). These and other observations (Rukwied et al., 2000; Horsmanheimo et al., 1994) suggest a role for mast cells in the disease, but the evidence is less compelling than that for experimental models of cutaneous late-phase reactions in animals (see, e.g., Togawa et al., 2001) or human subjects (Barata et al., 1998). That classic histamine H1 antogonists, but not the newer nonsedating antihistamines, attenuate itching in some patients has led to suggestions of a neurogenic component to the disease (Greaves, 2000) and the involvement of other mast cell products (Rukwied et al., 2000). Mast cell chymase (Mao et al., 1996) and IL-4 (Horsmanheimo et al., 1994) have been reported as potential candidates. The eosinophil chemoattractants RANTES and eotaxin (Morita et al., 2001) as well as T_H2 cytokines (Ong et al., 2002) have also been implicated in the disease process.

Conjunctivitis. Conjunctivitis is often associated with atopic rhinitis (rhinoconjunctivitis) or atopic dermatitis (keratoconjunctivitis) (Bielory, 2000). Nevertheless, the ocular symptoms can be the prominent component of the allergy in some patients. The clearly allergic forms of the disease, whether seasonal or perennial (Anderson, 2001), are mediated predominantly by conjunctival mast cells and recruited eosinophils (Graziano et al., 2001). As in rhinitis, the therapeutic experience with antihistamines indicates that histamine is a major inflammatory mediator for allergic conjunctivitis (Anderson, 2001). In atopic keratoconjunctivitis and other chronic forms of conjunctivitis, the involvement of mast cells is unclear and may be T cell mediated (Buckley, 1998).

Food Allergies. Unlike most atopic diseases, the inflammatory reactions to food allergens occur in multiple organs including skin, airways, gastrointestinal tract, and the cardiovascular system (Sampson, 1999a; Sampson, 1999b). The skin and respiratory airways are most commonly affected to produce urticaria, angioedema, skin flush, and asthma like symptoms. The oral cavity, nose, and larynx are often affected with pruritus, edema, and tingling of oral tissues and lips as well as rhinitis, hoarseness, and dry cough. Gastrointestinal symptoms maybe subclinical (i.e., loss of weight and appetite) or disabling (i.e., colic, vomiting, diarrhea). More rarely, but of grave clinical concern, the involvement of the cardiovascular system can lead to the classic symptoms of anaphylaxis, namely, syncope, arrythmia, hypotension, and vascular collapse (Sampson, 2000). The time course is variable. Symptoms may appear within seconds or develop over the course of hours. The reaction can be biphasic (in about 30% of patients), with symptoms reappearing after several hours.

Oral provocation tests indicate that the true prevalence of food allergies, although lower than widely assumed, ranges from 2% in adults to as high as 8% in children (Sampson, 1999a). However, food-induced anaphylaxis is the leading single cause of anaphylaxis in the U.S. and other developed countries (Sampson, 2000). Potentially any food is capable of inducing food allergy. However, a few items (peanuts, tree nuts, fish, shell fish in adults plus milk, eggs, and wheat in children) account for 80–90% of established food allergies (Sampson, 1999a).

The diagnostic challenge is to differentiate IgE-mediated from non-IgE-mediated inflammatory intestinal disorders. At this time, clinically acceptable serological/test reagents are limited and skin tests yield a high proportion of false positives. Apart from a personal or family history of atopic disease, the responses to allergen-deficient diet and placebo-controlled food challenge remain essential tests for diagnosis (see discussion in Sampson, 1999b). The characteristics of food allergy suggest that the access of ingested allergen to the circulation is due to an immature or dysfunctional gastrointestinal epithelial barrier (DeMeo et al., 2002). It is believed that the immaturity of this barrier accounts for the prevalence of food allergy in infants, especially in the first few months of life (Sampson, 1999a). The introduction of food solids by 4 months has been linked to later development of atopic disease in genetically predisposed infants. Even in adults, ~2% of ingested food antigens are absorbed immunologically intact. However, most adults develop "oral tolerance" to these ingested antigens.

As in other atopic diseases, clinical and animal studies support the T_H2/IgE paradigm for development of food allergy (Helm and Burks, 2000). Intestinal biopsies indicate increased numbers of intraepithelial lymphocytes, mucosal mast cells, and eosinophils associated with food allergy. The question of whether or not the disease starts as a localized inflammatory reaction in the gut and then becomes systemic is unclear (Helm and Burks, 2000). However, the rate and extent of allergen absorption are markedly increased during IgE-mediated reactions in gastrointestinal tissues, possibly as a consequence of increased production of IL-4 by activated T cells, mast cells, and circulating basophils (Sampson, 1999a).

Allergen-Independent Chronic Urticarias

Although release of inflammatory mediators from mast cells underlies most forms of urticaria, the majority of urticarias are not induced by allergens. A subgroup of chronic urticarias, about 40%, are induced by physical stimuli such as heat, exertion, sweating (cholinergic urticaria),

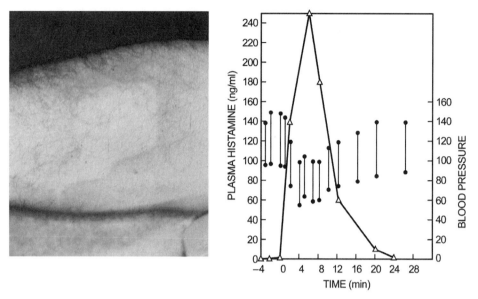

Figure 10.1. *Cold-Induced Urticaria and Histamine Release.* The left-hand panel shows the reaction after removal of an ice cube that had been on the arm of patient with cold-induced urticaria. This reaction has all the hallmarks of the classic "triple response" and develops as the exposed tissue rewarms. The right-hand panel shows the correlation between histamine release into the circulation (blood was collected from the brachial vein) and blood pressure after removal of the hand from ice- water (the hand was immersed for 4 min). The drop in blood pressure was also accompanied by flushing and tightness of chest. The patient had a history of such reactions after wading in cold seawater. Data from the study of Kaplan and Beaven (1976).

cold (cold induced urticaria) (Fig. 10.1), sunlight (solar urticaria), and pressure (symptomatic dermographism). These forms exhibit characteristic wheal-and-flare or angioedematous reactions, which may be localized or generalized or, in rare cases, result in fatal anaphylaxis (Greaves, 2002). The symptoms usually subside within 30 to 90 minutes. Cholinergic urticaria, as its name implies, is dependent on the sympathetic cholinergic pathway, and the rash is readily blocked by prior application of atropine to the skin. Early studies of some of these urticarias revealed that histamine release accompanied the onset of symptoms (Kaplan and Beaven, 1976; Metzger et al., 1976; Soter et al., 1976) (Fig. 10.1) and that antihistamines relieved the itching but not necessarily the rash (Greaves, 2002). Passive transfer studies also pointed to the involvement of circulating factors.

Another subgroup of patients, about 50%, have a chronic urticaria that is not attributable to physical or foreign agents and is appropriately called idiopathic chronic urticaria. In contrast to physically induced urticarias, the urticarial reactions persist for many hours and reappear periodically. Wheals and frequently angioedema may occur on any or extensive areas of the skin. The precipitating cause is unknown, although *Helicobacter pylori* and other infectious agents have been invoked (Greaves, 2002; Kaplan, 2002). As with physical urticarias, early studies showed that the disease was associated with release of histamine (Kaplan et al., 1978) and circulating mediating factors (Grattan et al., 1986) in a substantial number of patients.

It now appears that the chronic urticarias, physical and idiopathic, have an autoimmune basis in about half of patients (Kaplan, 2002). Of these patients, most have circulating IgG antibodies against the α-subunit of FcεRI (Hide et al., 1993; Fiebiger, 1995; Ferrer et al., 1998; Tong et al., 1997) and a smaller number have IgG antibody against IgE (Tong et al., 1997). Other patients without detectable IgG antibodies have unidentified circulating mast cell/basophil activating factors (Tong et al., 1997). Antibodies (primarily IgG2 and IgG4) against the FcεRIα subunit are also prevalent in connective tissue autoimmune diseases that can affect skin (e.g., pemiphigus vulgaris and systemic lupus erythematosis). In contrast, those found in chronic urticaria are uniquely mast cell activating and are of the complement fixing subtypes IgG1 and IgG3 (Fiebiger et al., 1998). These antibodies also activate rat RBL-2H3 mast cells made to express human FcεRIα (Tong et al., 1997), and they presumably induce aggregation of FcεRI to elicit the normal array of activating signals in these cells. The IgG-mediated mast cell activation is augmented by C5a produced as a consequence of complement fixation (Kukuchi and Kaplan, 2002). As in late-phase cutaneous allergic reactions, skin lesions have infiltrations of CD4-positive lymphocytes, monocytes, neutrophils, and eosinophils (Kaplan, 2002).

It is unclear why circulating IgG autoantibodies do not activate mast cells in tissues other than skin. Here again, mast cell heterogeneity could be a determining factor (Greaves, 2002). Pulmonary mast cells, in contrast to skin mast cells, are not activated by C5a (Schulman et al., 1983). Chronic urticaria is commonly associated with Hashimoto disease and, less frequently, Graves disease along with antithyroglobulin and antimicrosomal antibodies (Kaplan, 2002). These thyroid disorders are probably a concurrent autoimmune disease process.

Mastocytosis

The hallmark of mastocytosis is the abnormal growth and accumulation of mast cells in one or more organ systems. The manifestations are variable and range from a disease restricted to skin (cutaneous mastocytosis) to one that involves multiple organs (systemic mastocytosis). The primary pathology varies from mild mast cell hyperplasia to severe neoplastic mast cell proliferation. Mastocytosis can also be associated with disorders of myeloid proliferation and differentiation (Valent et al., 2001c).

Cutaneous mastocytosis occurs primarily in young children who exhibit maculopapular pigmented lesions in the skin, a condition originally referred to as urticaria pigmentosa. This condition is associated with urticaria and pruritis, with the latter symptom predominating at joints such as the wrist. Biopsies of the lesions indicate multifocal accumulations of mast cells. Less frequently, children exhibit a diffuse cutaneous form of mastocytosis with diffuse infiltrates of mast cells in the skin associated with thickened skin and perhaps occasional nodules with high populations of mast cells. Solitary skin mastocytomas with densely packed mast cells can occur in some children. The childhood variants of cutaneous mastocytosis are associated with relatively mature mast cells of the phenotype normally found in skin. Accordingly, these cells stain metachromatically and express tryptase and chymase. The patients show localized areas of edema and erythema when the affected skin is rubbed (the Darier sign). The predominant mediator is histamine, and the condition can be alleviated with topical application of antihistamine drugs. The prognosis for childhood cutaneous mastocytosis is good, as the disease usually resolves spontaneously before puberty.

In adults, the disease usually manifests itself systemically as multifocal lesions in bone marrow and other tissues. The variants of adult mastocytosis are many, but a recent consensus report (Valent et al., 2001a) proposes the following broad categories: indolent systemic mastocytosis (ISM), aggressive systemic mastocytosis (ASM), mast cell leukemia (MCL), and systemic mastocytosis that is associated with clonal hematologic non-mast cell disease (SM-AHNMD).

Patients with ISM typically have urticaria pigmentosa-like skin lesions and, in addition, diffusely scattered mast cells in bone marrow and sometimes organs such as liver, spleen, and the gastrointestinal tract. The infiltration of mast cells in extracutaneous tissue is low grade and insufficient to disrupt the organ architecture or function except for occasional visceromegaly. This form of the disease, as its name indicates, is clinically indolent, and its symptoms are mainly related to mediator release from mast cells.

In ASM the infiltration of mast cells in organs is aggressive and typically involves bone marrow, liver, spleen, and the gastrointestinal tract. Here the damage and functional impairment of organ function can be severe and life threatening. Diagnostic biochemical abnormalities accompany organ impairment. The mast cells themselves may exhibit significant abnormalities in morphology such as hypogranulation and atypical nuclei. The numbers of mast cells are sufficient to raise serum levels of mast cell tryptase. The progression of the disease is variable, ranging from slow to rapid.

MCL is a malignant disease marked by the presence of leukemic mast cells in bone marrow and blood. Usually the disease is only recognized once mast cells account for a significant portion of nucleated cells in bone marrow smears and blood. Onset of the disease is rapid, and the clinical symptoms are usually related to mediator release from mast cells such as hypotension, flushing, and diarrhea. At later stages as other organs become involved symptoms include bone pain, visceromegaly, weight loss, and internal bleeding due to defective coagulation through release of mast cell heparin. The disease progresses very rapidly, and there is no known therapy.

The primary disorder in SM-AHNMD, as the name implies, is not in the mast cell lineage but in dyscrasias primarily of myeloid origin such as myelodysplasic syndrome, myeloproliferative syndrome, acute myelocytic leukemia, and non-Hodgkin lymphoma. Myeloid progenitor cells have some capacity to differentiate into mast cells, and this capacity may account for the appearance of immature, blastlike mast cells in bone marrow and peripheral blood. Skin lesions are unusual in this group of disorders. In addition to the above, there are occasional reports of mast cell sarcomas and extracutaneous mastocytomas. Unlike the mast cell sarcomas, mastocytomas (primarily in lung) are not destructive and mast cells appear normal.

The clinical features in mastocytosis that can be attributed to release of histamine and eicosanoids from mast cells include, for example, pruritus and urticaria in skin, gastrointestinal symptoms (gastric hypersecretion, cramping, diarrhea), pulmonary distress (bronchoconstriction, pulmonary edema, mucus secretion), vascular instability, and increased vascular permeability. Release of mast cell proteases and proteoglycans such as heparin may lead to bone lesions, osteoporosis, and localized bleeding. Release of mast cell-derived cytokines and growth factors results in infiltration of other types of cells (lymphocytes, eosinophils), fibrosis, and cachexia.

Readers are referred to the consensus report (Valent et al., 2001a) for further details on the clinical/biochemical diagnostic features of the various categories of mastocytosis listed above.

POLYMORPHISMS/MUTATIONS LINKED TO MAST CELL-RELATED DISEASES

Atopic Diseases

Background Information. The literature increasingly points to a genetic basis for some features of atopic diseases with two qualifications. First, it is unlikely that genetic factors account for the growing prevalence of these diseases. Second, it is likely that multiple genes and different sets of genes are involved to account for the diversity of atopic phenotypes such as rhinitis, asthma, and eczema. Genetic mutations that result in, for example, elevated IgE production are probably accompanied by another set of mutations that lead to clinical manifestations in specific tissues such as airway hypersensitivity in asthmatic patients (Feijen et al., 2000). The genetic landscape must account for the overlap in atopic phenotypes. As noted above, some but not all asthmatic patients suffer from allergic rhinitis or atopic eczema and some patients with atopic eczema suffer from rhinitis. Also, not all asthmatic patients are atopic.

At present there is no coherent picture of the genetic basis for any atopic disease. A review of the Asthma Gene Database (Wjst and Immervoll, 1998) indicated more than 500 atopy and asthma loci throughout the human genome (Hakonarson and Wjest, 2001), yet few of these linkages were regarded as achieving reasonable levels of significance. Also, inconsistent replication of original findings suggested either that there is variability in the frequencies of the genes in question among different populations or that the genes are in linkage disequilibrium with other genes. There are, nevertheless, plausible genetic linkages and candidate genes that include those for IL-4, the IL-4 receptor α-chain, and FcεRI among others. Most of these involve single-nucleotide polymorphisms (SNPs) in the coding (and hence an amino acid substitution) and regulatory domains of the gene. A few have been described in intergene segments. In this section we have restricted our discussion to genetic polymorphisms that might affect mast cell function as shown in Table 10.3. For more complete accounts of the genetics of atopic disease the reader is referred to several recent reviews (Feijen et al., 2000; Ono, 2000; Hakonarson and Wjest, 2001).

Polymorphisms Associated with the T_H2 Response and IgE Production.
Several candidate genes for atopy are located in a so-called cytokine-gene cluster on chromosome 5q31–33 (Ono, 2000; Hakonarson and Wjest, 2001). This cluster includes genes for CD14 as well as IL-3, -4, -5, -9, and -13. Interest has focussed on IL-4, IL-13, and to some extent CD14 because of their role in regulating IgE production. IL-4 promotes maturation of the T_H cell to the T_H2 phenotype and a switch to IgE production in B cells. IL-13 acts directly on B cells to induce production of IgE. CD14 (a high-affinity receptor for lipopolysaccharide and other bacterial products) is though to polarize the immune response toward the T_H1 phenotype and depress IgE production. As noted in Table 10.3, polymorphisms in the genes for IL-4 and IL-13 as well as the sequence between these genes are found at higher frequency in atopic subjects than unaffected subjects (Noguchi et al., 2001 and citations therein). In the case of the CD14 gene, one haplotype (a combination of alleles derived from two or more polymorphisms) defined by three nucleotide polymorphisms is associated with low serum IgE levels and the complementary haplotype with high IgE levels (Vercelli et al., 2001). Homozygotes for one of these polymorphisms were also found to have higher incidence of allergic rhinitis or hay fever along with high serum IgE (Koppelmann et al., 2001).

Studies have also focused on the receptors and the downstream molecules in the IL-4/Il-13 signaling pathways (reviewed in Howard et al., 2002). Both IL-4 and IL-13 can act thorough the IL-4 receptor (IL4R) or the IL-13 receptor (IL13R), both of which share a common IL4Rα subunit (Shirakawa et al., 2000). Of the 16 polymorphisms reported for the IL4Rα gene (*IL4RA*, located on chromosome 16p12), several are associated with higher frequency of atopic asthma and eczema (Hershey et al., 1997; Rosa-Rosa et al., 1999; Howard et al., 2002) (Table 10.3). Two

TABLE 10.3. Polymorphisms, Atopic Diseases, and Mastocytosis

Gene	Location	Polymorphism	Effect	Phenotype	References
Atopic disease					
IL4	5q31–32	SNPs/promoter & between *IL4/IL13* genes	Promoter activity ↑	Atopy/asthma with certain haplotypes	Rosenwasser et al., 1995; Noguchi et al., 2001
IL13	5q31	R110Q (R130Q?)[a]	Binding to receptor expression?	Atopy/asthma	van der Pouw Kraan et al., 1999; Heinzmann et al., 2000; Graves et al., 2000; Noguchi et al., 2001
CD14	5q31–33	SNPs/promoter	sCD14-expression ↑	IgE according to haplotype; Asthma/rhinitis	Baldini et al., 1999; Vercelli et al., 2001; Koppelmann et al., 2001
B2AR (ADRB2)	5q31–33	R16G Q27E I164T; SNPs/promoter	Alters β_2AR[b] function; Alters β_2AR expression	haplotypes correlate with asthma & responses to β-agonists	Drysdale et al., 2000; Ulbrecht et al., 2000; Liggett, 2001; Summerhill et al., 2000; Holloway et al., 2000
ALOX5	10q11.2	SNPs/promoter	5-LO[b] expression ↓?[c]	Diminished response to antileukotrienes	Drazen et al., 1999b
T-bet	11	None described; Deletion in mice	T_H2 response; None detected	T-bet ↓ in asthmatics; Asthma in mice	Szabo et al., 2002
FcεR1β	11q13	I181L–V183L[d] E237G; SNPs/5'UTR & introns		Asthma; IgE ↑ or no correlation	Donnadieu et al., 2000; Furomoto et al., 2000; Palmer et al., 1997; Rohrbach et al., 1998; Dickson et al., 1999

TABLE 10.3. (*Continued*)

Gene	Location	Polymorphism	Effect	Phenotype	References
STAT6	12q15	SNPs/5′ Flank, Intron 18, & 3′UTR	Expression? ? Expression?	Eosinophilia Atopy Asthma	Duetsch et al., 2002 Tamura et al., 2001 Gao et al., 2000
MCC (MCA1)	14q11.2	SNPs/5′ flank	?	Atopy/eczema	Mao et al., 1998 Kawashima et al., 1998
IL4RA	16p12	V50I	Signaling ↑	Atopy/asthma	Izuhara et al., 2000 Mitsuyasu et al., 1999
		E375A C406R		Atopy when linked to other SNPs	Howard et al., 2002 Ober et al., 2000
		P478S (P503S)[a] Q551R (Q576R)[a]	Signaling ↑ Signaling ↑	Atopy/asthma Atopy/eczema Asthma	Kruse et al., 1999 Hershey et al., 1997 Rosa-Rosa et al., 1999
RANTES	17p12 –17q11.2 C-C cluster	G403A	New GATA Promoter activity ↑	Eczema	Nickel et al., 2000
Mastocytosis					
c-kit	4q11–12	V560G D816V	Activating Activating	Indolent M Systemic M	Buttner et al., 1998 Nagata et al., 1995 Longley et al., 1996 Metcalfe and Akin, 2001
		D816F/Y/H	Activating	Systemic M	Longley et al., 1999
		D820G E839K	Activating? Inactivating	Systemic M Pediatric M	Pullarkat et al., 2000 Pignon et al., 1977 Longley et al., 1999
IL4R	16p12	Q551R (Q576R)[a]	Activating	Indolent M cutaneous M	Daley et al., 2001

[a] Enumeration dependent on whether based on the coded or the mature protein.

[b] Abbreviations: β2-AR, β2-adrenergic receptor; 5LO, 5-lipoxygenase; M, mastocytosis; MCC, mast cell chymase.

[c] ? Indicates that functional effects were surmised rather than established.

[d] A double mutation.

variants, the substitution of arginine for glutamine-551 (Q551R, single-letter amino acid code) and the substitution of isoleucine for valine-50 (V50I), were shown by biochemical or expression studies to be gain of function, that is, they enhanced activity of IL4Rα or increased production of IgE in response to IL-4 (Hershey et al., 1997; Izuhara et al., 2000). The effects of these two mutations on mast cell function have not been assessed, although the IL-4R is expressed on human mast cells and can influence mast cell growth and differentiation (Table 10.2). Two other polymorphisms (E375A and C406R) have been linked to atopic disease in some populations, but the functional consequences have not been assessed (Howard et al., 2002; Ober et al., 2000). Polymorphisms in the genes for other molecules in the IL4/IL13 signaling pathways have been examined, and possible associations with elevated serum IgE have been identified in the *IL13Rα* and *BCL6* genes (discussed in Shirakawa et al., 2000; Steinke and Borish, 2001).

The prevailing belief is that the atopic predispostion is conferred not by a single allele but by a combination of polymorphisms within a gene or several genes. Indeed, one study suggests that susceptibility is dependent on additional variation outside the coding region of the *IL4Rα* gene (Ober et al., 2000). Also in support of the notion that asthma is a polygenic disease, a rare T/C variant in the promoter region of the IL-13 gene is associated with airway hypersensitivity but not elevated serum IgE. Subjects carrying this variant in addition to one of the *IL4RA* atopic variants showed severalfold greater risk of asthma than subjects carrying the *IL4RA* variant alone (Howard et al., 2002).

It is also possible that the true atopic allele, if any, remains undiscovered. Chromosomal mapping suggests the presence of atopic genes on chromosome 11. One gene on chromosome 11, *T-bet*, has been directly linked to asthma but has not been examined for polymorphisms. Asthmatic patients express low levels of the gene product, T-bet, in lung tissue. Moreover, deletion of the *T-bet* gene in mice resulted in a mouse phenotype with all the hallmarks of human asthma including airway hypersensitivity to methacholine. Histologically, the lungs of these mice resembled those of chronic asthmatic patients. This and an associated study (Szabo et al., 2002) cast light on a possible scenario for the pathogenesis of asthma. T-bet, a T_H1-specific T-box transcription factor, prompts immature immune cells to become T_H1 cells rather than T_H2 cells. Deletion of *T-bet* leads to low expression of T_H1 cells and their major product, interferon-γ, as well as enhanced levels of the T_H2 cytokines IL-4 and IL-5.

Polymorphisms in the FcɛRIβ Gene. Chromosome 11 also contains the gene for FcɛRI β-chain (FcɛRIβ) (at 11q13). Two polymorphisms have been described for this gene and are reported to be linked to positive IgE responses, rhinits, and asthma with a maternal pattern of inheritance. One is a double mutation in which leucine is substituted for isoleucine-181 and for valine-183 (I181L/V183L) and is located in the fourth transmembrane domain of the β-chain (Shirakawa et al., 1994). The second polymorphism is a substitution of glycine for glutamic acid at position 237 (E237G) and is located immediately downstream of the immunoreceptor tyrosine-based activation motif (ITAM) in the cytosolic tail of FcεRIβ (Hill and Cookson, 1996). Other polymorphisms exist in the regulatory and promoter region of the gene for FcεRIβ, but their relevance, if any, to receptor expression and function is unknown (Palmer et al., 1997; Rohrbach et al., 1998; Dickson et al., 1999).

The linkage of atopic disease to chromosome 11q13 and the FcεRIβ gene has substantial experimental support, primarily through the work of Cookson, Hopkin, and colleagues, although there are dissenting views (see discussions in Donnadieu et al., 2000 and Ono, 2000). The issue may be confounded by the probable polygenic nature of atopic diseases and the maternal inheritance of the FcεRIβ variants (Hill and Cookson, 1996).

Other Candidate Genes for Atopic Disease. Other genes under investigation that might impact on mast cell function include those encoding mast cell chymase (Mao et al., 1998; Mao et al., 1996; Kawashima et al., 1998), the chemokine RANTES (Nickel et al., 2000), the signal transducer and activation of transcription (STAT) 6 (Gao et al., 2000; Tamura et al., 2001; Duetsch et al., 2002), 5-lipoxygenase (Drazen et al., 1999b), and the β-adrenergic receptor (Ulbrecht et al., 2000). Some of these genes are located at chromosomal sites previously linked to atopic diseases, namely, chromosome 5q31–32 (the β-adrenergic receptor gene), chromosome 12q13 (STAT6 gene), chromosome 14q11 (the mast cell chymase gene), and the C-C chemokine cluster on chromosome 17q12–17q11.2 (the RANTES gene). Polymorphisms in the promoter regions of the genes for mast cell chymase (Mao et al., 1996) and RANTES (Nickel et al., 2000) are associated with atopic dermatitis but not with allergic rhinitis or asthma.

The findings with mast cell chymase are intriguing because the enzyme is abundant in skin mast cells but not in bronchial or pulmonary mast cells. Along with other genetic variations contributing to atopy, polymorphism in the mast cell chymase gene would provide the additional variable required for targeting skin in atopic eczema. However, the original reports (Mao et al., 1996; Mao et al., 1998) have not been replicated (see Kawashima et al., 1998), and the consequences of this polymorphism on levels of the enzyme by mast cells has not been assessed.

RANTES is a potent chemoattractant for eosinophils, basophils, lymphocytes, and monocytes and is produced in airway epithelial and mast cells. The mutation in *RANTES* that is associated with atopic eczema results in a new consensus binding site for GATA transcription factors and substantially augments promoter activity of the gene (Nickel et al., 2000). The functional consequences of the STAT6 gene polymorphisms (all 13 GT repeats in noncoding regions) have not been examined. The most recent work finds no association of any *STAT6* allele with asthma as such. Rather, one allele showed a strong association with elevated IgE levels and another with eosinophilia (Duetsch et al., 2002).

The remaining polymorphisms may have therapeutic consequences. About 6% of asthmatic patients lack a wild-type allele in the core promoter of the 5-lipoxygenase gene (*ALOX5*) in chromosome 10q11.2 and do not respond to treatment with 5-lipoxygenase inhibitors (Drazen et al., 1999b). Therefore, leukotriene production does not appear to be critical for expression of asthma in this particular subset of patients. The β-adrenergic receptor mediates airway smooth muscle contraction and inhibits release of mediators from human lung mast cells (Weston and Peachell, 1998). The gene encoding this receptor contains three SNPs that result in amino acid substitutions in the coding sequence and multiple SNPs in the promoter region. The former SNPs alter receptor function, whereas the latter alter receptor expression (Drysdale et al., 2000). Although studies of the effects of individual SNPs on bronchodilatory responses have provided inconsistent results, studies of combinations of SNPs show strong correlations in responses to β-adrenergic receptor agonists to different combinations of SNPs (Drysdale et al., 2000; Liggett, 2001).

Mastocytosis

Mutations of c-kit *Linked to Mastocytosis.* The polymorphisms that we have described so far involve variant alleles distributed differently in normal and atopic populations. Mastocytosis, in contrast, involves variant alleles that are absent or rare in normal populations and may represent true mutations. Mutations in *c-kit* (Fig. 10.2), which is located on chromosome 4q11–12 and maps to the *white spotting* locus (*W*) in mouse (D'Auriol et al., 1988), have been implicated in a number of human diseases. In animals, and sometimes humans, mutations leading to loss of function or expression of KIT or SCF result in decreased fertility, macrocytic anemia, hypopigmentation, and deficiency in mast cells. Gain-of-function mutations result in neoplastic transformation of mast cells (as described below) and are found in many gastrointestinal stromal cell tumors and some germ cell tumors in humans (Ashman, 1999; Longley et al., 2001; Smith et al., 2001; Boissan et al., 2000). Of the mutations associated with mastocytosis and mast cell neoplasia, the most common and the first described is a single point mutation that leads to the substitution of valine for aspartate-816 (D816V) in Kit and the constitutive activation of the receptor. This mutation is present in many patients with adult-onset systemic mastocytosis and is detectable in peripheral blood monocytes (Nagata et al., 1995) and tissue mast cells (Longley et al., 1996) as well as in the HMC-1 mast cell line (Furitsu et al., 1993), which was originally derived from a patient with a rare and aggressive mast cell leukemia (Butterfield et al., 1988). This mutation is also observed in the corresponding codon of *c-kit* in the rodent P815 and RBL-2H3 neoplastic mast cell lines.

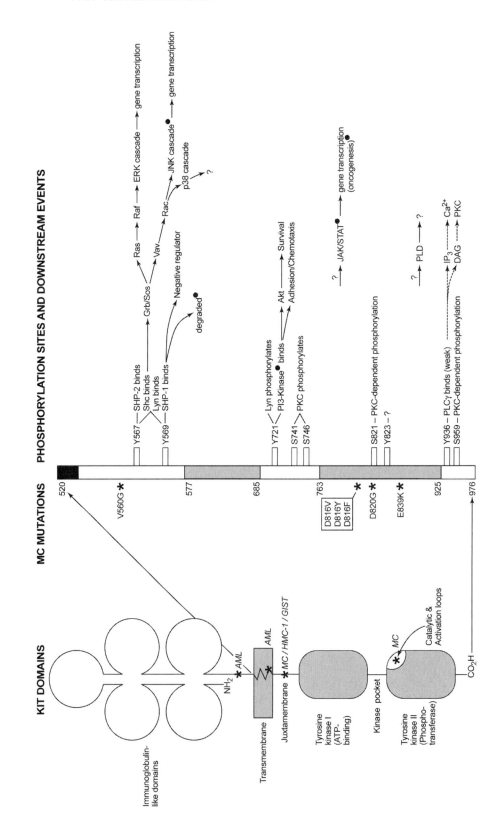

The D816V mutation has been detected in mast cells from different tissues in the same patient, thus demonstrating the clonal nature of mastocytosis (Longley et al., 1996; Nagata et al., 1998). This mutation is found in adult patients with indolent mastocytosis, although it is more common in patients with severe forms of mastocytosis associated with early to advanced myeloproliferative disease (Worobec et al., 1998). In some but not all cases, the mutation may be found in both neoplasms. Less frequently, D816 may be replaced by phenylalanine, tyrosine, or histidine (Longley et al., 1999; Pullarkat et al., 2000). All the described D816 mutations are activating mutations (Moriyama et al., 1996). A nearby mutation, D820G, has been described in one patient with aggressive systemic mastocytosis (Pignon et al., 1977). The only other mutation reported for adults is a V560G mutation in two patients in ISM (Buttner et al., 1998), which is also present in the HMC-1 cell line (Furitsu et al., 1993).

The D816V mutation is not expressed in the typical childhood form of mastocytosis (Buttner et al., 1998). Instead, another mutation in which lysine is substituted for glycine-839 (E839K) may be expressed (Longley et al., 1999). This is a dominant inactivating mutation, and it is puzzling why this mutation should be associated with mast cell malignancy in children. Rather than being part of the malignant process, the E839K mutation may instead help limit the severity of the disease in children. Certainly, the association of the D816V mutation with persistent disease in adults and the E839K mutation with the transient childhood forms of the disease suggest that these two mutations dictate the course of the mastocytosis.

Not all adult patients exhibit the D816V mutation (Longley et al., 1999), and it is possible that some other undetected mutation(s) contribute to the pathogenesis of the disease. However, a cluster of activating mutations in the cytosolic juxtamembrane coding region of *c-kit* (see

Figure 10.2. *A Schematic Diagram of KIT, Sites of Mutation, and Signaling Events.* The left-hand diagram shows the regions within KIT where mutations occur and diseases associated with these mutations (AML, acute myelogenous leukemia; MC, mastocytosis; HMC-1, the human mastocytosis cell line; GIST, gastrointestinal stromal tumour). The right-hand diagram shows the mutations associated with mastocytosis in the cytosolic domain of KIT. The sites of phosphorylation, the molecules interacting with these sites, and some of the downstream signaling events are also depicted. V560 D816 are activating mutations, and E839 is an inactivating mutation (see Table 10.3). Components of the signaling/functional pathways that are known to be constitutively activated (some are not—see text) as a result of the D816 mutation are noted (solid circles). Pathways leading to the activation of the JAK/STAT and PLD pathways are uncertain. The phospholipase C-mediated pathway is minimally stimulated via KIT. The diagram is based on published data (Linnekin, 1999; Taylor and Metcalfe, 2000; Boissan et al., 2000; and other authors) and our unpublished data. Abbreviations: PI3-kinase, phosphatidylinositol 3′-kinase; PK, protein kinase; PL, phospholipase; IP_3, inositol 1,4,5-trisphosphate; DAG, diacylglycerol; other abbreviations as defined in text.

next section) have been found in neoplastic mast cell lines and masto-cytomas from dog (and gastrointestinal stromal cell tumors). These include the previously noted V560G mutation in the human HMC-1 cell line (Furitsu et al., 1993) and the deletion of seven amino acids (T573–H579) in the mouse FMA3 mast cell line (Tsujimura et al., 1996). Both of these defects were found to be activating and oncogenic when the defective genes were introduced into *c-kit*-negative mast cells (Tsujimura et al., 1996; Hashimoto et al., 1996). Mastocytomas in dog, a common, often aggressive tumor in this species, may exhibit activating point mutations (W556R and L575P) or activating deletions (W556-K557 and V558) in the juxtamembrane coding region (Ma et al., 1996b). Other mutations in regions encoding the extracellular and transmembrane domain of KIT are associated with myeloproliferative disorders and acute myelocytic leukemia, which sometimes give rise to increased numbers of immature mast cells in bone marrow and peripheral blood (Longley et al., 2001). However, the mast cell dysplasia may be secondary to the underlying myeloproliferative disorder (Valent et al., 2001c) because myeloid progenitors have a certain capacity to differentiate into mast cells.

Appearance of the Atopic IL4Rα Gene in Mastocytosis. In addition to the association of the R576 allele of *IL4Rα* with atopic disease as noted above, mastocytosis patients bearing this allele usually have the less severe cutaneous form of mastocytosis (Daley et al., 2001). The gain in function produced by this polymorphism is thought to restrict extensive mast cell hyperplasia that might be induced by activating muations in *c-kit* or other predisposing cryptic events. The R576 allele is not directly linked to mastocytosis, however, because the proportion of patients bearing this allele is similar to that in the general population (~30%).

EFFECTS ON RECEPTOR SIGNALING

KIT-Mediated Signaling

Structure and Activation of KIT. KIT is a member of the type III receptor tyrosine kinase family of enzymes that include receptors for colony-stimulating factor 1 and platelet-derived growth factor among others (Ullrich and Schlessinger, 1990). In its unliganded state, KIT exists as au enzymatically inactive monomer that contains an extracellular immunoglobulin-like loop domain, a cytosolic juxtamembrane amphipathic helix, and two cytosolic enzymatic subdomains (ATP-binding and phosphotransferase subdomains) (Fig. 10.2A). The latter subdomains are connected by an activation loop to form an enzymatic pocket (Ullrich and Schlessinger, 1990). The amphipathic helix serves as an inhibitory regulator of Kit kinase activity in the absence of SCF (Ma et al., 1999a; Longley et al., 2001). The mutable sites in KIT and the effects on signaling events are depicted in Figure 10.2.

The KIT ligand, SCF, exists predominantly as a bivalent dimer. Binding of SCF to KIT induces dimerization of the receptor, phosphorylation of tyrosines in the juxtamembrane region (amino acids 544–577), and the binding/activation of Src tyrosine kinases as well as Src homology domain 2-containing protein tyrosine phosphatases, SHP-1 and SHP-2, to these phosphorylated residues (Linnekin, 1999). In turn, the activation of Src kinase leads to further phosphorylation of additional tyrosines including Y721 in KIT. The latter serves as docking site for the p85 regulatory subunit of phosphatidylinositol 3-kinase (p85^{PI3K}) and thus initiates activation of this enzyme and downstream substrates such as Akt. Other signaling moleules are also activated, although the mechanisms are not clear in all details. These include Gab2 (Nishida et al., 2002), phospholipase Cγ, the Janus kinase (JAK)/STAT pathway, and the MAP kinases (references in Linnekin, 1999; Boissan et al., 2000).

Perturbations Induced by Mutations at Sites Coding the Enzymatic Pocket of KIT.

Alterations in the vicinity of the enzymatic pocket of KIT can be either activating (i.e., D816V) or inactivating (i.e., E839K). With respect to the D816V variant, mutations of the equivalent codon (814) in mouse *c-kit* indicate that these mutations (D814V/F/Y/H) result in autophosphorylation of KIT, activation of KIT kinase activity, and degradation of KIT in the absence of SCF (Moriyama et al., 1996; Piao et al., 1996) without dimerization of the receptor (Kitayama et al., 1995). This constitutive activation of KIT leads to downstream perturbations. Such perturbations include enhanced degradation of SHP-1 (Piao et al., 1996), constitutive binding of phosphorylated p85^{PI3K} to KIT thorugh tyrosine-721 (Chian et al., 2001), and the constitutive activation of JNK 1, JNK 2 (Chian et al., 2001), STAT1, and STAT3 (Ning et al., 2001). However, not all the known KIT signaling pathways are activated. The ERK MAP kinases and Akt remain inactive (Chian et al., 2001). A plausible but untested explanation for the restricted set of activations may lie in the fact that the mutation of KIT promotes a slightly different pattern of autophosphorylation of the receptor from that induced by SCF activation of wild-type KIT (Piao et al., 1996).

The signals induced by KIT^{D816V} have several potential consequences. SHP-1 is a negative regulator of KIT and of other hematopoietic receptors. Its degradation in cells with KIT^{D816V} might lead to deregulation of both KIT and other growth-regulating receptors and thus account for the hematopoietic abnormalities sometimes associated with systemic mastocytosis (Piao et al., 1996; Boissan et al., 2000). Transfection experiments point to essential roles for p85^{PI3K} and STAT3 also. Oncogenic activity is lost when the p85^{PI3K}-binding site on KIT (i.e., Y721) is inactivated by mutation. The oncogenic activity of KIT^{D816V} is abolished in the double mutated D816V/Y721F KIT or by the PI 3-Kinase inhibitor wortmannin (Chian et al., 2001). Transfection studies indicate that STAT3 is also critical for KIT^{D816H} oncogenic activity in human cells as it appears to be for other receptor tyrosine kinases (Ning et al., 2001).

As alluded to above, the signals initiated by wild-type KIT in the presence of SCF and KIT^{D814Y} are not equivalent, and this is true of the functional consequences. The mutation not only induces spontaneous proliferation in the absence of SCF but also affects the maturation of cells. Expression of the mutated receptor in the murine mouse cell line IC2 leads to abundant expression of mRNA for the mouse mast cell proteases 4 and 6, whereas these transcripts are expressed at low levels in cells bearing wild-type KIT (Piao and Bernstein, 1996).

With respect to the inactivating E839K mutation, this is at a site that is believed to be essential for the formation of a salt bridge with R914. Both E839 and R914 are conserved completely in receptor tyrosine kinases and may be critical for preserving the tertiary structure of the enzymatic region of KIT and other tyrosine kinase receptors (Longley et al., 1999). KIT^{E839K} suppresses autophosphorylation of itself and of wild-type KIT when both receptors are coexpressed. Autophosphorylation is also suppressed when the E839K mutation coexists with the activating D816V mutation in the same molecule. The E839K mutation appears to interfere with intracellular processing because KIT^{E839K} is incompletely glycosylated and is retained largely in the Golgi (Longley et al., 1999). Thus the presence of KIT^{E839K} may help limit mast cell growth and severity of disease in children, especially if the primary cause of childhood mastocytosis lies outside of KIT.

Perturbations Induced by Alterations in the Juxtamembrane Region of KIT. Disruption of the helical conformation of the juxtamembrane region alleviates the inhibitory control of KIT kinase activity (Ma et al., 1999a). The juxtamembrane mutations described above and similar mutations found in human gastrointestinal stromal tumors are kinase-activating muations that result in spontaneous phosphorylation (Ma et al., 1996b) and, in contrast to the D816V mutation, the dimerization of the receptor (Kitayama et al., 1995). At this time, downstream signaling events have not been described (Boissan et al., 2000).

FcεRI-Mediated Signaling

The double I181L/V183L variants in FcεRIβ could conceivably interfere with the association of FcεRIβ with the other subunits of the receptor and therby alter expresssion and activity of FcεRI, whereas the single E237G variant could alter FcεRI activity (Donnadieu et al., 2000). However, neither of these variants was found to affect the functions ascribed to FcεRIβ, namely, amplification of FcεRI expression and of signaling/functional responses to antigen stimulation (Donnadieu et al., 2000; Furomoto et al., 2000). Specifically, expression and N-glycosylation of FcεRIα and, in antigen-stimulated cells, the amplitude of the Ca^{2+} response, the secretion of granules, and the generation of inflammatory mediators (IL-6, TNF-α, and LTC$_4$) were the same for wild-type and variant FcεRIβ when compared in transfected cell systems. It is suggested (Donnadieu et al., 2000) that these apparently innocuous variants could be in linkage disequilibrium with other polymorphisms in the regulatory and promoter region of the gene for FcεRIβ (Palmer et al., 1997; Rohrbach et al., 1998; Dickson et al., 1999). However, the impact of the latter polymorphisms on receptor expression and function awaits evaluation.

The Interleukin Receptor (IL4R) Pathway

Some populations of mast cells (e.g., those derived from human cord blood) express IL4R (Brown et al., 1987), and IL-4 itself can induce apoptosis of mature mast cells and have other effects during mast cell maturation in the presence of other growth factors (Osteritzian et al., 1999). However, the signaling pathways activated by IL-4 have not been specifically examined in mast cells. In other hematopoietic cells binding of IL-4 to IL4R results in activation of JAK-1, JAK3 among other kinases and subsequently STAT6 (Jiang et al., 2000). STAT6 activation is critical for mediating increased IgE production, T_H2 differentiation, and the proliferation of B cells and T cells. Recruitment of SHP-1 by IL4R terminates the signaling process. These responses are dependent on several key tyrosine residues on the IL4Rα subunit that when phosphorylated recruit signaling molecules such as JAK1 and STAT6 (Gessner and Rollinghoff, 2000). IL4R also contains a γc subunit common to several cytokine receptors that allows recruitment of JAK3 and amplification of the signaling functions of IL4Rα.

Four of the known IL4R polymorphisms (E375A, C406R, S478P, and Q551R) are close to the intracellular tyrosine-containing substrate binding sites. The less frequent alleles of two of them, P478 and R551, have been shown to perturb signaling mechanisms in functional studies in vitro and in cell-based systems. Binding assays with synthetic peptides that span the polymorphic regions as well as the tyrosine-containing binding sites suggest that R551 allele significantly diminishes binding of SHP-1 to phosphorylated Y550 peptide (Hershey et al., 1997). The potential consequences of diminished SHP-1 binding could be prolongation of signals transduced by phosphorylated JAKs, STAT6, phosphatidylinositol 3-kinase, and IL4R itself, all potential substrates for SHP-1. In addition, cell-based assays revealed a synergy between the R551 and P478 variants: Phosphorylation of IRS1 and IRS2 was enhanced but phosphorylation of STAT6 was reduced when both variants were present, whereas neither by itself had such an effect (Kruse et al., 1999). Also, expression of the R551 variant by itself is not associated with increased IL-4-dependent germline ε production (Mitsuyasu et al., 1999). If *IL4Rα* is an atopic gene, more than one allele is probably involved. The functional consequences of the E375A and C406R polymorphisms on IL-4-mediated signaling have not been investigated.

The other polymorphism of note, V50I, is located on the extracellular domain of IL4Rα. Expression of the I50 variant in B cell lines upregulates IL-4-induced STAT6 activation, germline ε transcriptional activity, and IgE production compared to the V50 variant (Izuhara et al., 2000;

Mitsuyasu et al., 1999). However, the mechanism is unclear because the affinity of receptor for IL-4 is the same for both variants.

The effects of the *STAT-6* atopy-related polymorphisms (Gao et al., 2000; Tamura et al., 2001) on STAT-6 function has not been determined. These SNPs could have relevance to mast cell function. In addition to mediating signals in response to IL-4 and IL-13 stimulation (Foster, 1999), gene deletion studies suggest that the JAK3/STAT6 pathway participates in the production and secretion of inflammatory cytokines in antigen-stimulated mast cells (Malaviya and Uckun; 1999, Malaviya and Uckun, 2002) and in late–stage allergic responses (Miyata et al., 1999; Malaviya and Uckun, 2002).

CURRENT THERAPIES FOR ALLERGIC DISEASES AND MASTOCYTOSIS

Atopic Diseases and Chronic Urticarias

Established therapies rely on use of mediator antagonists (antihistamines and antileukotrienes), immunosuppressants (corticosteroids), the enigmatic mast cell stabilizers (the cromones cromolyn and nedocromil), or, for asthma specifically, bronchodilators (β_2-adrenergic agents) and smooth muscle relaxants (theophylline). These drugs may be delivered by inhalers (corticosteroids, cromones, and bronchodilators for asthma and corticosteroids for hay fever), orally (antihistamines, theophylline, and the antileukotrienes), or by topical application for cutaneous disorders and conjunctivitis (corticosteroids). The treatment is usually tailored for the severity of disease and may require oral or even intravenous administration of corticosteroids for severe asthma and cutaneous disease. The abrogation of acute anaphylaxis requires immediate injection of epinephrine to resurrect circulatory function before any additional therapy.

Corticosteroids are effective in all atopic diseases, but they do not bring immediate relief because the effects are time dependent. Topical administration is preferred to minimize undesirable endocrine and metabolic side effects. The efficacy of mediator antagonists varies according to which mast cell mediator predominates in a given disease. We have mentioned a few key examples in previous sections. Antihistamines are effective in allergic rhinitis and alleviate itching in cutaneous disease but are ineffective in asthma. Antileukotrienes are effective in some patients with asthma but are ineffective by themselves in allergic rhinitis. Cromones, which minimize mediator release by what is described as "stabilizing" mast cells, are effective in some cases of mild allergic asthma, rhinitis, and conjuctivitis.

The reader is referred to the many excellent reviews for more specialized information on the treatment of atopic diseases in general (Turvey, 2001) and of asthma (Prussin and Metcalfe, 2001; Naureckas and Solway, 2001), rhinitis (Stempl and Woolf, 2002; Lipworth, 2001), conjunctivitis (Bielory, 2000; Buckley, 1998), food allergies (Sampson, 2000; Sicherer and Sampson, 1999), and cutaneous disease (Kaplan, 2002; Greaves and Sabroe, 1998).

Mastocytosis

There is no cure for mastocytosis. Current therapies rely on the same arsenal of drugs used for atopic diseases. A notable addition to this arsenal is aspirin and other NSAIDs, which are used to suppress episodic flushing, tachycardia, and syncope except for the few patients who have idiosyncratic reactions to these drugs (reviewed in Valent et al., 2001a; Worobec and Metcalfe, 2002), NSAIDs suppress prostaglandin production in mast cells by inhibiting the cyclooxygenases. Systemic corticosteroids are not effective for cutaneous mastocytosis but may be used topically with caution (Marone et al., 2001). If these drugs are introduced in step-wise fashion and are selected according to the mediator or organ involved, satisfactory responses can be achieved in patients with indolent mastocytosis (Valent et al., 2001a). For patients with aggressive systemic mastocytosis or mast cell leukemia the situation is grim. The only resort is ameliorative treatment perhaps coupled with experimental polychemotherapy and bone marrow transplantation (Valent et al., 2001a).

THERAPEUTIC POTENTIAL OF TRANSDUCTION-BASED DRUGS IN MAST CELL-RELATED DISEASES

Atopic Diseases and Chronic Urticarias

Conventional Approaches. Conventional pharmacologic approaches may still yield new drugs that help fine tune the management of allergic diseases. These include the search for antimediator drugs with improved therapeutic efficacy as well as drugs that disrupt signaling pathways in inflammatory cells such as phosphodiesterase, tyrosine kinases, MAP kinases, and NF-κB (Barnes, 1999; Wong and Koh, 2000). Leukotriene antagonists are one recent addition to the current repertoire of therapeutic agents (Hansel, 2001). There are, however, limitations to these approaches. New drugs must have a clear cost benefit ratio because current drugs, alone or in combinations, improve the quality of life for most patients who suffer from atopic disease. The problem with inhibitors of inflammatory mediators is the diversity of mediators released from mast cells. No one drug is likely to inhibit all major symptoms. Mediator inhibitors have utility in diseases where one mediator predominates such as the nonsedative antihistamines in rhinitis and itch of atopic eczema. The problem with drugs that disrupt FcεRI-mediated signaling pathways is that they are unlikely to have the desired specificity or even safety when administered systemically. These pathways are common to many other receptor systems. This does not preclude use of such drugs topically when systemic adsorption is minimal, as is the case, for example, with inhaled steroids for asthma. Finally, none of these drugs will cure atopic diseases. Allergen-specific immunotherapy remains the only approach that can reverse the underlying mechanisms in the allergic phenotype. Currently this approach is effective in specific instances, for example, desensitizing patients to anaphylactic hypersensitivity to insect stings and some patients with allergic rhinitis.

New therapeutic opportunities may be identified once the predisposing genetic factors for atopic disease become clearer. In the meantime, the accumulated information on the immunopathogenesis of atopic disease provides additional opportunities for investigation. These include approaches based on modulating expression of the T_H2 phenotype and, as a consequence, IgE production, suppression of binding of IgE to FcεRI, activation of inhibitory receptors on mast cells, and the development of glucocoriticoid drugs that dissociate between the antiinflammatory and side effects of glucocorticoids.

Potential Targets for Suppression of IL-4/-13 Signaling Pathways. IL-4 is critical for driving T_H0 lymphocytes into the T_H2 phenotype and, ultimately, production of IgE. Hence, the IL-4 system is one potential therapeutic target under consideration. In addition to its ability to upregulate IgE production by B lymphocytes and expression of FcεRI on mast cells and basophils (Pawankar et al., 2000), IL-4 also contributes to airway obstruction by induction of the mucin gene, secretion of mucus (Dabbagh et al., 1999), and recruitment of eosinophils in airways. Clinical trials with nebulized soluble IL4R lacking transmembrane and cytoplasmic activating domains (Nuvance™) show efficacy in moderate asthma (Steinke and Borish, 2001; Borish, 2001). The IL-4 mediator STAT6 is another potential target for drug design. Deletion of the STAT6 gene has the same antiasthmatic effects as deletion of the IL-4 gene (Miyata et al., 1999). An endogenous inhibitor of STAT activation is the recently described suppressor of cytokine signaling (SOCS-1), which potently inhibits IL-4 signaling pathways and thus provides another potential target (Chen et al., 2000).

DNA Vaccines. Synthetic oligonucleotide DNA vaccines containing CpG motifs are being developed to supplement or replace conventional protein-based desensitization immunotherapy (Horner et al., 2001). CpG motifs are abundant in many microbioal genomes but rare in mammalian DNA. These motifs stimulate the innate immune system and steer the immune system toward a T_H1 response. The DNA oligonucleotide can be used alone or mixed with or conjugated to specific small allergenic peptides. Interestingly, the efficacy of Freund's adjuvant (a mycobacterial extract mixed with paraffin oil) is now thought to be dependent on CpG motifs within palindromic sequences in the mycobacterial DNA that is present in Freund's extract. The recognition of CpG motifs by the innate immune system is thought to be of evolutionary benefit against microbial infections, and Toll-like receptor 9 has been identified as one receptor that mediates activation of immune cells (Takeshita et al., 2001). The extensive research on DNA vaccines has reached the preclinical phase (Verthelyi et al., 2002) and suggests potential use in IgE-related diseases (Horner et al., 2001). The therapeutic values of these vaccines should become evident over the next few years when current clinical trials are completed (Krieg, 2001). As with protein-based immunotherapy, DNA vaccines may have most value where the specific allergen can be identified, as in patients with seasonal rhinitis or a predisposition to insect- or drug-induced anaphylaxis.

Targeting IgE. In view of the key role played by IgE in atopic diseases, the most selective therapy for these diseases would be the use of agents that block IgE binding to FcεRI. The low circulating levels of IgE (~0.1 μM) compared to other immunoglobulins (~0.004% of total serum antibodies) hold prospect for high efficacy of this form of therapy. The problems faced are the high affinity of IgE for FcεRI and its extraordinarily slow detachment from the receptor (Kinet and Metzger, 1990). Apart from mast cells and basophils, few other types of cells express FcεRI. These include dendritic cells, in which the receptor promotes endocytosis and presentation of IgE-directed antigens to T cells (Shibaki, 1999), as well as eosinophils, which express low levels of the receptor (Kayaba et al., 2001). Strategies have been tested in model systems. These include use of nonanaphylactic anti-IgE antibodies (Coyle et al., 1996; Rudolf et al., 1996) and soluble forms of human FcεRI (Ohtsuka et al., 1999) or mutagenized FcεRI (Mackay et al., 2002). Clinical trials with a humanized monoclonal anti-IgE antibody (omalizumab, Genentech) show promise in the treatment of allergic rhinitis (Casale et al., 2001; Adelroth et al., 2000) and asthma (Arshad and Holgate, 2001) when administered subcutaneously and intravenously. This antibody is directed against an epitope on the F_c fragment of IgE and because of its long half-life in plasma can be administered every 1 or 2 weeks. The reduction in symptoms appears to correlate with reduction in the levels of free IgE (also see commentary in Plaut, 2001).

Targeting Inhibitory Receptors. Inhibitory receptors provide an additional but largely unexplored avenue for therapeutic intervention. FcεRI-induced signals are negatively regulated in mast cells by receptors bearing immunoreceptor tyrosine-based inhibition motifs (ITIMs). These include FcγRIIb, the glycoprotein gp49B1, paired Ig-like receptor(PIR)-B (p91), the signal regulatory protein α1 (SIRPα1), and the mast cell function-associated protein (MAFA) (Ott and Cambier, 2000) (Table 10.2). These receptors are single transmembrane peptides with two or more extracellular Ig-like domains except for MAFA, which contains an extracellular C-type lectin domain and may exist as a monomer or a disulfide-linked homodimer. The known ligands are IgG (for FcγRIIB), integrin-associated protein (IAP/CD47) (for SIRP), and oligosaccharides (for MAFA). The ligands for gp49B1 and PIR-B are not identified. Phosphorylation of the ITIMs enable these receptors to recruit the SH-2 domain-containing inositol 5-phosphatases (SHIP) 1 and 2 (by FcγRIIB and MAFA) or tyrosine phosphatases SHP-1 and SHP-2 (by gp49B2, PIR-B and SIRPα1). These phosphatases negatively regulate FcεRI-mediated formation of polyphosphorylated inositides or protein tyrosine phosphorylation cascades and downstream events (for more details, see Ott and Cambier, 2000; Xu et al., 2001). In general, inhibitory signals are initiated by coaggregation of the inhibitory receptor and FcεRI (or other relevant receptors in other immune cells) although MAFA is unique in that inhibitory signals can be initiated by cross-linking of MAFA alone in addition to cross-linking with FcεRI. The inhibitory receptors would be prime targets for the treatment of chronic urticarias because most of the therapeutic approaches discussed here are not relevant to this group of autoimmune diseases.

FcγRIIB provides a physiological example of the potential of inhibitory receptors as targets for therapy. This receptor is the only one known to coaggregate with FcεRI under physiological conditions. The benefit of allergen-specific immunotherapy has been attributed to the generation of allergen-specific IgG with increasing doses of allergen and the coaggregation of FcεRI and FcγRIIB by the allergen-IgG complex (Ebner, 1999). Indeed, administration of allergen-IgG complexes is reported to diminish symptoms of allergic asthma (Machiels et al., 1993). MAFA might be another therapeutic target because of its affinity for mannose-containing saccharides (Binsack and Pecht, 1997) and its inhibitory actions upon self-aggregation alone.

Glucocorticoids. Glucocorticoids mediate their effects by binding to the glucocorticoid receptor (GR), which when activated either dimerizes or remains as a monomer. Dimeric GR interacts with DNA glucocorticoid response elements (GREs) to increase gene expression (referred to as transactivation). Monomeric GR interacts with transcription factors to suppress gene expression (referred to as transrepression). These two types of interactions have been demonstrated by selective mutation of GR. Many of the antiinflammatory actions of the glucocorticoids are thought to be mediated via transrepression mechanisms that lead to inhibition of proinflammatory transcription factors such as activator protein (AP)-1, nuclear factor of activated T cells (NF-AT), NF-κB, and STATs (Adcock, 2001). These findings have led to the search for glucocorticoids that favor transrepression over transactivation. Drugs already in use, such as budesonide and fluticasone propionate, have such properties that might account for their efficacy in asthma (Adcock et al., 1999). Recently, glucocorticoids with little transactivating but potent transrepressive activities have been described and are called "dissociated" glucorticoids (Vayssiere et al., 1997; Berghe et al., 1999; Bamberger and Schulte, 2000). Theoretically, such drugs should have minimal endocrine side effects that are largely mediated through transactivation mechanisms. The promise of minimal side effects has not borne out in animal studies with one of the most potent of these drugs (Belvisi et al., 2001), but this line of research is still in its early days.

Atopic Dermatitis: Immunophilins

Much of the current research in this area is in the development of improved antagonists of inflammatory mediators for topical or systemic administration. Immunosuppressive/immunophilin-binding agents cyclosporin A and FK-506 (tacrolimus) are being investigated for topical use in allergic dermatitis. These two agents bind, respectively, to cyclophilins and FK-binding proteins (FKBPs). The immunophilin-drug complexes so formed inhibit the calcium/calmodulin-dependent phosphatase calcineurin (Nghiem et al., 2002), which is a critical component of signaling pathways in immune cells (Crabtree, 1999). In lymphoctes, calcineurin is thought to primarily regulate the interaction of NF-AT with cytokine genes. In mast cells, the role of calcineurin is not clearly defined and probably has multiple substrates. Cyclosporin A and tacrolimus inhibit not only the synthesis of cytokines but also the generation of eicosanoids and degranulation of mast cells (see citations in Marone et al., 2001). Tacrolimus has recently been approved for topical use in the treatment of atopic dermatitis. Its advantage over glucocortcoids is that does not cause dermal atrophy. A related and newer calcineurin inhibitor, pimecrolimus, which has better skin-penetrating abilities than tacrolimus, is also being developed for topical use (Nghiem et al., 2002). This compound also inhibits mediator release from human skin mast cells and, in fact, is said to be more effective in this regard than cyclosporine A and dexamethasone (Zuberbier et al., 2001). Therefore, the therapeutic efficacy of the immunophilins in cutaneous disease, which have been referred to as bacterial "smart bombs" (Nghiem et al., 2002), may be due to suppression of mast cell as well as T cell activation. The concerns with the immunophilins are whether use over large skin areas or over long periods of time results in systemic side effects and an increased risk for skin cancers, respectively.

Mastocytosis: KIT Inhibitors and Interferon-γ

The finding that activating mutations of KIT are associated with mastocytosis has provided the opportunity for devising new therapeutic approaches based on KIT inhibitors. Preliminary studies with indolinone derivatives designed to inhibit ATP binding in the pocket of receptor tyrosine kinases have shown that some of these derivatives inhibit activity of wild-type KIT. One of them, SU6577, inhibited KIT with activating mutations in the juxtamembrane or kinase domain and was lethal to neoplastic mast cells expressing these mutations (Ma et al., 2000).

Another KIT inhibitor is the drug imatinib mesylate (Gleevec, Novartis Pharmaceuticals; formerly ST1571), which was originally designed to selectively target the Bcr-Abl oncogene present in patients with chronic myelogenous leukemia. This drug is now in clinical use for treatment of chronic myeloid leukemia and is undergoing evaluation in the treatment of metastatic gastrointestinal stromal tumors with promising results (Demetri, 2001). An interesting facet has emerged from the work with KIT inhibitors. Some of the indoline derivatives (Ma et al., 2000) and imatinib mesylate (Ma et al., 2002) inhibit wild-type KIT and constitutively active KIT having mutations in the juxtamembrane regulatory domain of KIT (i.e., as found in gastrointestinal stromal tumors and dog mastocytomas) and are lethal to neoplastic mast cells expressing these forms of KIT. The same inhibitors do not inhibit constitutively active KIT^{V816} (i.e., as found in patients with systemic mastocytosis), nor are they lethal to neoplastic mast cells bearing this mutation. These findings suggest that drugs could be selectively designed to be effective against the tumorigenic effects of one type of mutation and not others. The present generation of KIT inhibitors offers hope for what was once a bleak prospect for patients with gastrointestinal stromal disease, but further research is required to optimize KIT inhibitors for treatment of systemic mastocytosis (Longley et al., 2000).

Other experimental therapies have been tested in humans including the use of interferon-γ, which has been shown to inhibit proliferation of mast cell progenitors in cultures of IL-3-dependent murine bone marrow-derived mast cells (Nafziger et al., 1990). However, the results of clinical trials with interferon in a limited number of patients are inconclusive (Worobec and Metcalfe, 2002), although further trials are desirable (Marone et al., 2001). The uncertainties with this form of therapy are that interferon-γ does not inhibit growth of mature mast cells and it is not known whether or not neoplastic mast cells express interferon-γ receptors.

Therapeutic Outlook

The most significant advance in the pharmacologic treatment of atopic diseases is the recognition that asthma is an inflammatory disease and that long-term control is best achieved by treating the underlying inflammatory disease. Recent advances in understanding the immunological basis of atopic diseases now point to possible therapies for treatment of these diseases on a broader immunological basis. The future challenge, rectification of the atopic process, must await further elucidation of the polygenetic nature of these disease. With respect to the treatment of allergic asthma and rhinitis specifically, the current armamentarium of drugs can adequately control, albeit not cure, these diseases in most cases. New drugs must have clear clinical benefit, because adequate control may be achieved with well-established therapies at much lower cost. The situation for mastocytosis is different. Cost is not a factor, because there is no established treatment for reversing the progression of disease in adults with more virulent forms of systemic mastocytosis. However, research into the genetic basis of mastocytosis has yielded, for at least certain forms of the disease, valuable clues that provide the real prospect of helping patients with an otherwise bleak prognosis.

REFERENCES

Adcock, I. M. (2001). Glucocorticoid-regulated transcription factors. Pulm Pharmacol Ther *2001*, 211–219.

Adcock, I. M., Nasuhara, Y., Stevens, D. A., and Barnes, P. J. (1999). Ligand-induced differentiation of glucocorticoid receptor (GR) trans-repression and transactivation: preferential targetting of NF-κB and lack of I-κB involvement. Br J Pharmacol *127*, 1003–1008.

Adelroth, E., Rak, S., Haahtela, T., Aasand, G., Rosenhall, L., Zetterstrom, O., Byrne, A., Champain, K., Thirlwell, J., Cioppa, G. D., and Sandstrom, T. (2000). Recombinant humanized mAb-E25, an anti-IgEmAb, in birch pollen-induced seasonal allergic rhinitis. J Allergy Clin Immunol *106*, 253–259.

Anderson, D. F. (2001). Management of seasonal allergic conjunctivitis (SAC): current therapeutic strategies. Clin Exp Allergy *31*, 823–826.

Arshad, S. H., and Holgate, S. (2001). The role of IgE in allergen-induced inflammation and the potential for intervention with humanized monoclonal anti-IgE antibody. Clin Exp Allergy *31*, 1344–1351.

Asai, K., Kitaura, I., Kawakami, Y., Yamagata, N., Tsai, M., Carbone, D. P., Liu, F.-T., Galli, S. J., and Kawakami, T. (2001). Regulation of mast cell survival by IgE. Immunity *14*, 791–800.

Ashman, L. K. (1999). The biology of stem cell factor and its receptor C-kit. Int J Biochem Cell Biol *31*, 1037–1051.

Austen, K. F., and Boyce, I. A. (2001). Mast cell lineage development and phenotypic regulation. Leukoc Res *25*, 511–518.

Baldini, M., Lohman, I. C., Holonen, M., Erickson, R. P., Holt, P. G., and Martinez, F. D. (1999). A polymorphism in the 5′ flanking region of the CD14 gene is associated with circulating soluble CD14 levels with total serum immunoglobulin E. Am J Respir Cell Mol Biol *20*, 976–983.

Bamberger, C. M., and Schulte, H. M. (2000). Molecular mechanisms of dissociative glucocorticoid activity. Eur J Clin Invest *30* Suppl. 3, 6–9.

Barata, L. T., Ying, S., Meng, O., Barkans, J., Rajakulasingam, K., Durham, S. R., and Kay, A. B. (1998). IL-4 and IL-5-positive T lymphocytes, eosinophils, and mast cells in allergen-induced late-phase cutaneous reactions in atopic subjects. J Allergy Clin Immunol *101*, 222–230.

Barnes, P. J. (1999). Therapeutic strategies for allergic diseases. Nature *402*, B31–38.

Baumgartner, R. A., and Beaven, M. A. (1996). Mediator release by mast cells and basophils. In: Herzenberg, L. A., Herzenberg, L., Weir, D. M., Blackwell, C. (eds). Weir's Handbook of Experimental Immunology, 5th ed. Cambridge, MA: Blackwell Science, Vol. 4, pp. 213.1–213.8.

Belvisi, M. G., Brown, T. L., Wicks, S., and Foster, M. L. (2001). New glucocorticoids with an improved therapeutic ratio? Pulm Pharmacol Ther *14*, 221–227.

Benyon, R. C., Robinson, C., and Church, M. K., (1989). Differential release of histamine and eicosanoids from human skin mast cells activated by IgE-dependent and non-immunological stimuli. Br J Pharmacol *97*, 898–904.

Berghe, W. V., Francesconi, E., de Bosscher, K., Resche-Rigon, M., and Haegeman, G. (1999). Dissociated glucocorticoids with anti-inflammatory potential repress interleukin-6 gene expression by a nuclear factor-κB-dependent mechanism. Mol Pharmacol *56*, 797–806.

Bielory, L. (2000). Allergic and immunologic disorders of the eye. Part II: ocular allergy. J Allergy Clin Immunol *106*, 1019–1032.

Bingham, C. O., and Austen, K. F. (2000). Mast-cell responses and asthma. J Allergy Clin Immunol *105*, S527–S534.

Binsack, R., and Pecht, I. (1997). The mast cell function-associated antigen exhibits saccharide binding capacity. Eur J Immunol *27*, 2557–2561.

Bochner, B. S., and Schleimer, R. P. (2001). Mast cells, basophils, and eosinophils: distinct but overlapping pathways for recruitment. Immunol Rev *179*, 5–15.

Boissan, M., Feger, F., Guillosson, J.-J., and Arock, M. (2000). C-Kit and c-kit mutations in mastocytosis and other hematological diseases. J Leukoc Biol *67*, 135–148.

Borish, L. C., Nelson, H. S., Corren, J., Bensch, G., Busse, W. W., Whitmore, J. B., Agosti, J. M., and IL-4R Asthma Study Group (2001). Efficacy of soluble IL-4 receptor for the treatment of adults with asthma. J Allergy Clin Immunol *107*, 963–970.

Brown, M. A., Pierce, J. H., Watson, C. J., Falco, J., Ihle, J. N., and Paul, W. E. (1987). B cell stimulatory factor-1/interleukin-4 mRNA is expressed by normal and transformed mast cells. Cell *50*, 809–818.

Buckley, R. J. (1998). Allergic eye disease—a clinical challenge. Clin Exp Allergy *28*, 39–43.

Busse, W. W., and Lemanske, R. F. (2001). Asthma. N Engl J Med *344*, 350–362.

Butterfield, J. H., Weiler, D., Dewald, G., and Gleich, G. J. (1988). Establishment of an immature mast cell line from a patient with mast cell leukemia. Leukoc Res *12*, 345–355.

Buttner, C., Beate, P. M., Sepp, N., and Grabbe, J. (1998). Identification of activating c-kit mutations in adult-, but not in childhood-onset indolent mastocytosis: A possible explanation for divergent clinical behavior. J Invest Dermatol *111*, 1227–1231.

Casale, T. R., Condemi, I., LaForce, C., Nayak, A., Rowe, M., Watrous, M., McArlary, M., Fowler-Taylor, A., Racine A., Gupta N., Fick, R., and Cioppa, G. D. (2001). Effect of omalizumab on symptoms of seasonal allergic rhinitis. A randomized controlled trial. JAMA *286*, 2956–2967.

CDC. (1998). Forecasted state-specific estimates of self reported asthma prevalence-United States. MMWR Morb Mortal Wkly Rep *47*, 1022–1025.

Chaves-Dias, C., Hundley, T. R., Gilfillan, A. M., Kirshenbaum, A. S., Cunha-Melo, J. R., Metcalfe, D. D., and Beaven, M. A. (2001). Induction of telomerase activity during development of human mast cells from peripheral blood CD34[+] cells: Comparisons with tumor mast cell lines. J Immunol *166*, 6647–6656.

Chen, P., Losman, J. A., and Rothman, P. (2000). SOCS proteins, regulators of intracellular signaling. Immunity *13*, 287–290.

Chian, R., Young, S., Danilkovitch, A., Ronnstrand, I., Leonard, E., Ferrao, P., Ashman, P., and Linnekin, D. (2001). Phosphatidylinositol 3 kinase contributes to the transformation of hematopoietic cells by the D816V c-Kit mutant. Blood *98*, 1365–1373.

Clark, J. M., Abrahamm, W. M., Fishman, C. E., Forteza, R., Ahmed, A., Cortes, A., Warne, R. L., Moore, W. R., and Tanaka, R. D. (1995). Tryptase inhibitors block allergen-induced airway and inflammatory responses in allergic sheep. Am J Respir Crit Care Med *152*, 2076–2083.

Colten, H. R. (1994). Drawing a double-edged sword. Nature *371*, 474–475.

Coyle, A. I., Wagner, K., Bertrand, C., Tsuyuki, S., Bews, J., and Heusser, C. (1996). Central role of immunoglobulin (Ig) E in the induction of lung eosinophil infiltration and T helper 2 cell cytokine production: inhibition by a non-anaphylactogenic anti-IgE antibody. J Exp Med *183*, 1303–1310.

Crabtree, G. R. (1999). Generic signals and specific outcomes: signaling through Ca^{2+}, calcineurin, and NF-AT. Cell *66*, 611–614.

D'Auriol, I., Mattei, M. G., Andre, C., and Galibert, F. (1988). Localization of the human c-kit protooncogene on the q11–q12 region of chromosome 4. Hum Genet *78*, 374–376.

Dabbagh, K., Takeyama, K., Lee, H. M., Ueki, I. F., Lausier, J. A., and Nadel, J. A. (1999). IL-4 induces mucin gene expression and goblet cell metaplasia in vitro and in vivo. J Immunol *162*, 6233–6237.

Daley, T., Metcalfe, D. D., and Akin, C. (2001). Association of the Q576R polymorphism in the interleukin-4α chain with indolent mastocytosis limited to skin. Blood *98*, 880–882.

DeMeo, M. T., Mutlu, E. A., Keshavarzian, A., and Tobin, M. C. (2002). Intestinal permeation and gastrointestinal disease. J Clin Gastroenterol *34*, 385–396.

Demetri, G. D. (2001). Targeting *c-kit* mutations in solid tumors: scientific rationale and novel therapeutic options. Semin Oncol *28*, 19–26.

Dickson, P. W., Wong, Z. Y., Harrap, S. B., Abramson, M. J., and Walters, E. H. (1999). Mutational analysis of the high affinity immunoglobulin E receptor β subunit gene in asthma. Thorax *54*, 409–412.

Donnadieu, E., Cookson, W. O., Jouvain, M.-H., and Kinet, J.-P. (2000). Allergy-associated polymorphisms of the FcεRIβ subunit do not impact its two amplification functions. J Immunol *165*, 3917–3922.

Drazen, J. M., Israel, E., and O'Brien, P. M. (1999). Treatment of asthma with drugs modifying the leukotriene pathway. N Engl J Med *340*, 197–206.

Drazen, I. M., Yandava, C. N., Dube, I., Szczerback, N., Hippensteel, R., Pillari, A., Israel, E., Schork, N., Silveerman, E. S., Katz, D. A., and Drajesk, J. (1999). Pharmacogenetic association between ALOX5 promoter genotype and the response to anti-asthma treatment. Nat Genet *22*, 168–170.

Drysdale, C. M., McGraw, D. W., Stack, C. B., Stephens, I. C., Judson, R. S., Nandabalan, K., Arnold, K., Ruano, G., and Liggett, S. B. (2000). Complex promotor and coding region β₂-adrenergic receptor haplotypes alter receptor expression and predict in vivo responsiveness. Proc Natl Acad Sci USA *97*, 10483–10488.

Duetsch, G., Illig, T., Loesgen, S., Rohde, K., Klopp, N., Herbon, N., Gohlke, H., Altmueller, J., and Wjst, M. (2002). *STAT6* as an asthma candidate gene: polymorphism-screening, association and haplotype analysis in a Caucasian sib-pair study. Hum Mol Genet *11*, 613–621.

Ebner, C. (1999). Immunological mechanisms operative in allergen-specific immunotherapy. Int Arch Allergy Immunol *119*, 1–5.

Echtenacher, B., Mannel, D. N., and Hultner, L. (1996). Critical protective role of mast cells in a model of acute septic peritonitis. Nature *381*, 75–77.

Eklund, K. K., Ghildyal, N., Austen, K. F., and Stevens, R. L. (1993). Induction by IL-9 and suppression by IL-3 and IL-4 of the levels of chromosome 14-derived transcripts that encode late-expressed mouse mast cell proteases. J Immunol *151*, 4266–4273.

Ennerback, L. (1986). Mast cells. In: Spicer S. S. (ed). Histochemistry in pathologic diagnosis. New York, NY: Marcel Dekker, pp. 695–728.

Feijen, M., Gerritsen, I., and Postma, D. S. (2000). Genetics of allergic disease. Br Med Bulletin *56*, 894–907.

Ferrer, M., Kinet, J. P., and Kaplan, A. P. (1998). Comparative studies of functional and binding assays for IgG anti-FcεRIα (α-subunit) in chronic urticaria. J Allergy Clin Immunol *101*, 672–676.

Fiebiger, E., Hammerschmid, F., Stingl, G., and Maurer, D. (1998). Anti-FcεRIα autoantibodies in autoimmune-mediated disorders. Identification of a structure-function relationship. J Clin Invest *101*, 243–251.

Fiebiger, E., Maurer, D., Holub, H., Reininger, B., Hartmann, G., Woisetschlager, M., Kinet, J. P., and Stingl, G. (1995). Serum IgG autoantibodies directed against the α chain of FcεRI: a selective marker and pathogenetic factor for a distinct subset of chronic urticaria patients. J Clin Invest *96*, 2606–2612.

Forsythe, P., and Ennis, M. (2000). Clinical consequences of mast cell heterogeneity. Inflamm Res *49*, 147–154.

Foster, P. S. (1999). STAT6: an intracellular target for the inhibition of allergic disease. Clin Exp Allergy *29*, 12–16.

Fureder, W., Agis, H., Willheim, M., Bankl, H. C., Maier, U., Kishi, K., Muller, M. R., Czerwenka, K., Radaszkiewicz, T., Butterfield, J. H., Klappacher, G. W., Sperr, W. R., Oppermann, M., Lechner, K., and Valent, P. (1995). Differential expression of complement receptors on human basophils and mast cells: Evidence for mast cell heterogeneity and CD88/C5aR expression on skin mast cells. J Immunol *155*, 3152–3160.

Furitsu, T., Tsujimura, T., Tono, T., Ikeda, H., Kitayama, H., Koshimizu, U., Sugahara, H., Butterfield, J. H., Ashman, L. K., Kanayama, Y., Matsuzawa, Y., Kitamura, Y., and Kanakura, Y. (1993). Identification of mutations in the coding sequence of the proto-oncogene c-kit in a human mast cell leukemia cell line causing ligand-independent activation of c-kit product. J Clin Invest *92*, 1736–1744.

Furomoto, Y., Hiraoka, S., Kawamoto, K., Masaki, S., Kitamura, T., Okumura, K., and Ra, C. (2000). Polymorphisms in FcεRIβ chain do not affect IgE-mediated mast cell activation. Biochem Biophys Res Commun *273*, 765–771.

Gao, P.-S., Mao, X. O., Roberts, M. H., Arinobu, Y., Akaiwa, M., Enomoto, T., Dake, Y., Kawai, M., Sasaki, S., Hamasaki, N., Izuhara, K., Shirakawa, T., and Hopkin, J. M. (2000). Variants of STAT6 (signal transducer and activator of transcription 6) in atopic asthma. J Med Genet *37*, 380–382.

Gessner, A., and Rollinghoff, M. (2000). Biologic functions and signaling of the interleukin-4 receptor complexes. Immunobiology *201*, 285–307.

Gordon, I. R. (1997). FcεRI-induced cytokine production and gene expression. In: Hamawy M. M. (ed). IgE Receptor (FcεRI) Function in Mast Cells and Basophils. Ausin, TX: R. G. Landes, pp. 209–242.

Grattan, C. E., Wallington, T. B., Warin, R. P., Kennedy, C. T., and Bradfield, J. W. (1986). A serological mediator in chronic idiopathic urticaria—a clinical, immunological and histological evaluation. Br J Dermatol *114*, 583–590.

Graves, P. E., Kabesch, M., Holonen, M., Holberg, C. J., Baldini, M., Fritzsch, C., Weiland, S. K., Erickson, R. P., von Mutius, E., and Martinez, F. D. (2000). A cluster of seven tightly linked polymorphisms in the IL-13 gene is associated with total serum IgE levels in three populations of white children. J Allergy Clin Immunol *105*, 506–513.

Graziano, F. M., Stahl, J. L., Cook, E. B., and Barney, N. P. (2001). Conjunctival mast cells in ocular allergic disease. Allergy Asthma Proc *22*, 121–126.

Greaves, M. (2000). Mast cell mediators other than histamine induce pruritis in atopic dermatitis patients—a dermal microdialysis study. Br J Dermatol *142*, 1079–1083.

Greaves, M. W. (2002). Pathophysiology of chronic urticaria. Int Arch Allergy Immunol *127*, 3–9.

Greaves, M. W., and Sabroe, R. A. (1998). ABC of allergies. Allergy and the skin. I—Urticaria. BMJ *316*, 1147–1150.

Hakonarson, H., and Wjest, M. (2001). Current concepts on the genetics of asthma. Curr Opin Pediatr *13*, 267–277.

Hansel, T. T. (2001). New treatments for asthma: current and future aspects. Curr Opin Pulm Med *7*, S3–3S6.

Hartmann, K., Henz, B. M., Kruger-Krasagakes, S., Kohl, J., Burger, R., Guhl, S., Haase, I., Lippert, U., and Zuberbier, T. (1997). C3a and C5a stimulate chemotaxis of human mast cells. Blood *89*, 2863–2870.

Hashimoto, K., Tsujimura, T., Moriyama, Y., Yamatodani, A., Kimura, M., Tohya, K., Morimoto, M., Kitayama, H., Kanakura, Y., and Kitamura, Y. (1996). Transforming and differentiation-inducing potential of constitutively activated *c-kit* mutant genes in the IC-2 murine interleukin-3-dependent mast cell line. Am J Pathol *148*, 189–200.

Heinzmann, A., Mao, X. O., Akaiwa, M., Kreomer, R. T., Gao, P. S., Ohshima, K., Umeshita, R., Abe, Y., Braun, S., Yamashita, T., Roberts, M. H., Sugimoto, R., Arima, K., Arinobu, Y., Yu, B., Kruse, S., Enomot, T., Dake, Y., Kawai, M., Shimazu, S., Sasaki, S., Adra, C. N., Kitaichi, M., Inoue, H., Yamauchi, K., Tomichi, N., Krimoto, F., Hamasaki, F., Hopkin, J. M., Izuhara, K., Shirakawa, T., and Deichman, K. A. (2000). Genetic variants of IL-13 signalling and human asthma and atopy. Hum Mol Genet *9*, 549–559.

Helm, R. M., and Burks, A. W. (2000). Mechanisms of food allergy. Curr Opin Immunol *12*, 647–653.

Hershey, G. K. K., Friedrich, M. F., Esswein, L. A., Thomas, M. L., and Chatila, T. A. (1997). The association of atopy with a gain-of-function mutation in the α subunit of the interleukin-4 receptor. N Engl J Med *337*, 1720–1725.

Hide, M., Francis, D. M., Grattan, C. E., Hakimi, J., Kochan, J. P., Greaves, M. W. (1993). Autoantibodies against the high-affinity IgE receptor as a cause of histamine release in chronic urticaria. N Engl J Med *328*, 1599–1604.

Hill, M. R., and Cookson, W. O. (1996). A new variant of the β subunit of the high affinity receptor for immunoglobulin E (FcεRI-β E237G) associations with measures of atopy and bronchial hyper-responsiveness. Hum Mol Genet *5*, 959–962.

Holgate, S. T. (1999). The epidemic of allergy and asthma. Nature *402*, B2–2B4.

Holloway, J. W., Dunbar, P. R., Riley, G. A., Sawyer, G. M., Fitzharris, P. F., Pearce, N., Le Gross, G. S., and Beasley, R. (2000). Association of β_2-adrenergic receptor polymorphisms with severe asthma. Clin Exp Allergy, *30*, 1097–1103.

Horner, A. A., Van Uden, J. H., Zubeldia, J. M., Broide, D., and Raz, E. (2001). DNA-based immunotherapeutics for the treatment of allergic disease. Immunol Rev *179*, 102–118.

Horsmanheimo, L., Harvima, I. T., Jarikallio, A., Naukkarinen, A., and Horsmanheimo, M. (1994). Mast cells are one major source of interleukin-4 in atopic dermatitis. Br J Dermatol *131*, 348–353.

Houtman, R., Koster, A. S., and Nijkamp, F. P. (2001). Integrin VLA-5: modulator and activator of mast cells. Clin Exp Allergy *31*, 817–822.

Howard, T. D., Koppelman, G. H., Xu, J., Zheng, S. L., Postma, D. J., Meyers, D. M., and Bleecker, E. R. (2002). Gene-gene interaction in asthma: *IL4RA* and *IL13* in a Dutch population with asthma. Am J Hum Genet *70*, 230–236.

Izuhara, K., Yanagihari, Y., Hamasaki, N., Shirakawa, T., and Hopkin, J. M. (2000). Atopy and the human IL-4 receptor α chain. J Allergy Clin Immunol *106*, S65–71.

Jiang, H., Harris, M. B., and Rothman, P. (2000). IL-4/IL-13 signaling beyond JAK/STAT. J Allergy Clin Immunol *105*, 1063–1070.

Kaplan, A. P. (2002). Chronic urticaria and angioedema. N Engl J Med *346*, 175–179.

Kaplan, A. P., and Beaven, M. A. (1976). In vivo studies of the pathogenesis of cold urticaria, cholinergic urticaria, and vibration-induced swelling. J Invest Dermatol *67*, 327–332.

Kaplan, A. P., Horakova, Z., and Katz, S. I. (1978). Assessment of tissue fluid histamine levels in patients with urticaria. J Allergy Clin Immunol *61*, 350–354.

Katz, H. R., and Lobell, R. B. (1995). Expression and function of FγR in mouse mast cells. Int Arch Allergy Immunol *107*, 76–78.

Kawashima, T., Noguchi, E., Areinami, T., Kobayashi, K., Otsuka, F., and Hamaguchi, H. (1998). No evidence for an association between a variant of the mast cell chymase gene and atopic dermatitis based on case-control and haplotype-relative-risk analyses. Hum Hered *48*, 271–274.

Kayaba, H., Dombrowicz, D., Woerly, G., Papin, J. P., Loiseau, S., and Capron, M. (2001). Human eosinophils and human high affinity IgE receptor transgenic mouse eosinophils express low levels of high affinity IgE receptor, but release IL-10 upon receptor activation. J Immunol *167*, 995–1003.

Kinet, J.-P., and Metzger, H. (1990). Genes, structure, and actions of the high-affinity Fc receptor for immunoglobulin E. In: Metzger H. (ed). Fc Receptors and the Action of Antibodies. Washington DC: American Society for Microbiology, pp. 239–259.

Kitayama, H., Kanakura, Y., Furitsu, T., Tsujimura, T., Oritani, K., Ikeda, H., Sugahara, H., Mitsui, H., Kanayama, Y., Kitamura, Y., and Matsuzawa, Y. (1995). Constitutively activating mutations of c-kit receptor tyrosine kinase confer factor-independent growth and tumorigenicity of factor-dependent hematopoietic cell lines. Blood *85*, 790–798.

Koppelmann, G. H., Reijmerink, N. E., Stine, O. C., Howard, T. D., Whittaker, P. A., Meyers, D. A., Postma, D. S., and Bleecker, E. R. (2001). Association of a promoter polymorphism of the CD14 gene and atopy. Am J Respir Crit Care Med *163*, 965–969.

Krieg, A. M. (2001). From bugs to drugs. Therapeutic immunomodulation with oligodeoxynucleotides containing CpG sequences from bacterial DNA. Antisense Nucleic Acid Drug Dev *11*, 181–188.

Krishna, M. T., Chauhan, A., Little, L., Sampson, K., Hawksworth, R., Mant, T., Djukanovic, R., Lee, T., and Holgate, S. (2001). Inhibition of mast cell tryptase by inhaled APC 366 attenuates allergen-induced late-phase airway obstruction in asthma. J Allergy Clin Immunol *107*, 1039—1045.

Kruse, S., Japha, T., Tedner, M., Sparholt, S. H., Forster, J., Kuehr, J., and Deichman, K. A. (1999). The polymorphisms S503P and Q576R in the interleukin-4 receptor α gene are associated with atopy and influence signal transduction. Immunology *96*, 365–371.

Kukuchi, Y., and Kaplan, A. P. (2002). A role for C5a in augmenting IgG-dependent histamine release from basophils in chronic urticaria. J Allergy Clin Immunol *109*, 114–148.

Lantz, C. S., Boesiger, J., Song, C. H., Mach, N., Kobayashi, T., Mulligan, R. C., Nawa, Y., Dranoff, G., and Galli, S. J. (1998). Role for interleukin-3 in mast cell and basophil development and in immunity to parasites. Nature *392*, 90–93.

Lappin, M. B., and Cambell, J. D. M. (2002). The Th1-Th2 classification of cellular immune responses: concepts, current thinking and applications in haematological malignancy. Blood Rev *14*, 228–239.

Leal-Berumen, I., Conlon, P., and Marshall, J. S. (1994). IL-6 production by rat peritoneal mast cells is not necessarily preceded by histamine release and can be induced by bacterial lipopolysaccharide. J Immunol *152*, 5468–5476.

Leong, K. P., and Huston, D. P. (2001). Understanding the pathogenesis of allergic asthma using mouse models. Ann Allergy Asthma Immunol *87*, 96–110.

Liggett, S. B. (2001). Pharmacogenetic applications of the Human Genome project. Nature Med *7*, 281–283.

Linnekin, D. (1999). Early signaling pathways activated by c-Kit in hematopoietic cells. Int J Biochem Cell Biol *31*, 1053–1074.

Lipworth, B. J. (2001). Emerging role of antileukotriene therapy in allergic rhinitis. Clin Exp Allergy *31*, 1813–1821.

Longley, B. J., Ma, Y., Carter, E., and McMahon, G. (2002). New approaches to therapy for mastocytosis. A case for treatment with kit kinase inhibitors. Hematol Oncol Clin North Am *14*, 689–695.

Longley, P. J., Metcalfe, D. D., Tharp, M., Wang, X., Tyrrell, I., Lu, S.-Z., Heitjan, D., and Ma, Y. (1999). Activating and dominant inactivating *c-KIT* catalytic domain mutations in distinct clinical forms of human mastocytosis. Proc Natl Acad Sci USA *96*, 1609–1614.

Longley, B. J., Reguera, M. J., and Ma, Y. (2001). Classes of *c-KIT* activating mutations: proposed mechanisms of action and implications for disease classification and therapy. Leuk Res *25*, 571–576.

Longley, B. J., Tyrell, L., Lu, S. Z., Ma, Y. S., Langley, K., Ding, T. G., Duffy, T., Jacobs, P., Tang, L. H., and Modlin, I. (1996). Somatic c-KIT activating mutation in urticaria pigmentosa and aggressive mastocytosis: establishment of clonality in human mast cell neoplasm. Nat Genet *12*, 321–314.

Lorentz, A., Schuppan, D., Gebert, A., Manns, M. P., and Bischoff, S. C. (2002). Regulatory effects of stem cell factor and interleukin-4 on adhesion of human mast cells to extracellular matrix proteins. Blood *99*, 966–972.

Louahed, J., Kermouni, A., Van Snick, J., and Renauld, J. C. (1995). IL-9 induces expression of granzymes and high-affinity IgE receptor in murine T helper clones. J Immunol *154*, 5061–5070.

Ma, Y., Carter, E., Wang, X., Shu, C., McMahon, G., and Longley, B. J. (2000). Indoline derivatives inhibit constitutively activated KIT mutants and kill neoplastic mast cells. J Invest Dermatol *114*, 392–394.

Ma, Y., Cunningham, M. E., Wang, A., Ghosh, I., Regan, I., and Longley, B. J. (1999). Inhibition of spontaneous receptor phosphorylation by residues in a putative α-helix in the KIT intracellular juxtamembrane region. J Biol Chem *274*, 13399–13402.

Ma, Y., Longley, B. J., Wang, X., Blount, I. L., Langley, K., and Caughey, G. H. (1999). Clustering of activating mutations in c-KIT's juxtamembrane coding region in canine mast cell neoplasms. J Invest Dermatol *112*, 165–170.

Ma, Y., Zeng, S., Metcalfe, D. D., Akin, C., Dimitrijevic, S., Butterfield, J. H., McMahon, G., and Longley, B. J. (2002). The c-KIT mutation causing human mastocytosis is resistant to ST1571 and other KIT kinase inhibitors; kinase with enzymatic site mutations show different inhibitor sensitivity profiles than wild-type kinases and those with regulatory-type mutations. Blood *99*, 1741–1744.

Machiels, J. J., Lebrun, P. M., Jacquemin, M. G., and Saint-Remy, J. M. (1993). Significant reduction of nonspecific bronchial reactivity in patients with *Dermatophagoides pteronyssinus*-sensitive allergic asthma under therapy with allergen-antibody complexes. Am Rev Respir Dis *147*, 1407–1412.

Mackay, G. A., Hulett, M. D., Cook, I. P., Trist, H. M., Henry, A. J., McDonnell, J. M., Beavil, A. J., Sutton, B. J., Hogarth, P. M., and Gould, H. J. (2002). Mutagenesis within human FcεRIα differentially affects human and murine IgE binding. J Immunol *168*, 1787–1795.

Malaviya, R., and Abraham, S. N. (2001). Mast cell modulation of immune responses to bacteria. Immunol Rev *179*, 16–24.

Malaviya, R., Gao, Z., Thankavel, K., van der Merwe, P. A., and Abraham, S. N. (1999). The mast cell tumor necrosis factor α response to FimH-expressing *Escherichia coli* is mediated by the glycosylphosphatidylinositol-anchored molecule CD48. Proc Natl Acad Sci USA *96*, 8110–8115.

Malaviya, R., Ikeda, T., Ross, E., and Abraham, S. N. (1996). Mast cell modulation of neutrophil influx and bacterial clearance at sites of infection through TNF-α. Nature *381*, 77–80.

Malaviya, R., and Uckun, F. M. (1999). Genetic and biochemical evidence for a critical role of Janus kinase (JAK)-3 in mast cell-mediated type I hypersensitivity reactions. Biochem Biophys Res Commun *257*, 807–813.

Malaviya, R., and Uckun, F. M. (2002). Role of STAT6 in IgE receptor/FcεRI-mediated late phase allergic responses of mast cells. J Immunol *168*, 421–426.

Mao, X. Q., Shirakawa, T., Enomoto, T., Shimazu, S., Kitano, H., Hagihara, A., and Hopkin, J. M. (1998). Association between variants of mast cell chymase gene and serum IgE levels in eczema. Hum Hered *48*, 38–31.

Mao, X. Q., Shirakawa, T., Yoshikawa, T., Kawai, M., Sasaki, S., Enomoto, T., Hashimoto, T., Furuyama, J., Hopkin, J. M., and Morimoto, K. (1996). Association beween genetic variants of mast-cell chymase and eczema. Lancet *348*, 581–583.

Mapp, C. E. (2001). Agents, old and new, causing occupational asthma. Occup Environ Med *58*, 354–360.

Marone, G., Spadaro, G., Granata, F., and Triggiani, M. (2001). Treatment of mastocytosis: pharmacologic basis and current concepts. Leuk Res *25*, 583–594.

McCurdy, J. D., Lin, T. J., and Marshall, J. S. (2001). Toll-like receptor 4-mediated activation of murine mast cells. J Leukoc Biol *70*, 977–984.

Mekori, Y. A., and Metcalfe, D. D. (2000). Mast cells in innate immunity. Immunol Rev *173*, 131–140.

Metcalfe, D. D., and Akin, C. (2001). Mastocytosis: molecular mechanisms and clinical disease heterogeneity. Leuk Res *25*, 577–582.

Metzger, W. J., Kaplan, A. P., Beaven, M. A., Irons, J. S., and Patterson, R. (1976). Hereditary vibratory angioedema: Confirmation of histamine release in a type of physical hypersensitivity. J Allergy Clin Immunol *57*, 605–608.

Meyer, J. D., Holt, D. I., Chen, Y., Cherry, N. M., and McDonald, J. C. (2001). SWORD '99: surveillance of work-related and occupational respiratory disease in the UK. Occup Med (Lond) *51*, 204–208.

Mitsuyasu, H., Yanagihara, Y., Mao, X., Gao, P.-S., Arinobu, Y., Ihara, K., Takabayashi, A., Hara, T., Enomoto, T., Sasaki, S., Kawai, M., Hamasaki, N., Shirakawa, T., Hopkin, J. M., and Izuhara, K. (1999). Dominant effect of Ile50Val variant of the human IL-4 receptor α-chain in IgE synthesis. J Immunol *162*, 1227–1231.

Miyata, S., Matsuyama, T., Kodama, Y., Nishioka, Y., Kuribayashi, K., Takeda, K., Akira, S., and Sugita, M. (1999). STAT6 deficiency in a mouse model of allergen-induced airways inflammation of CD8⁺ T cells. Clin Exp Allergy *29*, 114–123.

Morita, E., Kameyoshi, Y., Hiragun, T., Mihara, S., and Yamamoto, S. (2001). The C-C chemokines, RANTES and eotaxin, in atopic dermatitis. Allergy *56*, 194–195.

Moriyama, Y., Tsujimura, T., Hashimoto, K., Morimoto, M., Kitayama, H., Matsuzawa, Y., Kitamura, Y., and Kanakura, Y. (1996). Role of aspartic acid 814 in the function and expression of *c-kit* receptor tyrosine kinase. J Biol Chem *271*, 3347–3350.

Mousli, M., Bueb, I.-I., Bronner, C., Rouot, B., and Landry, Y. (1990). G protein activation: a receptor-independent mode of action for cationic amphiphilic neuropeptides and venom peptides. Trends Pharmacol Sci *11*, 358–362.

Nafziger, J., Arock, M., Guillosson, J. J., and Wietzerbin, J. (1990). Specific high-affinity receptors for interferon-γ on mouse bone marrow-derived mast cells: inhibitory effect of interferon-γ on mast cell precursors. Eur J Immunol *20*, 113–117.

Nagata, H., Worobec, A. S., Oh, C. K., Chowdhury, B. A., Tannenbaum, S., Suzuki, Y., and Metcalfe, D. D. (1995). Identification of a point mutation in the catalytic domain of the proto-oncogene *c-kit* in the peripheral blood mononuclear cells of patients with mastocytosis. Proc Natl Acad Sci USA *92*, 10565–10564.

Nagata, H., Worobec, A. S., Semere, T., and Metcalfe, D. D. (1998). Elevated expression of the proto-oncogene *c-kit* in patients with mastocytosis. Leukemia *12*, 175–181.

Naureckas, E. T., and Solway, J. (2001). Mild asthma. N Engl J Med *345*, 1257–1262.

Nghiem, P., Pearson, G., and Langley, R. G. (2002). Tacrolimus and pimecrolimus: from clever prokaryotes to inhibiting calcineurin and treating atopic dermatitis. J Am Acad Dermatol *46*, 228–241.

Nickel, R. G., Casolaro, V., Wahn, U., Beyer, K., Barnes, K. G., Plunkett, B. S., Freidhoff, L. R., Sengler, C., Plitt, J. R., Schleimer, R. P., Caraballo, L., Naidu,

R. P., Nevett, P. N., Beaty, T. H., and Huang, S.-K. (2000). Atopic dermatitis is associated with a functional mutation in the promoter of the C-C chemokine RANTES. J Immunol *164*, 1612–1616.

Ning, Z. Q., Li, J. McGuinness, M., and Arceci, R. J. (2001). STAT3 activation is required for Asp(816) mutant c-Kit induced tumorigenicity. Oncogene *20*, 4528–4536.

Nishida, K., Wang, L., Morii, E., Park, S. J., Narimatsu, N., Itoh, S., Yamasaki, S., Fusishima, M., Ishihara, K., Hibi, M., Kitamura, Y., and Hirano, T. (2002). Requirement of Gab2 for mast cell development and KitL/c-KIT signaling. Blood *99*, 1866–1869.

Noguchi, E., Nukaga-Nishio, Y., Jian, Z., Yokouchi, Y., Kamoika, M., Yamakawa-Kobayashi, K., Hamaguchi, H., Mastsui, A., Shibasaki, M., and Arinami, T. (2001). Haplotypes of the 5′ region of the IL-4 gene and SNPs in the intergene sequence between the IL-4 and IL-13 genes are associated with atopic asthma. Hum Immunol *62*, 1251–1257.

Novak, N., Kraft, S., and Bieber, T. (2001). IgE receptors. Curr Opin Immunol *13*, 721–726.

Ober, C., Leavitt, S. A., Tsalenko, A., Howard, T. D., Hoki, D. M., Daniel, R., Newman, D. L., Wu, X., Parry, R., Lester, I. A., Solway, J. Blumenthal, M. King, R. A., Xu, J., Meyers, D. A., Bleecker, E. R., and Cox, N. J. (2000). Variation in the interleukin 4-receptor α gene confers susceptibility to asthma and atopy in ethnically diverse populations. Am J Genet *66*, 517–526.

Ochi, H., De Jesus, N. H., Hsieh, F. H. Austin, K. E., and Boyce, J. A. (2000). IL-4 and IL-5 prime human mast cells for different profiles of IgE-dependent cytokine production. Proc Natl Acad Sci USA *97*, 10509–10513.

Ochi, H., Hirani, W. M., Yuan, Q., Friend, D. S., Austin, K. F., and Boyce, J. A. (1999). T helper cell type 2 cytokine-mediated comitogenic responses and CCR3 expression during differentiation of human mast cells in vitro. J Exp Med *190*, 267–280.

Oh, S.-W., Pae, C. I., Lee, D.-K., Jones, F., Chiang, G. K. S., Kim, H., Moon, S., Cao, B., Ogbu, C., Jeong, K., Kozu, G., Nakaishi, H., Kahn, M., Chi, E. Y., and Henderon, W. R., Jr. (2002). Tryptase inhibition blocks airway inflammation in mouse asthma model. J Immunol *168*, 1992–2000.

Ohtsuka, Y., Naito, K., Yamashiro, Y., Yabuta, K., Okumura, K., and Ra, C. (1999). Induction of anaphylaxis in mouse intestine by orally administered antigen and its prevention with soluble high affinity receptor for IgE. Pediatr Res *45*, 300–305.

Oliveira, S. H., and Lukacs, N. W. (2001). Stem cell factor and IgE-stimulated murine mast cells produce chemokines (CCL2, CCL17, CCL22) and express chemokine receptors. Inflamm Res *50*, 168–174.

Ong, P. Y., Hamid, Q. A., Travers, J. B., Strickland, I., Al Kerithy, M., Boguniewicz, M., and Leung, D. Y. (2002). Decreased IL-15 may contribute to elevated IgE and acute inflammation in atopic dermatitis. J Immunol *2002*, 505–510.

Ono, S. J. (2000). Molecular genetics of allergic disease. Annu Rev Immunol *18*, 347–366.

Osteritzian, C. A., Wang, Z., Kochan, J. P., Grimes, M., Du, Z, Chang, H.-W., Grant, S., and Schwartz, L. B. (1999). Recombinant human (rh)IL-4-mediated apoptosis and recombinant human IL-6-mediated protection of recombinant human stem cell factor-dependent human mast cells derived from cord blood mononuclear cell progenitors. J Immunol *163*, 5105–5115.

O'Sullivan, S. (1999). On the role of PGD2 metabolites as markers of mast cell activation in asthma. Acta Physiol Scand Suppl *644*, 1–74.

Ott, V. L., and Cambier, I. C. (2000). Activating and inhibitory signaling in mast cells: New opportunities for therapeutic intervention. J Allergy Clin Immunol *106*, 429–440.

Page, S., Ammit, A. J., Black, J. I., and Armour, C. I. (2001). Human mast cell and airway smooth muscle cell interactions: implications for asthma. Am J Physiol *281*, L1313–L1323.

Palmer, I. J., Pare, P. D., Faux, J. A., Moffatt, M. F., Daniels, S. E., LeSouef, P. N., Bremmer, P. R., Mockford, E., Gracey, M., Spargo, R., Musk, A. W., and Cookson, W. O. (1997). FcεRI-β polymorphism and total serum IgE levels in endemically parasitized Australian aborigines. Am J Hum Genet *61*, 182–188.

Pare, P. D., Bai, T. R., and Roberts, C. R. (1997). The structural and functional consequences of chronic allergic inflammation of the airways. Ciba Found Symp *206*, 71–86.

Pawankar, R., Yamagishi, S., and Yagi, T. (2000). Revisiting the roles of mast cells in allergic rhinitis and its relation to local IgE synthesis. Am J Rhinol *14*, 309–317.

Piao, X., and Bernstein, A. (1996). A point mutation in the catalytic domain of c-kit induces growth factor independence, tumorigenicity, and differentiation of mast cells. Blood *87*, 3117–3123.

Piao, X., Paulson, R., van der Geer, P., Pawson, T., and Bernstein, A. (1996). Onco-genic mutation in the Kit receptor tyrosine kinase alters substrate specificity and induces degradation of the protein tyrosine phosphatase. Proc Natl Acad Sci USA *93*, 14665–14669.

Pignon, J. M., Giraudier, S., Duquesnoy, P., Jouault, H., Imbert, M., Vainchenker, W., Vernant, J. P., and Tulliez, M. (1977). A new *c-kit* mutation in a case of aggressive mast cell disease. Br J Haematol *96*, 374–376.

Plaut, M. (2001). Immune-based, targeted therapy for allergic diseases. JAMA *286*, 3005–3006.

Prussin, C., and Metcalfe, D. D. (2001). Update on the management of asthma. Adv Intern Med *46*, 31–50.

Pullarkat, V. A., Pullarkat, S. T., Calverley, D. C., and Brynes, R. K. (2000). Mast cell disease associated with acute myeloid leukemia; Detection of a new *c-kit* mutation Asp816His. Am J Hematol *65*, 307–309.

Reddy, S. T., Tiano, H. E., Lagenbach, R., Morham, S. G., and Herschman, H. R. (1999). Genetic evidence for distinct roles of COX-1 and COX-2 in the imme-diate and delayed phases of prostaglandin synthesis in mast cells. Biochem Biophys Res Comm *265*, 205–210.

Rohrbach, M., Kraemer, R., and Liechti-Gallati, S. (1998). Screening of the FcεRI-β-gene in a Swiss population of asthmatic children: no association with E237G and identification of new sequence variations. Dis Markers *14*, 177–186.

Rosa-Rosa, L., Zimmermann, N., Bernstein, I. A., Rothenberg, M. E., and Hershey, G. K. K. (1999). The R576IL-4 receptor α allele correlates with asthma severity. J Allergy Clin Immunol *104*, 1008–1014.

Rosenwasser, L. J., Klemm, D. J., Dresback, J. K., Inamura, H., Mascali, J. J., Klinnert, M., and Borish, L. (1995). Promoter polymorphisms in chromosome 5 gene cluster in asthma and atopy. Clin Exp Allergy *25*, 74–78.

Rudolf, M. P., Furukawa, K., Meischer, S., Vogel, M., Kricek, E., and Stadler, B. M. (1996). Effect of anti-IgE antibodies on FcεRI-bound IgE. J Immunol *157*, 5646–5652.

Rukwied, R., Lischetzki, G., McGlone, E., Heyer, G., and Schmelz, M. (2000). Mast cell mediators other than histamine induce pruritus in atopic dermatitis patients: a dermal microdialysis study. Br J Dermatol *142*, 1114–1120.

Saito, H., Ebisawa, M., Tachimoto, H., Shichijo, M., Fukagawa, K., Matsumoto, K., Iikura, Y., Awaji, T., Tsujimoto, G., Yanagida, M., Uzumaki, H., Takahashi, G., Tsuji, K., and Nakahata, T. (1996). Selective growth of human mast cells induced by Steel factor, IL-6, and prostaglandin E2 from cord blood mononuclear cells. J Immunol *157*, 343–350.

Sampson, H. A. (1999a). Food allergy. Part 1: Immunopathogenesis and clinical disorders. J Allergy Clin Immunol *103*, 717–728.

Sampson, H. A. (1999b). Food allergy. Part 2: diagnosis and management. J Allergy Clin Immunol *103*, 981–989.

Sampson, H. A. (2000). Food anaphylaxis. Br Med Bull *56*, 925–935.

Schafer, T., Heinrich, J., Wjst, M., Adam, H., Ring, J., and Wichmann, H. E. (1999). Association between severity of atopic eczema and degree of sensitization to aeroallergens in schoolchildren. J Allergy Clin Immunol *104*, 1280–1284.

Schulman, E. S., Kagey-Sobotka, A., MacGlashan, D. W., Adkinson, N. E., Jr, Peters, S. P., Schleimer, R. P., and Lichtenstein, L. M. (1983). Heterogeneity of human mast cells. J Immunol *131*, 1936–1940.

Schwartz, L. B., and Kepley, C. (1994). Development of markers for human basophils and mast cells. J Allergy Clin Immunol *94*, 1231–1240.

Seymour, M. L., Rak, S., Aberg, D., Riise, G. C., Penrose, J. F., Kanaoka, Y., Austen, K. F., Holgate, S. T., and Sampson, A. P. (2001). Leukotriene and prostanoid pathway enzymes in bronchial biopsies of seasonal allergic asthmatics. Am J Respir Crit Care Med *164*, 2051–2056.

Shibaki, A. (1999). FcεRI on dendritic cells: a receptor, which links IgE allergic reaction and T cell mediated cellular response. J Dermatol Sci *20*, 29–38.

Shirakawa, T., Deichmann, K. A., Izuhara, K., Mao, X.-Q., Adra, C. N., and Hopkin, J. M. (2000). Atopy and asthma; genetic variants of IL-4 and IL-13 signalling. Trends Immunol Today *21*, 60–64.

Shirakawa, T., Li, A., Dubowitz, M., Dekker, J. W., Shaw, A. E., Faux, J. A., Ra, C., Cookson, W. O., and Hopkin, J. M. (1994). Association between atopy and variants of the β subunit of the high-affinity immunoglobulin E receptor. Nat Genet *7*, 125–129.

Sicherer, S. H., and Sampson, H. A. (1999). Food hypersensitivity and atopic dermatitis: pathophysiology, epidemiology, diagnosis, and management. J Allergy Clin Immunol *104*, S114–S122.

Skoner, D. P. (2001). Allergic rhinitis: definition, epidemiology, pathophysiology, detection, and diagnosis. J Allergy Clin Immunol *108*, S2–8.

Smith, M. A., Pallister, C. J., and Smith, J. G. (2001). Stem cell factor: biology and relevance to clinical practice. Acta Haematol *105*, 143–150.

Soter, N. A., Wasserman, S. I., and Austen, K. F. (1976). Cold urticaria: release into the circulation of histamine and eosinophil chemotactic factor of anaphylaxis during cold challenge. N Engl J Med *294*, 687–690.

Steinke, J. W., and Borish, L. (2001). Th2 cytokines and asthma. Interleukin-4: its role in the pathogenesis of asthma, and targeting it for asthma treatment with interleukin-4 receptor antagonists. Respir Res *2*, 66–70.

Stempl, D. A., and Woolf, R. (2002). The cost of treating allergic rhinitis. Curr Allergy Asthma Rep *2*, 223–230.

Stewar, A. S., Mitchell, E. A., Pearce, N., Strachan, D. P., and Weiland, S. K. (2001). The relationship of per capita gross national product to the prevalence of symptoms of asthma and other atopic diseases in children (ISAAC). Int J Epidemiology *30*, 173–179.

Summerhill, E., Leavitt, S. A., Gidley, H., Parry, R., Solway, J., and Ober, C. (2000). β-Adrenergic receptor Arg 16/Arg 16 genotype is associated with reduced lung function, but not with asthma, in the Hutterites. Am J Respir Crit Care Med *162*, 599–602.

Supajatura, V., Ushio, H., Nakao, A., Okumura, K., Ra, C., and Ogawa, H. (2001). Protective roles of mast cells against enterobacterial infection are mediated by Toll-like receptor 4. J Immunol *167*, 2250–2256.

Szabo, S. J., Sullivan, B. M., Stemmann, C., Satoskar, A. R., Sleckman, B. P., and Glimcher, L. H. (2002). Distinct effects of T-bet in T_H1 lineage commitment and IFN-γ production in CD4 and CD8 T cells. Science *295*, 338–342.

Takeshita, F., Leifer, C. A., Gursel I, Ishii, K., Takeshita, S., and Gursel, M. (2001). Role of Toll-like receptor 9 in CpG DNA-induced activation of human cells. J Immunol *167*, 3555–3558.

Tamura, K., Arakawa, H., Suzuki, M., Kobayashi, Y., Mochizuki, H., Kato, M., Tokuyama, K., and Morkawa, A. (2001). Novel dinucleotide repeat polymorphism in the first exon of the STAT-6 gene is associated with allergic disease. Clin Exp Allergy *31*, 1509–1514.

Taylor, M. L., and Metcalfe, D. D. (2000). Kit signal transduction. Hematol Oncol Clin North Am *14*, 517–535.

Thestrup-Pederson, K. (2000). Clinical aspects of atopic dermatitis. Clin Dermatol *25*, 535–543.

Tilley, S. L., Wagoner, V. A., Salvatore, C. A., Jacobsen, M. A., and Koller, B. H. (2000). Adenosine and inosine increase cutaneous vasopermeability by activating A_3 receptors on mast cells. J Clin Invest *105*, 361–367.

Togawa, M., Kiniwa, M., and Nagai, H. (2001). The roles of IL-4, IL-5 and mast cells in the accumulation of eosinophils during allergic late phase reaction in mice. Life Sci *69*, 699–705.

Tong, L. J., Balakrishnan, G., Kochan, J. P., Kinet, J.-P., and Kaplan, A. P. (1997). Assessment of autoimmunity in patients with chronic urticaria. J Allergy Clin Immunol *99*, 461–465.

Toren, K., Brisman, J., Olin, A.-C., and Blanc, P. D. (2000). Asthma on the job: work-related factors in new-onset asthma and in exacerbations of pre-existing asthma. Resp Med *94*, 529–535.

Tsujimura, T., Morimoto, M., Hashimoto, K., Moriyama, Y., Kitayama, H., Matsuzawa, Y., Kitamura, Y., and Kanakura, Y. (1996). Constitutive activation of c-kit in FMA3 murine mastocytoma cells caused by deletion of seven amino acids at the juxtamembrane domain. Blood *87*, 273–283.

Turvey, S. E. (2001). Atopic diseases of childhood. Curr Opin Pediat *13*, 487–495.

Ulbrecht, M., Hergeth, M. T., Wjst, M., Heinrich, J., Bickelboller, H., Wichmann, H.-E., and Weiss, E. H. (2000). Association of β$_2$-adrenoreceptor variants with bronchial hyperresponsiveness. Am J Respir Crit Care Med *161*, 469–474.

Ullrich, A., and Schlessinger, J. (1990). Signal transduction by receptors with tyrosine kinase activity. Cell *61*, 203–212.

Valent, P., Horny, H.-P., Escribano, L., Longley, B. J., Li, C. Y., Schwartz, L. B., Marone, G., Nunez, R., Akin, C., Sotlar, K., Sperr, W. R., Wolff, K., Brunning, R. D., Parwaresch, R. M., Austen, K. F., Lennert, K., Metcalfe, D. D., Vardiman, J. W., and Bennett, J. W. (2001a). Diagnostic criteria and classification of mastocytosis: a consensus proposal. Leuk Res *25*, 603–625.

Valent, P., Schernthaner, G.-H., Sperr, W. R., Fritsch, G., Agis, H., Wilheim, M., Buhring, H.-J., Orfao, A., and Escribano, L. (2001b). Variable expression of activation-linked surface antigens on human mast cells in health and disease. Immunol Rev *179*, 74–81.

Valent, P., Sperr, W. R., Samorapoompichit, P., Geisler, K., Lechner, K., Horny, H. P., and Bennett, J. M. (2001c). Myelomastocytic overlap syndromes: biology, criteria, and relationship to mastocytosis. Leuk Res *25*, 595–602.

van der Pouw Kraan, T. C., van Veen, A., Boeije, I. C., van Tuyl, S. A., de Groot, E. R., Stapel, S. O., Bakker, A., Verweij, C. L., Aarden, L. A., and van der Zee, J. S. (1999). An Il-13 promoter polymorphism associated with increased risk of allergic asthma. Genes Immun *1*, 61–65.

Vayssiere, B. M., Dupont, S., Choquart, A., Petit, F., Garcia, T., Marchandeau, C., Gronemeyer, H., and Resche-Rigon, M. (1997). Synthetic glucocorticoids that dissociate transactivation and AP-1 transrepression exhibit antiinflammatory activity in vivo. Mol Endocrinol *11*, 1245–1255.

Vercelli, D., Baldini, M., Stern, D., Lohman, I. C., Halonen, M., and Martinez, F. (2001). CD14: a bridge between innate immunity and adaptive IgE responses. J Endotoxin Res *7*, 45–48.

Verthelyi, D., Kenney, R. T., Seder, R. A., Gam, A. G., Friedag, B., and Klinman, D. M. (2002). CpG oligodeoxynucleotides as vaccine adjuvants in primates. J Immunol *168*, 1659–1663.

Wedemeyer, J., and Galli, S. J. (2000). Mast cells and basophils in acquired immunity. Brit Med Bull *56*, 936–955.

Wedemeyer, J., Tsai, M., and Galli, S. J. (2000). Roles of mast cells and basophils in innate and acquired immunity. Curr Opin Immunol *12*, 624–631.

Weston, M. C., and Peachell, P. T. (1998). Regulation of human mast cells and basophil function by cAMP. Gen Pharmacol *31*, 715–719.

Wjst, M., and Immervoll, T. (1998). Asthma Gene Database: An internal linkage and mutation database for the complex phenotype asthma. Bioinformatics *14*, 827–828.

Wong, W. S. E., and Koh, D. S. K. (2000). Advances in immunopharmacolgy of asthma. Biochem Pharmacol *59*, 1323–1355.

Woolhiser, M. R., Okayama, Y., Gilfillan, A. M., and Metcalfe, D. D. (2001). IgG-dependent activation of human mast cells following up-regulation of FcγRI by IFN-γ. Eur J Immunol *31*, 3298–3307.

Worobec, A. S., and Metcalfe, D. D. (2002). Mastocytosis: current treatment concepts. Int Arch Allergy Immunol *127*, 153–155.

Worobec, A. S., Semere, T., Nagata, H., and Metcalfe, D. D. (1998). Clinical correlates of the presence of the Asp816Val c-*Kit* mutation in the peripheral blood mononuclear cells of patients with mastocytosis. Cancer *83*, 2120–2129.

Xia, Z., Ghildyal, N., Austen, K. F., and Stevens, R. L. (1996). Post-translational regulation of chymase expression in mast cells. A cytokine-dependent mechanism for controlling the expression of granule neutral proteases of hematopoietic cells. J Biol Chem *271*, 8747–8753.

Xu, R., Abramson, J., Fridkin, M., and Pecht, I. (2001). SH2 domain-containing inositol 5′-phosphatase is the main mediator of the inhibitory action of the mast cell function-associated antigen. J Immunol *167*, 6394–6402.

Zuberbier, T., Chong, S. U., Grunov, K., Guhl, S., Welker, P., Grassberger, M., and Henz, B. M. (2001). The ascomycin macrolactam pimecrolimus (Elidel, SDZ ASM 981) is a potent inhibitor of mediator release from human dermal mast cells and peripheral blood basophils. J Allergy Clin Immunol *108*, 275–280.

CHAPTER II

RHEUMATOLOGY AND SIGNAL TRANSDUCTION

KEITH M. HULL and DANIEL L. KASTNER

Office of the Clinical Director, National Institute of Arthritis and Musculoskeletal and Skin Diseases, National Institutes of Health, Bethesda, Maryland; Genetics and Genomics Branch, National Institute of Arthritis and Musculoskeletal and Skin Diseases; National Institutes of Health, Bethesda, Maryland

INTRODUCTION

The field of rheumatology encompasses a broad group of clinical disorders, many of which involve inflammation of the joints. Inflammation is the body's normal initial defense mechanism against tissue injury and invading organisms and involves both innate and acquired immunological processes. The cytokines tumor necrosis factor-α (TNF-α) and interleukin-1 (IL-1) play a major role in mediating these effects, and their dysregulation can result in disease. This point is illustrated by two clinical disorders that serve as the focus for this chapter: TNF-α receptor-associated periodic syndrome (TRAPS), which directly demonstrates the profound consequences of isolated mutations in the receptors for TNF–α, and rheumatoid arthritis (RA), a more complex disorder implicating several cytokines through the final common pathway of nuclear factor-κB (NF-κB) signaling.

TNF-α RECEPTOR-ASSOCIATED PERIODIC SYNDROME

The term TRAPS was first proposed in 1999 (McDermott et al., 1999) to denote patients with recurrent fevers and localized inflammation and mutations in *TNFRSF1A*, which encodes the 55-kDa TNF-α receptor, one of two known receptors for TNF-α. Because TRAPS is the only human disease known to be caused by TNF-α receptor mutations, it provides insight into the function of TNF signaling in human health and disease.

Signal Transduction and Human Disease, Edited by Toren Finkel and J. Silvio Gutkind
ISBN 0-471-02011-7 Copyright © 2003 John Wiley & Sons, Inc.

TABLE 11.1. Clinical Features of TRAPS

Genetics
  Autosomal dominantly inherited
  Mutations in *TNFRSF1A*

Clinical features
  Signs and symptoms
    Attacks
      Duration: 5–21 days
      Frequency: Every 4–8 weeks
    Fever
    Myalgia
    Arthralgia/arthritis
    Abdominal PAIN
    Pleuritis
    Rash (migratory)
    Conjunctivitis
    Amyloidosis

Treatment
  Anti-TNF drugs (e.g., etanercept)
  Glucocorticosteroids
  Nonsteroidal antiinflammatory drugs (NSAIDs)

Clinical Manifestations and Genetics

TRAPS is an autosomal dominant disorder and as such affects both genders equally (Table 11.1). The majority of TRAPS patients present with symptoms in early childhood with attacks lasting between 3 and 21 days and recurring every 4 to 6 weeks (Hull et al., 2002a). Attacks typically begin with generalized fatigue and fever associated with the subtle onset of inflammatory symptoms that eventually crescendo over the course of 1 to 3 days, climax over the next several days and then slowly resolve. Most patients are asymptomatic during intercritical periods, and although it is not known what precipitates an attack, some patients note that physical or mental stress may play a role. Given the widespread distribution of TNFRSF1A throughout the body, it is not surprising that TRAPS comprises such a wide array of symptoms with the most common being fever, muscle pain (myalgia), rash, abdominal pain, and ocular inflammation.

Myalgia is almost always a feature of attacks and is frequently associated with an overlying erythematous rash. Patients describe the pain as cramplike in nature with waxing and waning severity that is often disabling. Symptoms typically affect a single area of the body, most commonly the extremities or torso, and migrate in a distinctive centrifugal pattern. Recent studies suggest that the myalgia and accompanying rash result from monocytic infiltration of the surrounding fascia that produces severe degradation of surrounding collagen and adipose tissue (Hull et al., 2002b). In addition to the macular, erythematous rash described above, a wide spectrum of cutaneous manifestations have been described. Skin biopsies from 10 patients at the National Institutes of Health revealed both deep and superficial interstitial infiltration of monocytes and lymphocytes (Toro et al., 2000). This finding is in contrast to the neutrophilic infiltrate often observed in other recurrent fever syndromes.

Abdominal pain is also a clinical hallmark of TRAPS that may represent inflammation within either the peritoneal cavity or the musculature of the abdominal wall itself. Symptoms are often severe and can include vomiting, constipation, and evidence of bowel obstruction. Consequently, a large proportion of TRAPS patients have undergone exploratory laparotomy and appendectomy, and although histologic examination of the appendix has been usually unrevealing, examination of the resected portions of bowel from TRAPS patients have demonstrated a mononuclear infiltrate, similar to what is observed in skin and muscle biopsies.

Conjunctivitis and periorbital edema are also regularly observed during attacks and may be either unilateral or bilateral in nature. More severe ocular manifestations, including uveitis, have been observed.

To date, 20 different disease-associated mutations in *TNFRSF1A* have been described, 19 single-nucleotide missense mutations occurring within exons 2, 3, and 4 and 1 splicing mutation (c.193-14G→A) that occurs within intron 3 (Hull et al., 2002a). Of the 19 missense mutations, 12 involve highly conserved cysteine residues and 7 involve other amino acid substitutions. The splicing mutation results in a four-amino acid insertion. All but one of the mutations affect the first two extracellular cysteine-rich domains of TNFRSF1A (Hull et al., 2002a).

Although there does not seem to be a distinct correlation between genotype and phenotype, two notable exceptions exist. First, patients with the R92Q mutation appear to have a more heterogeneous clinical presentation than do other TRAPS patients, and, second, individuals harboring cysteine mutations appear to have an increased risk of developing amyloidosis, a potentially life-threatening condition that affects about 15% of TRAPS patients.

Pathogenesis

TNF-α is a pleiotropic cytokine that helps to mediate inflammation, apoptosis, immunoregulation, and cellular proliferation/differentiation. The gene for TNF-α is located on the short arm of chromosome 6, encoding a 157-amino acid type II membrane protein that is anchored to the cell membrane (Spriggs et al., 1992). This unprocessed form of the protein can then be enzymatically cleaved by a metalloproteinase, termed TNF-α-associated cleavage enzyme (TACE), to produce the soluble form of TNF-α (sTNF-α), which exists in solution as a homotrimer and is the form responsible for binding and activating receptors (Loetscher et al., 1991; Smith and Baglioni, 1987).

TNF-α mediates its effect via two distinct receptors, the aforementioned TNFRSF1A, and the 75-kDa receptor, TNFRSF1B (Fig. 11.1a). TNFRSF1A represents the prototype of a larger family of receptors referred to as the TNF-α receptor superfamily (TNFRSF) that now includes 41 different members. This superfamily of receptors shares a common molecular structure consisting of two to six tandemly repeated extracellular regions termed cysteine-rich domains (CRDs) (Naismith and Sprang, 1998). Each CRD consists of between 30 and 40 amino acids, of which 6 are cysteine residues that form three intrachain disulfide bonds.

TNFRSF1A, which is ubiquitously expressed on almost all cell types, and TNFRSF1B, which is largely limited to cells of the hematopoietic system, only share limited homology (~28%) in the extracellular portion of the molecule and none within the intracellular regions, helping to explain their distinct signaling pathways (Armitage, 1994; Beutler and Bazzoni, 1998). As explained in greater detail below, TNFRSF1A is responsible for mediating TNF-α-induced apoptosis, as well as TNF-α-induced NF-κB activation, which prevents cell death. Signaling via TNFRSF1B results mainly in NF-κB activation, and although unable to induce apoptosis by itself, may under certain circumstances potentiate TNFRSF1A-mediated apoptosis. Although it is not entirely clear what ultimately determines whether a cell will undergo apoptosis or not after TNF-α stimulation, evidence suggests that it may depend on the individual cell type, relative population of receptors present on the cell surface, intracellular adaptor proteins, and other simultaneous cytokine signaling.

The structure of TNFRSF1A has been determined by crystallography, both with and without binding to ligand. These studies have demonstrated that homotrimers of TNF-α bind to homotrimer configurations of TNFRSF1A by interacting with amino acid residues within the second and third CRDs (Banner et al., 1993). Signaling is thought to occur when a homotrimer of TNF-α interacts with a homotrimer of its receptor, thereby inducing a conformational change of the receptor, which consequently brings the cytoplasmic domains in close proximity to one another and initiates a signaling cascade. It is interesting to note that crystallography of TNFRSF1A without ligand demonstrates the receptor to exist in a dimer form, thus leaving open the question as to what is the precise stoichiometry of the receptor when it is not bound to ligand (Naismith et al., 1996).

Both TNFRSF1A and TNFRSF1B possess a preligand-binding assembly domain (PLAD) within CRD1 that is distinct from the ligand-binding region of the receptors (Chan et al., 2000a). The PLAD allows for the formation of homotypic receptor complexes to form in the absence of TNF-α and appears to be necessary for efficient ligand binding. Preassembly of receptors may facilitate a rapid recruitment of intracellular signaling mediators after ligand binding, thereby inducing a signal immediately without the need to recruit monomeric molecules into the appropriate signaling trimer (Chan et al., 2000b). Spontaneous

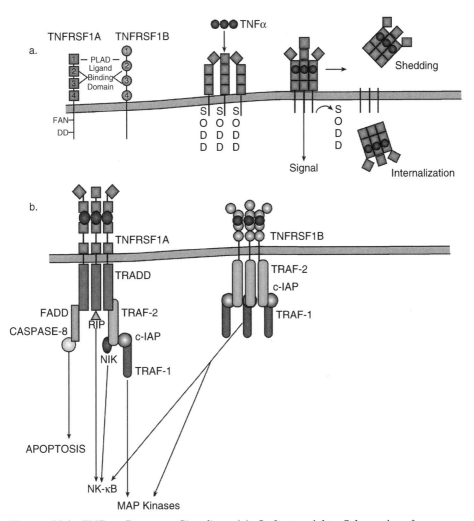

Figure 11.1. *TNF-α Receptor Signaling.* (a) Left to right. Schematic of TNFRSF1A and TNFRSF1B receptors with cysteine-rich domains 1–4 and intracellular signaling domains. A homotrimer of TNFRSF1A before binding of TNF-α. After ligand binding, the homotrimer of TNFRSF1A undergoes a three-dimensional conformational change allowing for the disassociation of SODD and propagation of intracellular signaling. TNFRSF1A is then either cleaved from the cell surface or internalized. (b) TNFRSF1A and TNFRSF1B intracellular signaling pathways. PLAD, preligand assembly domain; DD, death domain; SODD, silencer of death domain; TRADD, TNF receptor-associated kinase; TRAF, TNF receptor-associated death domain; FADD, FAS-associated death domain; RIP, receptor-interacting protein; TRAF, TNF receptor-associated factor; NIK, NF-κB-inducing kinase; c-IAP, cellular inhibitor of apoptosis protein.

intracellular signaling does not appear to occur despite the close proximity of the cytoplasmic domains before ligand binding because of the presence of adapter proteins that effectively inhibit transduction of the signal in the absence of ligand.

After ligand binding and receptor activation, TNFRSF1A and TNFRSF1B can undergo metalloproteinase-dependent cleavage (shedding) from the cell membrane (Crowe et al., 1995; Mullberg et al., 1995). Shedding is thought to contribute to the clearance of the receptor from the membrane and to produce a pool of soluble receptors that may attenuate the inflammatory response by competing for sTNF-α with membrane-bound receptors. Furthermore, the total number of receptors available to form homotrimeric complexes is decreased. In addition to shedding, TNFRSF1A has also been demonstrated to undergo internalization of the ligand-receptor complex after activation, again limiting the receptor pool that would be available for signaling.

In contrast to TNFRSF1B, the cytoplasmic portion of TNFRSF1A possesses a death domain (DD) motif, so named because of its role in mediating the effects of programmed cell death, commonly referred to as apoptosis. This structural motif is also found in a number of different TNFRSF receptors, including Fas/Apo-1, and consists of approximately 80 amino acids that form six amphipathic α-helical regions antiparallel to one another (Armitage, 1994). The DD serves as a protein-docking site for homotypic interactions. The silencer of death domain (SODD) interacts with the DD of TNFRSF1A to inhibit signaling in the absence of TNF-α but disassociates when the ligand binds the receptor, allowing other activator proteins to interact with the DD motif (Jiang et al., 1999). In addition to the DD, the cytoplasmic tail of TNFRSF1A also possesses a sequence that can bind the adaptor protein FAN, which is responsible for activation of neutral sphingomyelinase, which catalyzes the degradation of sphingolipids into smaller ceramide-containing molecules that are key signaling intermediates. Both TNFRSF1A and TNFRSF1B possess sequences that are capable of binding intracellular adaptor proteins that link TNF-α receptor stimulation to activation of many signaling processes. These TNF-α receptor-associating factor (TRAF) adaptors are what transduce the TNF-α signal from the cell surface (Wallach et al., 1999).

Currently, six mammalian TRAF molecules (TRAF1–6) have been identified and all contain a C terminus TRAF domain sequence, as well as a RING finger and zinc finger motif in their N terminus (Bradley and Pober, 2001). TRAF2 is thought to interact directly with the cytoplasmic tail of TNFRSF1B (Fig. 1b), as well as indirectly acting with TRAF1 and TRAF3. TRAF2 also interacts with TNFRSF1A, although indirectly via the adaptor protein TNF-α receptor-associated death domain (TRADD), which itself interacts directly through the DD of TNFRSF1A. TRADD recruits the downstream signaling adaptor molecules Fas-associated death domain (FADD) and receptor interacting protein (RIP), a serine/threonine kinase. FADD possesses a death effector domain (DED) sequence that can directly recruit caspase-8 or -10, initiating the enzymatic cascade of apoptosis. It was recently suggested that FADD and RIP can interact with TNFRSF1B indirectly via binding to TRAF2. TRAF2 can also interact with downstream signaling molecules such as NF-κB-inducing kinase (NIK), which is a member of the serine/threonine mitogen-activated protein kinase (MAPK) family. NIK directly phosphorylates the repressor protein inhibitor of κB (IκB) kinase (IKK), which subsequently triggers the NF-κB pathway transcription cascade (described in the next section).

TRAF2 also mediates binding of cellular inhibitor of apoptosis protein 2 (c-IAP2), which is capable of blocking caspase-8 activation and subsequently apoptosis. NIK and RIP are not the only kinases demonstrated to interact with TNF-α receptors. Apoptosis-stimulating kinase (ASK1), MEKK1, germinal center kinase (GCK), and c-Jun N-terminal kinase (JNK) have also been shown to mediate signaling.

TNF-α Receptor Signaling and TRAPS. The association of TNF receptor mutations with a disorder of excessive inflammation at first came as somewhat of a surprise. Assuming a predominantly proinflammatory role for TNFRSF1A signaling, the clinical features of TRAPS suggested either constitutive activation of the receptor or loss of some inhibitory function of the receptor. Early data supported the latter possibility. TRAPS patients had been shown to have lower serum levels of anti-inflammatory soluble TNFRSF1A compared with normal control subjects, and subsequent experiments demonstrated increased levels of cell surface TNFRSF1A (McDermott, 1997; McDermott et al., 1999). Mononuclear and polymorphonuclear (PMN) cells collected from patients with the C52F mutation exhibited decreased shedding of TNFRSF1A after PMA stimulation, whereas simultaneous analysis of TNFRSF1B shedding was found to be normal. These data were consistent with a model in which mutated receptors are retained on the cell surface and are consequently unable to sequester sTNF-α, thereby perpetuating TNF-α signaling and producing the observed inflammatory phenotype. In fact, impaired TNFRSF1A shedding has been demonstrated in patients with the H22Y, C30S, C33G, P46L, T50M, and C52F, but not R92Q or the splice mutation (Aksentijevich et al., 2001; Galon et al., 2000; Hull et al., 2002b). Because not all TRAPS patients have defective TNFRSF1A shedding, other mechanisms may also account for the clinical manifestations of TRAPS.

How else could the *TNFRSF1A* mutations found in TRAPS account for the observed phenotype? As noted above, almost all of the mutations known to date affect the first and second CRDs. The cysteine missense mutations result in an unpaired cysteine within the extracellular domain of TNFRSF1A that may cause improper intrachain and interchain disulfide bond formation, ultimately leading to improper folding and receptor dysfunction. Among the noncysteine mutations, T50M disrupts a highly conserved threonine that participates in an intrachain hydrogen bond. P46L, S86P, and R92P involve proline substitutions that may disrupt or introduce a bend in the receptor's secondary amino acid structure, thus interfering with proper three-dimensional folding. Possibly, these changes may interfere with proper homotypic interaction at the PLAD and thereby prevent the interaction of the receptors. Additionally, mutations occurring within amino acids 77–114 (S86P, C88R, R92P, R92Q, and F112I) may directly interfere with TNF-α binding to the receptor. Although initial studies with the C52F mutation failed to demonstrate altered TNF-α binding, other mutations may allow for either increased affinity for ligand, hence increasing signaling, or decreased affinity, perhaps leading to "shunting" down a TNFRSF1B-mediated proinflammatory pathway.

Treatment

Nonsteroidal antiinflammatory drugs (NSAIDs; e.g., aspirin, ibuprofen, naproxen) are beneficial in relieving symptoms of fever but are generally unable to resolve musculoskeletal and abdominal symptoms. This class of drugs is thought to mediate their effects primarily via inhibition of cyclooxygenase (COX), the enzyme responsible for the conversion of arachidonic acid to its prostaglandin derivatives. As discussed below, there is evidence to suggest that at least one member of this group, aspirin, may act to decrease inflammation directly by inhibiting NF-κB. Glucocorticosteroids are effective in decreasing the severity of symptoms, both acutely and chronically, but do not alter the frequency of the attacks. Also as discussed in the next section, glucocorticosteroids mediate their effect by a myriad of mechanisms within the cell including binding to glucocorticosteroid-responsive elements (GREs) present within the nucleus that regulate transcription of effector genes, some of which inhibit NF-κB. Although clinically effective, this class of drugs is associated with serious long-term adverse effects. Additionally, other immunosuppressive medications have been tried empirically, including colchicine, azathioprine, cyclosporin, thalidomide, cyclophosphamide, chlorambucil, intravenous immunoglobulin, dapsone, and methotrexate, but with no consistent benefit having been observed for any of these agents.

Within the past 5 years two anti-TNF biological agents have been approved for therapeutic use in RA. Etanercept consists of two TNFRSF1B molecules joined by an immunoglobulin (Ig)G1 Fc fragment, and infliximab is a chimeric anti-human TNF-α IgG$_1$ antibody. Both drugs have been shown to ameliorate clinical symptoms of RA by effectively binding TNF-α and therefore prohibiting its interaction with functional receptors. Evidence is accumulating that these medications may also be effective in TRAPS, especially given the evidence for defective TNFRSF1A shedding. A pilot study demonstrated that six of nine TRAPS patients had decreased frequency, severity, and duration of attacks while receiving etanercept at standard doses over a 6-month period. Additionally, etanercept has been reported to induce regression of renal amyloidosis in a TRAPS patient, although this does not appear to be universal to all TRAPS patients with amyloidosis.

RHEUMATOID ARTHRITIS

RA is one of the most common autoimmune disorders affecting humans, leading to chronic and progressive inflammation of the synovial lining of the joints. Although the underlying etiology is unknown, genetic, environmental, and immunologic factors probably contribute to its pathogenesis. TNF-α and IL-1 play a major role in the initiation and progression of the synovitis, and although they act at different receptors, these two cytokines have overlapping functions, which is not unexpected given that both receptors are able to activate the NF-κB signaling pathway. Moreover, NF-κB is relevant to the majority of rheumatic diseases, including RA, because it mediates the transcription of numerous genes that participate in inflammatory and immune responses.

Clinical Manifestations

RA is a systemic inflammatory disorder that most commonly affects the synovial membranes that line the joints (Table 11.2). If not effectively treated, chronic synovial inflammation ultimately results in joint destruction and, consequently, severely limited function. Statistically, females are

TABLE 11.2. Clinical Features of Rheumatoid Arthritis

Genetics
  Complex genetic underpinnings

Clinical Features
  Signs and symptoms
    Articular manifestations
      Morning stiffness
      Synovitis
        Hands
        Wrists
        Elbows
        Shoulders
        Knees
        Ankles
        Feet
    Extraarticular manifestations
      Rheumatoid nodules
      Pulmonary involvement
      Vasculitis
      Ocular involvement

Treatment
  Disease-modifying antirheumatic drugs (DMARDS; e.g., methotrexate)
  Glucocorticosteroids
  Nonsteroidal antiinflammatory drugs (NSAIDs)
  Anti-TNF drugs (e.g., etanercept)
  Anti-IL-1 drugs (e.g., anakinra)

2.5 times more likely to develop RA than males, and the peak incidence of onset occurs between 30 and 50 years of age.

Because there are no clinical or laboratory pathognomonic features of RA, diagnosis is based on a constellation of symptoms and diagnostic findings. The most commonly used set of criteria for the classification of RA was proposed by the American College of Rheumatology in 1987 (Arnett et al., 1988). These criteria have been shown to have a sensitivity and specificity of ~90%, making them useful in identifying patients early in the course of their disease. For this discussion, symptoms can be broadly grouped into articular and extraarticular manifestations.

Articular manifestations are the major clinical findings of RA and are characterized by morning stiffness and synovial inflammation. Morning stiffness is a common feature of many rheumatic diseases that does not actually reflect a direct function of the time of day but rather reflects prolonged immobilization of the joint as happens during sleep or periods of rest. Patients with active RA commonly experience morning stiffness lasting more than 1 hour and have improvements in their symptoms on warming or with movement. The duration of stiffness has been associated with the degree of synovitis and therefore can be useful in assessing disease activity.

The synovitis associated with RA commonly occurs in a symmetrical distribution that predominantly affects the joints of the hands, and the wrists, elbows, shoulders, feet, and ankles. Physical signs of synovitis are usually observed in more superficial joints that have an easily distensible capsule (e.g., the knee and wrist). Often, evidence of synovitis is subjective and clinical signs may not be readily apparent, but as the disease progresses, irreversible joint damage ensues, resulting in joint deformity and overall decreased function.

Although best known for synovitis, it is important to remember that RA is a systemic disease process that can produce significant inflammation in other organ systems. Extraarticular manifestations more commonly occur in RA patients who have rheumatoid factor (an Ig with antigen specificity to the Fc portion of IgG). Rheumatoid nodules are one of the most common extraarticular findings and are usually located over pressure points such as the elbow, extensor surface of the forearm, knuckles, ischial tuberosity, and Achilles tendon. Other extraarticular manifestations include dermatological, pulmonary, cardiovascular, renal, ocular, gastrointestinal, hematological, and neurological complications.

Pathogenesis

A combination of genetic and environmental factors appears likely to cause RA. Genetic predisposition of RA has been demonstrated in studies of monozygotic twins showing concordance rates of 15–30% versus 1% in dizygotic twins. Additionally, RA can be seen to cluster within families, although simple Mendelian inheritance cannot be demonstrated. The best-characterized evidence to date involves the major histocompatibility complex (MHC) located on chromosome 6

(Jawaheer and Gregersen, 2002). Studies have demonstrated that certain polymorphisms (commonly occurring single-nucleotide changes in the DNA) that encode a "shared epitope" at amino acids 67–74 of HLA-DRB1 are overrepresented in patients with RA and may influence repertoire selection in the thymus or antigen presentation. However, not all patients with RA possess these particular polymorphisms in HLA-DRB1, nor do all patients with the polymorphisms develop RA. This suggests that other genetic or environmental factors also contribute to the disease phenotype. It is likely that the genetic basis of RA, and other so-called "complex" diseases, results from a large number of polymorphisms that individually do not cause a phenotype but in combination, may cause clinical illness. Several collaborative efforts throughout the world are currently being undertaken to determine which other susceptibility polymorphisms outside the HLA coding area may contribute to RA.

Histologic examination of synovial biopsies from patients with RA demonstrate synovial membrane thickening. As the disease progresses, the synovium extends into the joint and causes resorption of cartilage and bone. The infiltration of lymphocytes (CD4$^+$ T cells and B cells) and macrophages are also observed, both of which can produce the "proinflammatory" cytokines, TNF-α (see above) and IL-1.

IL-1 injected into humans produces fevers, myalgia, joint pain (arthralgia), and headache. Its role in RA is supported by both human and animal data demonstrating its ability to induce inflammatory mediators such as nitric oxide, phospholipase A2, COX 2, and metalloproteinases, all of which are abundantly present in inflamed synovium.

The IL-1 family consists of three members: IL-1α, IL-1β, and IL-1 receptor antagonist (IL-1ra). IL-1α and IL-1β each act as agonists at the same receptor but differ in their regulation. Both are synthesized as precursor molecules, but only IL-1α is biologically active in its pro-form. Pro-IL-1β is retained in the cytoplasm and undergoes cleavage by caspase-1 to its active form as it is transported out of the cell. IL-1ra acts as a competitive antagonist to both IL-1α and IL-1β, binding membrane IL-1 receptors without triggering a response. There are two different IL-1 receptors, type I and type II (Auron, 1998; Dinarello, 1997). Only type I receptors have been found to be capable of transducing an intracellular signal, which is mediated by the Toll/IL-1 R (TIR) domain in its cytoplasmic tail that interacts with the adaptor proteins MyD88, IL-1 receptor associated kinase (IRAK), and TRAF-6 to ultimately activate NF-κB (Fig. 11.2). Conversely, type II receptors cannot signal and are hypothesized to act as "decoy" receptors in the intricate regulation of IL-1 signaling.

IL-1 and TNF-α act synergistically and are commonly found within the same sites of inflammation. Their importance in RA is underscored by a large number of trials in humans and animal models demonstrating that treatment with either anti-TNF-α or anti-IL-1 drugs attenuates the inflammatory symptoms and slows bone erosion. Although the receptors for IL-1 and TNF-α are different, they share common intracellular signaling pathways, primarily through NF-κB, which is rapidly activated

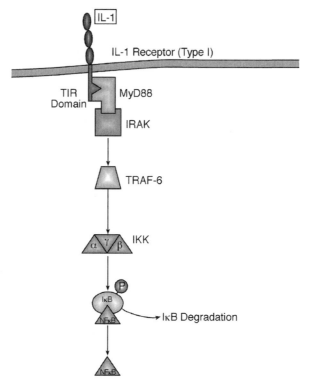

Figure 11.2. *IL-1 Receptor (Type I) Signaling.* TIR, Toll/IL-1R domain; IRAK, IL-1 receptor-associated kinase; TRAF, TNF receptor-associated factor, IKK, inhibitor of kappaB (IκB) kinase; MyD88, myeloid differentiation primary response gene 88; NF-κB, nuclear factor-kappaB.

after receptor activation and is responsible for regulating the transcription of a large number of genes that participate in inflammation and acute stress responses.

NF-κB Signaling

Overview. In the quiescent state, NF-κB is sequestered in the cytoplasm by IκB (Fig. 11.3). After cytokine or LPS receptor activation, IκB is phosphorylated by IKK, resulting in a conformational change that releases NF-κB and simultaneously targets the phosphorylated IκB for degradation. Release of IκB exposes a nuclear localization sequence (NLS) on NF-κB that permits it to translocate to the nucleus, where it binds to specific κB DNA consensus sequences and regulates gene transcription. Among many others, IκB genes are transcribed, the protein products of which enter the nucleus, bind NF-κB, and transport it from the nucleus to the cytoplasm in an inactive form (for more details, see Chapter 2).

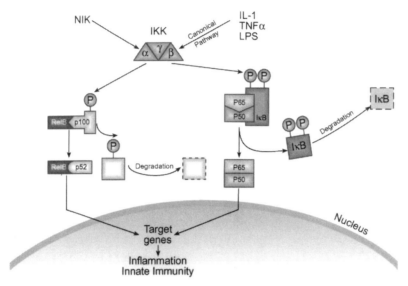

Figure 11.3. *NF-κB Signaling Overview.* NIK, NF-κB-inducing kinase; IκB, inhibitor of kappaB; IKK, IκB kinase; LPS, lipopolysaccharide.

NF-κB. NF-κB was first described as a transcription factor responsible for inducing the synthesis of the κ immunoglobulin light chain in B lymphocytes. Subsequently, NF-κB binding activity was found to occur in most cell types after treatment with several different proinflammatory molecules. NF-κB is not a single protein, but in fact, a group of dimers composed of members of the NF-κB/Rel family that share a 30-amino acid Rel homology domain that is responsible for dimerization, DNA binding, nuclear translocation, and binding of IκB (Karin and Ben-Neriah, 2000). To date, five members have been identified, NF-κB1 (p50 and its precursor p105), NF-κB2 (p52 and its precursor p100, which is largely restricted to expression in B lymphocytes), c-Rel, RelA (p65), and RelB. Heterodimers consisting of p65 and p50 represent the most abundant forms of NF-κB in most cell types, but other homo- and heterodimers exist and differ in their activation of target genes. Only p65 and c-Rel have potent transcriptional activation domains, and because p50, p52, and RelB are lacking this domain, they are believed to act as transcriptional repressors on κB sites.

NF-κB induces more than 150 genes, many of which are involved in the mediation of inflammatory and immune responses. Some of the genes perpetuate inflammation (e.g., TNF-α, IL-1, IL-6, IL-8), whereas others are antiinflammatory in nature (e.g., IL-10, IκBα). Additionally, NF-κB also works in concert with other transcription factors associated with the acute-phase response, for example, activator protein-1 (AP-1), and the nuclear factor protein of IL-6 (NF-IL6). Although not absolute, NF-κB activation has been repeatedly demonstrated to inhibit apoptosis, thus preventing cell death and perpetuating the inflammatory response.

NF-κB regulation. IκB is responsible for the sequestration and inactivation of NF-κB in the cytoplasm by binding to and blocking the NLS of NF-κB. Seven members of the mammalian IκB family have been identified, including IκBα, IκBβ, IκBγ, IκBε, Bcl-3, and the NF-κB precursor molecules, p105 and p100 (Karin and Ben-Neriah, 2000). Although p100 and p105 are precursor molecules of p52 and p50, respectively, they possess IκB-like domains in their C termini and therefore perform an IκB-like function of retaining heterodimers formed with p65, RelB, or c-Rel in the cytoplasm. p105 and p100 are processed to their "active" forms after phosphorylation of their C termini, which are subsequently ubiquitinated and degraded.

Of the IκB family, IκBα (the most thoroughly characterized), IκBβ, and IκBε appear to be the most important mediators in mammals because they contain an N-terminal regulatory domain that allows for its stimulus-induced degradation and subsequent activation of NF-κB. Phosphorylation of two serine residues on IκB has been demonstrated to immediately precede, and be required for, the translocation of NF-κB to the nucleus. After serine phosphorylation, IκB is polyubiquitinated and quickly degraded by the 26S proteosome.

All of the IκB proteins possess a sequence motif that specifically binds to NF-κB at its NLS, thus preventing its translocation to the nucleus. IκBα, whose transcription is regulated by NF-κB, also plays a role in terminating NF-κB signaling. Soon after translation of its mRNA, IκBα enters the nucleus, where it binds to NF-κB. Because there is a higher affinity of binding between IκBα and NF-κB than between NF-κB and its κB binding site on DNA, IκBα is able to bind and export NF-κB back to the cytoplasm via a nuclear export sequence found on IκBα.

IκB is itself regulated by IKK, which consists of three subunits: IKKα, IKKβ, and IKKγ. IKKγ [also known as NF-κB essential modulator (NEMO)] is a regulatory subunit consisting of a leucine zipper (LZ) and a zinc finger domain in its C terminus that mediates interactions with upstream activators of the IKK complex. Most studies support the notion that the IKK complex exists as heterodimer of IKKα and IKKβ associated with a dimer of IKKγ that mediates the formation of a complex containing two catalytic subunits. An intact IKKγ is required for functional activation of IKKα and IKKβ, which serve as the catalytic subunits of IKK and share a high degree of sequence homology. Both consist of an N-terminal protein kinase domain, as well as a LZ, and helix-loop-helix motif in the C terminus. IKKβ, but not IKKα, is necessary for NF-κB signaling via IL-1, TNF-α, and other inflammatory mediators and is referred to as the canonical pathway. A second pathway, specific for the conversion of p100 to p52 in B cells, is regulated by NIK and is independent from the canonical pathway. In this scheme, IKKα, but not IKKβ, is required for processing of p100 to p52 after the phosphorylation of IKKα by NIK. Because ubiquitination is under the control of constitutively expressed enzymes, phosphorylation-dependent degradation of IκB or processing of p100 represents a major regulatory step by

IKKβ and IKKα, respectively. Furthermore, because IKKβ regulates the canonical pathway of NF-κB activation, which is responsive to proinflammatory cytokines (i.e., TNF-α and IL-1), drugs targeted at IKKβ may prove to be therapeutically useful in diseases such as RA. Similarly, drugs targeted at IKKα may be useful in controlling specific B cell responses such as immunoglobulin production.

In summary, NF-κB represents a signaling pathway one of whose purposes is to coordinate and augment the inflammatory/acute-phase response by responding to proinflammatory signals (e.g., cytokines, LPS, dsRNA). This is executed by the coordinated transcription of a large number of genes with κB regulatory sequences. The regulation of this pathway occurs at several key points: 1) IKK phosphorylation of IκB, 2) the NF-κB-dependent transcription of IκBα that then sequesters NF-κB to the cytoplasm, and 3) NF-κB induction of pro- and antiinflammatory cytokines.

Treatment

Until the late 1970s, high doses of aspirin, in conjunction with prednisone, were the mainstay of early therapy for RA. This approach often failed to halt the progression of the disease. Newer agents have been proven to be clinically effective both in attenuating symptoms and in lessening the development of bone erosions. These medications are referred to as disease-modifying antirheumatic drugs (DMARDs) and include methotrexate, etanercept, infliximab, leflunomide, and sulfasalazine, all of which are commonly used today in the treatment of RA.

NSAIDs(e.g., aspirin, ibuprofen, naproxen, celocoxib, refocoxib) are still routinely used for the treatment of pain and inflammation. Additionally, glucocorticosteroids are still commonly prescribed, but with the aim of minimizing their dosage once the DMARDs have taken effect. Within the past decade, a new twist on the mechanism of action of NSAIDs and glucocorticosteroids has been reported involving the inhibition of NF-κB.

NSAID-Induced NF-κB Inhibition. The forerunner of aspirin, salicylic acid, was originally isolated from the bark of willow trees more than 100 years ago and found to be a useful antiinflammatory agent. Aspirin (acetylsalicylic acid) was derived from salicylic acid and is used therapeutically to reduce fever (antipyretic), inflammation (antiinflammatory), and pain (analgesic) and to prevent myocardial infarction. These functions have been demonstrated to result from the inhibition of COX, a key enzyme in prostaglandin synthesis. However, the doses that were commonly used to treat RA result in tissue concentrations that far exceed those needed to inhibit COX. Furthermore, other NSAIDs that are less effective at inhibiting COX are still able to adequately suppress inflammation. These observations suggested that other mechanisms of action might be mediating the effect of NSAIDs. In fact, salicylates have been demonstrated to modulate gene expression in both mammalian and plant cells, suggesting that this class of drugs may modulate gene transcription, either directly or indirectly.

Kopp and Ghosh (1994) first described the ability of salicylic acid and aspirin to inhibit NF-κB activation independent of their action on COX. Cells treated with therapeutically relevant concentrations of sodium salicylate or aspirin and stimulated with LPS demonstrated markedly reduced proteolysis of IκBα. These data suggested that NF-κB activation was inhibited either by inhibition of IκBα phosphorylation or by inhibition of the degradation of IκBα in the 26S proteosome. Subsequent studies confirmed these results and extended them to include cytokine-induced NF-κB activation (Pierce et al., 1996). NF-κB-mediated maturation of human dendritic cells was inhibited by the addition of salicylic acid or aspirin, at doses equivalent to those found in individuals taking antiinflammatory doses of salicylates, but not with ketoprofen, indomethacin, or NS398, a COX-2-selective agent (Matasic et al., 2000).

Inhibition of NF-κB by these NSAIDs does not appear to be selective because numerous cellular kinases are also inhibited (Frantz and O'Neill, 1995). Insight into this mechanism of action was gained when it was demonstrated that aspirin could directly interact with and inhibit the kinase activity of IKKβ by decreasing its ability to bind ATP (Yin et al., 1998). The nonselective nature of NSAID antagonism may be explained by the ability of these agents to act at the ATP-binding sites of kinases in an indiscriminate manner. Important to this discussion is the evidence demonstrating inhibition of other kinase pathways that interact at the transcription factor level with NF-κB. For example, salicylic acid and aspirin were shown to inhibit the activation of AP-1 in murine epidermal cells treated with epidermal growth factor and UV light by the inhibition of Erk-1 and Erk-2, two kinases that regulate the phosphorylation and ultimate activation of the AP-1 complex (Dong et al., 1997). Similarly, flurbiprofen was demonstrated to inhibit AP-1 activation in LPS-treated murine macrophages (Tegeder et al., 2001). Although this effect may be mediated through direct interaction with the AP-1 complex, evidence suggests that it is accomplished by inhibition of the upstream kinases Erk-1 and Erk-2 under certain circumstances.

In summary, at least several of the NSAIDs, including the former standard RA treatment, aspirin, appear to derive a portion of their effectiveness by inhibiting cellular kinases including IKKβ, a key regulating kinase in the canonical pathway of NF-κB activation. Additionally, they may inhibit AP-1 activity, which collaborates with NF-κB in the transcription of genes utilized during acute stress situations.

Glucocorticosteroid-Induced Inhibition of NF-κB. Glucocorticosteroids are widely utilized to suppress inflammation in a wide variety of diseases including RA. In fact, Philip Hench won the 1950 Nobel Prize in Medicine for his discovery of an adrenal-derived substance, later identified as cortisol, which seemed to cure RA. However, as the use of cortisol increased, its adverse effects were soon apparent.

Glucocorticosteroids freely traverse the cell membrane by nature of their lipophilic molecular structure and bind to the glucocorticosteroid receptor (GR) that is largely localized within the cytoplasm. After binding, the glucocorticosteroid-GR complex translocates to the nucleus and regulates gene transcription by directly binding as a homodimer (or heterodimer with other homologous steroid receptors) to GREs on the DNA, as well as interacting with specific transcription factors that link it to the basal transcription apparatus, including other transcription factors and RNA polymerase II. Both the absolute number of GREs and their proximity to the gene appear to be important in mediating the magnitude of the response.

Glucocorticosteroids are thought to mediate a portion of their anti-inflammatory properties by increasing the transcription of antiinflammatory proteins including IκBα. However, the primary effect appears to occur by their ability to directly repress inflammatory and immune-related gene expression. This is mediated by activated GR directly interacting with proinflammatory transcription factors such as the p65 component of NF-κB (Ray and Prefontaine, 1994). This interaction may result in competition between the GR and the binding sites of *trans*-acting elements for other transcription factors (e.g., NF-κB, AP-1, CREB) or perhaps by the GR binding to one of several transcription corepressor molecules (Adcock, 2001).

In summary, TRAPS and RA clearly illustrate the clinical consequences of dysregulation of inflammatory pathways and demonstrate the importance of TNF-α receptor signaling, as well as the role of NF-κB, a major common intracellular mediator of inflammation. Future research aimed at better understanding these pathways should allow us to ultimately develop medications specifically targeted to controlling inflammation.

REFERENCES

Adcock, I. M. (2001). Glucocorticoid-regulated transcription factors. Pulm Pharmacol Ther *14*, 211–219.

Aksentijevich, I., Galon, J., Soares, M., Mansfield, E., Hull, K., Oh, H. H., Goldbach-Mansky, R. et al. (2001). The tumor-necrosis-factor receptor-associated periodic syndrome: new mutations in TNFRSF1A, ancestral origins, genotype-phenotype studies, and evidence for further genetic heterogeneity of periodic fevers. Am J Hum Genet *69*, 301–314.

Armitage, R. J. (1994). Tumor necrosis factor receptor superfamily members and their ligands. Curr Opin Immunol *6*, 407–413.

Arnett, F. C., Edworthy, S. M., Bloch, D. A., McShane, D. J., Fries, J. F., Cooper, N. S., Healey, L. A. et al. (1988). The American Rheumatism Association 1987 revised criteria for the classification of rheumatoid arthritis. Arthritis Rheum *31*, 315–324.

Auron, P. E. (1998). The interleukin 1 receptor: ligand interactions and signal transduction. Cytokine Growth Factor Rev *9*, 221–237.

Banner, D. W., D'Arcy, A., Janes, W., Gentz, R., Schoenfeld, H. J., Broger, C., Loetscher, H. et al. (1993). Crystal structure of the soluble human 55 kd TNF receptor-human TNF β complex: implications for TNF receptor activation. Cell *73*, 431–445.

Beutler, B., and Bazzoni, F. (1998). TNF, apoptosis and autoimmunity: a common thread? Blood Cells Mol Dis *24*, 216–230.

Bradley, J. R., and Pober, J. S. (2001). Tumor necrosis factor receptor-associated factors (TRAFs). Oncogene *20*, 6482–6491.

Chan, F. K., Chun, H. J., Zheng, L., Siegel, R. M., Bui, K. L., and Lenardo, M. J. (2000a). A domain in TNF receptors that mediates ligand-independent receptor assembly and signaling. Science *288*, 2351–2354.

Chan, K. F., Siegel, M. R., and Lenardo, J. M. (2000b). Signaling by the TNF receptor superfamily and T cell homeostasis. Immunity *13*, 419–422.

Crowe, P. D., Walter, B. N., Mohler, K. M., Otten-Evans, C., Black, R. A., and Ware, C. F. (1995). A metalloprotease inhibitor blocks shedding of the 80-kD TNF receptor and TNF processing in T lymphocytes. J Exp Med *181*, 1205–1210.

Dinarello, C. A. (1997). Interleukin-1. Cytokine Growth Factor Rev *8*, 253–265.

Dong, Z., Huang, C., Brown, R. E., and Ma, W. Y. (1997). Inhibition of activator protein 1 activity and neoplastic transformation by aspirin. J Biol Chem *272*, 9962–9970.

Frantz, B., and O'Neill, E. A. (1995). The effect of sodium salicylate and aspirin on NF-κB. Science *270*, 2017–2019.

Galon, J., Aksentijevich, I., McDermott, M. F., O'Shea, J. J., and Kastner, D. L. (2000). TNFRSF1A mutations and autoinflammatory syndromes. Curr Opin Immunol *12*, 479–486.

Hull, K. M., Drewe, E., Aksentijevich, I., Singh, H. K., Wong, K., McDermott, E. M., Dean, J. et al. (2002a). The TNF receptor-associated periodic syndrome (TRAPS): emerging concepts of an autoinflammatory disorder. Medicine. *81*, 349–368.

Hull, K. M., Wong, K., Wood, G. M., Chu, W. S., and Kastner, D. L. (2002b). Monocytic fasciitis: a new clinical feature of TNF-receptor dysfunction. Arthritis Rheum. *46*, 2189–2194.

Jawaheer, D., and Gregersen, P. K. (2002). Rheumatoid arthritis. The genetic components. Rheum Dis Clin North Am *28*, 1–15, v.

Jiang, Y., Woronicz, J. D., Liu, W., and Goeddel, D. V. (1999). Prevention of constitutive TNF receptor 1 signaling by silencer of death domains. Science *283*, 543–546.

Karin, M., and Ben-Neriah, Y. (2000). Phosphorylation meets ubiquitination: the control of NF-κB activity. Annu Rev Immunol *18*, 621–663

Kopp, E., and Ghosh, S. (1994). Inhibition of NF-κB by sodium salicylate and aspirin. Science *265*, 956–959.

Loetscher, H., Gentz, R., Zulauf, M., Lustig, A., Tabuchi, H., Schlaeger, E. J., Brockhaus, M. et al. (1991). Recombinant 55-kDa tumor necrosis factor (TNF) receptor. Stoichiometry of binding to TNF α and TNF β and inhibition of TNF activity. J Biol Chem *266*, 18324–18329.

Matasic, R., Dietz, A. B., and Vuk-Pavlovic, S. (2000). Cyclooxygenase-independent inhibition of dendritic cell maturation by aspirin. Immunology *101*, 53–60.

McDermott, E. M., and Powell, R. J. (1997). Circulating cytokine concentrations in familial Hibernian fever. In: Sohar E. G. J., Pras M. (eds) Familial Mediterranean Fever. Freund, London

McDermott, M. F., Aksentijevich, I., Galon, J., McDermott, E. M., Ogunkolade, B. W., Centola, M., Mansfield, E. et al (1999). Germline mutations in the extracellular domains of the 55 kDa TNF receptor, TNFR1, define a family of dominantly inherited autoinflammatory syndromes. Cell 97, 133–144.

Mullberg, J., Durie, F. H., Otten-Evans, C., Alderson, M. R., Rose-John, S., Cosman, D., Black, R. A. et al (1995). A metalloprotease inhibitor blocks shedding of the IL-6 receptor and the p60 TNF receptor. J Immunol 155, 5198–5205.

Naismith, J. H., Devine, T. Q., Kohno, T., and Sprang, S. R. (1996). Structures of the extracellular domain of the type I tumor necrosis factor receptor. Structure 4, 1251–1262.

Naismith, J. H., and Sprang, S. R. (1998) Modularity in the TNF-receptor family. Trends Biochem Sci 23, 74–79.

Pierce, J. W., Read, M. A., Ding, H., Luscinskas, F. W., and Collins, T. (1996). Salicylates inhibit I κB-α phosphorylation, endothelial-leukocyte adhesion molecule expression, and neutrophil transmigration. J Immunol 156, 3961–3969.

Ray, A., and Prefontaine, K. E. (1994). Physical association and functional antagonism between the p65 subunit of transcription factor NF-κB and the glucocorticoid receptor. Proc Natl Acad Sci USA 91, 752–756.

Smith, R. A., and Baglioni, C. (1987). The active form of tumor necrosis factor is a trimer. J Biol Chem 262, 6951–6954.

Spriggs, D. R., Deutsch, S., and Kufe, D. W. (1992). Genomic structure, induction, and production of TNF-α. Immunol Ser 56, 3–34

Tegeder, I., Niederberger, E., Israr, E., Guhring, H., Brune, K., Euchenhofer, C., Grosch, S. et al. (2001). Inhibition of NF-κB and AP-1 activation by R- and S-flurbiprofen. FASEB J 15, 2–4.

Toro, J. R., Aksentijevich, I., Hull, K., Dean, J., and Kastner, D. L. (2000). Tumor necrosis factor receptor-associated periodic syndrome: a novel syndrome with cutaneous manifestations. Arch Dermatol 136, 1487–1494.

Wallach, D., Varfolomeev, E. E., Malinin, N. L., Goltsev, Y. V., Kovalenko, A. V., and Boldin, M. P. (1999). Tumor necrosis factor receptor and Fas signaling mechanisms. Annu Rev Immunol 17, 331–367.

Yin, M. J., Yamamoto, Y., and Gaynor, R. B. (1998). The anti-inflammatory agents aspirin and salicylate inhibit the activity of I(κ)B kinase-β. Nature 396, 77–80.

CHAPTER 12

MOLECULAR MECHANISMS OF NEURODEGENERATIVE DISORDERS

BENJAMIN WOLOZIN
Department of Pharmacology, Loyola University Medical Center,
Maywood, Illinois

INTRODUCTION

Neurodegeneration can occur slowly or rapidly. Neuronal death occurs within hours to days after a stroke, but neuronal death occurs over years in Alzheimer disease (AD). Awareness of the time frame for any particular type of neurodegeneration is critical for understanding the pathophysiology because the mechanisms of cell death differ greatly depending on the kinetics of neurodegeneration. After a traumatic injury or stroke, neurons die rapidly because of excitotoxicity (typically occurring within 48 h) and apoptosis (typically occurring within 4–14 days). There can also be extensive apoptosis of astrocytes and oligodendrocytes during this period. In contrast, neuronal death in diseases such as AD or Parkinson disease (PD) occurs over a period of years as a result of necrosis induced by oxidative stress and the accumulation of protein aggregates. The steady accumulation of protein aggregates in late-onset neurodegenerative diseases presents an ongoing insult to neurons that causes steady, progressive injury and steady, progressive cell death. Conversely, preventing the accumulation of toxic protein aggregates appears to be the most effective strategy for inhibiting neurodegeneration in these progressive, late-onset neurodegenerative diseases. Given the tremendous advances in our knowledge of the mechanisms of apoptosis, this focus on protein aggregation might seem surprising. However, the toxic aggregates cause ongoing injury, which differs from the time-delimited cell death stimulus typically associated with apoptosis. This chapter reviews current research into the mechanisms of late-onset neurodegeneration and, in the process, clarify why the field has focused to such a great extent on protein aggregation.

Signal Transduction and Human Disease, Edited by Toren Finkel and
J. Silvio Gutkind
ISBN 0-471-02011-7   Copyright © 2003 John Wiley & Sons, Inc.

DEFINING THE DISEASES

Late-onset neurodegenerative diseases constitute a diverse array of diseases that are all characterized by the loss of specific populations of neurons and the accumulation of particular types of inclusions in each disease. Table 12.1 lists some of the major neurodegenerative illnesses, the nerve populations that are affected in each illness, and the proteins that make up the inclusions.

The content of Table 12.1 highlights one of the essential features of neurodegenerative diseases: There is a great deal of overlap in the proteins that accumulate to form inclusions among the different diseases. For instance, inclusions containing tau proteins are present in both AD and frontotemporal dementias (FTDs), and inclusions containing α-synuclein are present in both PD and diffuse Lewy body disease (Fig. 12.1) (Lee et al., 2001; Spillantini et al., 1998b). The involvement of these proteins in multiple diseases suggests that different diseases utilize similar pathological mechanisms. On the other hand, the brain areas affected in each disease often differ despite the involvement of similar proteins in the diseases. The differences in sites of pathology among diseases that show similar protein inclusions likely reflect differences among the pathophysiologies of the disease. In some cases, some of the general pathophysiological mechanisms are known, but the specific factors that cause death of a selected population of neurons is poorly

TABLE 12.1. Major Neurodegenerative Diseases

Disease Name	Brain Area Affected	Pathological Hallmark	Major Proteins Comprising Inclusions
Alzheimer disease	Cortex, hippocampus, and nucleus basalis of Meynert	Neuritic plaque Neurofibrillary tangle	β-Amyloid Tau protein
Parkinson disease	Substantia nigra	Lewy body	α-Synuclein Ubiquitin
Frontotemporal dementia	Frontal cortex	Neurofibrillary tangle	Tau protein
Huntington disease	Striatum	Nuclear inclusion	Huntingtin
Amyotrophic lateral sclerosis	Spinal cord	Cytoplasmic inclusion Neurofibrillary tangle	Superoxide dismutase Neurofilament protein, α-Synuclein
Diffuse Lewy body disease	Cortex	Lewy body	α-Synuclein Ubiquitin

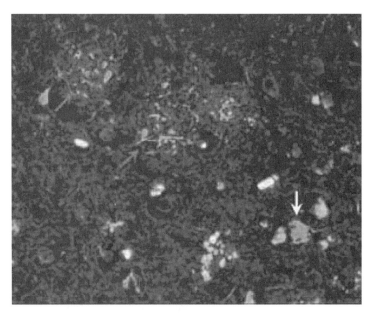

Figure 12.1. Neuritic plaques and neurifibrillary tangles in cortex from a patient with Alzheimer disease.

understood. For instance, aggregation of tau protein to form neurofibrillary tangles presumably occurs in response to toxicity of the β-amyloid peptide (Aβ) in AD. Neuritic plaques (composed of Aβ) and neurofibrillary tangles (composed of tau protein) selectively accumulate in the temporal, frontal, and parietal cortex, despite the presence of the Aβ peptide and tau protein throughout the central nervous system (Selkoe and Lansbury, 1999). Recent studies suggest that the distribution of Aβ and neurofibrillary pathology might reflect the distribution of an enzyme that degrades Aβ, termed neprilysin, but this hypothesis remains to be fully tested (Iwata et al., 2000). Similarly, α-synuclein is present throughout the central nervous system, but Lewy bodies are largely present in the substantia nigra in PD and cortex in diffuse Lewy body disease (Duda et al., 2000). To fully understand the pathophysiology of each disease, we need to understand both the general factors that cause the accumulation of each protein aggregate and the specific factors that lead to the spatial specificity characteristic of each disease.

OXIDATIVE STRESS: THE FIRST HIT IN THE ONE-TWO PUNCH CAUSING NEURODEGENERATION

There is a growing consensus that oxidative stress plays a key role in both aging and the pathophysiology of most neurodegenerative diseases.

Free radicals are continuously produced in living cells from many different sources. Mitochondria release free radicals from the electron transport chain, signal transduction agents such as nitrous oxide form reactive nitroso free radicals, and redox-active metals generate free radicals via the Fenton and Haber–Weiss reactions.

Haber–Weiss reactions: $O_2^- + Fe^{3+} \rightarrow O_2 + Fe^{2+}$

Fenton reaction: $2H_2O_2 + Fe^{2+} \rightarrow 2H_2O + O_2^- + O_2^\bullet + Fe^{3+}$

Once generated, free radicals rapidly react with many cellular constituents. Free radicals oxidize DNA to produce DNA adducts such as 8-hydroxy guanidine (Nunomura et al., 1999). Free radicals oxidize tyrosines to produce nitrotyrosine and tyrosine dimers, free radicals oxidize thiol groups on proteins to form disulfide bonds, and free radicals oxidize lipids. The oxidation of lipids is particularly problematic because the free electron associated with the oxidation is passed from one lipid to the next, leading to a chain of oxidized lipids resembling a zipper. The huge number of oxidative events every minute compounds the damage associated with oxidation. Each cell is estimated to be subject to millions of oxidation events per day. Each cell generates about 3×10^7 superoxide anions per day, which cause oxidation, and further oxidation is caused by free radicals, such as hydroxyl radicals and nitric oxide. Dividing cells constantly renew their organelles, which helps to dilute the affects of oxidation. However, neurons do not divide, so oxidation products accumulate with time. Lipofuscin is a type of inclusion that is produced as a by-product of lipid oxidation, which accumulates with age in neurons. By the time we are elderly, lipofuscin inclusions are present in most neurons in our brains.

DNA damage is also evident. The nucleus contains multiple mechanisms to repair oxidized DNA, but the mitochondria lack many of these mechanisms for DNA repair. Hence, damaged mitochondria accumulate with age, which presumably decreases the functioning of these neurons. Aged mitochondria in nondividing tissues contain mitochondria that have portions of the DNA deleted. By age 60, up to 1% of mitochondrial DNA in some regions contains any particular deletion, and by age 80, up to 10% of the mitochondria contain any particular deletion (Corral-Debrinski et al., 1992). The fraction of mitochondria carrying deletions in the brains of elderly patients, though, is likely much higher because there are many possible deletions (Corral-Debrinski et al., 1992). It seems possible that damage to mitochondrial DNA might impair mitochondrial function and pose a stress for neurons in the elderly. Under normal conditions this stress might not be evident, but the effects of oxidation likely render aged neurons more vulnerable to other insults and might account for the reduced resilience of aged neurons. Thus oxidation represents a major challenge to the survival of neurons as we age.

Neurodegenerative illness adds to this oxidative stress immensely by increasing metal levels and activating new pathways for producing free

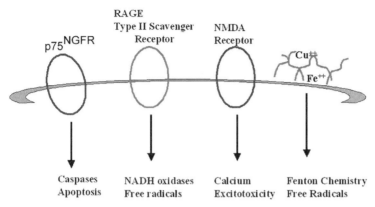

Figure 12.2. The Aβ peptide causes injury through many independent biochemical pathways. Some of the major pathways linked to Aβ toxicity are shown above.

radicals. The process of neurodegeneration increases oxidative stress through multiple independent mechanisms (Fig. 12.2). The protein aggregates that make up inclusions tend to bind metals, which produce free radicals via the Fenton and Haber–Weiss reactions. Protein aggregates also are "sticky" and tend to bind to many receptors. Some of the receptors, such as the RAGE receptor, are coupled to biochemical pathways that stimulate free radical production (Yan et al., 1996). Other receptors, such as the p75[NGF] receptor, are directly coupled to apoptotic pathways (Yaar et al., 1997). Activating apoptosis indirectly stimulates free radical production. Protein aggregates also stress the endoplasmic reticulum and inhibit the proteasome, all of which stimulate free radical production and promote apoptosis (Bence et al., 2001). Extracellular protein aggregates, such as Aβ, attract and activate macrophages. Activated macrophages generate large bursts of free radicals that further contribute to oxidative stress. Thus protein aggregates stimulate free radical production through a large number of different biochemical cascades. This pleiotropic response has important implications for the design of therapeutics because it suggests that blocking any single cell death mechanism will not significantly slow the progress of neurodegeneration.

Multiple neuropathological studies have demonstrated increased levels of oxidation products in brains from donors with neurodegenerative diseases. The levels of 8-hydroxy-guanidine are increased dramatically in brains of patients with AD (Nunomura et al., 1999; Smith et al., 1998). In PD, oxidized products, such as nitrosylated α-synuclein, accumulate in Lewy bodies and the levels of the antioxidants, such as glutathione and tetrahydrobiopterin (BH$_4$), are greatly decreased (Perry et al., 1982). In addition, the redox-active ferrous ion accumulates in Lewy bodies (Castellani et al., 2000). Increased levels of oxidation products have also been noted in brains of patients who died with Huntington disease (HD), multiple-systems atrophy, and FTD (Sayre et al., 2001).

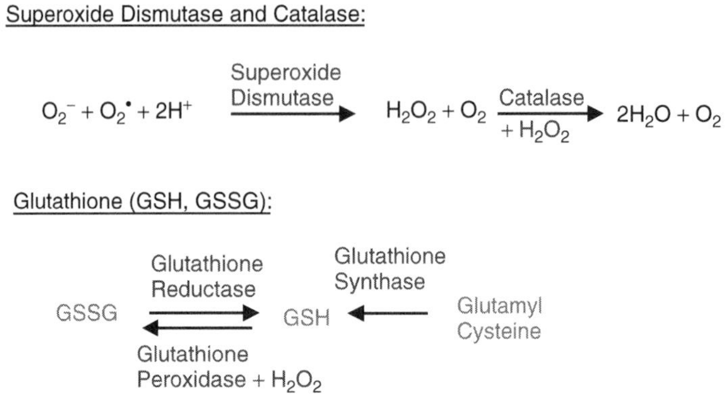

Figure 12.3. Biochemical mechanisms for scavenging free radicals. Two of the major enzymatic routes for scavenging free radicals are those mediated by super-oxide dismutase/catalase and glutathione.

Given all of these changes, there is no doubt that oxidative stress plays a key role in the process of neurodegeneration.

Oxidative stress, though, does not affect all tissues equally. Some tissues appear to be more vulnerable to oxidative stress than other tissues. One factor that controls vulnerability to oxidative stress is antioxidant capacity. Tissues have developed numerous mechanisms to scavenge free radicals. These mechanisms are summarized in Figure 12.3. One important mechanism to reduce free radical damage is the antioxidant enzyme system. The enzymes catalase and superoxide dismutase (SOD) eliminate superoxide and hydrogen peroxide and are a set of ubiquitous enzymes that are one important class of antioxidant proteins. Many of these enzymes are regulated by a master neuroprotective transcription factor, termed NF-κB (Baldwin, 1996). Activation of NF-κB increases transcription of these enzymes and thereby increases the ability to withstand oxidative insults. There are three different forms of SOD: cytoplasmic, mitochondrial, and extracellular. Of these, mitochondrial SOD appears to be the most important dismutase, because eliminating mitochondrial SOD in mice causes perinatal death (Lebovitz et al., 1996). The brain has a particularly high amount of oxidative stress because of the high metabolism of neurons of the central nervous system. To circumvent this problem, neurons generate large amounts of the antioxidant glutathione, using a regenerative enzyme system involving glutathione synthase, glutathione peroxidase, and glutathione reductase (Fig. 12.3) (Nakamura et al., 1997). The dopaminergic neurons of the substantia nigra are subject to added oxidative stress because dopamine is an oxidant. Surprisingly, dopaminergic neurons are not more sensitive to oxidative insult than other neurons (Nakamura et al., 1997). The reason appears to be that they also have high levels of an antioxidant, BH_4, that also functions as a cofactor in the mitochondrial electron transport chain (Nakamura et al., 2000). Loss of BH_4 renders dopaminergic neurons far more sensitive to oxidative insult (Nakamura et al., 2000). The endogenous oxidative load and requirement for BH_4 might also render dopaminergic neurons particularly susceptible to degeneration in response to mitochondrial toxins. Finally, another small molecule that has received attention for its scavenging powers is estrogen. Estrogen has a triple function (Behl and Holsboer, 1999). The molecule itself directly scavenges free radicals. Estrogen also acts as a neuroprotective transcription factor, and it activates the cytoprotective protein kinase B signal transduction cascade. Together, glutathione, BH_4, estrogen, and SOD constitute the major mechanisms for protecting against oxidative stress.

PROTEIN AGGREGATION: THE SECOND HIT IN THE ONE-TWO PUNCH CAUSING NEURODEGENERATION

A key factor determining cellular vulnerability to oxidative stress appears to be protein oxidation and protein aggregation. Many neurodegenerative diseases contain inclusions characteristic of that disorder. Each inclusion contains a particular type of protein (some examples are described in Table 12.1) that is present in an aggregated state—often as a fibrillar aggregate. Each of the major proteins that form pathological inclusions, such as β-amyloid, α-synuclein, tau protein, and huntingtin, appears to be characterized by a strong tendency to aggregate, and oxidation of these proteins generally increases their tendency to aggregate. Linkage studies using molecular genetics provide one of the strongest lines of evidence indicating that these protein aggregates contribute directly to neurodegeneration. Mutations in each of these proteins have been linked to neurodegenerative disease, and biochemical studies show that each of these mutations increases the tendency of the protein to aggregate. Mutations in two proteins, APP or presenilins, which increase production of rapidly aggregating Aβ42, cause familial AD (Selkoe and Lansbury, 1999). Expansions of polyglutamine stretches cause proteins such as huntingtin and ScaI to aggregate, which causes HD and spinocerebellar ataxia, respectively (Ross, 1995). Mutations in tau protein and α-synuclein increase the tendency of each of these proteins to aggregate and cause FTD of chromosome 17 and PD, respectively (Lee et al., 2001). Accumulating evidence shows that these protein aggregates cause toxicity through a number of different mechanisms. Most recently, many protein aggregates have been shown to inhibit the proteasome, which is a process that causes acute apoptosis in cells grown in culture (Bence et al., 2001; Snyder et al., 2003). The concordance of these different lines of evidence leads to the conclusion that protein aggregation is a process that is fundamentally important to the progression of neurodegenerative disease, and therefore investigating the pathophysiology of disease-related aggregates could provide fundamental insights into the pathophysiology of neurodegenerative disease.

The toxicity of protein aggregates appears to lie in their "stickiness" and their ability to activate receptors by cross-linking. To understand why aggregation might increase affinity, we need to consider receptor-binding characteristics. Most compounds have both high-affinity binding sites and low-affinity binding sites. The high-affinity binding site is designated as the specific receptor, whereas the low-affinity receptor can be thought of as resulting from nonspecific binding. The difference in affinity between specific and nonspecific receptors is usually high enough so that the monomeric protein (e.g., Aβ or α-synuclein) exhibits little binding to the "nonspecific" receptor under physiological concentrations. However, oligomerization, or aggregation, allows cross-linking of receptors, which can increase the affinity greatly. This increased affinity enables the aggregated protein to bind to more proteins than the monomeric form. For example, aggregated α-synuclein inhibits the proteasome with an IC50 at least 10,000-fold stronger than monomeric α-synuclein (1 nM for aggregated α-synuclein vs. 10 μM for monomeric α-synuclein) (Snyder et al., 2003). Cross-linking also activates receptors. For example, both monomeric and aggregated β-amyloid bind the p75 neurotrophin receptor (p75^{NTR}) with an affinity of less than 25 nM (Yaar et al., 1997). However, aggregated Aβ activates apoptosis through the p75^{NTR}, whereas monomeric Aβ is nontoxic. One way to understand why aggregation would allow Aβ to activate a receptor is to consider the stoichiometry of binding. Monomeric Aβ has only one binding site for the p75^{NTR}, but a complex of aggregated or oligomeric Aβ can bind more than one p75^{NTR} receptor molecules at the same time. Binding to multiple receptors leads to dimerization of the receptors, and this dimerization is a classic signal for activating growth factor receptors. In the case of p75^{NTR}, activation stimulates a proapoptotic cascade mediated by caspase-9 (Khursigara et al., 2001). Binding and activation by aggregated Aβ is true of other proteins that normally show low-affinity binding to monomeric Aβ. The result is that aggregated Aβ binds and activates many, many proteins. Binding of aggregated Aβ to most of these proteins is detectable but is irrelevant to the disease process because binding does not produce injury. The relevance appears when binding occurs to a protein that stimulates free radical production (e.g., RAGE) or is linked to cell death pathways (e.g., p75^{NTR}) or is required for normal cellular function (e.g., the proteasome) (Fig. 12.2). Aβ-induced changes in the function of such a protein stimulate cell death processes, which contribute to neurodegeneration. Thus protein aggregation increases "nonspecific binding," and this can lead to neurodegeneration when some of the protein targets are either linked to cell death cascades or necessary for cellular viability.

The specific mechanisms of degeneration vary from protein aggregate to protein aggregate. In the next section, we review several examples of how protein aggregates lead to neurodegeneration.

β-Amyloid (Aβ): The Paradigm for Protein Aggregation

The Aβ peptide is a cleavage product generated from its parent precursor protein, amyloid precursor protein (APP). Most APP is cleaved at a site termed the α-secretase site, which cuts in the middle of the Aβ domain at position 16 of Aβ and prevents production of Aβ. About

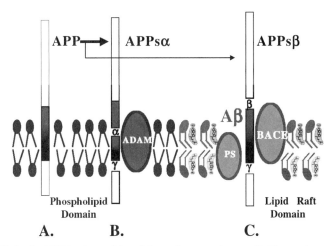

Figure 12.4. A. APP is cleaved by either of two pathways. **B.** The major pathway is via α-secretase, which occurs in a phospholipid-rich domain. **C.** A minor pathway is via β- and γ-secretases, which occurs in a sphingomyelin/cholesterol-rich lipid domain, termed a lipid raft. This pathway is significant for disease because it leads to formation of Aβ, which causes Alzheimer disease.

10% of the APP, though, is cleaved via a pathway that generates the Aβ peptide (Fig. 12.4) (Selkoe and Lansbury, 1999). Two cleavages are required to generate Aβ. The N terminus of APP is cleaved by the enzyme β-secretase, and the C terminus is cleaved by an enzyme termed γ-secretase, or presenilin. There are two homologs of each enzyme, BACE 1 and 2 and presenilin 1 and 2 (Levy-Lahad et al., 1995; Sherrington et al., 1995; Sinha et al., 1999; Vassar et al., 1999). The major enzymes cleaving APP in the brain appear to be BACE 1 and presenilin 1. Cleavage produces an Aβ peptide that is mostly 40 amino acids long, but a small amount of the peptide is 42 amino acids long. Mutations in presenilin 1 are the most common cause of mutations associated with familial AD, and these mutations increase production of the 42-amino acid Aβ peptide (Citron et al., 1997; Scheuner et al., 1996). The longer Aβ peptide aggregates much more rapidly than the 40-amino acid peptide that is normally produced (Jarrett and Lansbury, 1993). Aggregated Aβ is thought to be the cause of AD because every presenilin and APP mutation associated with AD results in increased production of either total Aβ or the Aβ42 peptide.

To understand why Aβ42 is important for the pathophysiology of PD, it is important to understand the kinetics of aggregation. The process of aggregation is analogous to the process of crystallization. Nucleation is the rate-determining step in crystallization. Once a nidus has formed, growth of the crystal occurs rapidly (Fig. 12.5) (Jarrett and Lansbury, 1993). This is why a solution of sugar can exist in a supersaturated state until a small nidus is added, whereupon crystallization occurs very rapidly. The process of Aβ aggregation also can be divided into two steps: nucleation and aggregation (Jarrett and Lansbury, 1993). Nucleation is the process of creating a nidus around which larger aggregates can form,

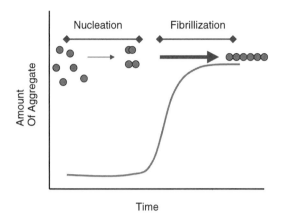

Figure 12.5. Aggregation of Aβ occurs in two steps. Nucleation of Aβ is a slow process that leads to formation of protofibrils. Once formed, fibrillization to make a large aggregate occurs rapidly.

and aggregation then occurs rapidly. Nucleation of Aβ depends on forming small multimers, such as dimers, trimers, and tetramers, which can be thought of as microaggregates or protofibrils. The critical feature of a microaggregate is that it contains several Aβ molecules aligned as β-pleated sheets. The kinetics of Aβ42 nucleation are much more rapid than the kinetics of Aβ40 nucleation. Once formed, these sheets of Aβ40 or Aβ42 appear to serve as templates that greatly facilitate binding of additional Aβ molecules in a β-pleated sheet structure. Thus nucleation occurs much more slowly than growth of the aggregate, and the ability of Aβ42 to nucleate rapidly greatly accelerates the kinetics of Aβ aggregation.

Multiple independent lines of evidence lead to the conclusion that aggregation of Aβ causes AD. Neuritic plaques, which are one of the pathological hallmarks of AD, contain abundant Aβ. All mutations associated with early-onset AD cause effects that would increase aggregation of Aβ. Mutations in presenilins associated with AD increase production of Aβ42. All mutations in APP associated with AD are near the Aβ domain. One type of mutation occurs at the β-secretase cleavage site, which increases total Aβ production (Citron et al., 1992). A second group of mutations flank the γ-secretase and specifically increase production of Aβ42 (Citron et al., 1997; Scheuner et al., 1996). A third group of mutations flank the α-secretase site and inhibit the α-secretase cleavage, which leads to an increase in Aβ production (Haass et al., 1994). Finally, a fourth genetically linked site affects apolipoprotein E (apo E) (Corder et al., 1993). The apo E4 polymorphism tends to increase aggregation of Aβ and leads to earlier appearance of AD (Ma et al., 1994; Strittmatter et al., 1993). Transgenic mice expressing AD-associated mutant forms of APP develop neuritic plaques, mild neurofibrillary pathology, and age-associated cognitive loss (Games et al., 1995; Hsiao et al., 1996). Crossing of these mice with mice expressing AD-associated presenilin

mutations increases production of Aβ42 and also accelerates both production of neuritic plaques and cognitive decline (Duff et al., 1996). Thus
all mutations appear to increase production of Aβ40 or -42 and correspondingly increase aggregation of Aβ in the brain.

One disappointing aspect of the transgenic mice has been the limited
amount of cell death and neurofibrillary tangles produced, but multiple
studies indicate that aggregated Aβ is highly toxic. Addition of Aβ is
toxic to primary cultures of hippocampal neurons, cortical neurons, cell
culture neurons, PC12 cells, and many neuroblastoma cell lines (Yankner
et al., 1990). Numerous biochemical pathways have all been shown to
mediate Aβ toxicity, which suggests that the mechanism of toxicity is
pleiotropic. The Aβ peptide is able to catalyze free radical production,
particularly when incubated with small amounts of metal (Hensley et al.,
1994). This is because Aβ has a histidine that can coordinate metals and
a cysteine that facilitates transfer of electrons from metals to Aβ
(Varadarajan et al., 1999). The tendency of Aβ to promote free radical
production is further increased by its strong ability to bind and concen-
trate redox-active metals that promote free radical production. Thus, Aβ
can promote free radical production independently of any direct interaction with cells.

The Aβ peptide also binds to numerous proteins associated with
neurons, and several of these proteins are linked to cell death cascades.
These proteins include the p75^{NGF} receptor, the RAGE receptor
(receptor for advanced glycation end products), the type II Scavenger
receptor, and ERAB (El Khoury et al., 1996; Yaar et al., 1997; Yan et al.,
1996; Yan et al., 1997). Each of these receptors has been shown to
bind to aggregated Aβ, with affinities in the high nanomolar range. In
each case, binding to aggregated Aβ is greater than binding to
monomeric Aβ. A great deal of information is known about the p75^{NGFR}
and the RAGE receptor. The p75^{NGFR} links the trk system with proapoptotic cascades (Khursigara et al., 2001; Rabizadeh et al., 1993). Consequently, binding of Aβ to p75^{NGFR} stimulates cell stress cascades mediated
by JNK and apoptotic cascades mediated by caspase-9 (Khursigara et
al., 2001). Activation of these cascades might explain why Aβ is toxic to
PC12 cells grown in culture, because these cells express the p75^{NGFR}. The
cholinergic neurons of the basal forebrain, which express abundant
p75^{NGFR}, are among the first neurons lost in AD and are affected to the
greatest extent (Mufson et al., 1989). Activation of cell death cascades
by binding of Aβ to the p75^{NGF} receptor likely explains why these
neurons are so vulnerable to Aβ exposure. Thus binding of Aβ to p75^{NGFR}
provides a mechanistic explanation for Aβ toxicity, both in cell culture
and *in vitro*.

The RAGE receptor has also been examined in detail. The RAGE
receptor appears to function as a scavenger for proteins that have undergone nonenzymatic glycation, which is a slow covalent reaction that
occurs between free glucose and proteins (Yan et al., 1996). Proteins with
slow turnover are particularly susceptible to nonenzymatic glycation
because they exist long enough for the slow nonenzymatic glycation to

occur. The RAGE receptor is present on macrophages and allows the macrophages and microglia to bind, endocytose, and degrade glycated proteins. Aggregated Aβ is also taken up by the RAGE receptor, perhaps because the β-pleated sheet structure of aggregated Aβ readily binds many complex sugar molecules. Degradation of molecules taken up via the RAGE receptor occurs via a burst of free radicals (Yan et al., 1994). It is this burst of free radicals that appears to render the RAGE receptor deleterious. For unknown reasons, neurons also express the RAGE receptor, which facilitates uptake of aggregated Aβ by neurons. The result is that exposing neurons to Aβ causes a burst of harmful free radical in neurons exposed to Aβ. A similar process likely occurs in response to binding of Aβ to the type II scavenge receptor. Unlike the RAGE receptor, the type II scavenger receptor is primarily located on microglial cells, which suggests that this receptor plays a greater role in uptake of Aβ by microglial cells than by neurons (El Khoury et al., 1996). In each case, specific receptors mediate the uptake of Aβ. Uptake of Aβ leads to an oxidative burst, which produces cell death and neurodegeneration in vulnerable cell populations.

Huntingtin: The Effects of Polyglutamine Expansions

An important reason that protein aggregation is thought to be important for disease is that many of the proteins that are associated with neurodegeneration aggregate readily and many of the disease-related mutations that occur in the proteins increase the rate of aggregation. The fact that disease-related mutations consistently accelerate protein aggregation strongly suggests that protein aggregation plays an important role in neurodegenerative disease.

Polyglutamine expansion underlies an increasing number of rare neurodegenerative diseases, including HD, Kennedy disease, spinocerebellar ataxias, fragile X syndrome, myotonic dystrophy, and others (Paulson and Fischbeck, 1996; Trottier et al., 1995). Each of these diseases is caused by mutations in a protein that increases the length of a polyglutamine stretch. For instance, huntingtin is the protein that normally contains a region with up to 22 glutamines in tandem (MacDonald et al., 1993). In HD, this polyglutamine region is expanded to a stretch of over 42 glutamines (MacDonald et al., 1993). This expanded polyglutamine stretch is "sticky," which results in new biochemical properties. One important property is that the expanded polyglutamine domain has a much greater tendency to aggregate, which accounts for the accumulation of nuclear aggregates of the brains of patients with HD (DiFiglia et al., 1997). One reason that accumulation of huntingtin aggregates might be harmful is that the aggregates tend to bind other proteins, such as the transcription factor CREB (Nucifora et al., 2001). Binding of CREB causes it to accumulate in the aggregate, which leads to the sequestration of CREB in the aggregate rather than in contact with either nuclear DNA or relevant signaling proteins. Because CREB is neuroprotective, sequestration of CREB might functionally deplete CREB and remove its neuroprotective function (Nucifora et al., 2001). Huntingtin protein containing polyglutamine expansions also interfere with BDNF-mediated gene transcription, which is another potential mechanism leading to cell death in HD (Zuccato et al., 2001). Extended polyglutamine repeats also bind other proteins. For instance, expanded polyglutamine repeats appear to bind caspases and stimulate their activity, which might promote apoptosis (Saudou et al., 1998). Expanded polyglutamine repeats therefore might promote neurodegeneration by removing CREB- and BDNF-mediated neuroprotection and increasing caspase-mediated apoptosis.

α-Synuclein: Oxidation Inducing Aggregation

Similar stories can be made for both α-synuclein in PD, and tau protein in FTDs. PD is characterized by a loss of dopaminergic neurons of the substantia nigra and the presence of eosinophilic intracellular inclusions, termed Lewy bodies, in the remaining dopaminergic neurons (Lang and Lozano, 1998). Lewy bodies contain a number of proteins, but the two main proteins are ubiquitin and α-synuclein (Duda et al., 2000; Spillantini et al., 1997; Spillantini et al., 1998). The presence of ubiquitin is not surprising because it is an 80-amino acid peptide tag that is added to damaged proteins that are targeted for degradation. Most intracellular proteins that accumulate in aggregated form are ubiquitinated; hence, ubiquitin cannot be considered to be specific to any one class of neurodegenerative illnesses. We examine the ubiquitin system in more detail in the next section. On the other hand, accumulation of aggregated α-synuclein is selective for a group of disorders, now referred to as synucleinopathies. Disorders in which α-synuclein represents a major component of the neuropathology include PD, diffuse Lewy body disease, multiple-systems atrophy, and brain iron accumulation disease type I (Hallervorden–Spatz syndrome) (Dickson et al., 1999; Duda et al., 2000; Galvin et al., 2000; Spillantini et al., 1997; Spillantini et al., 1998). α-Synuclein is an abundant intracellular protein, which is present in most cells but shows particular abundance in synapses. The function of α-synuclein is poorly understood, but it might play a role as a chaperone, analogous to 14-3-3, that regulates enzymes such as protein kinase C and phospholipase D (Jenco et al., 1998; Ostrerova et al., 1999). The association with lipids might regulate the activity of α-synuclein, because it changes from a disordered structure to a helical structure on contact with lipids (Clayton and George, 1998). The feature of α-synuclein that is perhaps most relevant to PD is that the protein aggregates readily, particularly after contact with free radicals and metals, such as iron (FeII) (Conway et al., 1998; Conway et al., 2000; Hashimoto et al., 1999; Ostrerova-Golts et al., 2000). Aggregation of α-synuclein is toxic to neurons, both in cell culture and *in vivo*, although the mechanism of toxicity is not yet known (Ostrerova-Golts et al., 2000). α-Synuclein is known to influence the proteasome and the proapoptotic protein BAD, and aggregated α-synuclein appears to bind metals (Golts et al., 2002; Ostrerova-Golts et al., 2000; Paik et al., 1999). Any of these mechanisms could contribute to the toxicity of α-synuclein, but the studies of α-synuclein are not far enough along to speculate with any confidence about why α-synuclein is harmful to dopaminergic neurons.

One of the intriguing aspects of the pathophysiology of PD is the association of PD with environmental factors, such as pesticides, herbicides, or heavy metals. Unlike most neurodegenerative diseases, PD is strongly associated with environmental factors. Farmers and metal workers exhibit an increased risk for PD (Langston, 1998). The association between toxins and PD has been further strengthened by the observations that specific toxins such as MPTP cause an acute Parkinsonian syndrome (Langston et al., 1983). MPTP is a by-product generated during heroin synthesis. It is metabolized to MPP^+, which is taken up by the dopamine transporter and concentrated in the outer membrane of the mitochondria (Tipton and Singer, 1993). The concentration of MPP^+ in the outer membrane disrupts the electron transport chain. Dopaminergic neurons are normally strongly protected from oxidative damage by an abundance of BH_4 and glutathione, but blockade of the mitochondria reduces BH_4 production, increases the free radical load, and kills the dopaminergic neurons (Nakamura et al., 2000). The toxicity of free radicals might be aggravated by a large pool of iron in the cells due to the sequestration of iron by neuromelanin (Zecca et al., 2001). Free radicals might enhance release of free iron, which then stimulates α-synuclein aggregation. Treatment of cells or animals with mitochondrial toxins such as MPTP or rotenone has been shown to stimulate aggregation of α-synuclein and production of free radicals, and cells overexpressing α-synuclein are more vulnerable to cell death mediated by such toxins (Betarbet et al., 2000; Uversky et al., 2001).

There is little doubt that α-synuclein plays an important role in the pathophysiology of PD, because two independent lines of evidence confirm this toxicity. Molecular genetics has identified two different kindreds in which mutations in α-synuclein, at positions A53T or A30P, are associated with familial PD (Kruger et al., 1998; Polymeropoulos et al., 1997). Both of these mutations increase the propensity of α-synuclein to aggregate (Conway et al., 1998; Conway et al., 2000). The pathogenicity of these mutations has been confirmed *in vivo*. Transgenic expression of A53T or wild-type α-synuclein in *Drosophila* or in cell lines produces dopaminergic dysfunction and degeneration (Feany and Bender, 2000; Masliah et al., 2000). In *Drosophila*, structures resembling Lewy bodies build up as the degeneration occurs (Feany and Bender, 2000). Thus, in PD, there appears to be an interaction between environmental toxins and aggregation-prone proteins, which together lead to dopaminergic neurodegeneration. It seems possible that inhibition of either process, either by removing the environmental stimulus or by preventing α-synuclein aggregation, might inhibit the progression of PD.

Tauopathies: The Intracellular Marker of Neurodegeneration

Diseases that accumulate neurofibrillary tangles are termed tauopathies. Tauopathies include AD, FTDs, FTDs associated with PD (FTDP-17),

Pick disease, and progressive supranuclear palsy (Spillantini et al., 1998). Each of these disorders is characterized by the accumulation of pathological inclusions that contain large amounts of the microtubule-associated protein tau (Spillantini et al., 1998). Although the causes of the accumulation of tau protein are unknown, molecular genetics has identified a group of tauopathies that are associated with mutations in the tau gene on chromosome 17 (Hong et al., 1998; Hutton et al., 1998). Each of the mutations present in the coding region results in a tau protein that increases the tendency of the protein to aggregate into filaments. The mutations also tend to inhibit the ability of tau to promote microtubule formation, but the effects on protein aggregation are likely the major pathophysiological effect of the mutation. Some cases of progressive supranuclear palsy also contain tau-associated mutations in an intron that alter the ratio of 3R to 4R tau and increase the tendency of tau to aggregate.

Tauopathies and synucleinopathies share similarities with respect to animal models, because transgenic mouse lines expressing a single transgene (either wild-type or mutant tau or α-synuclein) mimic disease only to a limited degree (Kahle et al., 2000; Lewis et al., 2000). Mice expressing these tau mutations develop tangles in the spinal cord, but not in the cortex, as is seen in human tauopathies (Lewis et al., 2000). However, crossing either transgene (α-synuclein or mutant tau) with the APP_{sw} overexpressing transgenic mouse produces a mouse model that mimics human neurodegenerative disease quite closely (Lewis et al., 2001; Masliah et al., 2001). Crossing the mutant tau mice with mice overexpressing APP, or exposing the animals to injected Aβ, produces a transgenic mouse that develops cortical neuritic plaques and neurofibrillary tangles, which resemble the distribution of neuropathology in AD (Lewis et al., 2001). The P301L tau X APP_{sw} mice also show increased cognitive loss compared to mice with either transgene alone (Lewis et al., 2001). Similarly, mice expressing both human α-synuclein and APP_{sw} develop cortical Lewy bodies and severe memory loss, which is a clinical phenotype that closely resembles diffuse Lewy body disease (Masliah et al., 2001). Thus mice carrying two disease-related transgenes mimic the pathophysiology of human diseases to a much greater degree than has been seen in any mouse model previously.

DEGRADING AGGREGATES: UBIQUITINATION TARGETS AGGREGATES FOR DESTRUCTION

An emerging concept in neurodegeneration is the key role played by proteasomal degradation development of the pathological inclusion. For many years, neuropathologists have noted that the inclusions that form in a neuron are highly ubiquitinated (Lennox et al., 1989; Mori et al., 1987). The significance of ubiquitination only recently became appreciated with the discovery that one of the key disease-related genes, termed parkin, is an E3 ubiquitin ligase (Kitada et al., 1998). To understand the role of the E3 ligase, one must examine the steps involved in proteaso-

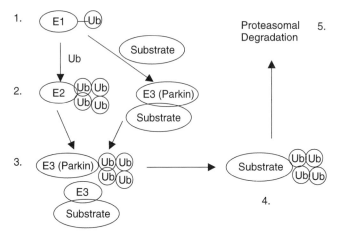

Figure 12.6. The ubiquitin proteasomal cascade proceeds via 5 steps. The first step is conjugation of ubiquitin by an E1 ligase. The second step occurs with two processes in parallel. Multiple ubiquitin peptides are covalently attached to the E2 ligase; meanwhile, the E3 ligase (which includes parkin) binds the substrate that will be targeted for degradation. In the third step, E2-ubiquitin complex and E3-substrate complex bind, which is followed by transfer of the ubiquitin complex to the substrate. In the final step, the ubiquitinated substrate is degraded by the proteasome and the ubiquitin peptides are recycled.

mal degradation. This multistep cascade begins with priming of ubiquitin by E1, then transfer of many ubiquitin molecules to the E2 conjugase, and finally the E2-ubiquitin complex transfers the ubiquitin tree to E3 and thence to the substrate, which is directed to the proteasome for degradation (Fig. 12.6) (Ciechanover et al., 2000). The proteasome is a massive complex consisting of more than 20 protein subunits and having a sedimentation coefficient of 26 (26S) (Ciechanover et al., 2000). It has a cap and barrel structure, with the cap (referred to as the 19S component of the proteasome) being the part that orients the ubiquitinated proteins in preparation for degradation by the 20S barrel component, which contains the proteases (Ciechanover et al., 2000).

Identification of which proteins are to be ubiquitinated is determined mostly by the E3 ligase and partially by the E2 conjugase. The levels of many, perhaps even most, proteins are regulated by the ubiquitin-proteasomal system (UPS). Until recently ubiquitination of the cyclins, which control cell cycle, and the IκB-NF-κB pathways have been investigated with the greatest depth (Alkalay et al., 1995; Hochstrasser, 2000; Tyers and Jorgensen, 2000). Ubiquitination normally keeps the levels of cyclins low until critical checkpoints in the cell cycle are reached, whereupon ubiquitination is inhibited and the cyclin levels increase (Tyers and Jorgensen, 2000). Similarly, ubiquitination regulates the IκB/NF-κB pathway (Alkalay et al., 1995). Under basal conditions, IκB is bound to NF-κB. Phosphorylation of IκB causes it to dissociate from NF-κB, whereupon it is rapidly ubiquitinated and degraded by the proteasome (Alkalay et al., 1995). Increasingly, ubiquitination is being shown to play a role in other signal transduction cascades as well. For instance, ubiquitination was recently shown to regulate MEK kinase kinase (MKK), a member of the c-Jun kinase cascade, by a ubiquitin-dependent kinase, Taf1 (Wang et al., 2001). Thus ubiquitination appears to be a posttranslational modification that regulates many normal cellular functions.

The UPS is particularly important in neurodegenerative disease because the UPS functions to eliminate damaged proteins, including denatured proteins, aggregated proteins, and oxidized proteins. Because oxidative stress plays a huge role in neurodegenerative disease, the cellular mechanisms for coping with oxidative damage are also important. The importance of ubiquitination in neurodegeneration is evident from the neuropathology. Virtually every inclusion that forms in neurodegenerative disease contains a large amount of ubiquitin, and immunostaining for ubiquitin represents one of the most sensitive methods of detecting inclusions (Lennox et al., 1989; Mori et al., 1987). One reason why ubiquitinated inclusions accumulate is that the UPS is impaired in many neurodegenerative diseases. The UPS in affected regions of cases of PD and AD shows a decrease in activity of between 30% and 40% (McNaught and Jenner, 2001). The causes of this decrease, though, are somewhat controversial. Proteasomal activity generally increases in response to oxidative stress. Oxidative stress combined with reduced ATP production (associated with cell injury) might reduce UPS activity (Conconi et al., 1996). The debate in this area centers on whether the protein aggregates themselves inhibit the UPS. Recent studies show that large protein aggregates that accumulate in the endoplasmic reticulum, termed aggresomes, inhibit the proteasome (Johnston et al., 1998). Aggregates of huntingtin and the CFTR both inhibit the UPS. However, whether UPS inhibition accounts for the mechanism of cell death associated with Aβ, α-synuclein, and tau is less clear. For instance, α-synuclein binds the TBP1 protein of proteasome, and overexpressing α-synuclein inhibits the proteasome, but there is no evidence *in vivo* showing that aggregated α-synuclein inhibits the proteasome more than monomeric α-synuclein (Ghee et al., 2000; Tanaka et al., 2001; Snyder et al, 2003). Because aggregation of α-synuclein is thought to play a key role in the pathophysiology of PD, any theory explaining the pathophysiology of PD must demonstrate selective actions of the aggregated form of α-synuclein. Similar problems can be raised for tauopathies. Together, this work suggests that oxidative stress does inhibit the UPS in many neurodegenerative diseases but that the contribution of protein aggregates to inhibition of the UPS might vary among different diseases.

In contrast to the ambiguous role of UPS inhibition in neurodegenerative disease, the role of protein ubiquitination in neurodegeneration has come to the forefront with the discovery of parkin, an E3 ligase (Kitada et al., 1998). Parkin was identified because of its association with the disease autosomal recessive juvenile Parkinsonism (ARJP) (Kitada et al., 1998). This disease presents in the teenage years and leads to a progressive Parkinsonism (tremor, bradykinesia, etc) that reflects progressive degeneration of the dopaminergic neurons of the substantia nigra. A wide variety of mutations in parkin cause ARJP. The mutations are located throughout the parkin protein. Some mutations create premature stop codons, other mutations are missense mutations, and other mutations are exon deletions (Abbas et al., 1999). Studies of parkin function indicate that the ARJP mutations lead to a loss of function, which suggests that maintenance of an active ubiquitinating system is essential for survival of dopaminergic neurons (Shimura et al., 2000). The reason that parkin is so important is that it appears to play a role in many diseases. Parkin binds α-synuclein, colocalizes with α-synuclein pathology, and ubiquitinates a glycosylated form of α-synuclein which contitutes about 1% of the total α-synuclein pool (Choi et al., 2001; Shimura et al., 2001). Our recent data show that parkin is also important for tauopathies because parkin binds tau, regulates the turnover of tau, and colocalizes with neurofibrillary tangles (which contain tau) in many tauopathies (Petrucelli et al., 2002). One difference between the interactions of parkin with tau versus parkin and α-synuclein is that enough parkin binds and ubiquitinates native tau to exert significant control over the turnover of tau and steady state levels of tau (Petrucelli et al., 2002). In contrast, parkin does not appear to regulate the turnover of native α-synuclein (Shimura et al., 2001). We have also observed that parkin colocalizes with huntingtin in the nuclear inclusions that accumulate in HD. The association of parkin with each of these proteins suggests that parkin plays a fundamentally important role in the turnover of proteins that are prone to aggregate or have aggregated.

Eliminating protein aggregates could represent an important mechanism of protection for neurons. Overexpressing parkin is appears to be protective. For instance, overexpressing parkin lowers the unfolded protein response, which is typically induced by cell stress (Imai et al., 2000). Conversely, loss of parkin causes dopaminergic neurodegeneration, such as occurs in ARJP (Kitada et al., 1998). The mechanism of this neurodegeneration is unknown and might depend on novel proteins that interact with parkin. However, study of parkin function already points to two potential mechanisms that might explain neurodegeneration associated with loss of parkin. One candidate parkin-interacting protein is the Pael receptor, which is a protein resident in the endoplasmic reticulum that is elevated in ARJP (Imai et al., 2001). This is significant because elevated levels of Pael receptor are known to be toxic to cells, causing acute apoptosis (Imai et al., 2001). The elevated levels of Pael receptor in ARJP might therefore acutely contribute to cell death (Imai et al., 2001). The weakness with this theory is that neurodegeneration in ARJP is a slow, chronic process that does not resemble the acute apoptosis induced by the Pael receptor. In contrast, tau is associated with many neurodegenerative illnesses, which suggests that the elevated levels of tau observed in ARJP could play a key role in the delayed neuro-degeneration occurring in that illness. Regardless of whether the Pael receptor, tau, or some other protein contributes to the neurodegeneration occurring in ARJP, the association of a member of the UPS with delayed neurodegeneration emphasizes the importance of the UPS in neurodegeneration.

THERAPEUTIC APPROACHES TO NEURODEGENERATION

Pharmaceutical Approaches

The advances in our knowledge of the mechanism of neurodegeneration have led to a clearer view of what might represent successful strategies for treating neurodegeneration, and some promising new medicines are in development. The first important concept arises from the pleiotropic mechanisms of cell death elicited by protein aggregates. This pleio-tropy suggests that preventing toxicity of aggregates by inhibiting any single biochemical pathway will be an unsuccessful therapeutic strategy because multiple biochemical pathways are activated at once. An example of this can be seen in studies of antioxidants and AD. Studies of Aβ toxicity using cell culture show that vitamin E and other antioxi-dants dramatically protect against Aβ-induced toxicity (Behl et al., 1994). However, clinical studies show that vitamin E induces only modest protection against AD—delaying admission to a nursing home by only a several months (Sano et al., 1997). In addition, despite the clear role of oxidative stress in PD, vitamin E exerts little to no benefit in pre-venting PD (Shoulson, 1998). These results are consistent with a hypoth-esis that once the protein aggregates are present neurodegeneration will likely proceed unless the underlying insult, often the protein aggregate, is eliminated.

On the basis of this hypothesis, drug development in neurodegen-erative diseases has focused on developing therapeutics that can either prevent formation of the protein aggregate or accelerate removal of the aggregate. One promising strategy being pursued in AD is to block the action of β or γ secretase, which are the enzymes that generate Aβ (Esler and Wolfe, 2001). Both secretases have at least two homologs. The two homologs of β-secretase are termed BACE 1 and BACE 2, and because BACE 1 is more abundant it is likely to be the major determinant of Aβ production. The two homologs of γ-secretase are termed presenilin 1 and presenilin 2, and presenilin 1 appears to be the major protease respon-sible for γ-secretase activity. The concept behind secretase inhibitors is to prevent Aβ from forming and thereby prevent further accumulation of Aβ aggregates (Esler and Wolfe, 2001). Without further accumulation of Aβ, endogenous secretase inhibitors might even be able to eliminate the aggregates of Aβ that have accumulated. Bristol Meyers Squibb is testing one γ-secretase inhibitor in phase II clinical trials. Although γ-secretase inhibitors are the first medicines capable of preventing Aβ formation to enter clinical trials, many investigators feel that β-secretase inhibitors have more promise. The reason is that γ-secretase is necessary for processing of Notch, which is a protein essential for normal neural development (De Strooper et al., 1999). Loss of γ-secretase activity is developmentally lethal. The role of Notch in the adult brain is unknown, but because of the important role played by Notch in development, it is conceivable that chronic inhibition of Notch activity in the adult might also be harmful. In contrast, knockout of BACE-1 or BACE-2 does not appear to be harmful (Roberds et al., 2001). Hence, inhibiting β-secretase appears less likely to produce side effects and more likely to be tolerated by humans.

Recent studies suggest an alternative method for reducing Aβ production that, although not as effective as secretase inhibitors, might be a highly valuable stop-gap measure until the secretase inhibitors become available. Multiple studies now demonstrate that Aβ production is sensitive to cholesterol levels. Cells treated with the cholesterol-lowering medicines, HMG CoA reductase inhibitors (also called statins), or subjected to treatments that remove cholesterol (cyclodextrans) show dramatic decreases in Aβ production (Simons et al., 1998). Studies of animal models also suggest that Aβ production is modulated by cholesterol levels. Feeding a high-cholesterol diet to transgenic mice over-expressing APP, or to rabbits, increases Aβ production and also increases production of neuritic plaques (Refolo et al., 2000). Conversely, treating guinea pigs with statins reduces Aβ production (Fassbender et al., 2001). This strategy has now been validated in humans. Patients treated with lovastatin show dose-dependent reductions in the peripheral levels of Aβ, and retrospective clinical studies show that patients taking statins have a prevalence of AD that is 70% lower than controls (Friedhoff et al., 2001; Jick et al., 2000; Wolozin et al., 2000). Statins have already been shown to prevent heart disease, reduce osteoporosis, and help the immune system. This research suggests that statins might also prevent AD. Perhaps it is not surprising that statins have been labeled "wonder drugs."

The Aβ peptide is unusual among protein aggregates because it is produced by cleavage of a precursor protein. Most protein aggregates differ from Aβ in that they are full-length proteins that tend to aggregate. For these proteins, it is generally not feasible to inhibit their production; hence, an alternate strategy that has become popular is to identify chemicals that prevent the proteins from aggregating. Protein aggregation tends to occur when two or more proteins align in a β-pleated sheet structure. It appears that the planar structure of β-pleated sheets facilitates formation of large aggregates. However, the planar nature of the β-pleated sheet provides a direct route to developing therapeutics that can prevent aggregation. One interesting strategy that, although unlikely to be relevant clinically, is still interesting is the concept of designing peptides that resemble the aggregates but have a β-sheet breaking mutation in the sequence, such as a proline. This type of strategy has been used successfully to prevent aggregation of Aβ, as well as of another amyloid that accumulates peripherally, transthyretin (Hammarstrom et al., 2001; Soto et al., 1998). Molecules like the dye Congo Red bind to β-pleated sheet structures, such as Aβ aggregates, and inhibit their formation. This same strategy has been used to identify molecules that can inhibit the aggregation of huntingtin and of the PrP^{sc} protein, which causes Creutzfeldt–Jakob disease. Analogs of Congo Red, a flat molecule that can intercalate in β-sheets, have been shown to prevent aggregation of huntingtin (Heiser et al., 2000). Similarly, analogs of the antipsychotic chlorpromazine have been shown to inhibit conversion of PrP^c, a normal cellular constituent, to PrP^{sc} (Korth et al., 2001). Correspondingly, these molecules prolonged the life of transgenic mice expressing human PrP that were inoculated with PrP^{sc} (Korth et al., 2001). Similar strategies might also be successful with α-synuclein, the protein that aggregates and accumulates in Lewy bodies.

Immunological Approaches

Finally, I end this discussion with a therapeutic strategy that is perhaps the most surprising and most promising strategy for preventing formation of protein aggregates, vaccines. Immunizing mice with Aβ elicits an immune response against Aβ (Morgan et al., 2000; Schenk et al., 1999). These antibodies seek out Aβ aggregates, coat the aggregates, and appear to promote removal of the aggregates by stimulating microglial cells to phagocytose and destroy the aggregates. The results are striking. Transgenic mice expressing APP$_{sw}$ normally develop large numbers of neuritic plaques by 11 months, but animals treated from birth show virtually no neuritic plaques (Schenk et al., 1999). The vaccines also appear effective when given to older animals, starting at 11 months of age (Schenk et al., 1999). Such a vaccine reduces the load of neuritic plaques by up to 90% and prevents cognitive decline in such mice. Interestingly, peripheral application of a vaccine also reduces plaque load in the brains of transgenic mice (DeMattos et al., 2001). One hypothesis explaining the success of peripheral antibodies is that Aβ in the brain is in equilibrium with peripheral Aβ, which means that reducing peripheral Aβ levels might also reduce central Aβ levels. This vaccine is currently being tested on humans (Hock et al., 2002). Although the vaccine was initially well tolerated, about 5% of patients receiving the vaccine developed an encephalitis, possibly because of T-cell infiltrates. This necessitated cessation of the trial. Current studies are focusing on the design of a vaccine that limits T-cell responses, which might limit adverse effects.

Increasing evidence suggests that vaccination is also valuable for treating other diseases. Abnormal prion protein causes a Creutzfeldt–Jakob disease, scrapie and bovine spongiform encephalitis (Prusiner, 1997). Antiprion protein antibodies slow the progression of scrapie and clear prions in cell culture (Heppner et al., 2001; Peretz et al., 2001).

Antibodies might also be useful to treat diseases with intracellular inclusions, such as HD. The basis of this strategy is to express antibodies that bind the relevant proteins in cells that are vulnerable to development of inclusions. For instance, coexpressing anti-huntingtin antibodies with huntingtin-containing polyglutamine expansions reduces formation of aggregates of huntingtin. To have this approach work in human disease requires both good antibodies and, more importantly, viral delivery of the genes. The requirement for viral delivery probably puts this kind of treatment farther in the future than more conventional pharmaceutical strategies. However, the concept remains promising.

Viruses and Stem Cells

Extensive research has been done to identify viruses that can safely and efficiently deliver genetic material to the central nervous system. The concept of viral delivery has been validated in many animal models, but perhaps the most extensive work has been done on PD. Most of this work has focused on glial-derived neurotrophic factor (GDNF), which is an important survival factor that promotes survival and neurite extension of dopaminergic neurons (Beck et al., 1995). GDNF infusion was shown to protect dopaminergic neurons from MPTP toxicity, but infusion of GDNF is not a feasible route for clinical administration of a medicine (Gash et al., 1996). Because of this, investigators have focused on viral delivery of GDNF. Initial studies using adenoviral therapy demonstrated the feasibility of using GDNF to protect dopaminergic neurons against toxic insults such as MPTP treatment (Choi-Lundberg et al., 1997). Although effective, adenoviruses are thought to have limited utility in humans because their strong antigenicity is likely to provoke an immune response, and because the viruses do not persist for long (<6 months) in the brain. Recent attention has shifted to lentivirus therapy after the demonstration that a lentivirus expressing GDNF protects against MPTP toxicity (Kordower et al., 2000). Lentiviruses are a family of viruses that include HIV. Because of the connection to HIV, clinical use of lentiviruses must proceed with caution. However, HIV is neurotropic and lentiviruses have many characteristics that make them highly appealing. They are neurotropic and nonantigenic, persist for extended periods of time (>6 months), and can be generated in high titers. Hence, the future of gene therapy in the central nervous system might reside in lentiviruses.

The final therapeutic approach to be addressed in this chapter is stem cells, which have garnered a tremendous amount of media attention. Stem cells are cells that can differentiate into more specialized cell types. They have been used to regenerate many cell types including heart cells, pancreatic cells, immune cells, and neuronal cells. Many tissues harbor stem cells that can differentiate into that tissue type, but the cells with the capability of differentiating into greatest number of cell types come from fetuses (Rietze et al., 2001). Multipotent neuronal stem cells, for instance, must be harvested from brain (Toma et al., 2001). Growth factors, morphogenic proteins, and cytokines all regulate the direction of differentiation of stem cells, and a great deal of attention has been devoted to identifying methods for purifying multipotent stem cells from peripheral tissues (Terskikh et al., 2001; Toma et al., 2001). Hence, it seems likely that future research will identify methods for generating neuronal stem cells from adult tissues.

The excitement over neuronal stem cells arises from their potential use in replacing damaged brain tissue in patients suffering from diseases such as stroke, spinal cord injury, and PD. Fetal neuronal cells have been shown to protect monkeys from MPTP toxicity, and they hold promise in treating patients with PD, although the most recent studies indicate a need to control the cells to prevent a debilitating overproduction of dopamine (Freed et al., 2001; Redmond et al., 1986). Fetal neurons are exceedingly difficult to obtain; hence, attention has shifted toward stem cells, which offer the possibility of a ready supply of dopaminergic neurons. A key hurdle is developing methods for generating large numbers of human dopaminergic form stem cells in the laboratory. Techniques are being developed to differentiate the cells into specific cell types, such as dopaminergic neurons, and it seems likely that future research will develop methods for differentiating stem cells obtained from adult human peripheral tissue into dopaminergic neurons or other neuronal types (Carvey et al., 2001). Once this has been done, neuronal stem cells can be used to replace damaged tissue in the central nervous system. These cells have a number of properties that are truly amazing. As mentioned above, they can differentiate into virtually any cell type. More surprisingly, they have a capacity to migrate to damaged areas, which suggests that stem cells given peripherally to patients might seek out damaged areas in the brain (Yandava et al., 1999). Stem cells might also be useful in therapy of gliomas because they also seek out glioblastoma cells, which suggests that stem cells tagged with inducible toxins could be used to search and destroy glioblastoma cells (Aboody et al., 2000).

Whether pharmacotherapy, gene therapy, or stem cell therapy will ultimately prevail in the clinic remains to be determined. It seems likely that the heterogeneity in clinical presentation will create room for application of each of these different types of therapies, depending on the particular situation. Conventional pharmacotherapy is the most obvious choice for many neurodegenerative illnesses because the medical system has the most experience with this mode of therapy. However, there is clearly room for application of more advanced therapies. The medical system is very familiar with vaccination, hence successful application to AD or Creutzfeldt–Jakob disease appears very likely. In addition, stem cells are particularly appealing for syndromes such as stroke and spinal cord injury, which are unlikely to be "cured" by pharmacotherapy.

REFERENCES

Abbas, N., Lücking, C., Ricard, S., Dürr, A., Bonifati, V., De Michele, G., Bouley, S., Vaughan, J., Gasser, T., Marconi, R., Broussolle, E., Brefel-Courbon, C., Harhangi, B., Oostra, B., Fabrizio, E., Böhme, G., Pradier, L., Wood, N., Filla, A., Meco, G., Denefle, P., Agid, Y., and Brice, A. (1999). A wide variety of mutations in the parkin gene are responsible for autosomal recessive parkinsonism in Europe. Hum Mol Gen 8, 567–574.

Aboody, K. S., Brown, A., Rainov, N. G., Bower, K. A., Liu, S., Yang, W., Small, J. E., Herrlinger, U., Ourednik, V., Black, P. M., Breakefield, X. O., and Snyder, E. Y. (2000). From the cover: neural stem cells display extensive tropism for pathology in adult brain: evidence from intracranial gliomas. Proc Natl Acad Sci USA 97, 12846–12851.

Alkalay, I., Yaron, A., Hatzubai, A., Orian, A., Ciechanover, A., and Ben-Neriah, Y. (1995). Stimulation-dependent IκBα phosphorylation marks the NF-κB inhibitor for degradation via the ubiquitin-proteasome pathway. Proc Natl Acad Sci USA 92, 10599–10603.

Baldwin, A. J. (1996). The NF-κB and IκB proteins: new discoveries and insights. Ann Rev Immunol 14, 649–683.

Beck, K. D., Valverde, J., Alexi, T., Poulsen, K., Moffat, B., Vandlen, R. A., Rosenthal, A., and Hefti, F. (1995). Mesencephalic dopaminergic neurons protected by GDNF from axotomy-induced degeneration in the adult brain. Nature 373, 339–341.

Behl, C., Davis, J., Lesley, R., and Schubert, D. (1994). Hydrogen peroxide mediates amyloid β protein toxicity. Cell 77, 817–827.

Behl, C., and Holsboer, F. (1999). The female sex hormone aestrogen as a neuroprotectant. TIPS 20, 441–444.

Bence, N. F., Sampat, R. M., and Kopito, R. R. (2001). Impairment of the ubiquitin-proteasome system by protein aggregation. Science 292, 1552–1555.

Betarbet, R., Sherer, T. B., MacKenzie, G., Garcia-Osuna, M., Panov, A. V., and Greenamyre, J. T. (2000). Chronic systemic pesticide exposure reproduces features of Parkinson's disease. Nat Neurosci 3, 1301–1306.

Carvey, P. M., Ling, Z. D., Sortwell, C. E., Pitzer, M. R., McGuire, S. O., Storch, A., and Collier, T. J. (2001). A clonal line of mesencephalic progenitor cells converted to dopamine neurons by hematopoietic cytokines: a source of cells for transplantation in Parkinson's disease. Exp Neurol 171, 98–108.

Castellani, R., Siedlak, S., Perry, G., and Smith, M. (2000). Sequestration of iron by Lewy bodies in Parkinson's Disease. Acta Neuropathologica 100, 111–121.

Choi, P., Golts, N., Snyder, H., Petrucelli, L., Chong, M., Hardy, J., Sparkman, D., Cochran, E., Lee, J., and Wolozin, B. (2001). Co-association of parkin and α-synuclein. NeuroReport 12, 2839–2844.

Choi-Lundberg, D. L., Lin, Q., Chang, Y. N., Chiang, Y. L., Hay, C. M., Mohajeri, H., Davidson, B. L., and Bohn, M. C. (1997). Dopaminergic neurons protected from degeneration by GDNF gene therapy. Science 275, 838–841.

Ciechanover, A., Orian, A., and Schwartz, A. L. (2000). Ubiquitin-mediated proteolysis: biological regulation via destruction. Bioessays 22, 442–451.

Citron, M., Oltersdorf, T., Haass, C., McConlogue, L., Hung, A., Seubert, P., Vigo-Pelfrey, C., Lieberburg, I., and Selkoe, D. (1992). Mutation of the β-amyloid precursor protein in familial Alzheimer's disease increases β-protein production. Nature 360, 672–674.

Citron, M., Westaway, D., Xia, W., Carlson, G., Diehl, T., Levesque, G., Johnson-Wood, K., Lee, M., Seubert, P., Davis, A., Kholodenko, D., Motter, R., Sherrington, R., Perry, B., Yao, H., Strome, R., Lieberburg, I., Rommens, J., Kim, S., Schenk, D., Fraser, P., St George Hyslop, P., and Selkoe, D. J. (1997). Mutant presenilins of Alzheimer's disease increase production of 42-residue amyloid β-protein in both transfected cells and transgenic mice. Nat Med *3*, 67–72.

Clayton, D., and George, J. (1998). The synucleins: a family of proteins involved in synaptic function, plasticity, neurodegeneration and disease. TINS *21*, 249–254.

Conconi, M., Szweda, L. I., Levine, R. L., Stadtman, E. R., and Friguet, B. (1996). Age-related decline of rat liver multicatalytic proteinase activity and protection from oxidative inactivation by heat-shock protein 90. Arch Biochem Biophys *331*, 232–240.

Conway, K., Harper, J., and Lansbury, P. (1998). Accelerated in vitro fibril formation by a mutant α-synuclein linked to early-onset Parkinson disease. Nat Med *4*, 1318–1320.

Conway, K. A., Lee, S. J., Rochet, J. C., Ding, T. T., Williamson, R. E., and Lansbury, P. T., Jr. (2000). Acceleration of oligomerization, not fibrillization, is a shared property of both alpha-synuclein mutations linked to early-onset Parkinson's disease: Implications for pathogenesis and therapy. Proc Natl Acad Sci USA *97*, 571–576.

Corder, E., Saunders, A., Strittmatter, W., Schmechel, D., Gaskell, P., Small, G., Roses, A., Haines, J., and Pericak-Vance, M. (1993). Gene dose of apolipoprotein E type 4 allele and the risk of Alzheimer's disease in late onset families. Science *261*, 921–923.

Corral-Debrinski, M., Horton, T., Lott, M. T., Shoffner, J. M., Beal, M. F., and Wallace, D. C. (1992). Mitochondrial DNA deletions in human brain: regional variability and increase with advanced age. Nat Genet *2*, 324–329.

DeMattos, R. B., Bales, K. R., Cummins, D. J., Dodart, J. C., Paul, S. M., and Holtzman, D. M. (2001). Peripheral anti-Aβ antibody alters CNS and plasma Aβ clearance and decreases brain Aβ burden in a mouse model of Alzheimer's disease. Proc Natl Acad Sci USA *98*, 8850–8855.

De Strooper, B., Annaert, W., Cupers, P., Saftig, P., Craessaerts, K., Mumm, J., Schroeter, E., Schrijvers, V., Wolfe, M., Ray, W., Goate, A., and Kopan, R. (1999). A presenilin-1-dependent γ-secretase-like protease mediates release of Notch intracellular domain. Nature *398*, 518–522.

Dickson, D., Lin, W.-L., Liu, W., and Yen, S. (1999). Multiple system atrophy: a sporadic synucleinopathy. Brain Pathol *9*, 721–732.

DiFiglia, M., Sapp, E., Chase, K., Davies, S., Bates, G., Vonsattel, J., and Aronin, N. (1997). Aggregation of huntingtin in neuronal intranuclear inclusions and dystrophic neurites in brain. Science *277*, 1990–1993.

Duda, J. E., Lee, V. M., and Trojanowski, J. Q. (2000). Neuropathology of synuclein aggregates. J Neurosci Res *61*, 121–127.

Duff, K., Eckman, C., Zehr, C., Yu, X., Prada, C. M., Pereztur, J., Hutton, M., Buee, L., Harigaya, Y., Yager, D., Morgan, D., Gordon, M. N., Holcomb, L., Refolo, L., Zenk, B., Hardy, J., and Younkin, S. (1996). Increased amyloid-β-42(43) in brains of mice expressing mutant presenilin 1. Nature *383*, 710–713.

El Khoury, J., Hickman, S., Thomas, C., Cao, L., Silverstein, S., and Loike, J. (1996). Scavenger receptor-mediated adhesion of microglia to β-amyloid fibrils. Nature *382*, 716–719.

Esler, W. P., and Wolfe, M. S. (2001). A portrait of Alzheimer secretases—new features and familiar faces. Science *293*, 1449–1454.

Fassbender, K., Simons, M., Bergmann, C., Stroick, M., Lutjohann, D., Keller, P., Runz, H., Kuhl, S., von Bergmann, K., Hennerici, M., Beyreuther, K., and Hartmann, T. (2001). Simvastatin strongly reduces Alzheimer's disease Aβ42 and Aβ40 levels in vitro and in vivo. Proc Natl Acad Sci USA *98*, 5856–5861.

Feany, M. B., and Bender, W. W. (2000). A *Drosophila* model of Parkinson's disease. Nature *404*, 394–398.

Freed, C. R., Greene, P. E., Breeze, R. E., Tsai, W. Y., DuMouchel, W., Kao, R., Dillon, S., Winfield, H., Culver, S., Trojanowski, J. Q., Eidelberg, D., and Fahn, S. (2001). Transplantation of embryonic dopamine neurons for severe Parkinson's disease. N Engl J Med *344*, 710–719.

Friedhoff, L. T., Cullen, E. I., Geoghagen, N. S., and Buxbaum, J. D. (2001). Treatment with controlled-release lovastatin decreases serum concentrations of human β-amyloid (Aβ) peptide. Int J Neuropsychopharmacol *4*, 127–130.

Galvin, J. E., Giasson, B., Hurtig, H. I., Lee, V. M., and Trojanowski, J. Q. (2000). Neurodegeneration with brain iron accumulation, type 1 is characterized by α-, β-, and γ-synuclein neuropathology. Am J Pathol *157*, 361–368.

Games, D., Adams, D., Alessandrini, R., Barbour, R., Berthelette, Blackwell, C., Carr, T., Clemens, J., Donaldson, T., Gillespie, F., Guido, T., Hagopian, S., Johnson, M., Wood, K., Khan, K., Lee, M., Leibowitz, P., Lieberburg, I., Little, S., Masliah, E., McConlogue, L., Montoya-Zavala, M., Mucke, L., Paganini, L., Penniman, E., Power, M., Schenk, D., Seubert, P., Snyder, B., Soriano, F., Tan, H., Vitale, J., Wadsworth, S., Wolozin, B., and Zhao, J. (1995). Development of neuropathology similar to Alzheimer's disease in transgenic mice overexpressing the 717_{V-F} β-amyloid precursor protein. Nature *373*, 523–527.

Gash, D. M., Zhang, Z., Ovadia, A., Cass, W. A., Yi, A., Simmerman, L., Russell, D., Martin, D., Lapchak, P. A., Collins, F., Hoffer, B. J., and Gerhardt, G. A. (1996). Functional recovery in parkinsonian monkeys treated with GDNF. Nature *380*, 252–255.

Ghee, M., Fournier, A., and Mallet, J. (2000). Rat α-synuclein interacts with Tat binding protein 1, a component of the 26S proteasomal complex. J Neurochem *75*, 2221–2224.

Golts, N., Snyder, H., Frasier, M., Theisler, C., Choi, P., and Wolozin, B. (2002). Magnesium inhibits spontaneous and iron-induced aggregation of α-synuclein. J Biol Chem *277*, 16116–16123.

Haass, C., Hung, A. Y., Selkoe, D. J., and Teplow, D. B. (1994). Mutations associated with a locus for familial Alzheimer's disease result in alternative processing of amyloid β-protein precursor. J Biol Chem *269*, 17741–17748.

Hammarstrom, P., Schneider, F., and Kelly, J. W. (2001). Trans-suppression of misfolding in an amyloid disease. Science *293*, 2459–2462.

Hashimoto, M., LJ, H., Xia, Y., Takeda, A., Sisk, A., Sundsmo, M., and Masliah, E. (1999). Oxidative stress induces amyloid-like aggregate formation of NACP/α-synuclein in vitro. NeuroReport *10*, 717–721.

Heiser, V., Scherzinger, E., Boeddrich, A., Nordhoff, E., Lurz, R., Schugardt, N., Lehrach, H., and Wanker, E. E. (2000). Inhibition of huntingtin fibrillogenesis by specific antibodies and small molecules: implications for Huntington's disease therapy. Proc Natl Acad Sci USA *97*, 6739–6744.

Hensley, K., Carney, J. M., Mattson, M. P., Aksenova, M., Harris, M., Wu, J. F., Floyd, R. A., and Butterfield, D. A. (1994). A model for β-amyloid aggregation and neurotoxicity based on free radical generation by the peptide: relevance to Alzheimer disease. Proc Natl Acad Sci US A *91*, 3270–3274.

Heppner, F. L., Musahl, C., Arrighi, I., Klein, M. A., Rulicke, T., Oesch, B., Zinkernagel, R. M., Kalinke, U., and Aguzzi, A. (2001). Prevention of scrapie pathogenesis by transgenic expression of anti-prion protein antibodies. Science *294*, 178–182.

Hochstrasser, M. (2000). Evolution and function of ubiquitin-like protein-conjugation systems. Nat Cell Biol *2*, E153–E157.

Hock, C., Konietzko, U., Papassotiropoulos, A., Wollmer, A., Streffer, J., von Rotz, R. C., Davey, G., Moritz, E., and Nitsch, R. M. (2002). Generation of antibodies specific for beta-amyloid by vaccination of patients with Alzheimer disease. Nat Med *8*, 1270–1275.

Hong, M., Zhukareva, V., Vogelsberg-Ragaglia, V., Wszolek, Z., Reed, L., Miller, B., Geschwind , D., Bird, T., McKeel, D., Goate, A., Morris, J., Wilhelmsen, K., Schellenberg, G., Trojanowski, J., and Lee, V. (1998). Mutation-specific functional impairments in distinct tau isoforms of hereditary FTDP-17. Science *282*, 1914–1917.

Hsiao, K., Chapman, P., Nilsen, S., Eckman, C., Harigaya, Y., Younkin, S., Yang, F., and Cole, G. (1996). Correlative memory deficits, Aβ elevation, and amyloid plaques in transgenic mice. Science *274*, 99–102.

Hutton, M., Lendon, C., Rizzu, P., Baker, M., Froelich, S., Houlden, H., Pickering-Brown, S., Chakraverty, S., Isaacs, A., Grover, A., Hackett, J., Adamson, J., Lincoln, S., Dickson, D., Davies, P., Petersen, R., Stevens, M., de Graaff, E., Wauters, E., van Baren, J., Hillebrand, M., Joosse, M., Kwon, J., Nowotny, P., Heutink, P., et al. (1998). Association of missense and 5′-splice-site mutations in tau with the inherited dementia FTDP-17. Nature *393*, 702–705.

Imai, Y., Soda, M., Inoue, H., Hattori, N., Mizuno, Y., and Takahashi, Y. (2001). An unfolded putative transmembrane polypeptide, which can lead to endoplasmic reticulum stress, is a substrate of parkin. Cell *105*, 891–902.

Imai, Y., Soda, M., and Takahashi, R. (2000). Parkin suppresses unfolded protein stress-induced cell death through its E3 ubiquitin-protein ligase activity. J Biol Chem *275*, 35661–35664.

Iwata, N., Tsubuki, S., Takaki, Y., Watanabe, K., Sekiguchi, M., Hosoki, E., Kawashima-Morishima, M., Lee, H. J., Hama, E., Sekine-Aizawa, Y., and Saido, T. C. (2000). Identification of the major Aβ1-42-degrading catabolic pathway in brain parenchyma: suppression leads to biochemical and pathological deposition. Nat Med *6*, 143–150.

Jarrett, J., and Lansbury, P. (1993). Seeding "one-dimensional crystallization" of amyloid: a pathogenic mechanism in Alzheimer's disease and scrapie? Cell *73*, 1055–1058.

Jenco, J., Rawlingson, A., Daniels, B., and Morris, A. (1998). Regulation of phospholipase D2: Selective inhibition of mammalian phospholipase D isoenzymes by α and β synucleins. Biochemistry *37*, 4901–4909.

Jick, H., Zornberg, G. L., Jick, S. S., Seshadri, S., and Drachman, D. A. (2000). Statins and the risk of dementia. Lancet *356*, 1627–1631.

Johnston, J., Ward, C., and Kopito, R. (1998). Aggresomes: A cellular response to misfolded proteins. J. Cell Biol *143*, 1883–1898.

Kahle, P. J., Neumann, M., Ozmen, L., Muller, V., Jacobsen, H., Schindzielorz, A., Okochi, M., Leimer, U., van Der Putten, H., Probst, A., Kremmer, E., Kretzschmar, H. A., and Haass, C. (2000). Subcellular localization of wild-type and Parkinson's disease- associated mutant α-synuclein in human and transgenic mouse brain. J Neurosci *20*, 6365–7363.

Khursigara, G., Bertin, J., Yano, H., Moffett, H., DiStefano, P. S., and Chao, M. V. (2001). A prosurvival function for the p75 receptor death domain mediated via the caspase recruitment domain receptor-interacting protein 2. J Neurosci *21*, 5854–5863.

Kitada, T., Asakawa, S., Hattori, N., Matsumine, H., Yamamura, Y., Minoshima, S., Yokochi, M., Mizuno, Y., and Shimizu, N. (1998). Mutations in the parkin gene cause autosomal recessive juvenile parkinsonism. Nature *392*, 605–608.

Kordower, J. H., Emborg, M. E., Bloch, J., Ma, S. Y., Chu, Y., Leventhal, L., McBride, J., Chen, E. Y., Palfi, S., Roitberg, B. Z., Brown, W. D., Holden, J. E., Pyzalski, R., Taylor, M. D., Carvey, P., Ling, Z., Trono, D., Hantraye, P., Deglon, N., and Aebischer, P. (2000). Neurodegeneration prevented by lentiviral vector delivery of GDNF in primate models of Parkinson's disease. Science *290*, 767–773.

Korth, C., May, B. C., Cohen, F. E., and Prusiner, S. B. (2001). Acridine and phenothiazine derivatives as pharmacotherapeutics for prion disease. Proc Natl Acad Sci USA *98*, 9836–9841.

Kruger, R., Kuhn, W., Muller, T., Woitalla, D., Graeber, M., Kosel, S., Przuntek, H., Epplen, J., Schols, L., and Riess, O. (1998). Ala30Pro mutation in the gene encoding α-synuclein in Parkinson's disease. Nat Genet *18*, 106–108.

Lang, A., and Lozano, A. (1998). Parkinson's Disease. N Engl J Med *339*, 1044–53.

Langston, J. (1998). Epidemiology versus genetics in Parkinson's disease: Progress in resolving an age-old debate. Ann Neurol *44*, S45–S52.

Langston, J., Ballard, P., Tetrud, J., and Irwin, I. (1983). Chronic Parkinsonism in humans due to a product of meperidine-analog synthesis. Science *219*, 979–980.

Lebovitz, R. M., Zhang, H., Vogel, H., Cartwright, J., Jr., Dionne, L., Lu, N., Huang, S., and Matzuk, M. M. (1996). Neurodegeneration, myocardial injury, and perinatal death in mitochondrial superoxide dismutase-deficient mice. Proc Natl Acad Sci USA *93*, 9782–9787.

Lee, V. M., Goedert, M., and Trojanowski, J. Q. (2001). Neurodegenerative tauopathies. Annu Rev Neurosci *24*, 1121–1159.

Lennox, G., Lowe, J., Morrell, K., Landon, M., and Mayer, R. J. (1989). Antiubiquitin immunocytochemistry is more sensitive than conventional techniques in the detection of diffuse Lewy body disease. J Neurol Neurosurg Psychiatry *52*, 67–71.

Levy-Lahad, E., Wasco, W., Poorkaj, P., Romano, D. M., Oshima, J., Pettingell, W. H., Yu, C. E., Jondro, P. D., Schmidt, S. D., Wang, K. et al. (1995). Candidate gene for the chromosome 1 familial Alzheimer's disease locus. Science *269*, 973–977.

Lewis, J., Dickson, D. W., Lin, W. L., Chisholm, L., Corral, A., Jones, G., Yen, S. H., Sahara, N., Skipper, L., Yager, D., Eckman, C., Hardy, J., Hutton, M., and McGowan, E. (2001). Enhanced neurofibrillary degeneration in transgenic mice expressing mutant tau and APP. Science *293*, 1487–1491.

Lewis, J., McGowan, E., Rockwood, J., Melrose, H., Nacharaju, P., Van Slegtenhorst, M., Gwinn-Hardy, K., Paul Murphy, M., Baker, M., Yu, X., Duff, K., Hardy, J., Corral, A., Lin, W. L., Yen, S. H., Dickson, D. W., Davies, P., and Hutton, M. (2000). Neurofibrillary tangles, amyotrophy and progressive motor disturbance in mice expressing mutant (P301L) tau protein. Nat Genet *25*, 402–405.

Ma, J., Yee, A., Brewer, H., Das, S., and Potter, H. (1994). Amyloid-associated proteins α₁-antichymotrypsin and apolipoprotein E promote assembly of Alzheimer β-protein into filaments. Nature *372*, 92–94.

MacDonald, M., Ambrose, C., Duyao, M., Myers, R., Lin, C., Srinidhi, L., Barnes, G., Taylor, S., James, M., Groot, N., MacFarlane, H., Jenkins, B., Anderson, M., Wexler, N., and Gusella, J. (1993). A novel gene containing a trinucleotide repeat that is expanded and unstable on Huntington's disease chromosomes. Cell *72*, 971–983.

Masliah, E., Rockenstein, E., Veinbergs, I., Mallory, M., Hashimoto, M., Takeda, A., Sagara, Y., Sisk, A., and Mucke, L. (2000). Dopaminergic loss and inclusion body formation in α-synuclein mice: Implications for neurodegenerative disorders. Science *287*, 1265–1269.

Masliah, E., Rockenstein, E., Veinbergs, I., Sagara, Y., Mallory, M., Hashimoto, M., and Mucke, L. (2001). β-Amyloid peptides enhance α-synuclein accumulation and neuronal deficits in a transgenic mouse model linking Alzheimer's disease and Parkinson's disease. Proc Natl Acad Sci USA *25*, 25.

McNaught, K. S., and Jenner, P. (2001). Proteasomal function is impaired in substantia nigra in Parkinson's disease. Neurosci Lett *297*, 191–194.

Morgan, D., Diamond, D. M., Gottschall, P. E., Ugen, K. E., Dickey, C., Hardy, J., Duff, K., Jantzen, P., DiCarlo, G., Wilcock, D., Connor, K., Hatcher, J., Hope, C., Gordon, M., and Arendash, G. W. (2000). A β peptide vaccination prevents memory loss in an animal model of Alzheimer's disease. Nature *408*, 982–985.

Mori, H., Kondo, J., and Ihara, Y. (1987). Ubiquitin is a component of paired helical filaments in Alzheimer's disease. Science *235*, 1641–1644.

Mufson, E. J., Bothwell, M., and Kordower, J. H. (1989). Loss of nerve growth factor receptor-containing neurons in Alzheimer's disease: a quantitative analysis across subregions of the basal forebrain. Exp Neurol *105*, 221–232.

Nakamura, K., Wang, W., and Kang, U. (1997). The role of glutathione in dopaminergic neuronal survival. J Neurochem *69*, 1850–1858.

Nakamura, K., Won, L., Heller, A., and Kang, U. J. (2000). Preferential resistance of dopaminergic neurons to glutathione depletion in a reconstituted nigrostriatal system. Brain Res *873*, 203–211.

Nucifora, F. C., Jr., Sasaki, M., Peters, M. F., Huang, H., Cooper, J. K., Yamada, M., Takahashi, H., Tsuji, S., Troncoso, J., Dawson, V. L., Dawson, T. M., and Ross, C. A. (2001). Interference by huntingtin and atrophin-1 with cbp-mediated transcription leading to cellular toxicity. Science *291*, 2423–2428.

Nunomura, A., Perry, G., Pappolla, M., Wade, R., Hirai, K. C., S, and Smith, M. (1999). Parkinson's disease is associated with oxidative damage to cytoplasmic DNA and RNA in substantia nigra neurons. Am J Pathol *154*, 1423–1429.

Ostrerova, N., Petrucelli, L., Farrer, M., Mehta, N., Alexander, P., Choi, P., Palacino, J., Hardy, J., and Wolozin, B. (1999). α-Synuclein shares physical and functional homology with 14-3-3 proteins. J Neurosci *19*, 5782–5791.

Ostrerova-Golts, N., Petrucelli, L., Hardy, J., Lee, J., Farrer, M., and Wolozin, B. (2000). The A53T α-synuclein mutation increases iron-dependent aggregation and toxicity. J Neurosci *20*, 6048–6054.

Paik, S., Shin, H., Lee, J., Chang, C., and Kim, J. (1999). Copper(II)-induced self-oligomerization of α-synuclein. Biochem J *340*, 821–828.

Paulson, H., and Fischbeck, K. (1996). Trinucleotide repeats in neurogenetic disorders. Annu Rev Neurosci *19*, 79–107.

Peretz, D., Williamson, R. A., Kaneko, K., Vergara, J., Leclerc, E., Schmitt-Ulms, G., Mehlhorn, I. R., Legname, G., Wormald, M. R., Rudd, P. M., Dwek, R. A., Burton, D. R., and Prusiner, S. B. (2001). Antibodies inhibit prion propagation and clear cell cultures of prion infectivity. Nature *412*, 739–743.

Perry, T., Godin, D., and Hansen, S. (1982). Parkinson's disease: a disorder due to nigral glutathione deficiency. Neurosc Lett *33*, 305–310.

Petrucelli, L., Ved, R., Kehoe, K., Lockhart, P., Lewis, J., Hernandez, D., Slegtenhorst, M., Dickson, D., Lee, J., Cochran, E., Choi, P., Cookson, M., Farrer, M., Hatori, N., Mizuno, Y., Dawson, T., Hardy, J., Hutton, M., and Wolozin, B. (2003). Tau is a parkin substrate: Implications for neurodegenerative disease. Nat Cell Biol. *(submitted)*.

Polymeropoulos, M. H., Lavedan, C., Leroy, E., Ide, S. E., Dehejia, A., Dutra, A., Pike, B., Root, H., Rubenstein, J., Boyer, R., Stenroos, E. S., Chandrasekharappa, S., Athanassiadou, A., Papapetropoulos, T., Johnson, W. G., Lazzarini, A. M., Duvoisin, R. C., Di Iorio, G., Golbe, L. I., and Nussbaum, R. L. (1997). Mutation in the α-synuclein gene identified in families with Parkinson's disease. Science *276*, 2045–2047.

Prusiner, S. (1997). Biology of prions. In The Molecular and Genetic Basis of Neurological Disease, R. Rosenberg, S. Prusiner, S. DiMauro and R. Barchi, eds. (Boston: Butterworth-Heinemann), pp. 103–143.

Rabizadeh, S., Oh, J., Zhong, L., Yang, J., Bitler, C., Butcher, L., and Bredesen, D. (1993). Induction of apoptosis by the low-affinity NGF receptor. Science *261*, 345–348.

Redmond, D. E., Sladek, J. R., Jr., Roth, R. H., Collier, T. J., Elsworth, J. D., Deutch, A. Y., and Haber, S. (1986). Fetal neuronal grafts in monkeys given methylphenyltetrahydropyridine. Lancet *1*, 1125–1127.

Refolo, L. M., Pappolla, M. A., Malester, B., LaFrancois, J., Bryant-Thomas, T., Wang, R., Tint, G. S., Sambamurti, K., and Duff, K. (2000). Hypercholesterolemia accelerates the Alzheimer's amyloid pathology in a transgenic mouse model. Neurobiol Dis *7*, 321–331.

Rietze, R. L., Valcanis, H., Brooker, G. F., Thomas, T., Voss, A. K., and Bartlett, P. F. (2001). Purification of a pluripotent neural stem cell from the adult mouse brain. Nature *412*, 736–739.

Roberds, S. L., Anderson, J., Basi, G., Bienkowski, M. J., Branstetter, D. G., Chen, K. S., Freedman, S. B., Frigon, N. L., Games, D., Hu, K., Johnson-Wood, K., Kappenman, K. E., Kawabe, T. T., Kola, I., Kuehn, R., Lee, M., Liu, W., Motter, R., Nichols, N. F., Power, M., Robertson, D. W., Schenk, D., Schoor, M., Shopp, G. M., Shuck, M. E., Sinha, S., Svensson, K. A., Tatsuno, G., Tintrup, H., Wijsman, J., Wright, S., and McConlogue, L. (2001). BACE knockout mice are healthy despite lacking the primary β-secretase activity in brain: implications for Alzheimer's disease therapeutics. Hum Mol Genet *10*, 1317–1324.

Ross, C. (1995). When more is less: Pathogenesis of glutamine repeat neurodegenerative disease. Neuron *15*, 493–496.

Sano, M., Ernesto, C., Thomas, R. G., Klauber, M. R., Schafer, K., Grundman, M., Woodbury, P., Growdon, J., Cotman, C. W., Pfeiffer, E., Schneider, L. S., and Thal, L. J. (1997). A controlled trial of selegiline, α-tocopherol, or both as treatment for Alzheimer's disease. The Alzheimer's Disease Cooperative Study. N Engl J Med *336*, 1216–1222.

Saudou, F., Finkbeiner, S., Devys, D., and Greenberg, M. (1998). Huntingtin acts in the nucleus to induce apoptosis but death does not correlate with formation of intranuclear inclusions. Cell 95, 55–66.

Sayre, L. M., Smith, M. A., and Perry, G. (2001). Chemistry and biochemistry of oxidative stress in neurodegenerative disease. Curr Med Chem 8, 721–738.

Schenk, D., Barbour, R., Dunn, W., Gordon, G., Grajeda, H., Guido, T., Hu, K., Huang, J., Johnson-Wood, K., Khan, K., Kholodenko, D., Lee, M., Liao, Z., Lieberburg, I., Motter, R., Mutter, L., Soriano, F., Shopp, G., Vasquez, N., Vandevert, C., Walker, S., Wogulis, M., Yednock, T., Games, D., and Seubert, P. (1999). Immunization with amyloid- attenuates Alzheimer-disease-like pathology in the PDAPP mouse. Nature 400, 173–178.

Scheuner, D., Eckman, C., Jensen, M., Song, X., Citron, M., Suzuki, N., Bird, T. D., Hardy, J., Hutton, M., Kukull, W., Larson, E., Levylahad, E., Viitanen, M., Peskind, E., Poorkaj, P., Schellenberg, G., Tanzi, R., Wasco, W., Lannfelt, L., Selkoe, D., and Younkin, S. (1996). secreted amyloid β-protein similar to that in the senile plaques of Alzheimers disease is increased in vivo by the presenilin 1 and 2 and APP mutations linked to familial Alzheimers disease. Nat Med 2, 864–870.

Selkoe, D., and Lansbury, P. (1999). Biochemistry of Alzheimer's and prion diseases. In Basic Neurochemistry, G. Siegel, B. Agranoff, R. Albers, S. Fisher, and M. Uhler, eds. (Philadelphia: Lippincott-Raven), pp. 949–968.

Sherrington, R., Rogaev, E., Liang, Y., Rogaeva, E., Levesque, G., Ideda, M., Chi, H., Lin, C., Li, G., Holman, K., Tsuda, T., Mar, L., Foncin, J., Bruni, A., Montesi, M., Sorbi, S., Rainero, I., Pinessi, L., Nee, L., Chumakov, I., Pollen, D., Brookes, A., Sanseau, P., Polinsky, R., Wasco, W., Da Silva, H., Haines, J., Pericak-Vance, M., Tanzi, R., Roses, A., Fraser, P., Rommens, J., and St George-Hyslop, P. (1995). Cloning of a gene bearing missense mutations in early-onset familial Alzheimer's disease. Nature 375, 754–760.

Shimura, H., Hattori, N., Kubo, S., Mizuno, Y., Asakawa, S., Minoshima, S., Shimizu, N., Iwai, K., Chiba, T., Tanaka, K., and Suzuki, T. (2000). Familial Parkinson disease gene product, parkin, is a ubiquitin-protein ligase. Nat Genet 25, 302–305.

Shimura, H., Schlossmacher, M. G., Hattori, N., Frosch, M. P., Trockenbacher, A., Schneider, R., Mizuno, Y., Kosik, K. S., and Selkoe, D. J. (2001). Ubiquitination of a new form of α-synuclein by parkin from human brain: implications for Parkinson's disease. Science 293, 263–269.

Shoulson, I. (1998). DATATOP: a decade of neuroprotective inquiry. Parkinson Study Group. Deprenyl and tocopherol antioxidative therapy of parkinsonism. Ann Neurol 44, S160–166.

Simons, M., Keller, P., De Strooper, B., Beyreuther, K., Dotti, C., and Simons, K. (1998). Cholesterol depletion inhibits the generation of β-amyloid in hippocampal neurons. Proc Natl Acad Sci USA 95, 6460–6464.

Sinha, S., Anderson, J. P., Barbour, R., Basi, G. S., Caccavello, R., Davis, D., Doan, M., Dovey, H. F., Frigon, N., Hong, J., Jacobson-Croak, K., Jewett, N., Keim, P., Knops, J., Lieberburg, I., Power, M., Tan, H., Tatsuno, G., Tung, J., Schenk, D., Seubert, P., Suomensaari, S. M., Wang, S., Walker, D., and John, V. (1999). Purification and cloning of amyloid precursor protein β-secretase from human brain. Nature 402, 537–540.

Smith, M., Sayre, L., Anderson, V., Harris, P., MF, B., Kowall, N., and Perry, G. (1998). Cytochemical demonstration of oxidative damage in Alzheimer disease by immunochemical enhancement of the carbonyl reaction with 2,4-dinitrophenylhydrazine. J Histochem Cytochem *46*, 731–735.

Snyder, H., Mensah, K., Theisler, C., Lee, J. M., Matouschek, A., and Wolozin, B. (2003). Aggregated and Monomeric α-Synuclein bind to the S6' Proteasomal Protein and Inhibit Proteasomal Function. J Biol Chem (epub ahead of print).

Soto, C., Sigurdsson, E., Morelli, L., Kumar, R., Castano, E., and Frangione, B. (1998). β-Sheet breaker peptides inhibit fibrillogenesis in a rat brain model of amyloidosis: implications for Alzheimer's therapy. Nat Med. *4*, 822–826.

Spillantini, M., Schmidt, M., VM-Y, L., Trojanowski, J., Jakes, R., and Goedert, M. (1997). α-Synuclein in Lewy bodies. Nature *388*, 839–840.

Spillantini, M. G., Bird, T. D., and Ghetti, B. (1998a). Frontotemporal dementia and Parkinsonism linked to chromosome 17: a new group of tauopathies. Brain Pathol *8*, 387–402.

Spillantini, M. G., Crowther, R. A., Jakes, R., Hasegawa, M., and Goedert, M. (1998b). α-Synuclein in filamentous inclusions of Lewy bodies from Parkinson's disease and dementia with Lewy bodies. Proc Natl Acad Sci USA *95*, 6469–6473.

Strittmatter, W., Weisgraber, K., Huang, D., Dong, L., Salvesen, G., Pericak-Vance, M., Schmechel, D., Saunders, A., Goldgaber, D., and Roses, A. (1993). Binding of human apolipoprotein E to synthetic amyloid β peptide: isoform-specific effects and implications for late-onset Alzheimer disease. Proc Natl Acad Sci USA *90*, 8098–8102.

Tanaka, Y., Engelender, S., Igarashi, S., Rao, R. K., Wanner, T., Tanzi, R. E., Sawa, A., V, L. D., Dawson, T. M., and Ross, C. A. (2001). Inducible expression of mutant α-synuclein decreases proteasome activity and increases sensitivity to mitochondria-dependent apoptosis. Hum Mol Genet *10*, 919–926.

Terskikh, A. V., Easterday, M. C., Li, L., Hood, L., Kornblum, H. I., Geschwind, D. H., and Weissman, I. L. (2001). From hematopoiesis to neuropoiesis: evidence of overlapping genetic programs. Proc Natl Acad Sci USA *98*, 7934–7939.

Tipton, K., and Singer, T. (1993). Advances in our understanding of the mechanisms of the neurotoxicity of MPTP and related compounds. J Neurochem *61*, 1191–1206.

Toma, J. G., Akhavan, M., Fernandes, K. J., Barnabe-Heider, F., Sadikot, A., Kaplan, D. R., and Miller, F. D. (2001). Isolation of multipotent adult stem cells from the dermis of mammalian skin. Nat Cell Biol *3*, 778–784.

Trottier, Y., Lutz, Y., Stevanin, G., Imbert, G., Devys, D., Cancel, G., Saudou, F., Weber, C., David, G., Tora, L. et al. (1995). Polyglutamine expansion as a pathological epitope in Huntington's disease and four dominant cerebellar ataxias. Nature *378*, 403–406.

Tyers, M., and Jorgensen, P. (2000). Proteolysis and the cell cycle: with this RING I do thee destroy. Curr Opin Genet Dev *10*, 54–64.

Uversky, V. N., Li, J., and Fink, A. L. (2001). Pesticides directly accelerate the rate of α-synuclein fibril formation: a possible factor in Parkinson's disease. FEBS Lett *500*, 105–108.

Varadarajan, S., Yatin, S., Kanski, J., Jahanshahi, F., and Butterfield, D. A. (1999). Methionine residue 35 is important in amyloid β-peptide-associated free radical oxidative stress. Brain Res Bull *50*, 133–141.

Vassar, R., Bennett, B., Babu-Khan, S., Kahn, S., Mendiaz, E., Denis, P., Teplow, D., Ross, S., Amarante, P., Loeloff, R., Luo, Y., Fisher, S., Fuller, J., Edenson,

S., Lile, J., Jarosinski, M., Biere, A., Curran, E., Burgess, T., Louis, J., Collins, F., Treanor, J., Rogers, G., and Citron, M. (1999). β-Secretase cleavage of Alzheimer's amyloid precursor protein by the transmembrane aspartic protease BACE. Science *286*, 735–741.

Wang, C., Deng, L., Hong, M., Akkaraju, G. R., Inoue, J., and Chen, Z. J. (2001). TAK1 is a ubiquitin-dependent kinase of MKK and IKK. Nature *412*, 346–351.

Wolozin, B., Kellman, W., Ruosseau, P., Celesia, G. G., and Siegel, G. (2000). Decreased prevalence of Alzheimer disease associated with 3-hydroxy-3-methyglutaryl coenzyme A reductase inhibitors. Arch Neurol *57*, 1439–1443.

Yaar, M., Zhai, S., Pilch, P., Doyle, S., Eisenhauer, P., Fine, R., and Ba, G. (1997). Binding of β-amyloid to the p75 neurotrophin receptor induces apoptosis. A possible mechanism for Alzheimer's disease. J Clin Invest *100*, 2333–2340.

Yan, S., Chen, X., Fu, J., Chen, M., Zhu, H., Roher, A., Slattery, T., Zhao, L., Nagashima, M., Morser, J., Migheli, A., Nawroth, P., Stern, D., and Schmidt, A. (1996). RAGE and amyloid-β peptide neurotoxicity in Alzheimer's disease. Nature *382*, 685–691.

Yan, S. D., Fu, J., Soto, C., Chen, X., Zhu, H., Al-Mohanna, F., Collison, K., Zhu, A., Stern, E., Saido, T., Tohyama, M., Ogawa, S., Roher, A., and Stern, D. (1997). An intracellular protein that binds amyloid-β peptide and mediates neurotoxicity in Alzheimer's disease. Nature *389*, 689–695.

Yan, S. D., Schmidt, A. M., Anderson, G. M., Zhang, J., Brett, J., Zou, Y. S., Pinsky, D., and Stern, D. (1994). Enhanced cellular oxidant stress by the interaction of advanced glycation end products with their receptors/binding proteins. J Biol Chem *269*, 9889–9897.

Yandava, B. D., Billinghurst, L. L., and Snyder, E. Y. (1999). "Global" cell replacement is feasible via neural stem cell transplantation: evidence from the dysmyelinated shiverer mouse brain. Proc Natl Acad Sci USA *96*, 7029–7034.

Yankner, B. A., Duffy, L. K., and Kirschner, D. A. (1990). Neurotrophic and neurotoxic effects of amyloid β protein: reversal by tachykinin neuropeptides. Science *250*, 279–282.

Zecca, L., Gallorini, M., Schunemann, V., Trautwein, A. X., Gerlach, M., Riederer, P., Vezzoni, P., and Tampellini, D. (2001). Iron, neuromelanin and ferritin content in the substantia nigra of normal subjects at different ages: consequences for iron storage and neurodegenerative processes. J Neurochem *76*, 1766–1773.

Zuccato, C., Ciammola, A., Rigamonti, D., Leavitt, B. R., Goffredo, D., Conti, L., MacDonald, M. E., Friedlander, R. M., Silani, V., Hayden, M. R., Timmusk, T., Sipione, S., and Cattaneo, E. (2001). Loss of huntingtin-mediated BDNF gene transcription in Huntington's disease. Science *293*, 493–498.

CHAPTER 13

NEUROTROPHIC SIGNALING IN MOOD DISORDERS

JING DU, TODD D. GOULD, and HUSSEINI K. MANJI

Laboratory of Molecular Pathophysiology, NIMH, National Institutes of Health, Bethesda, Maryland

Mood disorders (MDs) such as major depression and bipolar disorder (BD, also referred to as manic-depressive illness) are common, severe, and chronic illnesses. Additionally, increasing evidence suggests that major depression and BD are many times life-threatening illnesses as well. Suicide is the cause of death in up to ~15 % of individuals with both disorders, and in addition to suicide many other deleterious health-related effects are increasingly being recognized (Ciechanowski et al., 2000; Michelson et al., 1996; Musselman et al., 1998; Schulz et al., 2000). Indeed, there is a growing appreciation that, far from being diseases with purely psychological manifestations, major depression and BD are systemic diseases with deleterious ramifications on multiple organ systems (Ciechanowski et al., 2000; Michelson et al., 1996; Musselman et al., 1998; Schulz et al., 2000). For example, both major depression and BD represent major risk factors for both the development of cardiovascular disease as well as death after a myocardial infarction (Musselman et al., 1998). Furthermore, a recent study that controlled for physical illness, smoking, and alcohol consumption found that the magnitude of the increased mortality risk conferred by the presence of high depressive symptoms was similar to that of stroke and congestive heart failure (Schulz et al., 2000). The costs associated with disability and premature death represent an economic burden of tens of billions of dollars annually in the United States alone. It is thus not altogether surprising that the Global Burden of Disease Study has identified major MDs among the leading causes of disability worldwide and as illnesses that are likely to represent an increasingly greater health, societal, and economic problem in the coming years (Murray and Lopez, 1997).

Signal Transduction and Human Disease, Edited by Toren Finkel and J. Silvio Gutkind

ISBN 0-471-02011-7   Copyright © 2003 John Wiley & Sons, Inc.

MOOD DISORDERS: A CLINICAL OVERVIEW

Major depression has a lifetime incidence of about 15% (Blazer et al., 1994; Fava and Kendler, 2000), and affects twice as many women as men. There is a significant genetic component to the disorder; studies report an approximately threefold increased risk of major depression in first-degree relatives of probands with the disorder compared to the general population (Fava and Kendler, 2000; Sullivan et al., 2000). The course of the major depression is characterized by episodes of depression, separated by periods of euthymia (normal mood). Depressive episodes are characterized by depressed mood, suicidal ideation, anhedonia (an inability to experience pleasure), impaired sleep, changes in psychomotor activity, and impaired memory and concentration (Table 13.1).

BD has an approximate lifetime incidence of about 1% and is characterized by two seemingly opposite mood states: mania and depression. It is equally prevalent in men and women and can occur at any age, with the early twenties being the median age of onset (Angst and Sellaro, 2000; Goodwin and Jamison, 1990). The manic stages of BD are characterized by a hyperaroused state (either euphoric or dysphoric), increases in motor activity, racing thoughts, impaired judgment, decreased sleep, and an apparent decreased need for sleep (Goodwin and Jamison, 1990) (Table 13.1). The depressive phases of the illness present with symptomatology similar to that seen in major depression—that is, depressed mood, cognitive changes, psychomotoric changes, and a host of neurovegetative symptoms.

Individuals afflicted with severe MDs clearly experience considerable morbidity; what is less well appreciated is that these disorders are often fatal illnesses. As mentioned above, it is estimated that approximately 15% of patients who suffer from MDs eventually end their lives through suicide (Goodwin and Jamison, 1990; Jamison, 1986). Highlighting the importance of adequate treatment, it is noteworthy that multiple studies have analyzed the effects of lithium—a primary treatment for BD—on

TABLE 13.1. Diagnostic Criteria for Manic and Depressive Episodes

Mania	Depression
Euphoric or irritable mood	Depressed mood
Decreased need of sleep	Markedly diminished interest/pleasure
Increased energy and/or activity	Sleep, appetite, energy changes
Racing thoughts	Psychomotor changes
Increased talkativeness	Feelings of worthlessness or guilt
Disturbed ability to make decisions	Poor concentration or indecisiveness
Distractable	Recurrent thoughts of death/suicide
Increase in activities that can have painful consequences (e.g., gambling, promiscuity)	

the risk of suicide and suicide attempts (Baldessarini et al., 1999; Muller-Oerlinghausen et al., 1992; Schou, 1999; Tondo et al., 1997; Vestergaard and Aagaard, 1991). Tondo et al., in a review of the available data, found that patients with BD on lithium treatment have between a 5- and 10-fold lower risk of committing suicidal acts (Tondo et al., 1997). Furthermore, the frequency of suicidal acts and the number of completed suicides both increased over sevenfold when patients with BD discontinued lithium treatment (Tondo et al., 1997). In addition to a decreased risk for suicide, effective therapies for MDs are associated with longer life expectancy, improved socioeconomic measures, and a lower incidence of concomitant diseases such as heart disease (Muller-Oerlinghausen et al., 1992).

PHARMACOLOGIC TREATMENT OF SEVERE MOOD DISORDERS

Medications useful for treating depression and mania fall into two general classes; these are antidepressants and mood-stabilizing medications. Most antidepressants likely exert their initial biochemical effects by increasing the intrasynaptic levels of serotonin and/or norepinephrine. Tricyclic antidepressants and selective serotonin reuptake inhibitors (SSRIs) inhibit reuptake of serotonin, norepinephrine, and dopamine into the presynaptic nerve terminal. Monoamine oxidase inhibitors (MAOIs) inhibit monoamine oxidase, an enzyme responsible for the breakdown of monoamines after their reuptake (Stahl, 2000). However, a simple increase in the intrasynaptic levels of a "deficient neurotransmitter" likely does not explain the true mechanisms of action of antidepressants. Indeed, this seems unlikely because the therapeutic effects are only observed after weeks of administration. This observation, coupled with a number of negative and spurious findings, has led to the conclusion that increasing intrasynaptic levels of serotonin and/or norepinephrine is simply an initiating event, which induces a cascade of signaling and gene expression changes in critical neuronal circuits, effects that are ultimately responsible for the medications' therapeutic effects.

Although the depressive phases of BD also respond to antidepressants, the antidepressant medications can often "overshoot," thereby precipitating manic episodes. Thus patients with BD are generally treated with a class of medications referred to as "mood stabilizers" and, if necessary, with adjunctive antidepressants. Mood stabilizers are agents that have antimanic effects, exert prophylactic effects in preventing recurrent manic or depressive episodes, and may also possess some antidepressant properties. The prototypic agent of this class is lithium, a seemingly simple monovalent cation. More recently, a variety of anticonvulsant agents, most notably valproic acid (a substituted pentanoic acid) have also been utilized as mood-stabilizing agents. Other medications that may have some usefulness include other anticonvulsants, for example, carbamazepine, lamotrigine, and gabapentin. However, while valproic acid and other anticonvulsants are gaining notoriety, lithium remains the most common first-line treatment for BD worldwide. In fact, the introduction of lithium just over 50 years ago marked the beginning of a new era in treatment of the disorder, over time revolutionizing the treatment of patients with BD. Furthermore, the efforts to understand how a monovalent cation like lithium can exert such profound beneficial effect initially led investigators to examine the potential role of signal transduction pathways in severe MD. Indeed, as we discuss in greater detail below, it is hardly surprising that abnormalities in multiple neurotransmitter systems and physiological processes have been found in a disorder as complex as BD. Eventually, this led investigators to focus on pathways that are able to affect the functional balance between multiple neurotransmitter systems.

We now turn to a discussion of the traditional views of the pathophysiology of MDs, before reviewing the emerging evidence that signaling pathways regulating neuroplasticity and cellular resilience may play a role in the pathophysiology and treatment of MDs.

THE PATHOPHYSIOLOGY OF RECURRENT MOOD DISORDERS: TRADITIONAL CONCEPTS

Despite the devastating impact that major MDs have on the lives of millions worldwide, little is known for certain about their etiology or pathophysiology. The brain systems that have heretofore received the greatest attention in neurobiological studies of these illnesses have been the monoaminergic neurotransmitter systems, systems that were implicated by discoveries that effective antidepressant drugs exerted their primary biochemical effects by regulating intrasynaptic concentrations of serotonin and norepinephrine and that antihypertensives that depleted these monoamines sometimes precipitated depressive episodes in susceptible individuals (Goodwin and Jamison, 1990) (Fig. 13.1). Furthermore, the monoaminergic systems are extensively distributed throughout the network of limbic, striatal, and prefrontal cortical neuronal circuits thought to support the behavioral and visceral manifestations of MD (Drevets, 2000). Thus clinical studies over the past 40 years have attempted to uncover the specific defects in these neurotransmitter systems in major MDs by utilizing a variety of biochemical and neuroendocrine strategies. Indeed, assessments of cerebrospinal fluid (CSF) chemistry, neuroendocrine responses to pharmacologic challenge, and neuroreceptor and transporter binding have, in fact, demonstrated a number of abnormalities of the serotonergic, noradrenergic, and other neurotransmitter and neuropeptide systems in MD. Although such investigations have been heuristic over the years, they have been of limited value in elucidating the unique neurobiology of MD. Thus MDs arise from the complex interaction of multiple susceptibility (and likely protective) genes and environmental factors, and the phenotypic expression of the diseases includes not only episodic and often profound mood

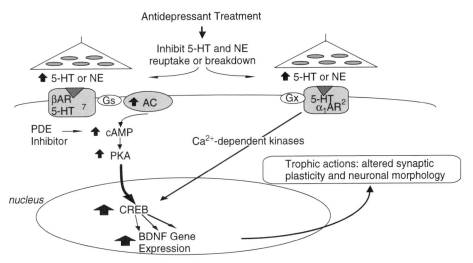

Figure 13.1. *Influence of Antidepressant Treatment on the Cyclic Adenosine Monophosphate (cAMP)-cAMP Response Element-Binding Protein (CREB) Cascade.* Antidepressant treatment increases synaptic levels of norepinephrine (NE) and serotonin (5-HT) via blocking the reuptake or breakdown of these monoamines. This results in activation of intracellular signal transduction cascades, one of which is the cAMP-CREB cascade. Chronic antidepressant treatment increases Gs coupling to adenylyl cyclase (AC), particulate levels of cAMP-dependent protein kinase (PKA), and CREB. CREB can also be phosphorylated by Ca^{2+}-dependent protein kinases, which can be activated by the phosphatidylinositol pathway (not shown) or by glutamate ionotropic receptors [e.g., *N*-methyl-D-aspartate (NMDA)]. Glutamate receptors and Ca^{2+}-dependent protein kinases are also involved in neural plasticity. One gene target of antidepressant treatment and the cAMP-CREB cascade is brain-derived neurotrophic factor (BDNF), which contributes to the cellular processes underlying neuronal plasticity and cell survival. BAR, β-adrenergic receptor; PDE, phosphodiesterase. Modified and reproduced with permission from Duman et al. (2000).

disturbance but also a constellation of cognitive, motoric, autonomic, endocrine, and sleep/wake abnormalities (see Table 13.1). Furthermore, as discussed above, although most antidepressants exert their initial effects by increasing the intrasynaptic levels of serotonin and/or norepinephrine, their clinical antidepressant effects are only observed after a lag period of 7–10 days and maximal benefits take upwards of 6 weeks, suggesting that a cascade of downstream effects are ultimately responsible for their therapeutic effects. These observations have led to the appreciation that although dysfunction within the monoaminergic neurotransmitter systems is likely to play important roles in mediating some facets of the pathophysiology of MD, they likely represent the downstream effects of other, more primary abnormalities (Manji et al., 2001a).

More recently, research into the pathophysiology and treatment of MD has focused on intracellular signaling pathways. These signaling

TABLE 13.2. Mood Disorders: A Putative Role for Signaling Pathways

- Amplify, attenuate, and integrate multiple signals—the basis of intracellular circuits and cellular modules
- Regulate multiple neurotransmitter and peptide systems—the basis of neuronal circuits and system modules
- Critical role in cellular memory and long-term neuroplasticity
- Dynamic regulation of complex signaling networks form the basis for higher-order brain function—mood and cognition
- Major targets for many hormones, including gonadal steroids, thyroid hormones, and glucocorticoids
- Abnormalities are indeed compatible with life—many human diseases arise from defects in signaling pathways
- Brain regional dysregulation and circumscribed symptoms are possible despite the relatively ubiquitous expression of signaling proteins
- Signaling proteins have been identified as targets for medications that are most effective in the treatment of mood disorders

From Manji et al. (2000a) with permission.

pathways are undoubtedly involved in neuroplastic events that regulate complex psychological and cognitive processes, as well as diverse vegetative functions such as appetite and wakefulness (Table 13.2). Consequently, recent evidence demonstrating that impairments of signaling pathways may play a role in the pathophysiology of MD, and that antidepressants and mood stabilizers exert major effects on signaling pathways that regulate neuroplasticity and cell survival, has generated considerable excitement among the clinical neuroscience community and are reshaping views about the neurobiological underpinnings of these disorders. We now review these data and discuss their implications not only for changing existing conceptualizations regarding the pathophysiology of MDs but also for the strategic development of improved therapeutics.

THE G$_S$/CAMP GENERATING SIGNALING PATHWAY IN THE PATHOPHYSIOLOGY AND TREATMENT OF MOOD DISORDERS

Pathophysiology

Several independent laboratories have now reported abnormalities in this signaling cascade in MDs (Garcia-Sevilla et al., 1997; Li et al., 2000; Manji et al., 1995b; Schreiber et al., 1991; Spleiss et al., 1998; Warsh et al., 2000; Young et al., 1993). Postmortem brain studies have reported increased levels of the stimulatory G protein (Gα_s) accompanied by increases in postreceptor stimulated adenylyl cyclase (AC) activity in BD (Warsh et al., 2000; Young et al., 1993). These observations, suggestive of elevated Gα_s levels and/or function, receive additional support from the demonstration of increased agonist-activated $[^{35}S]GTP\gamma S$ binding to G protein subunits in frontal cortical membranes from BD patients (Wang and Friedman, 1996). Additionally, several studies have also found elevated $G\alpha_s$ protein levels and mRNA levels in peripheral circulating cells in BD, although the dependence on clinical state remains unclear (Manji et al., 1995b; Spleiss et al., 1998; Wang et al., 1997; Warsh et al. 2000). It should be emphasized, however, that there is at present no evidence to suggest that the alterations in the levels of $G\alpha_s$ is due to a mutation in the $G\alpha_s$ gene itself (Li et al., 2000; Ram et al., 1997). There are numerous transcriptional and posttranscriptional mechanisms that regulate the levels of G protein α subunits, and the elevated levels of $G\alpha_s$ could potentially represent the indirect sequelae of alterations in any one of these other biochemical pathways (Li et al., 2000; Manji and Chen, 2000).

Considerable clinical research has focused on the activity of the cAMP generating system in readily accessible blood elements in patients with BD. Overall, the preponderance of the evidence suggests altered receptor and/or postreceptor sensitivity of the cAMP generating system in the absence of consistent alterations in the number of receptors themselves (Ebstein et al., 1988; Halper et al., 1988; Mann et al., 1985; Mork et al., 1992; Siever, 1987). Furthermore, a recent study looking more downstream in this signaling pathway found higher levels of cAMP-stimulated phosphorylation of a ~22-kDa protein (subsequently identified as Rap1) in platelets obtained from treated euthymic bipolar patients compared with healthy subjects (Perez et al., 2000). In recent postmortem studies, Warsh and associates (Warsh et al., 2000) found lower levels of cytosolic PKA regulatory subunits in various brain areas (Rahman et al., 1997). Compatible with lower levels of the regulatory subunits (which would theoretically "free up" more catalytic subunits), preliminary findings have revealed a higher basal kinase activity in BD in temporal cortex (Fields et al., 1999), and responsivity of the cAMP pathway (Garcia-Sevilla et al., 1999; Garcia-Sevilla et al., 1997). In addition, other studies have shown decreased basal, forskolin-, and GTP-stimulated AC activity (Gould and Manji, submitted).

Interestingly, diametric findings in this signaling pathway have been observed in MD compared to BD. For example, postmortem brain studies and peripheral cell studies have shown that antidepressant-free MD subjects have increased levels of the inhibitory G protein $(G\alpha_i)$, effects that would be expected to diminish cAMP activity in the frontal cortex of MD subjects. Other studies have found decreased stimulated AC activity in the frontal cortex of suicide victims compared to controls (Cowburn et al., 1994; Lowther et al., 1996). Moreover, studies of peripheral leukocytes have fairly consistently revealed blunting of the β-adrenergic response. Together, the data suggest that MD may be associated with downregulation of the cAMP pathway. Furthermore, as we discuss below, different classes of antidepressants appear to activate this signaling pathway, whereas antimanic/mood stabilizing agents attenuate it.

Treatment

Although it appears that lithium (at therapeutic concentrations) does not directly affect G protein, there is considerable evidence that chronic lithium administration indirectly affects G protein function (Manji and Lenox, 2000b; Manji et al., 1995b, Mork et al., 1992; Risby et al., 1991; Wang et al. 1999b). Although some studies have reported modest changes in the levels of G protein subunits, the preponderance of the data suggests that chronic lithium does not modify G protein levels per se, but rather modifies G protein function (Manji et al., 1995a; Manji and Lenox, 2000b; Manji et al., 1995b). Interestingly, for both G_s and G_i, lithium's major effects in both humans and rodents are most compatible with a stabilization of the heterotrimeric, undissociated ($\alpha\beta\gamma$) conformation of the G protein (Li et al., 2000; Manji et al., 1995a; Manji et al., 1995b; Stein et al., 1996; Warsh et al., 2000).

Lithium also exerts complex effects on the activity of AC, with the preponderance of the data demonstrating an elevation of basal AC activity but an attenuation of receptor-stimulated responses in both preclinical and clinical studies (Jope, 1999; Manji et al., 2000b; Mork et al., 1992; Wang et al. 1997). It has been postulated that these elevations of basal cAMP and dampening of receptor-mediated stimulated responses may play an important role in lithium's ability to prevent "excessive excursions from the norm" (Jope, 1999; Manji et al., 1995a). These complex effects likely represent the net effects of direct inhibition of AC, upregulation of certain AC subtypes, and effects on the stimulatory and inhibitory G proteins (Chen et al., 1996; Li et al., 2000; Manji and Lenox, 2000a). Consistent with a lithium-induced increase in basal cAMP and AC levels, platelets obtained from lithium treated euthymic bipolar patients exhibited enhanced basal and cAMP-stimulated phosphorylation of two PKA substrates (one of which was later identified as Rap1) (Perez et al., 2000). Most recently, lithium's effects on the phosphorylation and activity of cAMP response element binding protein (CREB) have been examined in rodent brain and in cultured human neuroblastoma cells, with somewhat conflicting results (Ozaki and Chuang, 1997; Wang et al., 1999a) However, Grimes and Jope recently reported results suggesting that lithium—through inhibition of the enzyme glycogen synthase kinase 3β—increases CREB DNA binding activity(Grimes and Jope, 2001).

The cAMP signaling cascade appears to be a major target for the actions of chronic antidepressants; recent studies have demonstrated an enhanced coupling between $G\alpha_s$ and the catalytic unit of AC (Rasenick et al., 2000) and activation of cAMP-dependent protein kinase enzyme activity (Nestler et al., 1989; Popoli et al., 2000). Antidepressants have also been demonstrated to activate cAMP-dependent and calcium/calmodulin-dependent protein kinases, effects that are accompanied by increases in the endogenous phosphorylation of selected substrates (microtubule-associated protein 2 and synaptotagmin) (Popoli et al., 2000). Duman and associates (Duman et al., 1997; Duman et al., 2000) have undertaken an elegant series of studies demonstrating that the chronic treatment of rats with a variety of antidepressants increases the levels of CREB mRNA, CREB protein, and CRE DNA binding activity in hippocampus. Furthermore, the same laboratory has demonstrated that chronic antidepressant treatment also increases the expression of two important genes known to be regulated by CREB, namely BDNF (brain-derived neurotrophic factor), and its receptor trkB (Duman et al., 1997; Duman et al., 2000) (Fig. 13.1). Preliminary postmortem human brain studies have also revealed increased CREB levels in patients treated with antidepressants, providing indirect support for the rodent and cell culture studies (Dowlatshahi et al., 1998).

THE PROTEIN KINASE C SIGNALING PATHWAY IN THE PATHOPHYSIOLOGY AND TREATMENT OF MOOD DISORDERS

The "inositol depletion hypothesis" posits that lithium, as an uncompetitive inhibitor of inositol-1-phosphatase (IMP), produces its therapeutic effects via a depletion of neuronal *myo*-inositol levels. Although this hypothesis has been of great heuristic value, numerous studies have examined the effects of lithium on receptor mediated PI responses, and although some report a reduction in agonist-stimulated PIP_2 hydrolysis in rat brain slices after acute or chronic lithium, these findings have often been small and inconsistent and subject to numerous methodological differences (Jope and Williams, 1994). However, more recently, a magnetic resonance spectroscopy study demonstrated that lithium-induced *myo*-inositol reductions are observed in the frontal cortex of BD patients after only 5 days of lithium administration, at a time when the patients' clinical state is completely unchanged (Moore et al., 1999). Consequently, these and other studies suggest that although inhibition of IMP may represent an initiating lithium effect, reducing *myo*-inositol levels per se is not associated with therapeutic response. This has led to the working hypothesis that some of the initial actions of lithium may occur as a result of a relative reduction of *myo*-inositol; putatively, this reduction of *myo*-inositol initiates a cascade of secondary changes in the PKC signaling pathway and gene expression in the CNS, effects which are ultimately responsible for lithium's therapeutic efficacy.

PKC in the Pathophysiology of Bipolar Disorder

To date, there have only been a limited number of studies directly examining PKC in BD (Hahn and Friedman, 1999). Although undoubtedly an oversimplification, particulate (membrane) PKC is sometimes viewed as the more active form of PKC, and thus an examination of the subcellular partitioning of this enzyme can be used as an index of the degree of activation. Friedman et al. investigated PKC activity and PKC translocation in response to serotonin in platelets obtained from BD subjects before and during lithium treatment (Friedman et al., 1993). They reported that the ratios of platelet membrane-bound to cytosolic PKC activities were elevated in the manic subjects. In addition, serotonin-elicited platelet PKC translocation was enhanced in those subjects. In postmortem brain tissue from BD patients, Wang and Friedman (Wang and Friedman, 1996) measured PKC isozyme levels, activity, and translocation; they reported increased PKC activity and translocation in BD brains compared to controls, effects that were accompanied by elevated levels of selected PKC isozymes in the cortex of BD subjects. Two recent studies have utilized phorbol dibutyrate ([^{3}H]PDBu, a radioligand that binds to PKC) to investigate particulate and cytosolic PKC in postmortem brain samples obtained from depressed patients and/or individuals who committed suicide. Pandey and associates (Pandey et al., 1997) found that the B_{max} of [^{3}H]PDBu binding sites was significantly decreased in both membrane and cytosolic fractions from Brodmann's areas 8–9 in teenage suicide subjects compared to matched controls. Coull and associates (Coull et al., 2000) found increased [^{3}H]PDBu binding in the soluble fraction (suggesting less in the active membrane fraction) in antidepressant-free suicides compared to controls in the frontal cortex. The results of these two studies could potentially be interpreted as reflecting reduced PKC function, either by a reduction in the absolute levels or by a reduction in the particulate/soluble fractions. However, considerable additional research is required to adequately justify such a conclusion.

PKC in the Treatment of Bipolar Disorder

Evidence accumulating from various laboratories has clearly demonstrated that lithium, at therapeutically relevant concentrations, exerts major effects on the PKC signaling cascade (Table 13.3). Currently available data suggest that acute lithium exposure facilitates a number of PKC-mediated responses whereas longer-term exposure results in an attenuation of phorbol ester-mediated responses, which is accompanied by a downregulation of specific PKC isozymes (Manji and Lenox, 1999).

TABLE 13.3. Protein Kinase C and the Pathophysiology of Bipolar Affective Disorder

- Amphetamine produces increases in PKC activity and GAP-43 phosphorylation (implicated in neurotransmitter release)
- PKC inhibitors block the biochemical and behavioral responses to amphetamine and cocaine and also block cocaine-induced sensitization
- Increased membrane/cytosol PKC partitioning in platelets from manic subjects; normalized with lithium treatment
- Increased PKC activity and translocation in BD brains compared to controls
- Lithium and valproate regulate PKC activity, PKCα, PKCε, and MARCKS
- Preliminary data suggest that PKC inhibitors may have efficacy in the treatment of acute mania

Studies in rodents have demonstrated that chronic (but not acute) lithium produces an isozyme-selective reduction in PKC α and ε in frontal cortex and hippocampus, in the absence of significant alterations in the β, γ, δ, or ζ isozymes (Chen et al., 2000b; Manji et al., 1993; Manji and Lenox, 1999). Concomitant studies carried out in immortalized hippocampal cells in culture exposed to chronic lithium show a similar reduction in the expression of both the PKC α and ε isozymes in the cell as determined by immunoblotting (Manji and Lenox, 1999). Furthermore, chronic lithium has been demonstrated to dramatically reduce the hippocampal levels of a major PKC substrate, myristoylated alanine-rich C kinase substrate (MARCKS), a protein that has been implicated in regulating long-term neuroplastic events (Lenox et al., 1992). Although these effects of lithium on PKC isozymes and MARCKS are striking, a major problem inherent in neuropharmacologic research is the difficulty in attributing therapeutic relevance to any observed biochemical finding. It is thus noteworthy that the structurally dissimilar antimanic agent VPA produces very similar effects to those of lithium on PKC α and ε isozymes and MARCKS protein (Chen et al., 1994; Lenox and Hahn, 2000; Manji and Chen, 2000; Manji and Lenox, 2000b; Manji et al., 1999c; Watson et al., 1998). Interestingly, lithium and VPA appear to bring about their effects on the PKC signaling pathway by distinct mechanisms (Lenox and Hahn, 2000; Manji and Lenox, 1999); these biochemical observations are consistent with the clinical observations that some patients show preferential response to one or other of the agent, and that one often observes additive therapeutic effects in patients when the two agents are coadministered.

In view of the pivotal role of the PKC signaling pathway in the regulation of neuronal excitability, neurotransmitter release, and long-term synaptic events (Chen et al., 1997; Conn and Sweatt, 1994; Hahn and Friedman, 1999), it was postulated that the attenuation of PKC activity may play a role in the antimanic effects of lithium and VPA. Recently, a pilot study found that tamoxifen [a nonsteroidal antiestrogen known to be a PKC inhibitor at higher concentrations (Baltuch et al., 1993)] may, indeed, possess antimanic efficacy (Bebchuk et al., 2000). Clearly, these results must be considered preliminary because of the small sample size thus far. However, in view of the preliminary data suggesting the involvement of the PKC signaling system in the pathophysiology of BD (vide supra), these results suggest that PKC inhibitors may be very useful agents in the treatment of mania. Thus larger double-blind placebo-controlled studies of tamoxifen and novel selective PKC inhibitors in the treatment of mania are warranted.

Overall, considerable data suggest that alterations in both the PKA and PKC signaling cascades may play a role in mediating some of the signs and symptoms of MDs; it should be noted, however, that there is presently no evidence that mutations in these pathways are etiologically responsible for MDs. The data are much more, compelling, however, that that these signaling pathways do represent therapeutically relevant targets for many of the beneficial effects of antidepressants and mood stabilizers. In addition to the identification of the involvement of these signaling pathways in MDs, there is now considerable evidence suggesting that severe MDs are associated with impairments of structural plasticity and cellular resilience and that neurotrophic signaling cascades may also be involved in the pathophysiology and treatment of MD and BD. We now review the recent advances in our understanding of cellular events regulating neurotrophic signaling cascades and the possible involvement of these signaling cascades in MDs.

NEUROTROPHIC SIGNALING CASCADES: A FOCUS ON BRAIN-DERIVED NEUROTROPHIC FACTOR

Neurotrophins are a family of regulatory factors that mediate the differentiation and survival of neurons, as well as the modulation of synaptic transmission and synaptic plasticity(Patapoutian and Reichardt, 2001; Poo, 2001). The neurotrophin family now include—among others— nerve growth factor (NGF), brain-derived neurotrophic factor (BDNF), and neurotrophin 3 (NT3), NT4/5, and NT6 (Patapoutian and Reichardt, 2001). These various proteins are closely related in terms of sequence homology and receptor specificity. They bind to and activate specific receptor tyrosine kinases belonging to the Trk family of receptors, including TrkA, TrkB, TrkC, and a pan-neurotrophin receptor, p75 (Patapoutian and Reichardt, 2001; Poo, 2001). Neurotrophins can be secreted constitutively or transiently and often in an activity-dependent manner. Recent observations support a model in which neurotrophins are secreted from the dendrite and act retrogradely at presynaptic terminals, where they act to induce long-lasting modifications (Poo, 2001) .

Within the neurotrophin family, BDNF is a potent physiological survival factor that has also been implicated in a variety of pathophysiological conditions, such as Parkinson disease, Alzheimer disease, and diabetic peripheral neuropathy (Nagatsu et al., 2000; Pierce and Bari, 2001; Salehi et al., 1998). The cellular actions of BDNF are mediated through two types of receptors: a high-affinity tyrosine receptor kinase (Trk B) and a low-affinity pan-neurotrophin receptor (p75). TrkB is preferentially activated by BDNF and NT4/5 and appears to mediate most of the cellular responses to these neurotrophins. Additionally, there are two isoforms of TrkB receptors: the full-length TrkB and the truncated form of TrkB, which does not contain the intracellular tyrosine kinase domain (Fryer et al., 1996). The truncated form of TrkB can thus function as a dominant-negative inhibitor for the TrkB receptor tyrosine kinase, thereby providing another mechanism to regulate BDNF signaling in the CNS (Eide et al., 1996; Gonzalez et al., 1999). Indeed, under certain pathological conditions, such as Alzheimer disease, the immunoreactivity of the full-length TrkB receptor isoform is selectively lost in both temporal lobe and frontal cortex (Allen et al., 1999). The functions of the p75 receptor are more diverse and complex than those of the Trks. In vitro studies have shown that p75 enhances the sensitivity of TrkA-expressing neurons to the survival-promoting effect of NGF, while decreasing their sensitivity to neurotrophin NT3 (Bibel et al., 1999; Kaplan and Miller, 2000). Direct interaction between p75 and Trk receptors, together with changes in ligand affinity and Trk signaling, account (at least in part) for these effects of p75 (Bibel et al., 1999; Kaplan and Miller, 2000). Recent studies has shown that p75-mediated activation of NF-κB plays a role in enhancing the survival response of developing sensory neurons to NGF (Hamanoue et al., 1999).

Binding of BDNF initiates TrkB dimerization and transphosphorylation of tyrosine residues in its cytoplasmic domain (Patapoutian and Reichardt, 2001). The phosphotyrosine residues of Trk B receptor function as binding sites for recruiting specific cytoplasmic signaling and scaffolding proteins. Binding of cytoplasmic src-homology 2 (SH2) domain-containing scaffolding proteins, including shc and Grb2, which recognize specific phosphotyrosine residues on the receptor, can thus result in the recruitment of a variety of effector molecules. This recruitment of effector molecules generally occurs via interaction of proteins with modular binding domains SH2, SH3 (named after homology to the src oncogenes—src homology domains); SH2 domains are a stretch of ~100 aa that allows high-affinity interactions with certain phosphotyrosine motifs. The ability of multiple effectors to interact with phosphotyrosines is undoubtedly one of the keys to the pleiotropic effects that neurotrophins can exert. These pleiotropic and yet distinct effects of growth factors are mediated by varying degrees of activation of three major signaling pathways—the Ras/MAP kinase (MAPK) pathway, the PI-3 kinase pathway (PI3K), and the phospholipase C-γ1 pathways. Among these pathways, the effects of the PI3K pathway and the MAPK pathway have been most directly linked to the cell survival effects of neurotrophins (Patapoutian and Reichardt, 2001).

Signaling Through the MAPK Cascade

Shc recruitment and phosphorylation results in recruitment to the membrane of a complex of the adaptor Grb-2 and the Ras exchange factor son of sevenless (SOS), thereby stimulating transient activation of Ras. Ras, in turn, activates PI3K, the p38 MAPK/MAPK-activating protein kinase 2 pathway, and the c-Raf/ERK pathway. Among the targets of ERK are the ribosomal S6 kinases (RSKs). Both RSK and MAPK-activating protein kinase 2 phosphorylate CREB and other transcription factors. Recent studies demonstrated that the activation of the MAPK pathway can inhibit apoptosis by inducing the phosphorylation of BAD (Bcl-xl/Bcl-2 associated death promoter) and increasing the expression of the antiapoptotic protein Bcl-2, the latter effect likely involves CREB (Bonni et al., 1999; Riccio et al., 1997). Phosphorylation of BAD occurs via activation of ribosomal S-6 kinase (Rsk). Rsk phosphorylates BAD and thereby promotes its inactivation. Activation of Rsk also mediates the actions of the MAPK cascade and neurotrophic factors on the expression of Bcl-2. Rsk can phosphorylate CREB, leading to induction of Bcl-2 gene expression (Fig. 13.2).

MAPKs are abundantly present in brain and in recent years have been postulated to play a major role in a variety of long-term CNS functions, both in the developing and mature CNS (Fukunaga and Miyamoto, 1998; Kornhauser and Greenberg ,1997; Matsubara et al., 1996; Robinson et al., 1998). With respect to their actions in the mature CNS, MAPKs have been implicated in mediating neurochemical processes associated with long term-facilitation (Martin et al., 1997), long-term potentiation (English and Sweatt, 1996; English and Sweatt, 1997), associative learning (Atkins et al., 1998), one-trial and multitrial classic conditioning (Crow et al., 1998), and long-term spatial memory (Blum et al., 1999) and

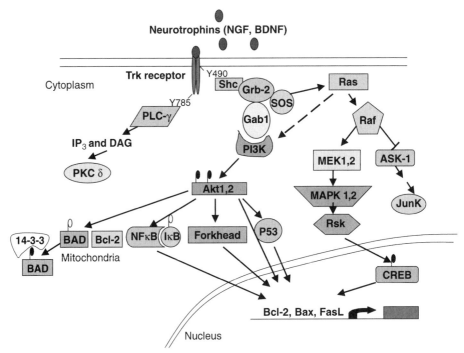

Figure 13.2. *Neurotrophin Signaling Cascade.* Neurotrophins, for example, nerve growth factor (NGF) and BDNF, signal through extracellular receptors (such as Trk), thereby stimulating signaling cascades within cells. Stimulation of Trk receptors results in activation of multiple pathways involved in gene regulation and cellular survival. One of these pathways is the Ras MAP kinase pathway that regulates; among other proteins, phosphorylation and activation of the transcription factor CREB. Activation of CREB mediates transcription of many genes including Bcl-2. Trk receptor signaling also activates PI3-kinase, a mediator of the actions of Akt, which phosphorylates and inactivates BAD. A third signaling pathway activated by Trk receptors involves phospholipase C-γ, resulting in IP$_3$- and diacylglycerol (DAG)-mediated signals. Both of these activate multiple pathways involved in cellular protection from apoptosis.

have also been postulated to integrate information from multiple, infrequent bursts of synaptic activity (Murphy et al., 1994). Importantly for the present discussion, MAPK pathways were recently demonstrated to regulate the responses to environmental stimuli and stressors in rodents (Xu et al., 1997), and to couple PKA and PKC to CREB phosphorylation in area CA1 of hippocampus (Roberson et al., 1999; Roberson et al., 1996). These recent studies suggest the possibility of a broad role for the MAPK cascade in regulating gene expression in long-term forms of synaptic plasticity (Roberson et al., 1999). Thus, overall, the data suggest that MAPKs play important physiological roles in the mature CNS and, furthermore, may represent important targets for the actions of CNS-active agents (Nestler, 1998; Yuan et al., 1998).

The PI3K-Akt Pathway: A Major Pathway Mediating Neuronal Survival

The PI3K-Akt pathway is also particularly important for mediating neuronal survival under a wide variety of circumstances. Trk receptors can activate PI3K through at least two distinct pathways, the relative importance of which differs among neuronal subpopulations. In many neurons, Ras-dependent activation of PI3K is the most important pathway through which neurotrophins promote cell survival. In some cells, however, PI3K can also be activated through three adaptor proteins, Shc, Grb-2 ,and Gab-1. Binding to phosphorylated tyrosine 490 of Shc results in recruitment of Grb-2. Phosphorylated Grb-2 provides a docking site for Gab-1, which in turn is bound by PI3K (Brunet et al., 2001).

PI3K directly regulates certain cytoplasmic apoptotic pathways. Akt has been proposed to act both before the release of cytochrome c by proapoptotic Bcl-2 family members and after the release of cytochrome c, by regulating components of the apoptosome. Akt phosphorylates the proapoptotic Bcl-2 family member BAD, thereby inhibiting BAD's proapoptotic functions (Datta et al., 1997).

Akt may also promote survival in an indirect fashion by regulating another major signaling enzyme—glycogen synthase kinase 3β (GSK-3β) (Woodgett, 2001). Thus elevated GSK-3β has been shown to promote apoptosis in cultured neurons (Bijur et al., 2000). Furthermore, neurotrophin withdrawal increases, whereas phosphorylation by Akt decreases, GSK-3β activity (Hetman et al., 2000). Moreover, a series of studies indicates that Akt controls a major class of transcriptional factors—the Forkhead box transcription factor, class O (FOXO) subfamily of Forkhead transcriptional regulators (FKHR, FKHRL1 and AFX). Several groups have independently shown that Akt directly phosphorylates FOXOs and inhibits their ability to induce the death genes (Brunet et al., 1999; Dijkers et al., 2000). Finally, activation of Akt also results in phosphorylation of NF-κB. Transcription activated by NF-κB was shown recently to promote neuronal survival (Maggirwar et al., 1998). Thus PI3K acting through Akt may promote survival by variety of mechanisms; precisely which of these mechanisms is operative in the actions of neurotrophins, and under what circumstances, is the focus of extensive current research (Fig. 13.2).

Signaling Through PLC-γ

Phosphorylated Trk receptors also recruit PLC-γ1. The Trk kinase then phosphorylates and activates PLC-γ1, which acts to hydrolyze phosphatidylinositides to generate diacylglycerol (DAG) and inositol 1,4,5-trisphosphate (IP_3). IP_3 induces the release of Ca^{2+} stores, increasing levels of cytoplasmic Ca^{2+} and thereby activating many pathways controlled by Ca^{2+}. In recent work, it has been shown that neurotrophins activate protein kinase C (PKC) δ, which is required for activation of the ERK cascade and for neurite outgrowth (Patapoutian and Reichardt, 2001) (Fig. 13.2).

Neurotrophins Also Function as Synaptic Modulators

It is now clear that neurotrophins not only support cell survival but also play important roles as "synaptic modulators" by regulating synapse development, synaptic transmission, and, indirectly, the formation of long-term potentiation (LTP) (discussed by Poo, 2001). Neurotrophic factors secreted by either pre-or postsynaptic cells are important in synapse development and normal maintenance; furthermore, overexpression of BDNF in transgenic mice increases the number of synapses in sympathetic ganglia and accelerates the maturation of inhibitory pathways in the developing visual cortex (Huang et al., 1999). In addition to more long-term effects, neurotrophic factors also acutely regulate synaptic transmission. In this context, BDNF specifically induces potentiation of glutamatergic synapses, with the potentiation only being observed when the postsynaptic neuron uses glutamate as a transmitter (and is not seen with GABA) (Lessmann and Heumann, 1998).

In addition to regulating synaptic efficacy, BDNF appears to function as a modulator that is required for the induction, expression, and/or maintenance of LTP. Genetic deletion of BDNF in mice disrupts normal induction of LTP, which can be rescued by reintroducing BDNF either by transfection of hippocampal slices with BDNF-expressing adenovirus or exogenous administration of BDNF (Korte et al., 1996; Patterson et al., 1996). In contrast, exogenous application of BDNF does not potentiate basal synaptic transmission. BDNF does, however, reduce the tetanus-induced depression of transmitter release at CA3–CA1 synapse of young rats, allowing sufficient postsynaptic activation for the induction of LTP (Pozzo-Miller et al., 1999). These data suggested that BDNF is a permissive factor required for formation of LTP rather than mediating LTP directly.

Retrograde Transportation of the Neurotrophin Receptors as a Signal to the Cell Body

Unlike most other internalized receptors, which are usually degraded after internalization, neurotrophin-Trk complexes in endocytic vesicles function as signal transducers and provide a mechanism for long-range signaling in the neuronal cytoplasm. Several studies have provided support for the retrograde transportation model of neurotrophin-Trk complexes. For sympathetic ganglionic neurons, internalization of NGF-TrkA complexes at axon terminals and retrograde transport of these complexes to the cell body is responsible for the NGF-dependent effects on neuronal survival (Riccio et al., 1997). The tyrosine kinase activity of TrkA is required to maintain the complex in an autophosphorylated state on its arrival in the cell body and for propagation of the signal to the transcription factor CREB within the nucleus (Riccio et al., 1997). Similarly, in the isthmooptic nucleus (ION) of chick embryos, transport of BDNF alone does not promote the survival of ION neurons when axonal TrkB is inactivated (von Bartheld et al., 1996). These results indicate that endocytotic vesicles containing neurotrophin-Trk complexes may be functionally active and should be viewed as activated signaling complexes that spread the cytosolic signaling of neurotrophin-Trk complexes to distant parts of the neuron via active transport mechanisms. Intriguingly, as has recently been shown with another tyrosine kinase (ErbB-4 receptor tyrosine kinase), other hitherto unappreciated methods, such as cleavage of receptor fragments, may also be operative in trafficking signals from extracellular receptors to intracellular and nuclear targets (Ni et al., 2001). Whether such novel signaling mechanisms are also utilized by neurotrophin receptors will undoubtedly be the focus of considerable future research.

BDNF Signaling Is Regulated by Neuronal Activity

The neurotrophic functions of neurotrophins depend in large part on a cytoplasmic signal transduction cascade, whose efficacy may be influenced by the presence of electrical activity in the neuron. Seizure activity, as well as nonseizure activity of a frequency/intensity capable of inducing LTP, have been shown to elevate BDNF mRNA levels and facilitate the release of BDNF from hippocampal and cortical neurons (reviewed by Poo, 2001). Although BDNF was originally considered to only be transported retrogradely, recent evidence indicates that BDNF can also act anterogradely to modulate synaptic plasticity (reviewed by Poo, 2001). High-frequency neuronal activity and synaptic transmission have also been shown to elevate the number of TrkB receptors on the surface of cultured hippocampal neurons through activation of the CAM kinase II pathway and may therefore facilitate the synaptic action of BDNF (Du et al., 2000). Thus electrically active nerve terminals may be more susceptible to synaptic potentiation by secreted neurotrophins compared to inactive terminals. Neuronal or synaptic activity is also known to promote the effects of neurotrophins on the survival of cultured retinal ganglion cells; here, neuronal or synaptic activity elevates cAMP levels to enhance the responsiveness of the neuron to neurotrophins, apparently by recruiting extra TrkB receptors to the plasma membrane (Meyer-Franke et al., 1998). Moreover, the internalization of BDNF receptor TrkB is also upregulated by activity as a retrograde signal to the cell body in cultured hippocampal neurons (Du et al., 2003). The activity-dependent regulation of BDNF signaling on BDNF synthesis and release, TrkB insertion onto neuronal surfaces, and activated TrkB tyrosine kinase internalization are crucial for its influence on synaptic plasticity and neuronal survival.

The information reviewed here clearly shows that neurotrophin signaling cascades play a major role in regulating various forms of neuronal and synaptic plasticity, as well as neuronal survival. As alluded to already, there is now evidence from a variety of sources demonstrating significant reductions in regional CNS volume, as well as regional reductions in the numbers and/or sizes of glia and neurons in MD. One line of evidence comes from structural imaging studies, which have recently begun to provide important clues about the neuroanatomic basis of MDs. These volumetric neuroimaging studies demonstrate an enlargement of third and lateral ventricles, as well as reduced gray matter volumes in parts of the orbital and medial prefrontal cortex, the ventral striatum, and the mesiotemporal cortex (Drevets et al., 1999). In addition to the accumulating neuroimaging evidence, several postmortem brain studies are now providing direct evidence for reductions in regional CNS volume and cell number as well (Manji et al., 2000c; Manji et al., 2001b; Rajkowska, 2000b). We now turn to a discussion of the evidence that neurotrophic signaling cascades are long-term targets for antidepressants and mood stabilizers.

INFLUENCE OF ANTIDEPRESSANT TREATMENT ON CELL SURVIVAL PATHWAYS

In an extensive series of studies, Duman and associates have demonstrated that the cAMP-CREB cascade—an important pathway involved in cell survival and plasticity—is upregulated by chronic antidepressant treatment, in a time frame that parallels clinical response (Duman et al., 1999; Nestler et al., 1989; Nibuya et al., 1996). Thus chronic antidepressant treatment increases CREB phosphorylation and also increases the expression of a major gene regulated by CREB, namely BDNF (Nibuya et al., 1995). A role for the cAMP-CREB cascade and BDNF in the actions of antidepressant treatment is also supported by studies demonstrating that upregulation of these pathways increases performance in behavioral models of depression (Siuciak et al., 1997). Consistent with the cellular effects, several reports support the hypothesis that chronic antidepressant treatment produces neurotrophic-like effects. Thus antidepressant treatment induces greater regeneration of catecholamine axon terminals in the cerebral cortex, enhances hippocampal synaptic plasticity, and may attenuate stress-induced atrophy of hippocampal CA3 pyramidal neurons (discussed in (Duman et al., 2000; Duman et al., 1999; Manji and Duman 2001)).

MOOD STABILIZERS REGULATE THE MAPK SIGNALING CASCADE

As discussed above, several endogenous growth factors—including NGF and BDNF—exert many of their neurotrophic effects via the MAPK signaling cascade. In view of the important role of MAPKs in mediating long-term neuroplastic events, it is noteworthy that lithium and VPA, at therapeutically relevant concentrations, were recently demonstrated to robustly activate the ERK MAPK cascade in human neuroblastoma SH-SY5Y cells (Chen et al., 2002; Yuan et al., 2001).

Because the ERK MAPKs are known to mediate many of the effects of various neurotrophic factors and to promote neurite outgrowth (Finkbeiner, 2000; Segal and Greenberg, 1996), VPA's effects on the morphology of human neuroblastoma cells have been examined in detail. Human neuroblastoma SH-SY5Y cells exposed to VPA (1.0 mM) in serum-free medium for 5 days exhibited prominent growth cones and dramatic neurite outgrowth. Growth cone associated protein-43 (GAP-43) is a protein expressed at elevated levels during neurite growth during development or regeneration, and a greater than threefold increase in GAP-43 levels was observed after 5-day VPA exposure (Yuan et al., 2001). Follow-up studies have recently shown that, similar to the effects observed in neuroblastoma cells in vitro, chronic lithium or VPA also robustly increases the levels of activated ERK in areas of brain that have been implicated in the pathophysiology and treatment of BD—the frontal cortex and hippocampus (Chen et al., 2002). Interestingly, neurotrophic factors are now known to promote cell survival by activating MAPKs to suppress intrinsic, cellular apoptotic machinery, not by inducing cell survival pathways (Pettmann and Henderson, 1998; Thoenen, 1995). Thus a downstream target of the MAPK cascade, Rsk, phosphorylates CREB, and this leads to induction of Bcl-2 gene expression (see Fig. 13.2). Our group therefore undertook studies to determine whether lithium or VPA regulated the expression of bcl-2.

Chronic treatment of rats with "therapeutic" doses of lithium and VPA produced a doubling of bcl-2 levels in FCx, effects that were primarily due to a marked increase in the number of bcl-2 immunoreactive cells in layers II and III of FCx (Chen et al., 1999; Manji et al., 1999a; Manji et al., 2000a). Interestingly, the importance of neurons in layers II–IV of the FCx in MDs has recently been emphasized, because primate studies indicate that these areas are important for providing connections with other cortical regions and that they are targets for subcortical input (Rajkowska, 2000a). Chronic lithium also markedly increased the number of bcl-2 immunoreactive cells in the dentate gyrus and striatum (Manji et al., 1999a); detailed immunohistochemical studies following chronic VPA treatment are currently under way. Subsequent to these findings, it has been demonstrated that lithium also increases bcl-2 levels in C57BL/6 mice (Chen et al., 1999), in human neuroblastoma SH-SY5Y cells in vitro (Manji and Chen, 2000), and in rat cerebellar granule cells in vitro (Chen and Chuang, 1999). This latter study was undertaken as part of studies investigating the molecular and cellular mechanisms underlying the neuroprotective actions of lithium against glutamate excitotoxicity (vide infra). These investigators found that lithium produced a remarkable increase in bcl-2 protein and mRNA levels. Moreover, lithium was recently demonstrated to reduce the levels of the proapoptotic protein p53 both in cerebellar granule cells (Chen and Chuang, 1999) and SH-SY5Y cells (Lu et al., 1999). Thus, overall, the data clearly show that chronic lithium robustly increases the levels of the neuroprotective protein bcl-2 in areas of rodent FCx, hippocampus, and striatum in vivo and in cultured cells of both rodent and human neuronal origin in vitro; furthermore, at least in cultured cell systems, lithium has also been demonstrated to reduce the levels of the proapoptotic protein p53.

Consistent with bcl-2's known cytoprotective effects, lithium, at therapeutically relevant concentrations, has been shown to exert neuroprotective effects in a variety of preclinical paradigms. Thus lithium has been demonstrated to protect against the deleterious effects of glutamate, NMDA receptor activation, aging, serum/nerve growth factor deprivation, ouabain, thapsigargin (which mobilizes intracellular MPP^+, Ca^{2+}) and β-amyloid in vitro (Chuang et al., 2002, in press, Manji et al. 2000a). More importantly, lithium's neurotrophic and cytoprotective effects have also been demonstrated in rodent brain in vivo. Thu, lithium treatment has been shown to attenuate the biochemical deficits produced by kainic acid infusion, ibotenic acid infusion, and forebrain cholinergic system lesions (Manji et al., 2000a; Manji et al., 1999b; Manji et al. 1999c), to exert dramatic protective effects against middle cerebral artery occlusion (Nonaka and Chuang, 1998), and to enhance hippocampal neurogenesis in the adult rodent hippocampus (Chen et al., 2000a). The potential therapeutic relevance of these preclinical findings is discussed below.

Human Evidence for the Neurotrophic Effects of Mood Stabilizers

Although the body of preclinical data demonstrating neurotrophic and neuroprotective effects of lithium is striking, considerable caution must clearly be exercised in extrapolating to the clinical situation with humans. In view of lithium and VPA's robust effects on the levels of the cytoprotective protein bcl-2 in the frontal cortex, Drevets and associates (Drevets et al., 1997) reanalyzed older data demonstrating ~40% reductions in subgenual PFC volumes in familial MD subjects. Consistent with neurotrophic/neuroprotective effects of lithium and VPA, they found that the patients treated with chronic lithium or VPA exhibited subgenual PFC volumes that were significantly higher than the volumes in non-lithium- or VPA-treated patients and not significantly different from control (Drevets, 2000).

Although the results of the study by Drevets (Drevets, 2000) suggests that mood stabilizers may have provided neuroprotective effects during naturalistic use, considerable caution is warranted in view of the small sample size and cross-sectional nature of the study. To investigate the potential neurotrophic effects of lithium in humans more definitively, a longitudinal clinical study was recently undertaken using proton magnetic resonance spectroscopy (MRS) to quantitate *N*-acetyl-aspartate (NAA) levels. NAA is a putative neuronal marker localized to mature neurons and not found in mature glial cells, CSF, or blood (Tsai and Coyle, 1995). A number of studies have now shown that initial abnormally low brain NAA measures may increase and even normalize with remission of CNS symptoms in disorders such as demyelinating disease, amyotrophic lateral sclerosis, mitochondrial encephalopathies, and HIV dementia. (Tsai and Coyle, 1995). NAA is synthesized within mitochondria, and inhibitors of the mitochondrial respiratory chain decrease NAA concentrations, effects that correlate with reductions in ATP and oxygen consumption (Manji et al., 2000a). Thus NAA is now generally regarded as a measure of neuronal viability and function, rather than strictly a marker for neuronal loss per se (for an excellent review on NAA see Tsai and Coyle, 1995). It was found that chronic lithium administration at therapeutic doses increases NAA concentration in the human brain in vivo (Moore et al., 2000a). These findings provide intriguing indirect support for the contention that, similar to the findings observed in the rodent brain and in human neuronal cells in culture, chronic lithium increases neuronal viability/function in the human brain. Furthermore, a striking ~0.97 correlation between lithium-induced NAA increases and regional voxel gray matter content was observed (Moore et al., 2000a), providing evidence for colocalization with the regional specific bcl-2 increases observed (e.g., gray vs. white matter) in the rodent brain cortices. These results suggest that chronic lithium may not only exert robust neuroprotective effects (as demonstrated in a variety of preclinical paradigms) but also exert neurotrophic effects in humans.

In follow-up studies to the NAA findings, it was hypothesized that in addition to increasing functional neurochemical markers of neuronal viability, lithium-induced increases in bcl-2 would also lead to neuropil increases and thus to increased brain gray matter volume in BD patients. In this clinical research investigation, brain tissue volumes were examined with high-resolution three-dimensional MRI and validated quantitative brain tissue segmentation methodology to identify and quantify the various components by volume, including total brain white and gray matter content. Measurements were made at baseline (medication free, after a minimum 14-day washout) and then repeated after 4 weeks of lithium at therapeutic doses. This study revealed that chronic lithium significantly increases total gray matter content in the human brain of patients with BD (Moore et al., 2000b) (Fig. 13.3). No significant changes were observed in brain white matter volume or in quantitative measures of regional cerebral water content, providing strong evidence that the observed increases in gray matter content are likely due to neurotrophic

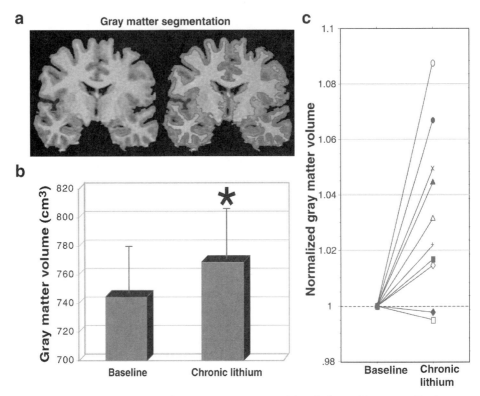

Figure 13.3. *Brain Gray Matter Increase After Lithium Treatment.* Brain gray matter volume is increased after 4 weeks of lithium administration at therapeutic levels in patients with BD. (a) A slice of three-dimensional volumetric magnetic resonance imaging (MRI) data that was segmented by tissue type with quantitative methodology to determine tissue volumes at each scan time point. Brain tissue volumes were examined with high-resolution three-dimensional MRI (124 1.5-mm-thick coronal T_1-weighted spoiled gradient echo images) and validated quantitative brain tissue segmentation methodology to identify and quantify the various components by volume, including total brain white and gray matter content. Measurements were made at baseline (medication free, after a minimum 14-day washout) and then repeated after 4 weeks of lithium at therapeutic doses. (b) & (c) Chronic lithium significantly increases total gray matter content in the human brain of patients with BD. No significant changes were observed in brain white matter volume or in quantitative measures of regional cerebral water. Modified and reproduced with permission from Moore et al. (2000b).

effects as opposed to any possible cell swelling and/or osmotic effects associated with lithium treatment. A finer-grained subregional analysis of this brain imaging data is ongoing and suggests that lithium produces a regionally selective increase in gray matter, with prominent effects being observed in hippocampus and caudate (unpublished observations).

CONCLUDING REMARKS: IMPLICATIONS FOR NEW MEDICATION DEVELOPMENT

As discussed above, there is a considerable body of evidence both conceptually and experimentally in support of the regulation of the PKC, PKA, and neurotrophic signaling cascades in the treatment (and potentially pathophysiology) of MD. Regulation of signal transduction within critical regions of the brain affects the intracellular signal generated by multiple neurotransmitter systems; these effects thus represent attractive putative mediators of the pathophysiology of MD and the therapeutic actions of antidepressants and mood stabilizers.

It is also becoming increasingly clear that for many refractory MD patients, new drugs simply mimicking many "traditional" drugs that directly or indirectly alter neurotransmitter levels and those that bind to cell surface receptors may be of limited benefit (Nestler, 1998). This is because such strategies implicitly assume that the target receptor(s)—and downstream signal mediators—are functionally intact and that altered synaptic activity will thus be transduced to modify the postsynaptic "throughput" of the system. However, the possible existence of abnormalities in signal transduction pathways suggests that for patients refractory to conventional medications improved therapeutics may only be obtained by the direct targeting of postreceptor sites. Recent discoveries concerning a variety of mechanisms involved in the formation and inactivation of second messengers offer the promise for the development of novel pharmacological agents designed to target signal transduction pathways (Guo et al., 2000) (Fig. 13.4).

Although clearly more complex than the development of receptor-specific drugs, it may be possible to design novel agents to selectively affect second messenger systems because they are quite heterogeneous at the molecular and cellular level, are linked to receptors in a variety of ways, and are expressed in different stoichiometries in different cell types (Manji and Duman, 2001). Additionally, because signal transduction pathways display certain unique characteristics depending on their activity state, they offer built-in targets for relative specificity of action, depending on the "set point" of the substrate. It is also noteworthy that a variety of strategies to enhance neurotrophic factor signaling are currently under investigation. An increasing number of strategies are being investigated to develop small molecule switches for protein-protein interactions, which have the potential to regulate the activity of growth factors, MAPK cascades, and interactions between homo- and heterodimers of the bcl-2 family of proteins (Guo et al., 2000); these developments hold much promise for the development of novel therapeutics for the long-term treatment of severe MDs, and for improving the lives of millions.

ACKNOWLEDGMENTS

The authors' research is supported by National Institute of Mental Health, the Theodore and Vada Stanley Foundation, and National Alliance for Research for Schizophrenia and Depression (NARSAD).

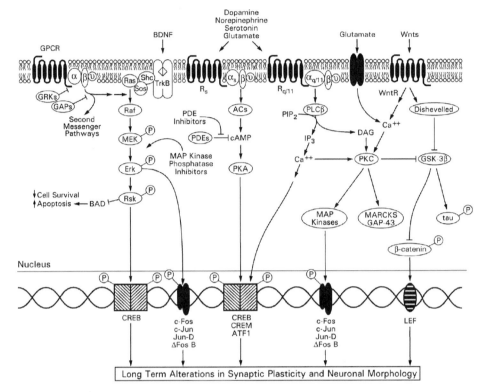

Figure 13.4. *Major Intracellular Signaling Pathways in Brain That Are Potential Future Targets for Medication Development.* The figure depicts some of the major intracellular signaling pathways involved in neural and behavioral plasticity. Cell surface receptors transduce extracellular signals such as neurotransmitters and neuropeptides into the interior of the cell. Most neurotransmitters and neuropeptides communicate with other cells by activating seven transmembrane spanning G protein-coupled receptors (GPCRs). As their name implies, GPCRs activate selected G proteins, which are composed of α and βγ subunits. Two families of proteins turn off the GPCR signal and may therefore represent attractive targets for new medication development. GPCR kinases (GRKs) phosphorylate GPCRs and thereby uncouple them from their respective G proteins. GTPase-activating proteins (GAPs, also called RGS or regulators of G protein signaling proteins) accelerate the G protein turn-off reaction (an intrinsic GTPase activity). Two major signaling cascades activated by GPCRs are the cAMP generating second messenger system and the phosphoinositide (PI) system. cAMP activates protein kinase A (PKA), a pathway that has been implicated in the therapeutic effects of antidepressants. Among the potential targets for the development of new antidepressants are certain phosphodiesterases (PDEs). PDEs catalyze the breakdown of cAMP; thus PDE inhibitors would be expected to sustain the cAMP signal and may represent an antidepressant augmenting strategy. Activation of receptors coupled to PI hydrolysis results in the breakdown of phosphoinositide 4,5-bisphosphate (PIP$_2$) into two second messengers—inositol 4,5-trisphosphate (IP$_3$) and DAG. IP$_3$ mobilizes Ca^{2+} from intracellular stores, whereas DAG is an endogenous activator of protein kinase C (PKC), which is also directly activated by Ca^{2+}. PKC, PKA, and other Ca^{2+}-dependent kinases directly or indirectly activate several important transcription

factors, including CREB, CREM, ATF-1, c-Fos, c-Jun, Jun-D, and ΔFos B. Endogenous growth factors such as BDNF utilize different types of signaling pathways. BDNF binds to and activates its tyrosine kinase receptor (TrkB); this facilitates the recruitment of other proteins (SHC, SOS), which results in the activation of the ERK-MAP kinase cascade (via sequential activation of Ras, Raf, MEK, Erk, and Rsk). In addition to regulating several transcription factors, the ERK-MAP kinase cascade, via Rsk, downregulates BAD, a proapoptotic protein. Enhancement of the ERK-MAP kinase cascade may have effects similar to those of endogenous neurotrophic factors; one potential strategy is to utilize inhibitors of MAP kinase phosphatases (which would inhibit the turn-off reaction) as potential drugs with neurotrophic properties. In addition to utilizing GPCRs, many neurotransmitters (e.g., glutamate and GABA) produce their responses via ligand-gated ion channels. Although these responses are very rapid, they also bring about more stable changes via regulation of gene transcription. One pathway gaining increasing recent attention in adult mammalian neurobiology is the Wnt signaling pathway. Wnts are a group of glycoproteins active in development but now known to play important roles in the mature brain. Binding of Wnts to the Wnt receptor (WntR) activates an intermediary protein, Disheveled, which regulates a glycogen synthase kinase (GSK-3β). GSK-3β exerts many cellular effects; it regulates cytoskeletal proteins, including tau and also plays an important role in determining cell survival/cell death decisions. GSK-3β has recently been identified as a target for lithium's actions. GSK-3β also regulates phosphorylation of β-catenin, a protein that when dephosphorylated acts as a transcription factor at LEF (lymphoid enhancer factor) sites. R_q and R_s, extracellular GPCRs coupled to stimulation or inhibition of adenylyl cyclases (ACs), respectively. Rq/11, GPCR coupled to activation of phospholipase C (PLC), MARCKS, myristoylated alanine-rich C kinase substrate, a protein associated with several neuroplastic events.

REFERENCES

Allen, S. J., Wilcock, G. K., and Dawbarn, D. (1999). Profound and selective loss of catalytic TrkB immunoreactivity in Alzheimer's disease. Biochem Biophys Res Commun *264*, 648–651.

Angst, J., and Sellaro, R. (2000). Historical perspectives and natural history of bipolar disorder. Biol Psychiatry *48*, 445–457.

Atkins, C. M., Selcher, J. C., Petraitis, J. J., Trzaskos, J. M., and Sweatt, J. D. (1998). The MAPK cascade is required for mammalian associative learning. Nat Neurosci *1*, 602–609

Baldessarini, R. J., Tondo, L., and Hennen, J. (1999). Effects of lithium treatment and its discontinuation on suicidal behavior in bipolar manic-depressive disorders. J Clin Psychiatry *60*, 77–84; discussion 111–116.

Baltuch, G. H., Couldwell, W. T., Villemure, J. G., and Yong, V. W. (1993). Protein kinase C inhibitors suppress cell growth in established and low-passage glioma cell lines. A comparison between staurosporine and tamoxifen. Neurosurgery *33*, 495–501.

Bebchuk, J. M., Arfken, C. L., Dolan-Manji, S., Murphy, J., Hasanat, K., and Manji, H. K. (2000). A preliminary investigation of a protein kinase C inhibitor in the treatment of acute mania. Arch Gen Psychiatry *57*, 95–97.

Bibel, M., Hoppe, E., and Barde, Y. A. (1999). Biochemical and functional inter-actions between the neurotrophin receptors trk and p75NTR. EMBO J *18*, 616–622.

Bijur, G. N., De Sarno, P., and Jope, R. S. (2000). Glycogen synthase kinase-3β facilitates staurosporine- and heat shock-induced apoptosis. Protection by lithium. J Biol Chem *275*, 7583–7590.

Blazer, D. G., Kessler, R. C., McGonagle, K. A., and Swartz, M. S. (1994). The prevalence and distribution of major depression in a national community sample: the National Comorbidity Survey. Am J Psychiatry *151*, 979–986.

Blum, S., Moore, A. N., Adams, F., and Dash, P. K. (1999). A mitogen-activated protein kinase cascade in the CA1/CA2 subfield of the dorsal hippocampus is essential for long-term spatial memory. J Neurosci *19*, 3535–3544

Bonni, A., Brunet, A., West, A. E., Datta, S. R., Takasu, M. A., and Greenberg, M. E. (1999). Cell survival promoted by the Ras-MAPK signaling pathway by tran-scription-dependent and -independent mechanisms. Science *286*, 1358–1362

Brunet, A., Bonni, A., Zigmond, M. J., Lin, M. Z., Juo, P., Hu, L. S., Anderson, M. J., Arden, K. C., Blenis, J., and Greenberg, M. E. (1999). Akt promotes cell sur-vival by phosphorylating and inhibiting a Forkhead transcription factor. Cell *96*, 857–868.

Brunet, A., Datta, S. R., and Greenberg, M. E. (2001). Transcription-dependent and -independent control of neuronal survival by the PI3K-Akt signaling pathway. Curr Opin Neurobiol *11*, 297–305.

Chen, G., Manji, H. K., Hawver, D. B., Wright, C. B., and Potter, W. Z. (1994). Chronic sodium valproate selectively decreases protein kinase C α and ε in vitro. J Neurochem *63*, 2361–2364.

Chen, G., Manji, H. K., Wright, C. B., Hawver, D. B., and Potter, W. Z. (1996). Effects of valproic acid on β-adrenergic receptors, G-proteins, and adenylyl cyclase in rat C6 glioma cells. Neuropsychopharmacology *15*, 271–280.

Chen, G., Rajkowska, G., Du, F., Seraji-Bozorgzad, N., and Manji HK. (2000a). Enhancement of hippocampal neurogenesis by lithium. J Neurochem *75*, 1729–1734.

Chen, G., Zeng, W. Z., Yuan, P. X., Huang, L. D., Jiang, Y. M., Zhao, Z. H., and ,Manji H. K. (1999). The mood-stabilizing agents lithium and valproate robustly increase the levels of the neuroprotective protein bcl-2 in the CNS. J Neurochem *72*, 879–882.

Chen, G. H., Einat, Yuan, P., and Manji, H. K. (2002). Evidence for the involve-ment of the ERK MAP kinase signaling cascade in mood modulation. Biol Psychiatry.

Chen, R. H., Ding, W. V., and McCormick, F. (2000b). Wnt signaling to β-catenin involves two interactive components. Glycogen synthase kinase-3β inhibition and activation of protein kinase C. J Biol Chem *275*, 17894–17899.

Chen, R. W., and Chuang, D. M. (1999). Long term lithium treatment suppresses p53 and Bax expression but increases Bcl-2 expression. A prominent role in neuroprotection against excitotoxicity. J Biol Chem *274*, 6039–6042.

Chen, S. J., Sweatt, J. D., and Klann, E. (1997). Enhanced phosphorylation of the postsynaptic protein kinase C substrate RC3/neurogranin during long-term potentiation. Brain Res *749*, 181–187.

Chuang, D. M., Chen, R., Chalecka-Franaszek, E., Ren, M., Hashimoto, R., Senatorov, V., Kanai, H., Hough, C., Hiroi, T., and Leeds, P. (2002).

Neuroprotective effects of lithium in cultured cells and animal model of disease. Bipolar Disorders *4*(2), 129–136.

Ciechanowski, P. S., Katon, W. J., and Russo, J. E. (2000). Depression and diabetes: impact of depressive symptoms on adherence, function, and costs. Arch Intern Med *160*, 3278–3285.

Conn, P. J., and Sweatt, J. D. (1994). Protein kinase C in the nervous system. In *Protein Kinase C*, New York: Oxford University Press. pp. 199–235.

Coull, M. A., Lowther, S., Katona, C. L., and Horton, R. W. (2000). Altered brain protein kinase C in depression: a post-mortem study. Eur Neuropsychopharmacol *10*, 283–288.

Cowburn, R. F., Marcusson, J. O., Eriksson, A., Wiehager, B., and O'Neill, C. (1994). Adenylyl cyclase activity and G-protein subunit levels in postmortem frontal cortex of suicide victims. Brain Res *633*, 297–304.

Crow, T., Xue-Bian, J. J., Siddiqi, V., Kang, Y., and Neary, J. T. (1998). Phosphorylation of mitogen-activated protein kinase by one-trial and multi-trial classical conditioning. J Neurosci *18*, 3480–3487.

Datta, S. R., Dudek, H., Tao, X., Masters, S., Fu, H., Gotoh, Y., and Greenberg, M. E. (1997). Akt phosphorylation of BAD couples survival signals to the cell-intrinsic death machinery. Cell *91*, 231–241.

Dijkers, P. F., Medema, R. H., Lammers, J. W., Koenderman, L., and Coffer, P. J. (2000). Expression of the pro-apoptotic Bcl-2 family member Bim is regulated by the forkhead transcription factor FKHR-L1. Curr Biol *10*, 1201–1204.

Dowlatshahi, D., MacQueen, G. M., Wang, J. F., and Young, L. T. (1998). Increased temporal cortex CREB concentrations and antidepressant treatment in major depression. Lancet *352*, 1754–1755.

Drevets, W. C. (2000). Functional anatomical abnormalities in limbic and prefrontal cortical structures in major depression. Prog Brain Res *126*, 413–431.

Drevets, W. C., Gadde, K., and Krishnan, R. (1999). Neuroimaging studies of depression. In *Neurobiology of Mental Illness*. New York, NY: Oxford University Press. pp. 394–418.

Drevets, W. C., Price, J. L., Simpson, J. R., Jr., Todd, R. D., Reich, T., Vannier, M., and Raichle, M. E. (1997). Subgenual prefrontal cortex abnormalities in mood disorders. Nature *386*, 824–827.

Du, J., Feng, L., Yang, F., and Lu, B. (2000). Activity- and Ca^{2+}-dependent modulation of surface expression of brain-derived neurotrophic factor receptors in hippocampal neurons. J Cell Biol *150*, 1423–1434.

Du, J., Feng, L., Zaizev, E., Liu, X., and Lu, B. (2003). Activity- and tyrosine kinase-dependent regulation of trkb receptors internalization in hippocampal neurons. J Cell Biol (under revision).

Du, J. F., Linyin, Zaisev, E., Liu, X., and Lu, B. (2003). Activity- and tyrosine kinase-dependent regulation of trkb receptors internalization in hippocampal neurons. In preparation.

Duman, R. S., Heninger, G. R., and Nestler, E. J. (1997). A molecular and cellular theory of depression. Arch Gen Psychiatry *54*, 597–606.

Duman, R. S., Malberg, J., Nakagawa, S., and D'Sa, C. (2000). Neuronal plasticity and survival in mood disorders. Biol Psychiatry *48*, 732–739.

Duman, R. S., Malberg, J., and Thome, J. (1999). Neural plasticity to stress and antidepressant treatment. Biol Psychiatry *46*, 1181–1191.

Ebstein, R. P., Lerer, B., Shapira, B., Shemesh, Z., Moscovich, D. G., and Kindler, S. (1988). Cyclic AMP second-messenger signal amplification in depression. Br J Psychiatry *152*, 665–669.

Eide, F. F., Vining, E. R., Eide, B. L., Zang, K., Wang, X. Y., and Reichardt, L. F. (1996). Naturally occurring truncated trkB receptors have dominant inhibitory effects on brain-derived neurotrophic factor signaling. J Neurosci *16*, 3123–3129.

English, J. D., and Sweatt, J. D. (1996). Activation of p42 mitogen-activated protein kinase in hippocampal long term potentiation. J Biol Chem *271*, 24329–24332.

English, J. D., and Sweatt, J. D. (1997). A requirement for the mitogen-activated protein kinase cascade in hippocampal long term potentiation. J Biol Chem *272*, 19103–19106.

Fava, M., and Kendler, K. S. (2000). Major depressive disorder. Neuron *28*, 335–341.

Fields, A., Li, P. P., Kish, S. J., and Warsh, J. J. (1999). Increased cyclic AMP-dependent protein kinase activity in postmortem brain from patients with bipolar affective disorder. J Neurochem. *73*, 1704–1710.

Finkbeiner. S. (2000). CREB couples neurotrophin signals to survival messages. Neuron *25*, 11–14.

Friedman. E., Hoau Yan. W., Levinson, D., Connell, T. A., and Singh, H. (1993). Altered platelet protein kinase C activity in bipolar affective disorder, manic episode. Biol Psychiatry *33*, 520–525.

Fryer, R. H., Kaplan, D. R., Feinstein, S. C., Radeke, M. J., Grayson, D. R., and Kromer, L. F. (1996). Developmental and mature expression of full–length and truncated TrkB receptors in the rat forebrain. J Comp Neurol *374*, 21–40.

Fukunaga, K., and Miyamoto, E. (1998). Role of MAP kinase in neurons. Mol Neurobiol *16*, 79–95.

Garcia-Sevilla, J. A., Escriba, P. V., and Guimon, J. (1999). Imidazoline receptors and human brain disorders. Ann N Y Acad Sci *881*, 392–409.

Garcia-Sevilla, J. A., Walzer, C., Busquets, X., Escriba, P. V., Balant, L., and Guimon, J. (1997). Density of guanine nucleotide-binding proteins in platelets of patients with major depression: increased abundance of the G α i2 subunit and down-regulation by antidepressant drug treatment. Biol Psychiatry *42*, 704–712.

Gonzalez, M., Ruggiero, F. P., Chang, Q., Shi, Y. J., Rich, M. M., Kraner, S., and Balice-Gordon, R. J. (1999). Disruption of Trkb-mediated signaling induces disassembly of postsynaptic receptor clusters at neuromuscular junctions. Neuron *24*, 567–583.

Goodwin, F. K., and Jamison, K. R. (1990). Manic-Depressive Illness. New York: Oxford University Press.

Gould, T. D., and Manji, H. K. (2002). Signaling networks in the pathophysiology & treatment of mood disorders. J Psychosom Res *53*(2), 687–697.

Grimes, C. A, and Jope, R. S. (2001). The multifaceted roles of glycogen synthase kinase 3β in cellular signaling. Prog Neurobiol *65*, 391–426.

Guo, Z., Zhou, D., and Schultz P. G. (2000). Designing small-molecule switches for protein-protein interactions. Science *288*, 2042–2045.

Hahn, C. G., and Friedman, E. (1999). Abnormalities in protein kinase C signaling and the pathophysiology of bipolar disorder. Bipolar Disord *1*, 81–86.

Halper, J. P., Brown, R. P., Sweeney, J. A., Kocsis, J. H., Peters, A., and Mann, J. J. (1988). Blunted β-adrenergic responsivity of peripheral blood mononuclear cells in endogenous depression. Isoproterenol dose-response studies. Arch Gen Psychiatry *45*, 241–244.

Hamanoue, M., Middleton, G., Wyatt, S., Jaffray, E., Hay, R. T., and Davies, A. M. (1999). p75-mediated NF-κB activation enhances the survival response of developing sensory neurons to nerve growth factor. Mol Cell Neurosci *14*, 28–40.

Hetman, M., Cavanaugh, J. E., Kimelman, D., and Xia, Z. (2000). Role of glycogen synthase kinase-3β in neuronal apoptosis induced by trophic withdrawal. J Neurosci *20*, 2567–2574.

Huang, Z. J., Kirkwood, A., Pizzorusso, T., Porciatti, V., Morales, B., Bear, M. .F, Maffei, L., and Tonegawa, S. (1999). BDNF regulates the maturation of inhibition and the critical period of plasticity in mouse visual cortex. Cell *98*, 739–755.

Jamison, K. R. (1986). Suicide and bipolar disorders. Ann NY Acad Sci *487*, 301–315.

Jope, R. S. (1999). Anti-bipolar therapy: mechanism of action of lithium. Mol Psychiatry *4*, 117–128.

Jope, R. S., and Williams, M. B. (1994). Lithium and brain signal transduction systems. Biochem Pharmacol *47*, 429–441.

Kaplan, D. R., and Miller, F. D. (2000). Neurotrophin signal transduction in the nervous system. Curr Opin Neurobiol *10*, 381–391.

Kornhauser, J. M., and Greenberg, M. E. (1997). A kinase to remember: dual roles for MAP kinase in long-term memory. Neuron *18*, 839–842.

Korte, M., Griesbeck, O., Gravel, C., Carroll, P., Staiger, V., Thoenen, H., and Bonhoeffer, T. (1996). Virus-mediated gene transfer into hippocampal CA1 region restores long- term potentiation in brain-derived neurotrophic factor mutant mice. Proc Natl Acad Sci USA *93*, 12547–12552.

Lenox, R. H., and Hahn C. G. (2000). Overview of the mechanism of action of lithium in the brain: fifty-year update. J Clin Psychiatry *61* Suppl 9, 5–15.

Lenox, R. H., Watson, D. G., Patel, J., and Ellis, J. (1992). Chronic lithium administration alters a prominent PKC substrate in rat hippocampus. Brain Res *570*, 333–340.

Lessmann, V., and Heumann, R. (1998). Modulation of unitary glutamatergic synapses by neurotrophin-4/5 or brain-derived neurotrophic factor in hippocampal microcultures: presynaptic enhancement depends on pre-established paired-pulse facilitation. Neuroscience *86*, 399–413.

Li, M., Wang, X., Meintzer, M. K., Laessig, T., Birnbaum, M. J., and Heidenreich, K. A. (2000). Cyclic AMP promotes neuronal survival by phosphorylation of glycogen synthase kinase 3β. Mol Cell Biol *20*, 9356–9363.

Lowther, S., Crompton, M. R., Katona, C. L., and Horton, R. W. (1996). GTP γS and forskolin-stimulated adenylyl cyclase activity in post-mortem brain from depressed suicides and controls. Mol Psychiatry *1*, 470–477.

Lu, R., Song, L., and Jope, R. S. (1999). Lithium attenuates p53 levels in human neuroblastoma SH-SY5Y cells. Neuroreport *10*, 1123–1125.

Maggirwar, S B., Sarmiere, P. D., Dewhurst, S., and Freeman, R, S. (1998). Nerve growth factor-dependent activation of NF-κB contributes to survival of sympathetic neurons. J Neurosci *18*, 10356–10365.

Manji, H. K., Bebchuk, J. M., Moore, G. J., Glitz, D., Hasanat, K. A., and Chen, G. (1999a). Modulation of CNS signal transduction pathways and gene expression by mood-stabilizing agents: therapeutic implications. J Clin Psychiatry 60, 27–39; discussion 40–41, 113–116.

Manji, H. K., and Chen, G. (2000). Post-receptor signaling pathways in the pathophysiology and treatment of mood disorders. Curr Psychiatry Rep 2, 479–489.

Manji, H. K., Chen, G., Hsiao, J. K., Masana, M. I., Moore, G. J., and Potter, W. Z. (2000a). Regulation of signal transduction pathways by mood stabilizing agents: implications for the pathophysiology and treatment of bipolar affective disorder. In Bipolar Medications: Mechanisms of Action. Washington, DC: American Psychiatric Press. pp. 129–177.

Manji, H. K., Chen, G., Shimon, H., Hsiao, J. K., Potter, W. Z., and Belmaker, R. H. (1995a). Guanine nucleotide-binding proteins in bipolar affective disorder. Effects of long-term lithium treatment. Arch Gen Psychiatry 52, 135–144.

Manji, H. K., Drevets, W. C., and Charney, D. S. (2001a). The cellular neurobiology of depression. Nat Med 7, 541–547.

Manji, H. K., and Duman, R. S. (2001). Impairments of neuroplasticity and cellular resilience in severe mood disorders: implications for the development of novel therapeutics. Psychopharmacol Bull 35, 5–49.

Manji, H. K., Etcheberrigaray, R., Chen, G., and Olds, J. L. (1993). Lithium decreases membrane-associated protein kinase C in hippocampus: selectivity for the α isozyme. J Neurochem 61, 2303–2310.

Manji, H. K., and Lenox, R. H. (1999). Ziskind-Somerfeld Research Award. Protein kinase C signaling in the brain: molecular transduction of mood stabilization in the treatment of manic-depressive illness. Biol Psychiatry 46, 1328–1351.

Manji, H. K., and Lenox, R. H. (2000a). The nature of bipolar disorder. J Clin Psychiatry 61, 42–57.

Manji, H. K., and Lenox, R. H. (2000b). Signaling: cellular insights into the pathophysiology of bipolar disorder. Biol Psychiatry 48, 518–530.

Manji, H. K., McNamara, R., Chen, G., and Lenox, R. H. (1999b). Signalling pathways in the brain: cellular transduction of mood stabilisation in the treatment of manic-depressive illness. Aust NZ J Psychiatry 33 Suppl, S65–S83.

Manji, H. K., Moore, G. J., and Chen, G. (1999c). Lithium at 50: have the neuroprotective effects of this unique cation been overlooked? Biol Psychiatry 46, 929–940.

Manji, H. K., Moore, G. J., and Chen, G. (2000b). Clinical and preclinical evidence for the neurotrophic effects of mood stabilizers: implications for the pathophysiology and treatment of manic-depressive illness. Biol Psychiatry 48, 740–754.

Manji, H. K., Moore, G. J., and Chen, G. (2000c). Lithium up-regulates the cytoprotective protein Bcl-2 in the CNS in vivo: a role for neurotrophic and neuroprotective effects in manic depressive illness. J Clin Psychiatry 61, 82–96.

Manji, H. K., Moore, G. J., and Chen, G. (2001b). Bipolar disorder: leads from the molecular and cellular mechanisms of action of mood stabilisers. Br J Psychiatry 178, S107–S119.

Manji, H. K., Potter, W. Z,, and Lenox, R. H. (1995b). Signal transduction pathways. Molecular targets for lithium's actions. Arch Gen Psychiatry 52, 531–543.

Mann, J. J., Brown, R. P., Halper, J. P., Sweeney, J A., Kocsis, J. H., Stokes, P. E., and Bilezikian, J. P. (1985). Reduced sensitivity of lymphocyte β-adrenergic receptors in patients with endogenous depression and psychomotor agitation. N Engl J Med *313*, 715–720.

Martin, K. C., Michael, D., Rose, J. C., Barad, M., Casadio, A., Zhu, H., and Kandel, E. R. (1997). MAP kinase translocates into the nucleus of the presynaptic cell and is required for long-term facilitation in *Aplysia*. Neuron *18*, 899–912.

Matsubara, M., Kusubata, M., Ishiguro, K., Uchida, T., Titani, K., and Taniguchi, H. (1996). Site-specific phosphorylation of synapsin I by mitogen-activated protein kinase and Cdk5 and its effects on physiological functions. J Biol Chem *271*, 21108–21113.

Meyer-Franke, A., Wilkinson, G. A, Kruttgen, A., Hu, M., Munro, E., Hanson, M. G., Jr., Reichardt, L. F., and Barres, B. A. (1998). Depolarization and cAMP elevation rapidly recruit TrkB to the plasma membrane of CNS neurons. Neuron *21*, 681–693.

Michelson, D., Stratakis, C., Hill, L., Reynolds, J., Galliven, E., Chrousos, G., and Gold, P. (1996). Bone mineral density in women with depression. N Engl J Med *335*, 1176–1181.

Moore, G. J., Bebchuk, J. M., Hasanat, K., Chen, G., Seraji-Bozorgzad, N., Wilds, I. B., Faulk, M. W., Koch, S., Glitz, D. A., Jolkovsky, L., and Manji, H. K. (2000a). Lithium increases *N*-acetyl-aspartate in the human brain: in vivo evidence in support of bcl-2's neurotrophic effects? Biol Psychiatry *48*, 1–8.

Moore, G. J., Bebchuk, J. M., Parrish, J. K., Faulk, M. W., Arfken, C. L., Strahl-Bevacqua, J., and Manji HK. (1999). Temporal dissociation between lithium-induced changes in frontal lobe myo-inositol and clinical response in manic-depressive illness. Am J Psychiatry *156*, 1902–1908.

Moore, G. J., Bebchuk, J. M., Wilds, I. B., Chen, G., and Manji, H. K. (2000b). Lithium-induced increase in human brain grey matter. Lancet 356, 1241–2.

Mork, A., Geisler, A., and Hollund, P. (1992). Effects of lithium on second messenger systems in the brain. Pharmacol Toxicol *71*, 4–17.

Muller-Oerlinghausen, B., Ahrens, B., Grof, E., Grof, P., Lenz, G., Schou, M., Simhandl, C,, Thau, K., Volk, J., Wolf, R. et al. (1992). The effect of long-term lithium treatment on the mortality of patients with manic-depressive and schizoaffective illness. Acta Psychiatr Scand *86*, 218–222.

Murphy, T. H., Blatte,r L. A., Bhat, R. V., Fiore, R. S., Wier, W. G., and Baraban, J. M. (1994). Differential regulation of calcium/calmodulin-dependent protein kinase II and p42 MAP kinase activity by synaptic transmission. J Neurosci *14*, 1320–1331.

Murray, C. J., and Lopez, A. D. (1997). Alternative projections of mortality and disability by cause 1990–2020: Global Burden of Disease Study. Lancet *349*, 1498–1504.

Musselman, D. L., Evans, D. L., and Nemeroff, C. B. (1998). The relationship of depression to cardiovascular disease: epidemiology, biology, and treatment. Arch Gen Psychiatry *55*, 580–592.

Nagatsu, T., Mogi, M., Ichinose, H., and Togari, A. (2000). Changes in cytokines and neurotrophins in Parkinson's disease. J Neural Transm Suppl, 277–290.

Nestler, E. J. (1998). Antidepressant treatments in the 21st century. Biol Psychiatry *44*, 526–533.

Nestler, E. J, Terwilliger, R. Z, and Duman, R. S. (1989). Chronic antidepressant administration alters the subcellular distribution of cyclic AMP-dependent protein kinase in rat frontal cortex. J Neurochem *53*, 1644–1647.

Ni, C., Murphy, P., Golde, T. E., and Carpenter, G. (2001). γ-Secretase cleavage and nuclear localization of ErbB-4 receptor tyrosine kinase. Science *294*, 2179–2181.

Nibuya, M., Morinobu, S., and Duman RS. (1995). Regulation of BDNF and trkB mRNA in rat brain by chronic electroconvulsive seizure and antidepressant drug treatments. J Neurosci *15*, 7539–7547.

Nibuya, M., Nestler, E. J., and Duman, R S. (1996). Chronic antidepressant administration increases the expression of cAMP response element binding protein (CREB) in rat hippocampus. J Neurosci *16*, 2365–2372.

Nonaka, S., and Chuang, D. M. (1998). Neuroprotective effects of chronic lithium on focal cerebral ischemia in rats. Neuroreport *9*, 2081–2084.

Ozaki, N., and Chuang, D. M. (1997). Lithium increases transcription factor binding to AP-1 and cyclic AMP-responsive element in cultured neurons and rat brain. J Neurochem *69*, 2336–2344.

Pandey, G. N., Conley, R. R., Pandey, S. C., Goel, S., Roberts, R. C., Tamminga, C. A., Chute, D., and Smialek, J. (1997). Benzodiazepine receptors in the postmortem brain of suicide victims and schizophrenic subjects. Psychiatry Res *71*, 137–149.

Patapoutian. A., and Reichardt, L. F. (2001). Trk receptors: mediators of neurotrophin action. Curr Opin Neurobiol *11*, 272–280.

Patterson, S. L., Abel, T., Deuel, T A., Martin, K. C., Rose, J. C., and Kandel, E. R. (1996). Recombinant BDNF rescues deficits in basal synaptic transmission and hippocampal LTP in BDNF knockout mice. Neuron *16*, 1137–1145.

Perez, J., Tardito, D., Mori, S., Racagni, G., Smeraldi, E., and Zanardi, R. (2000). Abnormalities of cAMP signaling in affective disorders: implication for pathophysiology and treatment. Bipolar Disord *2*, 27–36.

Pettmann, B., and Henderson, C. E. (1998). Neuronal cell death. Neuron *20*, 633–647.

Pierce, R. C., and Bari, A. A. (2001). The role of neurotrophic factors in psychostimulant-induced behavioral and neuronal plasticity. Rev Neurosci *12*, 95–110.

Poo, M. M. (2001). Neurotrophins as synaptic modulators. Nat Rev Neurosci *2*, 24–32.

Popoli, M., Brunello, N., Perez, J., and Racagni, G. (2000). Second messenger-regulated protein kinases in the brain: their functional role and the action of antidepressant drugs. J Neurochem *74*, 21–33.

Pozzo-Miller, L. D., Gottschalk, W., Zhang, L., McDermott, K., Du, J., Gopalakrishnan, R., Oho, C., Sheng, Z. H., and Lu, B. (1999). Impairments in high-frequency transmission, synaptic vesicle docking, and synaptic protein distribution in the hippocampus of BDNF knockout mice. J Neurosci *19*, 4972–4983.

Rahman, S., Li, P. P., Young, L. T., Kofman, O., Kish, S. J., and Warsh, J. J. (1997). Reduced [^{3}H]cyclic AMP binding in postmortem brain from subjects with bipolar affective disorder. J Neurochem *68*, 297–304.

Rajkowska, G. (2000a). Histopathology of the prefrontal cortex in major depression: what does it tell us about dysfunctional monoaminergic circuits? Prog Brain Res *126*, 397–412.

Rajkowska, G. (2000b). Postmortem studies in mood disorders indicate altered numbers of neurons and glial cells. Biol Psychiatry *48*, 766–777.

Ram, A., Guedj, F., Cravchik, A., Weinstein, L., Cao, Q., Badner, J. A., Goldin, L. R., Grisaru, N., Manji, H. K., Belmaker, R. H., Gershon, E. S., and Gejman, P. V. (1997). No abnormality in the gene for the G protein stimulatory α subunit in patients with bipolar disorder. Arch Gen Psychiatry *54*, 44–48.

Rasenick, M. M., Chen, J., and Ozawa, H. (2000). Effects of antidepressant treatments on the G protein-adenylyl cyclase axis as the possible basis of therapeutic action. In Bipolar Medications: Mechanisms of Action, ed. Washington, DC: American Psychiatric Press. pp. 87–108.

Riccio, A., Pierchala, B. A., Ciarallo, C. L., and Ginty, D. D. (1997). An NGF-TrkA-mediated retrograde signal to transcription factor CREB in sympathetic neurons. Science *277*, 1097–1100.

Risby, E. D., Hsiao, J. K., Manji, H. K., Bitran, J., Moses, F., Zhou, D. F., and Potter, W. Z. (1991). The mechanisms of action of lithium. II. Effects on adenylate cyclase activity and β-adrenergic receptor binding in normal subjects. Arch Gen Psychiatry *48*, 513–524.

Roberson, E. D., English, J. D., Adams, J. P., Selcher, J. C., Kondratick, C., and Sweatt, J. D. (1999). The mitogen-activated protein kinase cascade couples PKA and PKC to cAMP response element binding protein phosphorylation in area CA1 of hippocampus. J Neurosci *19*, 4337–4348.

Roberson, E. D., English, J. D., and Sweatt, J. D. (1996). A biochemist's view of long-term potentiation. Learn Mem *3*, 1–24.

Robinson, M. J., Stippec, S. A., Goldsmith, E., White, M. A., and Cobb, M. H. (1998). A constitutively active and nuclear form of the MAP kinase ERK2 is sufficient for neurite outgrowth and cell transformation. Curr Biol *8*, 1141–1150

Salehi, A., Verhaagen, J., and Swaab, D. F. (1998). Neurotrophin receptors in Alzheimer's disease. Prog Brain Res *117*, 71–89.

Schou, M. (1999). Perspectives on lithium treatment of bipolar disorder: action, efficacy, effect on suicidal behavior. Bipolar Disord *1*, 5–10.

Schreiber, G., Avissar, S., Danon, A., and Belmaker, R. H. (1991). Hyperfunctional G proteins in mononuclear leukocytes of patients with mania. Biol Psychiatry *29*, 273–280.

Schulz, C., Mavrogiorgou, P., Schroter, A., Hegerl, U., and Juckel, G. (2000). Lithium-induced EEG changes in patients with affective disorders. Neuropsychobiology *42* Suppl 1, 33–37.

Segal, R. A., and Greenberg, M. E. (1996). Intracellular signaling pathways activated by neurotrophic factors. Annu Rev Neurosci *19*, 463–489.

Siever, L J. (1987). Role of noradrenergic mechanisms in the etiology of the affective disorders. In Psychopharmacology : The Third Generation of Progress. New York: Raven Press. pp. 493–504.

Siuciak, J. A., Lewis, D. R., Wiegand, S. J., and Lindsay, R. M. (1997). Antidepressant-like effect of brain-derived neurotrophic factor (BDNF). Pharmacol Biochem Behav *56*, 131–137.

Spleiss, O., van Calker, D., Scharer, L., Adamovic, K., Berger, M., Gebicke-Haerter, P. J. (1998). Abnormal G protein α_s- and α_{i2}-subunit mRNA expression in bipolar affective disorder. Mol Psychiatry *3*, 512–520.

Stahl, S. M. (2000). Essential Psychopharmacology: Neuroscientific Basis and Practical Applications. Cambridge, U.K. Cambridge University Press.

Stein, M. B., Chen, G., Potter, W. Z., and Manji, H. K. (1996). G-protein level quantification in platelets and leukocytes from patients with panic disorder. Neuropsychopharmacology *15*, 180–186.

Sullivan, P. F., Neale, M. C., and Kendler, K. S. (2000). Genetic epidemiology of major depression: review and meta-analysis. Am J Psychiatry *157*, 1552–1562.

Thoenen, H. (1995). Neurotrophins and neuronal plasticity. Science *270*, 593–598.

Tondo, L., Jamison, K. R., and Baldessarini, R. J. (1997). Effect of lithium maintenance on suicidal behavior in major mood disorders. Ann NY Acad Sci *836*, 339–351.

Tsai, G., and Coyle, J. T. (1995). *N*-acetylaspartate in neuropsychiatric disorders. Prog Neurobiol *46*, 531–540.

Vestergaard, P., and Aagaard, J. (1991). Five-year mortality in lithium-treated manic-depressive patients. J Affect Disord *21*, 33–38.

von Bartheld, C. S., Williams, R., Lefcort, F., Clary, D. O., Reichardt, L. F., and Bothwell, M. (1996). Retrograde transport of neurotrophins from the eye to the brain in chick embryos: roles of the p75NTR and trkB receptors. J Neurosci *16*, 2995–3008.

Wang, H. Y., and Friedman, E. (1996). Enhanced protein kinase C activity and translocation in bipolar affective disorder brains. Biol Psychiatry *40*, 568–575.

Wang, H Y., Markowitz, P., Levinson, D., Undie, A. S., and Friedman, E. (1999a). Increased membrane-associated protein kinase C activity and translocation in blood platelets from bipolar affective disorder patients. J Psychiatr Res *33*, 171–179.

Wang, J.-F., Young, L. T., Li, P. P., and Warsh, J. J. (1997). Signal transduction abnormalities in bipolar disorder. In Bipolar Disorder: Biological Models and Their Clinical Application. New York: Dekker. pp. 41–79.

Wang, J. F., Asghari, V., Rockel, C., and Young, L. T. (1999b). Cyclic AMP responsive element binding protein phosphorylation and DNA binding is decreased by chronic lithium but not valproate treatment of SH-SY5Y neuroblastoma cells. Neuroscience *91*, 771–776.

Warsh, J. J., Young, L. T., and Li, P. P. (2000). Guanine nucleotide binding (G) protein disturbances in bipolar affective disorder. In Bipolar Medications : Mechanisms of Action. Washington, DC: American Psychiatric Press. pp. 299–329.

Watson, D. G., Watterson, J. M., and Lenox, R. H. (1998). Sodium valproate downregulates the myristoylated alanine-rich C kinase substrate (MARCKS) in immortalized hippocampal cells: a property of protein kinase C-mediated mood stabilizers. J Pharmacol Exp Ther *285*, 307–316.

Woodgett, J. R. (2001). Judging a protein by more than its name: gsk-3. Sci STKE *2001*, RE12.

Xu, Q., Fawcett, T. W., Gorospe, M., Guyton, K. Z., Liu, Y., and Holbrook, N. J. (1997). Induction of mitogen-activated protein kinase phosphatase-1 during acute hypertension. Hypertension *30*, 106–111.

Young, L. T., Li, P. P., Kish, S. J., Siu, K. P., Kamble, A., Hornykiewicz, O., and Warsh, J. J. (1993). Cerebral cortex Gs α protein levels and forskolin-stimulated cyclic AMP formation are increased in bipolar affective disorder. J Neurochem *61*, 890–898.

Yuan, D., Komatsu, K., Tani, H., Cui, Z., and Kano, Y. (1998). Pharmacological properties of traditional medicines. XXIV. Classification of antiasthmatics based on constitutional predispositions. Biol Pharm Bull *21*, 1169–1173.

Yuan, P. X., Huang, L. D., Jiang, Y. M., Gutkind, J. S., Manji, H. K., and Chen, G. (2001). The mood stabilizer valproic acid activates mitogen-activated protein kinases and promotes neurite growth. J Biol Chem *276*, 31674–31683.

CHAPTER 14

INHIBITING SIGNALING PATHWAYS THROUGH RATIONAL DRUG DESIGN

JAMES N. TOPPER and NEILL A. GIESE

Millennium Pharmaceuticals, Inc., South San Francisco, California

INTRODUCTION

Two of the biological disciplines that have had the most profound impact on the process of drug development are genomics and signal transduction research, the former by providing a comprehensive list of defined targets toward which putative agonists or antagonists may be directed and the latter by providing an understanding of how signals and information flow within cells and, by inference, how these may be manipulated to alter the physiology of the cell. Despite the rapid progress being achieved in both of these areas, the process of drug discovery and development continues to pose significant challenges. This chapter outlines some of the general issues that surround this process at its various stages and how these have been addressed successfully. This is followed by several examples of ongoing drug development efforts that are attempting to exploit advances in signal transduction research in specific pathways that have been implicated in human disease.

RATIONAL DRUG DESIGN

The phrase "rational drug design" has been used in a variety of contexts, but for the purposes of this discussion we will define it as the rational choice of a target toward which a drug will be directed, followed by a rational approach toward the development of this therapeutic reagent. The choice of the target is the initial and most important step in this process. In this context, scientists often refer to entities as "validated targets." In the strictest sense a validated target is a gene, or protein,

Signal Transduction and Human Disease, Edited by Toren Finkel and J. Silvio Gutkind

whose function or dysfunction can be selectively modulated by a therapeutic intervention (such as a drug), resulting in a favorable clinical response. An example would be the angiotensin-converting enzyme (ACE), whose specific inhibition by several drugs has resulted in marked improvements in the clinical outcomes of patients suffering from congestive heart failure. Although there are many validated targets in a variety of human diseases, by definition, novel therapeutics are developed against targets for which no such validation exists. A much more typical scenario will involve the implication of a specific target, or signaling pathway, in the pathogenesis of a disease state because of its behavior in an experimental model of the disease. For example, a growth factor receptor whose overexpression, or activation in cultured cells, results in oncogenic transformation and tumor formation when injected into experimental animals may be an attractive target for the treatment of human tumors that overexpress this same receptor.

In addition to the biology of the target, the "drugability" of the target must be considered as part of the selection process. In the case of small-molecule drugs (defined here as the products of synthetic medicinal chemistry efforts), there is a relatively short list of proteins whose functions have been successfully modulated. These include entities such as G protein-coupled receptors, protein kinases, proteases, enzymes of intermediary metabolism, and nuclear receptors. In contrast to these proteins, there are many protein classes such as transcriptional regulators, DNA-modifying enzymes, and structural proteins that represent attractive targets because of their roles in pathophysiological processes but whose function has not yet been effectively modulated by synthetic small molecules.

Once a target is selected, the most widely utilized approach toward the development of an inhibitor is the screening of large chemical libraries. This strategy involves the creation of an assay for the function of the target and the application of hundreds of thousands of test compounds to this assay to assess the ability of these to inhibit the function of the target. An assay may be as simple as a recombinantly expressed purified protein acting on a synthetic substrate or as complex as an engineered cell line designed to overexpress a particular protein. Many considerations go into the design and selection of a screening system. Two of the more important considerations are the validity of the assay (i.e., does it really assess the true function of the target as it may exist in vivo) and the throughput and efficiency of the screening process. Enormous progress has been made in this area, and with modern automation technologies almost any target can be effectively screened against large collections of compounds and inhibitors identified. An inhibitor identified by a screen (typically known as a "hit" or "lead") rarely represents a viable drug, however. This is because it typically does not demonstrate the requisite specificity (i.e., it may inhibit many other targets in addition to the one of interest) or have the required potency against the target that would enable the compound to be safe and effective in humans. Thus the next step in the process is known as lead or hit optimization. Although a meaningful discussion of this process is beyond the scope of this chapter, in its simplest form, this process represents an iterative process whereby the medicinal chemists systematically modify the structure of the lead compound to optimize its potency, specificity, and pharmaceutical properties (e.g., oral bioavailability and effective half-life in vivo). This process can take years to accomplish and may not always be successful. However, recent advances in structural biology and associated computational approaches are promising to enhance the speed and efficiency of this process.

TRANSFORMING GROWTH FACTOR β SIGNALING AS A MODEL FOR A SIGNALING CASCADE AMENABLE TO DRUG DEVELOPMENT

Transforming growth factor β (TGF-β) is a prototypical member of a larger family of growth factors in humans that include TGF-βs, bone morphogenetic proteins (BMPs) and activins. The actions of these growth factors have been implicated in a wide variety of pathophysiological processes such as fibroproliferative diseases of the lung, kidney, and liver, arterial vascular disease, and certain malignancies (Massague, 1996). For example, TGF-β is thought to play a central role in the pathogenesis of diabetic nephropathy. This disorder is thought to arise, in part, from the induction of TGF-β production by glomerular mesangial cells of the kidney in response to excessive hyperglycemia. This growth factor then acts as a potent stimulus for the production of excessive extracellular matrix by cells within the glomerulus, which ultimately leads to glomerular sclerosis/fibrosis and a progressive decline in renal function. Inhibition of the actions of TGF-β in experimental models of diabetic renal disease (via the application of decoy receptors or inhibitory antibodies) has resulted in significant improvements in renal function in several of these models. Thus inhibition of TGF-β may be an attractive strategy for the therapy of diabetic renal disease and other disorders in humans.

Over the last several years the detailed intracellular signaling mechanisms utilized by this family of effectors have begun to be elucidated (Massague, 1996; Baker and Harland, 1997). In the case of TGF-β_1, its cellular effects appear to be transduced via at least three types of cell surface receptors (types I, II, III), two of which are serine/threonine kinases. The active form of soluble TGF-β binds to the type II receptor at the cell surface, and this complex subsequently interacts with and transphosphorylates the cytoplasmic domain of the type I receptor. This phosphorylation event activates the type I receptor kinase domain, which then propagates downstream signals within the cell. This process can be modulated by the presence of a type III receptor in most cells. Although the precise role of the type III receptor(s) is unclear, the function of the other receptors is well supported by in vitro data demonstrating that the type II receptor is required for ligand (TGF-β) binding whereas the type I receptor, once activated, is responsible for initiating intracellular signaling.

Once activated, the type I receptor can specifically interact with a class of intracellular proteins known as Smad proteins. These proteins act as substrates for the kinase domain present within the type I receptor, and phosphorylation of the Smads results in their activation and translocation to the nucleus, where they act as transcriptional effectors on a variety of target genes. For example, in response to TGF-β_1, the type 1 receptor known as TBR1 or ALK5 is activated and this activated kinase specifically phosphorylates a set of proteins known as Smad2 and Smad3. These proteins then can dissociate from the activated receptor kinase, interact with a variety of other signaling proteins (including another Smad known as Smad4). and ultimately translocate to the nucleus. There they modulate the expression of a variety of genes including those encoding extracellular matrix components such as several collagen isoforms and fibronectin. Although a variety of other intracellular signaling cascades have been implicated in TGF-β signaling, the importance of the receptor kinases and the Smad proteins is highlighted by the fact that mutations in the TGF-β type II receptor as well as Smad2 and Smad4 have been causally linked to specific human malignancies and that disruption of many of these genes in the mouse results in early embryonic lethality (Zhou et al., 1998).

Importantly, there are a variety of lines of evidence that point to the activation of the type I receptor kinase activity as a critical step in this signaling system (Huse et al., 1999). For example, overexpression of a mutated type I receptor that is constitutively activated (and thus does not require ligand or type II receptor activity) has been shown to effectively initiate signaling and mimic the actions of TGF-β itself. Conversely, there is a unique class of Smad proteins (characterized by Smad6 and Smad7) that are capable of selectively binding to, and inhibiting the action of, activated type I receptors (Hayashi et al., 1997; Topper et al., 1997; Ulloa et al., 1999). When these proteins are induced, or overexpressed, virtually all of the actions of TGF- β on cells are inhibited. Thus, given the importance of excessive or dysregulated actions of TGF-β in many disease states and the data supporting the critical role for the type I receptor kinase in this signaling pathway, this receptor represents a very attractive molecular target for therapeutic inhibition.

To screen for inhibitors of TGF-β signaling we have set up a series of assay systems. For example, we have created several cell lines that harbor TGF-β/Smad-responsive promoters linked to reporter genes such as luciferase. When these cells are treated with TGF-β, they activate the transcription of these constructs and an increase in luciferase activity can be easily detected. A strength of this assay as a screen for the detection of putative inhibitors is that it is dependent on the endogenous TGF-β signaling components within the cell that are present in their native state. However, because this assay represents a complex system it also has the potential to identify inhibitors that target general cellular processes, many of which may not be selective to the TGF-β pathway. In fact, this screening assay did result in the identification of numerous inhibitory compounds that did not appear to be selective to the TGF-β signaling cascade (such as nonspecific inhibitors of transcription or protein synthesis). For this reason, a second set of assays designed to focus on the activity of the type I receptor itself were developed. These assays utilize recombinantly expressed TBR1 (ALK5) in a purified kinase assay that monitors the ability of this kinase to phosphorylate specific substrates such as Smad proteins. This assay represents a more straightforward assay and has allowed us to screen a large number of compounds for inhibitory activity. To date, these activities have resulted in the identification of a series of small molecules that are capable of inhibiting TBR1 kinase activity. When these compounds are tested for their ability to inhibit kinases of several other classes, such as non-receptor tyrosine kinases (e.g., fyn kinase) or serine threonine kinases (e.g., PKC) they are not active. Furthermore, when several of these inhibitors are applied to cultured cells, they are capable of inhibiting endogenous Smad protein phosphorylation in response to exogenous TGF-β. These results suggest that a selective inhibitor of the TGF-β type I receptor kinase can be identified and may be an effective means of modulating this signaling pathway in vivo.

GROWTH FACTOR SIGNALING THROUGH RECEPTOR TYROSINE KINASES: A RATIONAL TARGET FOR DEVELOPMENT OF THERAPEUTIC INHIBITORS

In humans, one of the most common signaling mechanisms utilized by extracellular growth factors is activation of a specific member or members of a class of cell surface proteins known as receptor tyrosine kinases (RTKs) (Fambrough et al., 1999; Schlessinger, 2000; Blume-Jensen and Hunter, 2001). These are characterized by an extracellular ligand (growth factor)-binding region, a single transmembrane spanning region, and an intracellular tyrosine kinase domain. At present, 58 human genes encoding RTKs have been identified and are distributed into 20 subfamilies. For many of these receptors, the basic signaling mechanisms have been described. Signaling by RTKs involves ligand-induced receptor oligomerization that results in the induction of autophosphorylation on specific tyrosines within the intracellular domain of the receptors themselves. Autophosphorylation on these tyrosines creates a unique site on the proteins that then recruits, and physically interacts with, a number of proteins that contain Src homology region 2 (SH2) domains and protein tyrosine-binding (PTB) domains. These interactions result in the activation of signaling pathways that are mediated by small G proteins, lipid kinases, cytoplasmic kinases, and phospholipases and, as a result, modulate a variety of cellular processes ranging from the synthesis and elaboration of bioactive effectors to proliferation and survival. Given the fundamental role that many of these growth factor signaling pathways subserve, it is not surprising that dysregulation of these signaling pathways has been implicated in a wide variety of disease states. For example, dysregulated RTK signaling has been shown to arise via a number of distinct mechanisms in a growing number of human malignancies (Bejcek et al., 1989; Guha et al., 1995; Carroll et al., 1996; Rusch et al., 1996; Todo et al., 1996; Uhrbom et al., 1998). The most common alteration leading to altered RTK signaling is receptor overexpression, an event that has been identified in tumor tissues for ~30 RTKs. The importance of this overexpression in cancer has been most robustly demonstrated for the epidermal growth factor receptor (EGFR) family. A specific member of this receptor family (EGFR-2/erbB2) is overexpressed in a significant portion of breast cancers, and the application of a monoclonal antibody that is capable of targeting this receptor has been demonstrated to favorably impact the progression of these tumors in humans (Baselga, 2001; Vogel et al., 2001). Aberrant RTK signaling leading to oncogenic transformation may also occur when a receptor and its ligand are expressed by the same or closely adjacent cells, thus establishing an autocrine/paracrine loop. Autocrine signaling by both the EGFR and platelet-derived growth factor receptor (PDGFR) has been shown to be a common event in certain human malignancies and is sufficient to cause tumor formation in some animal models. Finally, somatic mutations resulting in structural alterations in several RTK genes and their cognate proteins can lead to a constitutive (and presumably dysregulated) activation of the signaling pathways mediated by this receptor (Hirota et al., 1998; Hayakawa et al., 2000). By examining many tumor types, somatic, activating mutations have been identified in at least 13 RTK family members. These constitutively activated receptor kinases represent especially attractive targets for therapeutic inhibition. On the basis of this rationale, the members of the PDGFR family represent attractive targets for kinase inhibitor therapy because many of them are overexpressed in a wide array of diseased tissues and four of the five family members have been shown to harbor activating mutations in specific human tumors (Levitzki and Gazit, 1995; Levitzki, 1999).

DEVELOPMENT OF SELECTIVE INHIBITORS OF THE PDGFR FAMILY OF RTKS

The members of the PDGFR family include α PDGFR, β PDGFR, colony stimulating factor 1 receptor (CSF-1R), Flt-3, and c-kit. Within this RTK family, c-kit gain-of-function mutations have been the most extensively studied. Somatic mutations of the c-kit gene that result in a receptor protein that is constitutively activated have been identified in several disorders, including systemic mastocytosis, a syndrome characterized by an uncontrolled proliferation of mast cells, a subset of patients with acute myelogenous leukemia (AML), and a rare and very aggressive tumor of the gastrointestinal tract known as gastrointestinal stromal tumors (GISTs) (Hirota et al., 1998). The strategy of targeted inhibition of c-kit for the treatment of GIST has been validated clinically by treating patients with Gleevec. Gleevec is a small-molecule kinase inhibitor that is capable of inhibiting the kinase activity of c-kit (as well as the bcr-abl kinase, which is the basis for its dramatic efficacy as a therapeutic agent in patients with chronic myelogenous leukemia). Clinical trials using Gleevec to inhibit c-kit in GIST patients have shown positive (and occasionally dramatically positive) results indicating that c-kit is a valid target for treating this disease (Druker et al., 2001a; Druker et al., 2001b; Joensuu et al., 2001). Flt-3, the newest and least studied member of the PDGFR family, is expressed in myeloid cells and has recently been implicated in AML. Approximately 30% of patients who are diagnosed with AML can be shown to harbor a somatic, clonal, gain-of-function mutation in their Flt-3 gene (Rombouts et al., 2000). Although these vary among patients, they often involve either an internal tandem duplication in the juxtamembrane region or a point mutation in the kinase domain activation loop, both of which result in a receptor that is constitutively activated. Patients with these Flt-3 mutations have higher white cell counts, a shorter time to relapse after chemotherapy, and reduced long-term survival. These clinical observations suggest strongly that this aberrant activation of Flt-3 signaling plays an important role in the leukemic process and may be an attractive therapeutic strategy in these patients. Another member of the PDGFR family, CSF-1R, has also been implicated in AML. In this case, a point mutation in the C-terminal region of the gene has been described, but the frequency of this mutation in patients and its importance to tumor formation have not been well characterized to date. The founding member of the PDGFR family, β PDGFR, has been studied extensively for its role in cancer. The seminal observation that the v-sis oncogene of simian sarcoma virus encoded the PDGF B chain (the ligand for β PDGFR) and that when this was expressed in fibroblasts or glial cells in vivo it gave rise to sarcomas and brain tumors, respectively, established for the first time that RTK autocrine signaling could play a key role in the oncogenic process (Waterfield et al., 1983; Eriksson et al., 1992). More recent studies have identified chromosomal translocations involving the β PDGFR in chronic myelomonocytic leukemia (CMML) (Carroll et al., 1996). A subset of CMML patients have a t(5;12) (q33;p13) or a t(5;7) (q33;q11.2) chromosomal translocation that results in the N-terminal ligand-binding domain of the PDGFR being replaced by the N terminus of the transcription factor Tel or the Huntingtin interacting protein 1. Each of the fusion partners contain a self-association domain that mediates the formation of oligomers (in the absence of ligand), leading to constitutive activation of the β PDGFR kinase domain. Finally, PDGF B as well as β PDGFR are overexpressed in a variety of inflammatory and fibroproliferative disorders such as hepatic cirrhosis, restenosis after vascular injury, and diabetic kidney disease, suggesting that inhibition of this signaling pathway may favorably impact these diseases.

To develop inhibitors of the PDGFR class of RTKs, we developed a high-throughput cell-based two-site ELISA to screen compounds for inhibition of PDGFR tyrosine autophosphorylation. This system utilizes a series of cell lines that selectively express fusions between the extracellular ligand-binding domain of the β PDGFR and the unique kinase domains of each of the respective family members. The ELISA is designed to read out the tyrosine phosphorylation (and thus activation) of each of the kinase domains after signaling is initiated with a single ligand (PDGF B), thus allowing a uniform and standardized assay format. In addition, the use of a whole cell assay offers a distinct advantage over a more simplified purified kinase assay in that only compounds capable of effectively entering a cell and inhibiting these RTKs in their native states would be detected. Because RTK tyrosine phosphorylation in response to a specific extracellular ligand represents the most proximal step in the intracellular signaling pathway it was hypothesized that inhibition at this proximal step would provide the greatest level of specificity and efficacy in blocking the biological effects mediated by these signaling pathways.

These assays allowed us to screen more than 100,000 compounds and identify several hits or lead molecules that were capable of inhibiting PDGFR family signaling. As expected, none of these initial leads demonstrated the requisite pharmaceutical properties, and thus the process of lead optimization was pursued. A team of medicinal chemists proceeded to systematically produce hundreds of structurally related analogs, and these were retested in the assays. These efforts resulted in the identification of several compounds that incorporated a core quinazoline structure that began to demonstrate remarkable levels of potency and specificity. For example, several of the compounds synthesized were able to inhibit the α and β PDGFRs as well as Flt-3 and c-kit with IC_{50} values (the concentration of the compound that results in a 50% inhibition of activity) in the 50–100 nM range. Furthermore, when these compounds were tested for their ability to inhibit a variety of other kinases such as non-receptor tyrosine kinases (e.g., src, abl), mitogen-activated protein kinases (e.g., MAPK, JNK, MEKK1), and other RTKs (VEGFR, FGFR, IR), they were inactive.

The successful development of inhibitors with these levels of potency and specificity suggested that they may have clinical utility, and thus two strategies were pursued to understand whether these compounds could be developed into viable drugs for use in human disease. The first was to utilize these more selective and potent compounds in preclinical models of specific human diseases to assess the validity of the target and confirm the efficacy of the inhibitor. The second was to examine the pharmaceutical properties (e.g., oral bioavailability and half-life) of these compounds in experimental animals. The former is critical to ensure the rationale of the therapeutic and the latter to allow effective and safe administration to humans. In several initial analyses, lead compounds were examined for their ability to inhibit cellular proliferation and migration in response to PDGF-BB or bFGF. These assays demonstrated that the compounds could selectively inhibit the PDGF-BB-mediated responses (and not the bFGF-mediated responses) in cell fibroblasts and vascular smooth muscle cells, which are normally responsive to both growth factors. In parallel analyses, several of these compounds were found to demonstrate excellent pharmaceutical properties. That is, when administered to experimental animals such as rats or dogs, they were rapidly absorbed from the gastrointestinal tract and appeared in the blood at levels that would be predicted to be effectively inhibitory toward their target(s) (e.g., they were bioavailable) and these levels persisted in the blood for reasonable periods of time after dosing (i.e., they demonstrated good effective half-lives in vivo). The studies thus demonstrated that these drugs could putatively be administered to humans via oral administration and could effect inhibition of their target RTKs in vivo.

These types of compounds thus represent putative drugs and could now be tested in more definitive preclinical models. To this end, several models of neoplastic or proliferative disease mediated by activated RTKs of the PDGFR family were developed. One of these involved the introduction of a mutated, constitutively activated form of the Flt-3 receptor into an IL-3-dependent murine myeloid cell line. As a direct result of the presence of an activated form of the Flt-3 RTK, these cells are capable of growing in an IL-3-independent manner in vitro. Furthermore, when these cells are introduced into nude mice they induce a lethal myeloproliferative disorder. To test the effect of target inhibition of Flt-3 signaling, the ability of several of the inhibitory compounds to ameliorate this Flt-3-driven myeloproliferative disorder was evaluated. Chronic administration of these compounds proved highly efficacious at delaying, and in some cases apparently arresting, the progression of the myeloproliferative disease.

In summary, a focused drug development effort directed at identifying inhibitors of the PDGFR family of RTKs has successfully developed several potent and selective inhibitors that show remarkable efficacy in a variety of in vivo models. These inhibitors are now poised to begin clinical trials in humans with several forms of malignancy that are thought to be mediated by this family of RTKs.

CONCLUSIONS

This chapter has attempted to outline the most basic steps in the process of drug development. The explosion of knowledge in signal transduction and genomics has provided an unprecedented opportunity to identify the molecular mechanisms underlying virtually all of the significant human diseases. Coupled with the development of modern drug development strategies, these findings are leading to the identification of potent and selective modulators of many of the major signal transduction pathways. We have illustrated this process in two distinct contexts, and efforts like these may ultimately revolutionize the therapy of many human pathologies.

ACKNOWLEDGMENTS

We are indebted to Dr. Robert Scarborough and Dr. Scott Wasserman for assistance with this manuscript and to Ms. Elizabeth Park for editorial assistance.

REFERENCES

Baker, J. C., and Harland, R. M. (1997). From receptor to nucleus: the Smad pathway. Curr Opin Genet Dev 7(4), 467–473.

Baselga, J. (2001). Clinical trials of Herceptin® (trastuzumab). Eur J Cancer 37 Suppl 1, 18–24.

Bejcek, B. E., Li, D. Y. et al. (1989). Transformation by v-sis occurs by an internal autoactivation mechanism. Science 245(4925), 1496–1499.

Blume-Jensen, P., and Hunter, T. (2001). Oncogenic kinase signalling. Nature 411(6835), 355–365.

Carroll, M., Tomasson, M. H. et al. (1996). The TEL/platelet-derived growth factor beta receptor (PDGF βR) fusion in chronic myelomonocytic leukemia is a transforming protein that self-associates and activates PDGF βR kinase-dependent signaling pathways. Proc Natl Acad Sci USA 93(25), 14845–14850.

Druker, B. J., Sawyers, C. L. et al. (2001a). Activity of a specific inhibitor of the BCR-ABL tyrosine kinase in the blast crisis of chronic myeloid leukemia and acute lymphoblastic leukemia with the Philadelphia chromosome. N Engl J Med 344(14), 1038–1042.

Druker, B. J., Talpaz, M. et al. (2001b). Efficacy and safety of a specific inhibitor of the BCR-ABL tyrosine kinase in chronic myeloid leukemia. N Engl J Med 344(14): 1031–1037.

Eriksson, A., Rorsman, C. et al. (1992). Ligand-induced homo- and heterodimerization of platelet-derived growth factor α- and β-receptors in intact cells. Growth Factors 6(1), 1–14.

Fambrough, D., McClure, K. et al. (1999). Diverse signaling pathways activated by growth factor receptors induce broadly overlapping, rather than independent, sets of genes [see comments]. Cell 97(6), 727–741.

Guha, A., Dashner, K. et al. (1995). Expression of PDGF and PDGF receptors in human astrocytoma operation specimens supports the existence of an autocrine loop. Int J Cancer 60(2), 168–173.

Hayakawa, F., Towatari, M. et al. (2000). Tandem-duplicated Flt3 constitutively activates STAT5 and MAP kinase and introduces autonomous cell growth in IL-3-dependent cell lines. Oncogene 19(5), 624–631.

Hayashi, H., Abdollah, S. et al. (1997). The MAD-related protein Smad7 associates with the TGFβ receptor and functions as an antagonist of TGFβ signaling. Cell 89(7), 1165–1173.

Hirota, S., Isozaki, K. et al. (1998). Gain-of-function mutations of c-kit in human gastrointestinal stromal tumors. Science 279(5350), 577–580.

Huse, M., Chen, Y. G. et al. (1999). Crystal structure of the cytoplasmic domain of the type I TGF β receptor in complex with FKBP12. Cell 96(3), 425–436.

Joensuu, H., Roberts, P. J. et al. (2001). Effect of the tyrosine kinase inhibitor STI571 in a patient with a metastatic gastrointestinal stromal tumor. N Engl J Med *344*(14), 1052–1056.

Levitzki, A. (1999). Protein tyrosine kinase inhibitors as novel therapeutic agents. Pharmacol Ther *82*(2–3), 231–239.

Levitzki, A., and Gazit, A. (1995). Tyrosine kinase inhibition: an approach to drug development. Science *267*(5205), 1782–1788.

Massague, J. (1996). TGFβ signaling: receptors, transducers, and Mad proteins. Cell *85*(7), 947–950.

Rombouts, W. J., Blokland, I. et al. (2000). Biological characteristics and prognosis of adult acute myeloid leukemia with internal tandem duplications in the Flt3 gene. Leukemia *14*(4), 675–683.

Rusch, V., Mendelsohn, J. et al. (1996). The epidermal growth factor receptor and its ligands as therapeutic targets in human tumors. Cytokine Growth Factor Rev *7*(2), 133–141.

Schlessinger, J. (2000). Cell signaling by receptor tyrosine kinases. Cell *103*(2), 211–225.

Todo, T., Adams, E. F. et al. (1996). Autocrine growth stimulation of human meningioma cells by platelet-derived growth factor. J Neurosurg *84*(5), 852–858; discussion 858–859.

Topper, J. N., Cai, J. et al. (1997). Vascular MADs: two novel MAD-related genes selectively inducible by flow in human vascular endothelium. Proc Natl Acad Sci USA *94*(17), 9314–9319.

Uhrbom, L., Hesselager, G. et al. (1998). Induction of brain tumors in mice using a recombinant platelet-derived growth factor B-chain retrovirus. Cancer Res *58*(23), 5275–5279.

Ulloa, L., Doody, J. et al. (1999). Inhibition of transforming growth factor-beta/SMAD signalling by the interferon-γ/STAT pathway. Nature *397*(6721), 710–713.

Vogel, C., Cobleigh, M. A. et al. (2001). First-line, single-agent Herceptin® (trastuzumab) in metastatic breast cancer. a preliminary report. Eur J Cancer *37 Suppl 1*, 25–29.

Waterfield, M. D., Scrace, G. T. et al. (1983). Platelet-derived growth factor is structurally related to the putative transforming protein p28sis of simian sarcoma virus. Nature *304*(5921), 35–39.

Zhou, S., Buckhaults, P. et al. (1998). Targeted deletion of smad4 shows it is required for transforming growth factor β and activin signaling in colorectal cancer cells. Proc Natl Acad Sci USA *95*(5), 2412–2416.

INDEX